Fundamental Equations of Statics

Cartesian Vector

$$\mathbf{A} = A_x\mathbf{i} + A_y\mathbf{j} + A_z\mathbf{k}$$

Magnitude

$$A = \sqrt{A_x^2 + A_y^2 + A_z^2}$$

Directions

$$\mathbf{u}_A = \frac{\mathbf{A}}{A} = \frac{A_x}{A}\mathbf{i} + \frac{A_y}{A}\mathbf{j} + \frac{A_z}{A}\mathbf{k}$$

$$= \cos\alpha\mathbf{i} + \cos\beta\mathbf{j} + \cos\gamma\mathbf{k}$$

$$\cos^2\alpha + \cos^2\beta + \cos^2\gamma = 1$$

Dot Product

$$\mathbf{A}\cdot\mathbf{B} = AB\cos\theta$$

$$= A_xB_x + A_yB_y + A_zB_z$$

Cross Product

$$\mathbf{C} = \mathbf{A}\times\mathbf{B} = \begin{vmatrix} \mathbf{i} & \mathbf{j} & \mathbf{k} \\ A_x & A_y & A_z \\ B_x & B_y & B_z \end{vmatrix}$$

Cartesian Position Vector

$$\mathbf{r} = (x_2 - x_1)\mathbf{i} + (y_2 - y_1)\mathbf{j} + (z_2 - z_1)\mathbf{k}$$

Cartesian Force Vector

$$\mathbf{F} = F\mathbf{u} = F\left(\frac{\mathbf{r}}{r}\right)$$

Moment of a Force

$$M_O = Fd$$

$$\mathbf{M}_O = \mathbf{r}\times\mathbf{F} = \begin{vmatrix} \mathbf{i} & \mathbf{j} & \mathbf{k} \\ r_x & r_y & r_z \\ F_x & F_y & F_z \end{vmatrix}$$

Moment of a Force About a Specified Axis

$$M_a = \mathbf{u}\cdot\mathbf{r}\times\mathbf{F} = \begin{vmatrix} u_x & u_y & u_z \\ r_x & r_y & r_z \\ F_x & F_y & F_z \end{vmatrix}$$

Simplification of a Force and Couple System

$$\mathbf{F}_R = \Sigma\mathbf{F}$$

$$(\mathbf{M}_R)_O = \Sigma\mathbf{M} + \Sigma\mathbf{M}_O$$

Equilibrium

Particle

$$\Sigma F_x = 0, \Sigma F_y = 0, \Sigma F_z = 0$$

Rigid Body-Two Dimensions

$$\Sigma F_x = 0, \Sigma F_y = 0, \Sigma M_O = 0$$

Rigid Body-Three Dimensions

$$\Sigma F_x = 0, \Sigma F_y = 0, \Sigma F_z = 0$$

$$\Sigma M_{x'} = 0, \Sigma M_{y'} = 0, \Sigma M_{z'} = 0$$

Friction

Static (maximum) $\quad F_s = \mu_s N$

Kinetic $\quad F_k = \mu_k N$

Center of Gravity

Particles or Discrete Parts

$$\bar{r} = \frac{\Sigma\tilde{r}W}{\Sigma W}$$

Body

$$\bar{r} = \frac{\displaystyle\int \tilde{r}\,dW}{\displaystyle\int dW}$$

Area and Mass Moments of Inertia

$$I = \int r^2\,dA \qquad I = \int r^2\,dm$$

Parallel-Axis Theorem

$$I = \bar{I} + Ad^2 \qquad I = \bar{I} + md^2$$

Radius of Gyration

$$k = \sqrt{\frac{I}{A}} \qquad k = \sqrt{\frac{I}{m}}$$

Virtual Work

$$\delta U = 0$$

SI Prefixes

Multiple	Exponential Form	Prefix	SI Symbol
1 000 000 000	10^9	giga	G
1 000 000	10^6	mega	M
1 000	10^3	kilo	k

Submultiple			
0.001	10^{-3}	milli	m
0.000 001	10^{-6}	micro	μ
0.000 000 001	10^{-9}	nano	n

Conversion Factors (FPS) to (SI)

Quantity	Unit of Measurement (FPS)	Equals	Unit of Measurement (SI)
Force	lb		4.4482 N
Mass	slug		14.5938 kg
Length	ft		0.3048 m

Conversion Factors (FPS)

1 ft = 12 in. (inches)
1 mi. (mile) = 5280 ft
1 kip (kilopound) = 1000 lb
1 ton = 2000 lb

STATICS

ENGINEERING MECHANICS
Statics

Tenth Edition in SI Units

R. C. Hibbeler

SI conversion by S. C. Fan

Singapore London New York Toronto Sydney Tokyo Madrid
Mexico City Munich Paris Capetown Hong Kong Montreal

Published in 2004 by
Prentice Hall
Pearson Education South Asia Pte Ltd
23/25 First Lok Yang Road
Singapore 629733

Pearson Education offices in Asia: *Bangkok, Beijing, Hong Kong, Jakarta, Kuala Lumpur, Manila, New Delhi, Seoul, Shanghai, Singapore, Taipei, Tokyo,*

Original edition published by:
Prentice-Hall, Inc.
Copyright © 2004
R. C. Hibbeler

Printed in Singapore

5 4

07 06 05

ISBN 013-124-844-8 (pbk)
ISBN 013-127-844-4 (hbk)

PEARSON
Prentice
Hall

TO THE STUDENT

With the hope that this work will stimulate an interest in Engineering Mechanics and provide an acceptable guide to its understanding.

The main purpose of this book is to provide the student with a clear and thorough presentation of the theory and applications of engineering mechanics. To achieve this objective, the author has by no means worked alone; to a large extent, this book has been shaped by the comments and suggestions of hundreds of reviewers in the teaching profession as well as many of the author's students.

New Features

Some unique features used throughout this third edition include the following:

- **Illustrations.** Throughout the book, new photorealistic illustrations have been added that provide a strong connection to the 3-D nature of engineering. In addition, particular attention has been placed on providing a view of any physical object, its dimensions, and the vectors applied to it in a manner that can be easily understood.
- **Problems.** The problems sets have been revised so that instructors can select both design and analysis problems having a wide range of difficulty. Apart from the author, two other professionals have checked all the problems for clarity and accuracy of the solutions. At the end of some chapters, design projects are included.
- **Review Material.** New end-of-chapter review sections have been added to help students recall and study key chapter points.

Of course, the hallmarks of the book remain the same: Where necessary, a strong emphasis is placed on drawing a free-body diagram, and the importance of selecting an appropriate coordinate system, and associated sign convention for vector components is stressed when the equations of mechanics are applied.

Contents

The book is divided into 11 chapters, in which the principles are applied first to simple, then to more complicated situations. Most often, each principle is applied first to a particle, then to a rigid body subjected to a coplanar system of forces, and finally to a general case of three-dimensional force systems acting on a rigid body.

Chapter 1 begins with an introduction to mechanics and a discussion of units. The notation of a vector and the properties of a concurrent force system are introduced in Chapter 2. This theory is then applied to the equilibrium of a particle in Chapter 3. Chapter 4 contains a general discussion of both concentrated and distributed force systems and the methods used to simplify them. The principles of rigid-body equilibrium are developed in Chapter 5 and then applied to specific problems involving the equilibrium of trusses, frames, and machines in Chapter 6, and to the analysis of internal forces in beams and cables in Chapter 7. Applications to problems involving frictional forces are discussed in Chapter 8, and topics related to the center of gravity and centroid are treated in Chapter 9. If time permits, sections concerning more advanced topics, indicated by stars (★) may be covered. Most of these topics are included in Chapter 10 (area and mass moments of inertia) and Chapter 11 (virtual work and potential energy). Note that this material also provides a suitable reference for basic principles when it is discussed in more advanced courses.

Alternative Coverage. At the discretion of the instructor, some of the material may be presented in a different sequence with no loss of continuity. For example, it is possible to introduce the concept of a force and all the necessary methods of vector analysis by first covering Chapter 2 and Section 4.2. Then after covering the rest of Chapter 4 (force and moment systems), the equilibrium methods of Chapters 3 and 5 can be discussed.

Special Features

Organization and Approach. The contents of each chapter are organized into well-defined sections that contain an explanation of specific topics, illustrative example problems, and a set of homework problems. The topics within each section are placed into subgroups defined by boldface titles. The purpose of this is to present a structured method for introducing each new definition or concept and to make the book convenient for later reference and review.

Chapter Contents. Each chapter begins with an illustration demonstrating a broad-range application of the material within the chapter. A bulleted list of the chapter contents is provided to give a general overview of the material that will be covered.

Free-Body Diagrams. The first step to solving most mechanics problems requires drawing a diagram. By doing so, the student forms the habit of tabulating the necessary data while focusing on the physical aspects of the problem and its associated geometry. If this step is performed correctly, applying the relevant equations of mechanics becomes somewhat methodical since the data can be taken directly from the diagram. This step is particularly important when solving equilibrium problems, and for this reason drawing free-body diagrams is strongly emphasized throughout the book. In particular, special sections and examples are devoted to show

how to draw free-body diagrams, and specific homework problems in many sections of the book have been added to develop this practice.

Procedures for Analysis.   Found after many of the sections of the book, this unique feature provides the student with a logical and orderly method to follow when applying the theory. The example problems are solved using this outlined method in order to clarify its numerical application. It is to be understood, however, that once the relevant principles have been mastered and enough confidence and judgment have been obtained, the student can then develop his or her own procedures for solving problems.

Photographs.   Many photographs are used throughout the book to explain how the principles of mechanics apply to real-world situations. In some sections, photographs have been used to show how engineers must first make an idealized model for analysis and then proceed to draw a free-body diagram of this model in order to apply the theory.

Important Points.   This feature provides a review or summary of the most important concepts in a section and highlights the most significant points that should be realized when applying the theory to solve problems.

Conceptual Understanding.   Through the use of photographs placed throughout the book, theory is applied in a simplified way in order to illustrate some of its more important conceptual features and instill the physical meaning of many of the terms used in the equations. These simplified applications increase interest in the subject matter and better prepare the student to understand the examples and solve problems.

Example Problems.   All the example problems are presented in a concise manner and in a style that is easy to understand.

Homework Problems

- **Free-Body Diagram Problems.** Many sections of the book contain introductory problems that only require drawing the free-body diagram for the specific problems within a problem set. These assignments will impress upon the student the importance of mastering this skill as a requirement for a complete solution of any equilibrium problem.

- **General Analysis and Design Problems.** The majority of problems in the book depict realistic situations encountered in engineering practice. Some of these problems come from actual products used in industry and are stated as such. It is hoped that this realism will both stimulate the student's interest in engineering mechanics and provide a means for developing the skill to reduce any such problem from its physical description to a model or symbolic representation to which the principles of mechanics may be applied.

  Throughout the book, all problems use SI units. Furthermore, in any set, an attempt has been made to arrange the problems in order of increasing difficulty. (Review problems at the end of each chapter are presented in random order.) The answers to all but every fourth problem are listed in the back of the book.

- **Student Study Materials:** New chapter-by-chapter review includes key points, equations, and check up questions.
- **Free-Body Diagram Workbook:** 75-page workbooks that step students through numerous free-body diagram problems. Full explanations and solutions are provided.

Student Study Guide *Statics*; *Dynamics*

Students may purchase a further Study Guide containing more worked problems. Problems are partially solved and designed to help guide students through difficult topics.

Companion Website www.pearsoned-asia.com/hibbeler. This password-protected website provides additional statics/dynamics problems with solutions. Problems and solutions are supplemented and do not appear in the third edition. Solutions contain both math and associated free-body diagrams. Students can use these for practice before quizzes and tests, as well as self-drill. For instructors: Downloadable version of the powerpoint slides and Instructor's Solutions Manual are available.

Acknowledgments

The author has endeavored to write this book so that it will appeal to both the student and instructor. Through the years, many people have helped in its development, and I will always be grateful for their valued suggestions and comments. Specifically, I wish to personally thank the following individuals who have contributed their comments to the *Statics* and *Dynamics* series:

Paul Heyliger, *Colorado State University*
Kenneth Sawyers, *Lehigh University*
John Oyler, *University of Pittsburgh*
Glenn Beltz, *University of California*
Johannes Gessler, *Colorado State University*
Wilfred Nixon, *University of Iowa*
Jonathan Russell, *U.S. Coast Guard Academy*
Robert Hinks, *Arizona State University*
Cap. Mark Orwat, *U.S. Military Academy, West Point*
Cetin Cetinyaka, *Clarkson University*
Jack Xin, *Kansas State University*
Pierre Julien, *Colorado State University*
Stephen Bechtel, *Ohio State University*
W. A. Curtain, *Brown University*
Robert Oakberg, *Montana State University*
Richard Bennett, *University of Tennessee*

A particular note of thanks is also given to Professors Will Liddell, Jr. and Henry Kuhlman for their specific help. A special note of thanks is given to the accuracy checkers, Scott Hendricks of VPI and Karim Nohra of the University of South Florida, who diligently checked all of the text and problems. I should also like to acknowledge the proofreading assistance of my wife, Conny (Cornelie), during the time it has taken to prepare this manuscript for publication.

Lastly, many thanks are extended to all my students and to members of the teaching profession who have freely taken the time to send me their suggestions and comments. Since this list is too long to mention, it is hoped that those who have given help in this manner will accept this anonymous recognition.

I would greatly appreciate hearing from you if at any time you have any comments, suggestions, or problems related to any matters regarding this edition.

Russell Charles Hibbeler
hibbeler@bellsouth.net

C O N T E N T S

5

**Equilibrium of a Rigid Body
193**

6

Structural Analysis   257

7

Internal Forces   325

8

Friction   379

9

Center of Gravity and Centroid 437

10

Moments of Inertia 499

11

Virtual Work 551

Appendices

S T A T I C S

The design of this rocket and gantry structure requires a basic knowledge of both statics and dynamics, which forms the subject matter of engineering mechanics.

CHAPTER
1

General Principles

- To provide an introduction to the basic quantities and idealizations of mechanics.
- To give a statement of Newton's Laws of Motion and Gravitation.
- To review the principles for applying the SI system of units.
- To examine the standard procedures for performing numerical calculations.
- To present a general guide for solving problems.

1.1 Mechanics

Mechanics can be defined as that branch of the physical sciences concerned with the state of rest or motion of bodies that are subjected to the action of forces. In general, this subject is subdivided into three branches: *rigid-body mechanics, deformable-body mechanics*, and *fluid mechanics*. This book treats only rigid-body mechanics since it forms a suitable basis for the design and analysis of many types of structural, mechanical, or electrical devices encountered in engineering. Also, rigid-body mechanics provides part of the necessary background for the study of the mechanics of deformable bodies and the mechanics of fluids.

Rigid-body mechanics is divided into two areas: statics and dynamics. *Statics* deals with the equilibrium of bodies, that is, those that are either at rest or move with a constant velocity; whereas *dynamics* is concerned with the accelerated motion of bodies. Although statics can be considered as a special case of dynamics, in which the acceleration is zero, statics deserves separate treatment in engineering education since many objects are designed with the intention that they remain in equilibrium.

Historical Development.   The subject of statics developed very early in history because the principles involved could be formulated simply from measurements of geometry and force. For example, the writings of Archimedes (287–212 B.C.) deal with the principle of the lever. Studies of the pulley, inclined plane, and wrench are also recorded in ancient writings—at times when the requirements of engineering were limited primarily to building construction.

Since the principles of dynamics depend on an accurate measurement of time, this subject developed much later. Galileo Galilei (1564–1642) was one of the first major contributors to this field. His work consisted of experiments using pendulums and falling bodies. The most significant contributions in dynamics, however, were made by Issac Newton (1642–1727), who is noted for his formulation of the three fundamental laws of motion and the law of universal gravitational attraction. Shortly after these laws were postulated, important techniques for their application were developed by Euler, D'Alembert, Lagrange, and others.

1.2   Fundamental Concepts

Before we begin our study of engineering mechanics, it is important to understand the meaning of certain fundamental concepts and principles.

Basic Quantities.   The following four quantities are used throughout mechanics.

Length.   *Length* is needed to locate the position of a point in space and thereby describe the size of a physical system. Once a standard unit of length is defined, one can then quantitatively define distances and geometric properties of a body as multiples of the unit length.

Time.   *Time* is conceived as a succession of events. Although the principles of statics are time independent, this quantity does play an important role in the study of dynamics.

Mass.   *Mass* is a property of matter by which we can compare the action of one body with that of another. This property manifests itself as a gravitational attraction between two bodies and provides a quantitative measure of the resistance of matter to a change in velocity.

Force.   In general, *force* is considered as a "push" or "pull" exerted by one body on another. This interaction can occur when there is direct contact between the bodies, such as a person pushing on a wall, or it can occur through a distance when the bodies are physically separated. Examples of the latter type include gravitational, electrical, and magnetic forces. In any case, a force is completely characterized by its magnitude, direction, and point of application.

Idealizations. Models or idealizations are used in mechanics in order to simplify application of the theory. A few of the more important idealizations will now be defined. Others that are noteworthy will be discussed at points where they are needed.

Particle. A *particle* has a mass, but a size that can be neglected. For example, the size of the earth is insignificant compared to the size of its orbit, and therefore the earth can be modeled as a particle when studying its orbital motion. When a body is idealized as a particle, the principles of mechanics reduce to a rather simplified form since the geometry of the body will not be involved in the analysis of the problem.

Rigid Body. A *rigid body* can be considered as a combination of a large number of particles in which all the particles remain at a fixed distance from one another both before and after applying a load. As a result, the material properties of any body that is assumed to be rigid will not have to be considered when analyzing the forces acting on the body. In most cases the actual deformations occurring in structures, machines, mechanisms, and the like are relatively small, and the rigid-body assumption is suitable for analysis.

Concentrated Force. A *concentrated force* represents the effect of a loading which is assumed to act at a point on a body. We can represent a load by a concentrated force, provided the area over which the load is applied is very small compared to the overall size of the body. An example would be the contact force between a wheel and the ground.

Newton's Three Laws of Motion. The entire subject of rigid-body mechanics is formulated on the basis of Newton's three laws of motion, the validity of which is based on experimental observation. They apply to the motion of a particle as measured from a nonaccelerating reference frame. With reference to Fig. 1–1, they may be briefly stated as follows.

First Law. A particle originally at rest, or moving in a straight line with constant velocity, will remain in this state provided the particle is *not* subjected to an unbalanced force.

Second Law. A particle acted upon by an *unbalanced force* **F** experiences an acceleration **a** that has the same direction as the force and a magnitude that is directly proportional to the force.* If **F** is applied to a particle of mass m, this law may be expressed mathematically as

$$\mathbf{F} = m\mathbf{a} \qquad (1\text{–}1)$$

Third Law. The mutual forces of action and reaction between two particles are equal, opposite, and collinear.

Equilibrium

Accelerated motion

Action – reaction

Fig. 1–1

---

*Stated another way, the unbalanced force acting on the particle is proportional to the time rate of change of the particle's linear momentum.

Newton's Law of Gravitational Attraction.   Shortly after formulating his three laws of motion, Newton postulated a law governing the gravitational attraction between any two particles. Stated mathematically,

$$F = G\frac{m_1 m_2}{r^2} \tag{1-2}$$

where F = force of gravitation between the two particles

   G = universal constant of gravitation; according to experimental evidence, $G = 66.73(10^{-12})$ m^3/(kg·s^2)

   m_1, m_2 = mass of each of the two particles

   r = distance between the two particles

Weight.   According to Eq. 1–2, any two particles or bodies have a mutual attractive (gravitational) force acting between them. In the case of a particle located at or near the surface of the earth, however, the only gravitational force having any sizable magnitude is that between the earth and the particle. Consequently, this force, termed the *weight*, will be the only gravitational force considered in our study of mechanics.

From Eq. 1–2, we can develop an approximate expression for finding the weight W of a particle having a mass $m_1 = m$. If we assume the earth to be a nonrotating sphere of constant density and having a mass $m_2 = M_e$, then if r is the distance between the earth's center and the particle, we have

$$W = G\frac{mM_e}{r^2}$$

Letting $g = GM_e/r^2$ yields

$$\boxed{W = mg} \tag{1-3}$$

By comparison with $\mathbf{F} = m\mathbf{a}$, we term g the acceleration due to gravity. Since it depends on r, it can be seen that the weight of a body is *not* an absolute quantity. Instead, its magnitude is determined from where the measurement was made. For most engineering calculations, however, g is determined at sea level and at a latitude of 45°, which is considered the "standard location."

1.3   Units of Measurement

The four basic quantities—force, mass, length and time—are not all independent from one another; in fact, they are *related* by Newton's second law of motion, $\mathbf{F} = m\mathbf{a}$. Because of this, the *units* used to measure these quantities cannot *all* be selected arbitrarily. The equality $\mathbf{F} = m\mathbf{a}$ is maintained only if three of the four units, called *base units*, are *arbitrarily defined* and the fourth unit is then *derived* from the equation.

SI Units. The International System of units, abbreviated SI after the French "Système International d'Unités," is a modern version of the metric system which has received worldwide recognition. As shown in Table 1–1, the SI system specifies length in meters (m), time in seconds (s), and mass in kilograms (kg). The unit of force, called a newton (N), is *derived* from $\mathbf{F} = m\mathbf{a}$. Thus, 1 newton is equal to a force required to give 1 kilogram of mass an acceleration of 1 m/s^2 ($N = kg \cdot m/s^2$).

If the weight of a body located at the "standard location" is to be determined in newtons, then Eq. 1–3 must be applied. Here $g = 9.806\ 65 \text{ m/s}^2$; however, for calculations, the value $g = 9.81 \text{ m/s}^2$ will be used. Thus,

$$W = mg \qquad (g = 9.81 \text{ m/s}^2) \qquad (1\text{–}4)$$

Therefore, a body of mass 1 kg has a weight of 9.81 N, a 2-kg body weighs 19.62 N, and so on, Fig. 1–2.

1 kg

9.81 N

Fig. 1–2

TABLE 1–1 • System of Units

Name	Length	Time	Mass	Force
International System of Units (SI)	meter (m)	second (s)	kilogram (kg)	newton* (N) $\left(\dfrac{kg \cdot m}{s^2}\right)$

*Derived unit.

1.4 The International System of Units

The SI system of units is used extensively in this book since it is intended to become the worldwide standard for measurement. Consequently, the rules for its use and some of its terminology relevant to mechanics will now be presented.

Prefixes. When a numerical quantity is either very large or very small, the units used to define its size may be modified by using a prefix. Some of the prefixes used in the SI system are shown in Table 1–2. Each represents a multiple or submultiple of a unit which, if applied successively, moves the decimal point of a numerical quantity to every third place. * For example, $4\ 000\ 000 \text{ N} = 4\ 000 \text{ kN(kilo-newton)} = 4\text{MN}$ (mega-newton), or $0.005\text{m} = 5 \text{ mm (milli-meter)}$. Notice that the SI system does not include the multiple deca (10) or the submultiple centi (0.01), which form part of the metric system. Except for some volume and area measurements, the use of these prefixes is to be avoided in science and engineering.

TABLE 1–2 • Prefixes

	Exponential Form	Prefix	SI Symbol
Multiple			
1 000 000 000	10^9	giga	G
1 000 000	10^6	mega	M
1 000	10^3	kilo	k
Submultiple			
0.001	10^{-3}	milli	m
0.000 001	10^{-6}	micro	μ
0.000 000 001	10^{-9}	nano	n

*The kilogram is the only base unit that is defined with a prefix.

Rules for Use.   The following rules are given for the proper use of the various SI symbols:

1. A symbol is *never* written with a plural "s," since it may be confused with the unit for second (s).

2. Symbols are always written in lowercase letters, with the following exceptions: symbols for the two largest prefixes shown in Table 1–3, giga and mega, are capitalized as G and M, respectively; and symbols named after an individual are also capitalized, e.g., N.

3. Quantities defined by several units which are multiples of one another are separated by a *dot* to avoid confusion with prefix notation, as indicated by $N = kg \cdot m/s^2 = kg \cdot m \cdot s^{-2}$. Also, $m \cdot s$ (meter-second), whereas ms (milli-second).

4. The exponential power represented for a unit having a prefix refers to both the unit *and* its prefix. For example, $\mu N^2 = (\mu N)^2 = \mu N \cdot \mu N$. Likewise, mm^2 represents $(mm)^2 = mm \cdot mm$.

5. Physical constants or numbers having several digits on either side of the decimal point should be reported with a *space* between every three digits rather than with a comma; e.g., 73 569.213 427. In the case of four digits on either side of the decimal, the spacing is optional; e.g., 8537 or 8 537. Furthermore, always try to use decimals and avoid fractions; that is, write 15.25 *not* $15\frac{1}{4}$.

6. When performing calculations, represent the numbers in terms of their

base or derived units by converting all prefixes to powers of 10. The final result should then be expressed using a *single prefix*. Also, after calculation, it is best to keep numerical values between 0.1 and 1000; otherwise, a suitable prefix should be chosen. For example,

$$(50 \text{ kN})(60 \text{ nm}) = [50(10^3) \text{ N}][60(10^{-9}) \text{ m}]$$
$$= 3000(10^{-6}) \text{ N} \cdot \text{m} = 3(10^{-3}) \text{ N} \cdot \text{m} = 3 \text{ mN} \cdot \text{m}$$

7. Compound prefixes should not be used; e.g., kμs (kilo-micro-second) should be expressed as ms (milli-second) since $1 \text{ k}\mu\text{s} = 1(10^3)(10^{-6}) \text{ s} = 1(10^{-3}) \text{ s} = 1 \text{ ms}$.

8. With the exception of the base unit the kilogram, in general avoid the use of a prefix in the denominator of composite units. For example, do not write N/mm, but rather kN/m; also, m/mg should be written as Mm/kg.

9. Although not expressed in multiples of 10, the minute, hour, etc., are retained for practical purposes as multiples of the second. Furthermore, plane angular measurement is made using radians (rad). In this book, however, degrees will often be used, where $180° = \pi$ rad.

1.5 Numerical Calculations

Numerical work in engineering practice is most often performed by using handheld calculators and computers. It is important, however, that the answers to any problem be reported with both justifiable accuracy and appropriate significant figures. In this section we will discuss these topics together with some other important aspects involved in all engineering calculations.

Dimensional Homogeneity. The terms of any equation used to describe a physical process must be *dimensionally homogeneous;* that is, each term must be expressed in the same units. Provided this is the case, all the terms of an equation can then be combined if numerical values are substituted for the variables. Consider, for example, the equation $s = vt + \frac{1}{2}at^2$, where, in SI units, s is the position in meters, m, t is time in seconds, s, v is velocity in m/s, and a is acceleration in m/s^2. Regardless of how this equation is evaluated, it maintains its dimensional homogeneity. In the form stated, each of the three terms is expressed in meters [m, $(m/\cancel{s})\cancel{s}$, $(m/\cancel{s}^2)\cancel{s}^2$,] or solving for a, $a = 2s/t^2 - 2v/t$, the terms are each expressed in units of m/s^2 [m/s^2, m/s^2, $(m/s)/s$].

Since problems in mechanics involve the solution of dimensionally homogeneous equations, the fact that all terms of an equation are represented by a consistent set of units can be used as a partial check for algebraic manipulations of an equation.

Significant Figures. The accuracy of a number is specified by the number of significant figures it contains. A *significant figure* is any digit, including a zero, provided it is not used to specify the location of the decimal point for the number. For example, the numbers 5604 and 34.52 each have four significant figures. When numbers begin or end with zeros, however, it is difficult to tell how many significant figures are in the number. Consider the number 400. Does it have one (4), or perhaps two (40), or three (400) significant figures? In order to clarify this situation, the number should be reported using powers of 10. Using *engineering notation*, the exponent is displayed in multiples of three in order to facilitate conversion of SI units to those having an appropriate prefix. Thus, 400 expressed to one significant figure would be $0.4(10^3)$. Likewise, 2500 and 0.00546 expressed to three significant figures would be $2.50(10^3)$ and $5.46(10^{-3})$.

Computers are often used in engineering for advanced design and analysis.

Rounding Off Numbers. For numerical calculations, the accuracy obtained from the solution of a problem generally can never be better than the accuracy of the problem data. This is what is to be expected, but often handheld calculators or computers involve more figures in the answer than the number of significant figures used for the data. For this reason, a calculated result should always be "rounded off" to an appropriate number of significant figures.

To convey appropriate accuracy, the following rules for rounding off a number to n significant figures apply:

- If the $n + 1$ digit is *less than 5*, the $n + 1$ digit and others following it are dropped. For example, 2.326 and 0.451 rounded off to $n = 2$ significant figures would be 2.3 and 0.45.

- If the $n + 1$ digit is equal to 5 with zeros following it, then round off the nth digit to an *even number*. For example, $1.245(10^3)$ and 0.8655 rounded off to $n = 3$ significant figures become $1.24(10^3)$ and 0.866.

- If the $n + 1$ digit is *greater than 5* or equal to 5 with any nonzero digits following it, then increase the nth digit by 1 and drop the $n + 1$ digit and others following it. For example, 0.723 87 and 565.500 3 rounded off to $n = 3$ significant figures become 0.724 and 566.

Calculations. As a general rule, to ensure accuracy of a final result when performing calculations on a pocket calculator, always retain a greater number of digits than the problem data. If possible, try to work out the computations so that numbers which are approximately equal are not subtracted since accuracy is often lost from this calculation.

In engineering we generally round off final answers to *three* significant figures since the data for geometry, loads, and other measurements are often reported with this accuracy.* Consequently, in this book the intermediate calculations for the examples are often worked out to four significant figures and the answers are generally reported to *three* significant figures.

---

*Of course, some numbers, such as π, e, or numbers used in derived formulas are exact and are therefore accurate to an infinite number of significant figures.

E X A M P L E   **1.1**

Evaluate each of the following and express with SI units having an appropriate prefix: (a) (50 mN)(6 GN), (b) (400 mm)(0.6 MN)^2, (c) 45 MN^3/900 Gg.

Solution

First convert each number to base units, perform the indicated operations, then choose an appropriate prefix (see Rule 6 on p. 9).

Part (a)

$$(50 \text{ mN})(6 \text{ GN}) = [50(10^{-3}) \text{ N}][6(10^9) \text{ N}]$$

$$= 300(10^6) \text{ N}^2$$

$$= 300(10^6) \text{ N}^2\left(\frac{1 \text{ kN}}{10^3 \text{ N}}\right)\left(\frac{1 \text{ kN}}{10^3 \text{ N}}\right)$$

$$= 300 \text{ kN}^2 \qquad\qquad\qquad Ans.$$

Note carefully the convention kN^2 = (kN)^2 = 10^6 N^2 (Rule 4 on p. 9).

Part (b)

$$(400 \text{ mm})(0.6 \text{ MN})^2 = [400(10^{-3}) \text{ m}][0.6(10^6) \text{ N}]^2$$

$$= [400(10^{-3}) \text{ m}][0.36(10^{12}) \text{ N}^2]$$
$$= 144(10^9) \text{ m} \cdot \text{N}^2$$
$$= 144 \text{ Gm} \cdot \text{N}^2 \qquad\qquad\qquad Ans.$$

We can also write

$$144(10^9)\text{m} \cdot \text{N}^2 = 144(10^9)\text{m} \cdot \text{N}^2\left(\frac{1 \text{ MN}}{10^6 \text{ N}}\right)\left(\frac{1 \text{ MN}}{10^6 \text{ N}}\right)$$

$$= 0.144 \text{ m} \cdot \text{MN}^2$$

Part (c)

$$45 \text{ MN}^3/900 \text{ Gg} = \frac{45(10^6 \text{ N})^3}{900(10^6) \text{ kg}}$$

$$= 0.05(10^{12}) \text{ N}^3/\text{kg}$$

$$= 0.05(10^{12}) \text{ N}^3\left(\frac{1 \text{ kN}}{10^3 \text{ N}}\right)^3 \frac{1}{\text{kg}}$$

$$= 0.05(10^3) \text{ kN}^3/\text{kg}$$
$$= 50 \text{ kN}^3/\text{kg} \qquad\qquad\qquad Ans.$$

Here we have used Rules 4 and 8 on p. 9.

1.6 General Procedure for Analysis

The most effective way of learning the principles of engineering mechanics is to *solve problems*. To be successful at this, it is important to always present the work in a *logical* and *orderly manner*, as suggested by the following sequence of steps:

1. Read the problem carefully and try to correlate the actual physical situation with the theory studied.
2. Draw any necessary diagrams and tabulate the problem data.
3. Apply the relevant principles, generally in mathematical form.
4. Solve the necessary equations algebraically as far as practical, then, making sure they are dimensionally homogeneous, use a consistent set of units and complete the solution numerically. Report the answer with no more significant figures than the accuracy of the given data.
5. Study the answer with technical judgment and common sense to determine whether or not it seems reasonable.

When solving problems, do the work as neatly as possible. Being neat generally stimulates clear and orderly thinking, and vice versa.

IMPORTANT POINTS

- Statics is the study of bodies that are at rest or move with constant velocity.
- A particle has a mass but a size that can be neglected.
- A rigid body does not deform under load.
- Concentrated forces are assumed to act at a point on a body.
- Newton's three laws of motion should be memorized.
- Mass is a property of matter that does not change from one location to another.
- Weight refers to the gravitational attraction of the earth on a body or quantity of mass. Its magnitude depends upon the elevation at which the mass is located.
- In the SI system the unit of force, the newton, is a derived unit. The meter, second, and kilogram are base units.
- Prefixes G, M, k, m, μ, n are used to represent large and small numerical quantities. Their exponential size should be known, along with the rules for using the SI units.
- Perform numerical calculations to several significant figures and then report the final answer to three significant figures.
- Algebraic manipulations of an equation can be checked in part by verifying that the equation remains dimensionally homogeneous.
- Know the rules for rounding off numbers.

PROBLEMS

1-1. Round off the following numbers to three significant figures: (a) 4.65735 m, (b) 55.578 s, (c) 4555 N, (d) 2768 kg.

1-2. Represent each of the following quantities in the correct SI form using an appropriate prefix: (a) 0.000431 kg, (b) $35.3(10^3)$ N, (c) 0.00532 km.

*1-3. Represent each of the following combinations of units in the correct SI form using an appropriate prefix: (a) m/ms, (b) μkm, (c) ks/mg, and (d) km·μN.

1-4. Evaluate each of the following and express with an appropriate prefix: (a) $(430 \text{ kg})^2$, (b) $(0.002 \text{ mg})^2$, and (c) $(230 \text{ m})^3$.

*1-5. Represent each of the following combinations of units in the correct SI form: (a) kN/μs, (b) Mg/mN, and (c) MN/(kg·ms).

1-6. What is the weight in newtons of an object that has a mass of: (a) 10 kg, (b) 0.5 g, (c) 4.50 Mg? Express the result to three significant figures. Use an appropriate prefix.

1-7. Evaluate each of the following to three significant figures and express each answer in SI units using an appropriate prefix: (a) 354 mg(45 km)/(0.035 6 kN), (b) (.004 53 Mg)(201 ms), (c) 435 MN/23.2 mm.

*1-8. Two particles have a mass of 8 kg and 12 kg, respectively. If they are 800 mm apart, determine the force of gravity acting between them. Compare this result with the weight of each particle.

1-9. Determine the mass of an object that has a weight of (a) 20 mN, (b) 150 kN, (c) 60 MN. Express the answer to three significant figures.

1-10. Using the base units of the SI system, show that Eq. 1–2 is a dimensionally homogeneous equation which gives F in newtons. Determine to three significant figures the gravitational force acting between two spheres that are touching each other. The mass of each sphere is 200 kg and the radius is 300 mm.

*1-11. Evaluate each of the following to three significant figures and express each answer in SI units using an appropriate prefix: (a) $(0.631 \text{ Mm}) / (8.60 \text{ kg})^2$, (b) $(35 \text{ mm})^2 (48 \text{ kg})^3$.

This communications tower is stabilized by cables that exert forces at the points of connection. In this chapter, we will show how to determine the magnitude and direction of the resultant force at each point.

Force Vectors

- To show how to add forces and resolve them into components using the Parallelogram Law.
- To express force and position in Cartesian vector form and explain how to determine the vector's magnitude and direction.
- To introduce the dot product in order to determine the angle between two vectors or the projection of one vector onto another.

2.1 Scalars and Vectors

Most of the physical quantities in mechanics can be expressed mathematically by means of scalars and vectors.

Scalar. A quantity characterized by a positive or negative number is called a *scalar*. For example, mass, volume, and length are scalar quantities often used in statics. In this book, scalars are indicated by letters in italic type, such as the scalar A.

Vector. A *vector* is a quantity that has both a magnitude and a direction. In statics the vector quantities frequently encountered are position, force, and moment. For handwritten work, a vector is generally represented by a letter with an arrow written over it, such as $\vec{A}$. The magnitude is designated $|\vec{A}|$ or simply A. In this book vectors will be symbolized in boldface type; for example, **A** is used to designate the vector "A." Its magnitude, which is always a positive quantity, is symbolized in italic type, written as $|A|$, or simply A when it is understood that A is a positive scalar.

A vector is represented graphically by an arrow, which is used to define

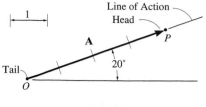

Fig. 2–1

its magnitude, direction, and sense. The *magnitude* of the vector is the length of the arrow, the *direction* is defined by the angle between a reference axis and the arrow's line of action, and the *sense* is indicated by the arrowhead. For example, the vector **A** shown in Fig. 2–1 has a magnitude of 4 units, a direction which is 20° measured counterclockwise from the horizontal axis, and a sense which is upward and to the right. The point *O* is called the *tail* of the vector, the point *P* the *tip* or *head*.

2.2 Vector Operations

Vector **A** and its negative counterpart

Fig. 2–2

Multiplication and Division of a Vector by a Scalar. The product of vector **A** and scalar *a*, yielding *a***A**, is defined as a vector having a magnitude $|aA|$. The *sense* of *a***A** is the *same* as **A** provided *a* is *positive*; it is *opposite* to **A** if *a* is *negative*. In particular, the negative of a vector is formed by multiplying the vector by the scalar (-1), Fig. 2–2. Division of a vector by a scalar can be defined using the laws of multiplication, since $\mathbf{A}/a = (1/a)\mathbf{A}, a \neq 0$. Graphic examples of these operations are shown in Fig. 2–3.

Vector Addition. Two vectors **A** and **B** such as force or position, Fig. 2–4a, may be added to form a "resultant" vector $\mathbf{R} = \mathbf{A} + \mathbf{B}$ by using the *parallelogram law*. To do this, **A** and **B** are joined at their tails, Fig. 2–4b. Parallel lines drawn from the head of each vector intersect at a common point, thereby forming the adjacent sides of a parallelogram. As shown, the resultant **R** is the diagonal of the parallelogram, which extends from the tails of **A** and **B** to the intersection of the lines.

We can also add **B** to **A** using a *triangle construction*, which is a special case of the parallelogram law, whereby vector **B** is added to vector **A** in a "head-to-tail" fashion, i.e., by connecting the head of **A** to the tail of **B**, Fig. 2–4c. The resultant **R** extends from the tail of **A** to the head of **B**. In a similar manner, **R** can also be obtained by adding **A** to **B**, Fig. 2–4d. By comparison, it is seen that vector addition is commutative; in other words, the vectors can be added in either order, i.e., $\mathbf{R} = \mathbf{A} + \mathbf{B} = \mathbf{B} + \mathbf{A}$.

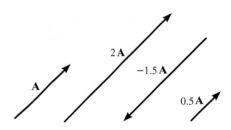

Scalar Multiplication and Division

Fig. 2–3

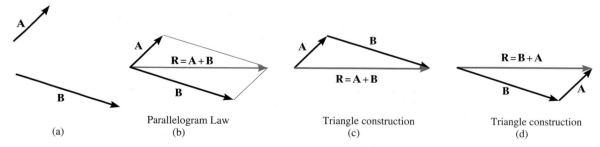

| (a) | Parallelogram Law (b) | Triangle construction (c) | Triangle construction (d) |

Vector Addition

Fig. 2–4

As a special case, if the two vectors **A** and **B** are *collinear*, i.e., both have the same line of action, the parallelogram law reduces to an *algebraic* or *scalar addition* $R = A + B$, as shown in Fig. 2–5.

Vector Subtraction.   The resultant *difference* between two vectors **A** and **B** of the same type may be expressed as

$$R' = A - B = A + (-B)$$

This vector sum is shown graphically in Fig. 2–6. Subtraction is therefore defined as a special case of addition, so the rules of vector addition also apply to vector subtraction.

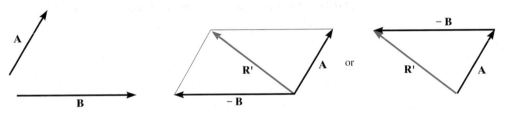

$R = A+B$

Addition of collinear vectors

Fig. 2–5

Parallelogram law

Vector Subtraction

Triangle construction

Fig. 2–6

Resolution of Vector. A vector may be resolved into two "components" having known lines of action by using the parallelogram law. For example, if **R** in Fig. 2–7a is to be resolved into components acting along the lines a and b, one starts at the *head* of **R** and extends a line *parallel* to a until it intersects b. Likewise, a line parallel to b is drawn from the *head* of **R** to the point of intersection with a, Fig. 2–7a. The two components **A** and **B** are then drawn such that they extend from the tail of **R** to the points of intersection, as shown in Fig. 2–7b.

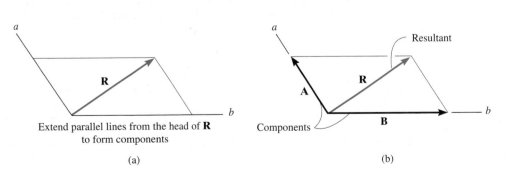

Extend parallel lines from the head of **R** to form components

(a)

Components

(b)

Resolution of a vector

Fig. 2–7

2.3 Vector Addition of Forces

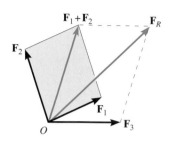

Fig. 2–8

Experimental evidence has shown that a force is a vector quantity since it has a specified magnitude, direction, and sense and it adds according to the parallelogram law. Two common problems in statics involve either finding the resultant force, knowing its components, or resolving a known force into two components. As described in Sec. 2.2, both of these problems require application of the parallelogram law.

If more than two forces are to be added, successive applications of the parallelogram law can be carried out in order to obtain the resultant force. For example, if three forces $\mathbf{F}_1$, $\mathbf{F}_2$, $\mathbf{F}_3$ act at a point O, Fig. 2–8, the resultant of any two of the forces is found—say, $\mathbf{F}_1 + \mathbf{F}_2$—and then this resultant is added to the third force, yielding the resultant of all three forces; i.e., $\mathbf{F}_R = (\mathbf{F}_1 + \mathbf{F}_2) + \mathbf{F}_3$. Using the parallelogram law to add more than two forces, as shown here, often requires extensive geometric and trigonometric calculation to determine the numerical values for the magnitude and direction of the resultant. Instead, problems of this type are easily solved by using the "rectangular-component method," which is explained in Sec. 2.4.

If we know the forces $\mathbf{F}_a$ and $\mathbf{F}_b$ that the two chains a and b exert on the hook, we can find their resultant force $\mathbf{F}_c$ by using the parallelogram law. This requires drawing lines parallel to a and b from the heads of $\mathbf{F}_a$ and $\mathbf{F}_b$ as shown thus forming a parallelogram.

In a similar manner, if the force $\mathbf{F}_c$ along chain c is known, then its two components $\mathbf{F}_a$ and $\mathbf{F}_b$, that act along a and b, can be determined from the parallelogram law. Here we must start at the head of $\mathbf{F}_c$ and construct lines parallel to a and b, thereby forming the parallelogram.

PROCEDURE FOR ANALYSIS

Problems that involve the addition of two forces can be solved as follows:

Parallelogram Law.

- Make a sketch showing the vector addition using the parallelogram law.
- Two "component" forces add according to the parallelogram law, yielding a *resultant* force that forms the diagonal of the parallelogram.
- If a force is to be resolved into *components* along two axes directed from the tail of the force, then start at the head of the force and construct lines parallel to the axes, thereby forming the parallelogram. The sides of the parallelogram represent the components.
- Label all the known and unknown force magnitudes and the angles on the sketch and identify the two unknowns.

Trigonometry.

- Redraw a half portion of the parallelogram to illustrate the triangular head-to-tail addition of the components.
- The magnitude of the resultant force can be determined from the law of cosines, and its direction is determined from the law of sines, Fig. 2–9.
- The magnitude of two force components are determined from the law of sines, Fig. 2–9.

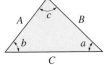

Sine law:
$$\frac{A}{\sin a} = \frac{B}{\sin b} = \frac{C}{\sin c}$$

Cosine law:
$$C = \sqrt{A^2 + B^2 - 2AB \cos c}$$

Fig. 2–9

IMPORTANT POINTS

- A scalar is a positive or negative number.
- A vector is a quantity that has magnitude, direction, and sense.
- Multiplication or division of a vector by a scalar will change the magnitude of the vector. The sense of the vector will change if the scalar is negative.
- As a special case, if the vectors are collinear, the resultant is formed by an algebraic or scalar addition.

E X A M P L E 2.1

The screw eye in Fig. 2–10a is subjected to two forces, $\mathbf{F}_1$ and $\mathbf{F}_2$. Determine the magnitude and direction of the resultant force.

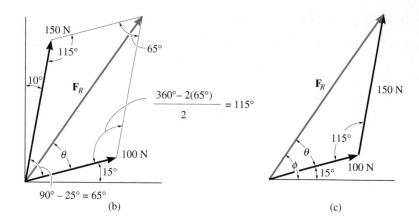

(a)

(b)

(c)

Fig. 2-10

Solution

Parallelogram Law. The parallelogram law of addition is shown in Fig. 2–10b. The two unknowns are the magnitude of $\mathbf{F}_R$ and the angle θ (theta).

Trigonometry. From Fig. 2–10b, the vector triangle, Fig. 2–10c, is constructed. F_R is determined by using the law of cosines:

$$F_R = \sqrt{(100 \text{ N})^2 + (150 \text{ N})^2 - 2(100 \text{ N})(150 \text{ N}) \cos 115°}$$
$$= \sqrt{10\,000 + 22\,500 - 30\,000(-0.4226)} = 212.6 \text{ N}$$
$$= 213 \text{ N} \qquad \qquad Ans.$$

The angle θ is determined by applying the law of sines, using the computed value of F_R.

$$\frac{150 \text{ N}}{\sin \theta} = \frac{212.6 \text{ N}}{\sin 115°}$$

$$\sin \theta = \frac{150 \text{ N}}{212.6 \text{ N}} (0.9063)$$

$$\theta = 39.8°$$

Thus, the direction ϕ (phi) of $\mathbf{F}_R$, measured from the horizontal, is

$$\phi = 39.8° + 15.0° = 54.8° \measuredangle^{\phi} \qquad \qquad Ans.$$

EXAMPLE **2.2**

Resolve the 1000-N ($\approx$ 100-kg) force acting on the pipe, Fig. 2–11a, into components in the (a) x and y directions, and (b) x' and y directions.

(a)

(b)

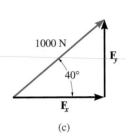

(c)

Fig. 2–11

Solution

In each case the parallelogram law is used to resolve **F** into its two components, and then the vector triangle is constructed to determine the numerical results by trigonometry.

Part (a). The vector addition $\mathbf{F} = \mathbf{F}_x + \mathbf{F}_y$ is shown in Fig. 2–11b. In particular, note that the length of the components is scaled along the x and y axes by first constructing lines from the tip of **F** parallel to the axes in accordance with the parallelogram law. From the vector triangle, Fig. 2–11c,

$$F_x = 1000 \text{ N} \cos 40° = 766 \text{ N} \qquad Ans.$$
$$F_y = 1000 \text{ N} \sin 40° = 643 \text{ N} \qquad Ans.$$

Part (b). The vector addition $\mathbf{F} = \mathbf{F}_{x'} + \mathbf{F}_y$ is shown in Fig. 2–11d. Note carefully how the parallelogram is constructed. Applying the law of sines and using the data listed on the vector triangle, Fig. 2–11e, yields

$$\frac{F_{x'}}{\sin 50°} = \frac{1000 \text{ N}}{\sin 60°}$$

$$F_{x'} = 1000 \text{ N} \left(\frac{\sin 50°}{\sin 60°} \right) = 884.6 \text{ N} \qquad Ans.$$

$$\frac{F_y}{\sin 70°} = \frac{1000 \text{ N}}{\sin 60°}$$

$$F_y = 1000 \text{ N} \left(\frac{\sin 70°}{\sin 60°} \right) = 1085 \text{ N} \qquad Ans.$$

(d)

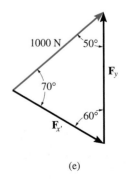

(e)

E X A M P L E   2.3

The force **F** acting on the frame shown in Fig. 2–12a has a magnitude of 500 N and is to be resolved into two components acting along members AB and AC. Determine the angle θ, measured *below* the horizontal, so that the component $\mathbf{F}_{AC}$ is directed from A toward C and has a magnitude of 400 N.

(b)

(c)

$F = 500$ N

(a)

Fig. 2–12

Solution

By using the parallelogram law, the vector addition of the two components yielding the resultant is shown in Fig. 2–12b. Note carefully how the resultant force is resolved into the two components $\mathbf{F}_{AB}$ and $\mathbf{F}_{AC}$, which have specified lines of action. The corresponding vector triangle is shown in Fig. 2–12c.

The angle ϕ can be determined by using the law of sines:

$$\frac{400 \text{ N}}{\sin \phi} = \frac{500 \text{ N}}{\sin 60°}$$

$$\sin \phi = \left(\frac{400 \text{ N}}{500 \text{ N}}\right) \sin 60° = 0.6928$$

$$\phi = 43.9°$$

Hence,

$$\theta = 180° - 60° - 43.9° = 76.1° \; \diagdown_\theta \qquad \textit{Ans.}$$

Using this value for θ, apply the law of cosines or the law of sines and show that $\mathbf{F}_{AB}$ has a magnitude of 561 N.

Notice that **F** can also be directed at an angle θ *above* the horizontal, as shown in Fig. 2–12d, and still produce the required component $\mathbf{F}_{AC}$. Show that in this case $\theta = 16.1°$ and $F_{AB} = 161$ N.

(d)

E X A M P L E   **2.4**

The ring shown in Fig. 2–13*a* is subjected to two forces, $\mathbf{F}_1$ and $\mathbf{F}_2$. If it is required that the resultant force have a magnitude of 1 kN and be directed vertically downward, determine (a) the magnitudes of $\mathbf{F}_1$ and $\mathbf{F}_2$ provided $\theta = 30°$, and (b) the magnitudes of $\mathbf{F}_1$ and $\mathbf{F}_2$ if F_2 is to be a minimum.

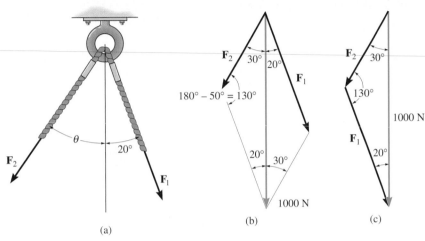

(a)

(b)

(c)

Fig. 2–13

(d)

(e)

Solution

Part (a). A sketch of the vector addition according to the parallelogram law is shown in Fig. 2–13*b*. From the vector triangle constructed in Fig. 2–13*c*, the unknown magnitudes F_1 and F_2 are determined by using the law of sines:

$$\frac{F_1}{\sin 30°} = \frac{1000 \text{ N}}{\sin 130°}$$

$$F_1 = 653 \text{ N} \qquad\qquad Ans.$$

$$\frac{F_2}{\sin 20°} = \frac{1000 \text{ N}}{\sin 130°}$$

$$F_2 = 446 \text{ N} \qquad\qquad Ans.$$

Part (b). If θ is not specified, then by the vector triangle, Fig. 2–13*d*, $\mathbf{F}_2$ may be added to $\mathbf{F}_1$ in various ways to yield the resultant 1000-N force. In particular, the *minimum* length or magnitude of $\mathbf{F}_2$ will occur when its line of action is *perpendicular to* $\mathbf{F}_1$. Any other direction, such as *OA* or *OB*, yields a larger value for F_2. Hence, when $\theta = 90° - 20° = 70°$, F_2 is minimum. From the triangle shown in Fig. 2–13*e*, it is seen that

$$F_1 = 1000 \sin 70°\text{N} = 940 \text{ N} \qquad\qquad Ans.$$

$$F_2 = 1000 \cos 70°\text{N} = 342 \text{ N} \qquad\qquad Ans.$$

PROBLEMS

2-1. Determine the magnitude of the resultant force $\mathbf{F}_R = \mathbf{F}_1 + \mathbf{F}_2$ and its direction, measured counterclockwise from the positive x axis.

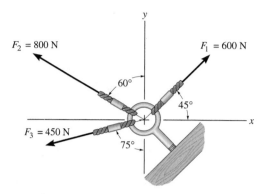

$F_2 = 800$ N

$F_1 = 600$ N

$60°$

$45°$

x

$F_3 = 450$ N

$75°$

Prob. 2–1

2-2. Determine the magnitude of the resultant force if:
(a) $\mathbf{F}_R = \mathbf{F}_1 + \mathbf{F}_2$; (b) $\mathbf{F}'_R = \mathbf{F}_1 - \mathbf{F}_2$.

$F_1 = 100$ N

$60°$

$F_2 = 80$ N

$45°$

Prob. 2–2

2-3. Determine the magnitude of the resultant force $\mathbf{F}_R = \mathbf{F}_1 + \mathbf{F}_2$ and its direction, measured counterclockwise from the positive x axis.

y

$F_1 = 250$ N

$30°$

x

$45°$

$F_2 = 375$ N

Prob. 2–3

*2-4. Determine the magnitude of the resultant force $\mathbf{F}_R = \mathbf{F}_1 + \mathbf{F}_2$ and its direction, measured clockwise from the positive u axis.

2-5. Resolve the force $\mathbf{F}_1$ into components acting along the u and v axes and determine the magnitudes of the components.

2-6. Resolve the force $\mathbf{F}_2$ into components acting along the u and v axes and determine the magnitudes of the components.

$70°$

u

$30°$

$45°$

$F_1 = 300$ N

$F_2 = 500$ N

v

Probs. 2–4/5/6

2-7. The plate is subjected to the two forces at A and B as shown. If $\theta = 60°$, determine the magnitude of the resultant of these two forces and its direction measured from the horizontal.

*2-8. Determine the angle θ for connecting member A to the plate so that the resultant force of F_A and F_B is directed horizontally to the right. Also, what is the magnitude of the resultant force.

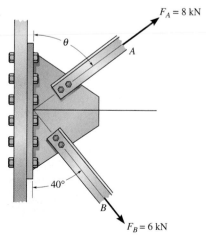

$F_A = 8$ kN

θ

A

$40°$

B

$F_B = 6$ kN

Probs. 2–7/8

2-9. The vertical force **F** acts downward at *A* on the two-membered frame. Determine the magnitudes of the two components of **F** directed along the axes of *AB* and *AC*. Set *F* = 500 N.

2-10. Solve Prob. 2–9 with *F* = 350 N.

2-13. The 500-N force acting on the frame is to be resolved into two components acting along the axis of the struts *AB* and *AC*. If the component of force along *AC* is required to be 300 N, directed from *A* to *C*, determine the magnitude of force acting along *AB* and the angle θ of the 500-N force.

Probs. 2–9/10

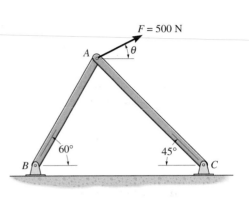

Prob. 2–13

2-11. The force acting on the gear tooth is *F* = 20 N. Resolve this force into two components acting along the lines *aa* and *bb*.

*2-12. The component of force **F** acting along line *aa* is require to be 30 N. Determine the magnitude of **F** and its component along line *bb*.

2-14. The post is to be pulled out of the ground using two ropes *A* and *B*. Rope *A* is subjected to a force of 600 N and is directed at 60° from the horizontal. If the resultant force acting on the post is to be 1200 N, vertically upward, determine the force *T* in rope *B* and the corresponding angle θ.

Probs. 2–11/12

Prob. 2–14

2-15. Determine the design angle $\theta(0° \le \theta \le 90°)$ for strut AB so that the 400-N horizontal force has a component of 500-N directed from A towards C. What is the component of force acting along member AB? Take $\phi = 40°$.

*2-16. Determine the design angle $\phi(0° \le \phi \le 90°)$ between struts AB and AC so that the 400-N horizontal force has a component of 600-N which acts up to the left, in the same direction as from B towards A. Take $\theta = 30°$.

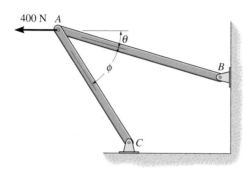

Probs. 2–15/16

2-17. The chisel exerts a force of 200 N on the wood dowel rod which is turning in a lathe. Resolve this force into components acting (a) along the n and t axes and (b) along the x and y axes.

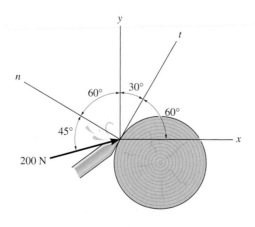

Prob. 2–17

2-18. Two forces are applied at the end of a screw eye in order to remove the post. Determine the angle θ ($0° \le \theta \le 90°$) and the magnitude of force **F** so that the resultant force acting on the post is directed vertically upward and has a magnitude of 750 N.

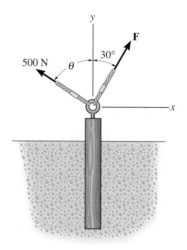

Prob. 2–18

2-19. If $F_1 = F_2 = 300$ N determine the angles θ and ϕ so that the resultant force is directed along the positive x axis and has a magnitude of $F_R = 200$ N.

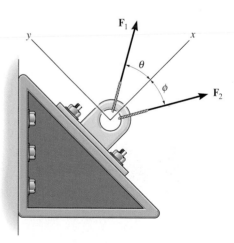

Prob. 2–19

*2-20. The truck is to be towed using two ropes. Determine the magnitude of forces F_A and F_B acting on each rope in order to develop a resultant force of 950 N directed along the positive *x* axis. Set $\theta = 50°$.

2-22. Determine the magnitude and direction of the resultant $\mathbf{F}_R = \mathbf{F}_1 + \mathbf{F}_2 + \mathbf{F}_3$ of the three forces by first finding the resultant $\mathbf{F}' = \mathbf{F}_1 + \mathbf{F}_2$ and then forming $\mathbf{F}_R = \mathbf{F}' + \mathbf{F}_3$.

2-23. Determine the magnitude and direction of the resultant $\mathbf{F}_R = \mathbf{F}_1 + \mathbf{F}_2 + \mathbf{F}_3$ of the three forces by first finding the resultant $\mathbf{F}' = \mathbf{F}_2 + \mathbf{F}_3$ and then forming $\mathbf{F}_R = \mathbf{F}' + \mathbf{F}_1$.

Prob. 2–20

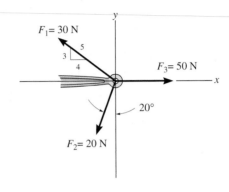

Probs. 2–22/23

2-21. The truck is to be towed using two ropes. If the resultant force is to be 950 N, directed along the positive *x* axis, determine the magnitudes of forces F_A and F_B acting on each rope and the angle of θ of F_B so that the magnitude of F_B is a *minimum*. F_A acts at 20° from the *x* axis as shown.

*2-24. Resolve the 500-N force into components acting along (a) the *x* and *y* axes, and (b) the *x* and *y'* axes.

Prob. 2–21

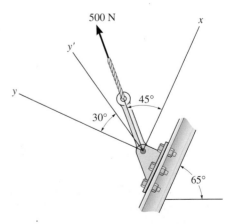

Prob. 2–24

2-25. The log is being towed by two tractors A and B. Determine the magnitude of the two towing forces $\mathbf{F}_A$ and $\mathbf{F}_B$ if it is required that the resultant force have a magnitude $F_R = 10$ kN and be directed along the x axis. Set $\theta = 15°$.

2-26. If the resultant $\mathbf{F}_R$ of the two forces acting on the log is to be directed along the positive x axis and have a magnitude of 10 kN, determine the angle θ of the cable, attached to B such that the force $\mathbf{F}_B$ in this cable is minimum. What is the magnitude of the force in each cable for this situation?

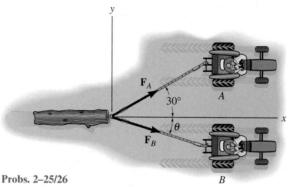

Probs. 2–25/26

2-27. The beam is to be hoisted using two chains. Determine the magnitudes of forces $\mathbf{F}_A$ and $\mathbf{F}_B$ acting on each chain in order to develop a resultant force of 600 N directed along the positive y axis. Set $\theta = 45°$.

*2-28. The beam is to be hoisted using two chains. If the resultant force is to be 600 N, directed along the positive y axis, determine the magnitudes of forces $\mathbf{F}_A$ and $\mathbf{F}_B$ acting on each chain and the orientation θ of $\mathbf{F}_B$ so that the magnitude of $\mathbf{F}_B$ is a minimum. $\mathbf{F}_A$ acts at 30° from the y axis as shown.

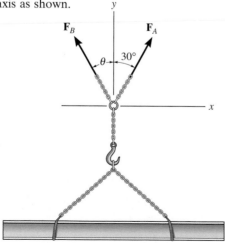

Probs. 2–27/28

2-29. Three chains act on the bracket such that they create a resultant force having a magnitude of 500 N. If two of the chains are subjected to known forces, as shown, determine the orientation θ of the third chain, measured clockwise from the positive x axis, so that the magnitude of force $\mathbf{F}$ in this chain is a *minimum*. All forces lie in the x–y plane. What is the magnitude of $\mathbf{F}$? *Hint*: First find the resultant of the two known forces. Force $\mathbf{F}$ acts in this direction.

Prob. 2–29

2-30. Three cables pull on the pipe such that they create a resultant force having a magnitude of 900 N. If two of the cables are subjected to known forces, as shown in the figure, determine the direction θ of the third cable so that the magnitude of force $\mathbf{F}$ in this cable is a *minimum*. All forces lie in the x–y plane. What is the magnitude of $\mathbf{F}$? *Hint*: First find the resultant of the two known forces.

Prob. 2–30

2.4 Addition of a System of Coplanar Forces

When the resultant of more than two forces has to be obtained, it is easier to find the components of each force along specified axes, add these components algebraically, and then form the resultant, rather than form the resultant of the forces by successive application of the parallelogram law as discussed in Sec. 2.3.

In this section we will resolve each force into its rectangular components F_x and F_y, which lie along the x and y axes, respectively, Fig. 2–14a. Although the axes are horizontal and vertical, they may in general be directed at any inclination, as long as they remain perpendicular to one another, Fig. 2–14b. In either case, by the parallelogram law, we require

$$\mathbf{F} = \mathbf{F}_x + \mathbf{F}_y$$

and

$$\mathbf{F}' = \mathbf{F}'_x + \mathbf{F}'_y$$

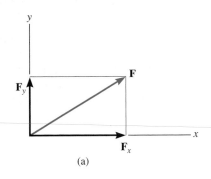

(a)

As shown in Fig. 2–14, the sense of direction of each force component is represented *graphically* by the *arrowhead*. For *analytical* work, however, we must establish a notation for representing the directional sense of the rectangular components. This can be done in one of two ways.

Scalar Notation. Since the x and y axes have designated positive and negative directions, the magnitude and directional sense of the rectangular components of a force can be expressed in terms of *algebraic scalars*. For example, the components of $\mathbf{F}$ in Fig. 2–14a can be represented by positive scalars F_x and F_y since their sense of direction is along the *positive x* and *y* axes, respectively. In a similar manner, the components of $\mathbf{F}'$ in Fig. 2–14b are F'_x and $-F'_y$. Here the y component is negative, since $\mathbf{F}'_y$ is directed along the negative y axis.

It is important to keep in mind that this scalar notation is to be used only for computational purposes, not for graphical representations in figures. Throughout the book, the *head of a vector arrow* in any figure indicates the sense of the vector *graphically*; algebraic signs are not used for this purpose. Thus, the vectors in Figs. 2–14a and 2–14b are designated by using boldface (vector) notation.* Whenever italic symbols are written near vector arrows in figures, they indicate the *magnitude* of the vector, which is *always* a *positive* quantity.

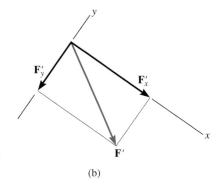

(b)

Fig. 2–14

*Negative signs are used only in figures with boldface notation when showing equal but opposite pairs of vectors as in Fig. 2–2.

Cartesian Vector Notation. It is also possible to represent the components of a force in terms of Cartesian unit vectors. When we do this the methods of vector algebra are easier to apply, and we will see that this becomes particularly advantageous for solving problems in three dimensions.

In two dimensions the *Cartesian unit vectors* **i** and **j** are used to designate the *directions* of the x and y axes, respectively, Fig. 2–15a.* These vectors have a dimensionless magnitude of unity, and their sense (or arrowhead) will be described analytically by a plus or minus sign, depending on whether they are pointing along the positive or negative x or y axis.

As shown in Fig. 2–15a, the *magnitude* of each component of **F** is *always a positive quantity*, which is represented by the (positive) scalars F_x and F_y. Therefore, having established notation to represent the magnitude and the direction of each vector component, we can express **F** in Fig. 2–15a as the *Cartesian vector*,

$$\mathbf{F} = F_x\mathbf{i} + F_y\mathbf{j}$$

And in the same way, **F'** in Fig. 2–15b can be expressed as

$$\mathbf{F'} = F'_x\mathbf{i} + F'_y(-\mathbf{j})$$

or simply

$$\mathbf{F'} = F'_x\mathbf{i} - F'_y\mathbf{j}$$

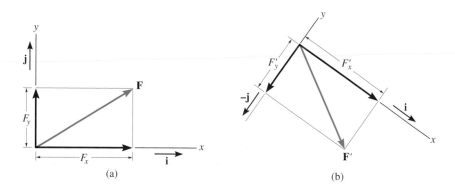

(a)

(b)

Fig. 2–15

---

*For handwritten work, unit vectors are usually indicated using a circumflex, e.g., $\hat{i}$ and $\hat{j}$.

Coplanar Force Resultants. Either of the two methods just described can be used to determine the resultant of several *coplanar forces*. To do this, each force is first resolved into its x and y components, and then the respective components are added using *scalar algebra* since they are collinear. The resultant force is then formed by adding the resultants of the x and y components using the parallelogram law. For example, consider the three concurrent forces in Fig. 2–16a, which have x and y components as shown in Fig. 2–16b. To solve this problem using *Cartesian vector notation*, each force is first represented as a Cartesian vector, i.e.,

$$\mathbf{F}_1 = F_{1x}\mathbf{i} + F_{1y}\mathbf{j}$$
$$\mathbf{F}_2 = -F_{2x}\mathbf{i} + F_{2y}\mathbf{j}$$
$$\mathbf{F}_3 = F_{3x}\mathbf{i} - F_{3y}\mathbf{j}$$

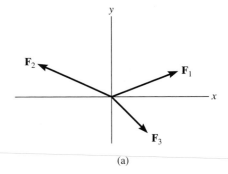
(a)

The vector resultant is therefore

$$\begin{aligned}
\mathbf{F}_R &= \mathbf{F}_1 + \mathbf{F}_2 + \mathbf{F}_3 \\
&= F_{1x}\mathbf{i} + F_{1y}\mathbf{j} - F_{2x}\mathbf{i} + F_{2y}\mathbf{j} + F_{3x}\mathbf{i} - F_{3y}\mathbf{j} \\
&= (F_{1x} - F_{2x} + F_{3x})\mathbf{i} + (F_{1y} + F_{2y} - F_{3y})\mathbf{j} \\
&= (F_{Rx})\mathbf{i} + (F_{Ry})\mathbf{j}
\end{aligned}$$

If *scalar notation* is used, then, from Fig. 2–16b, since x is positive to the right and y is positive upward, we have

$$(\xrightarrow{+}) \qquad\qquad F_{Rx} = F_{1x} - F_{2x} + F_{3x}$$
$$(+\uparrow) \qquad\qquad F_{Ry} = F_{1y} + F_{2y} - F_{3y}$$

These results are the *same* as the $\mathbf{i}$ and $\mathbf{j}$ components of $\mathbf{F}_R$ determined above.

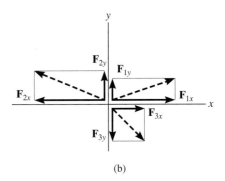
(b)

In the general case, the x and y components of the resultant of any number of coplanar forces can be represented symbolically by the algebraic sum of the x and y components of all the forces, i.e.,

$$\boxed{\begin{aligned} F_{Rx} &= \Sigma F_x \\ F_{Ry} &= \Sigma F_y \end{aligned}} \qquad (2\text{–}1)$$

When applying these equations, it is important to use the *sign convention* established for the components; and that is, components having a directional sense along the positive coordinate axes are considered positive scalars, whereas those having a directional sense along the negative coordinate axes are considered negative scalars. If this convention is followed, then the signs of the resultant components will specify the sense of these components. For example, a positive result indicates that the component has a directional sense which is in the positive coordinate direction.

(c)

Fig. 2–16

Fig. 2–16

The resultant force of the four cable forces acting on the supporting bracket can be determined by adding algebraically the separate x and y components of each cable force. This resultant $\mathbf{F}_R$ produces the *same pulling effect* on the bracket as all four cables.

Once the resultant components are determined, they may be sketched along the x and y axes in their proper directions, and the resultant force can be determined from vector addition, as shown in Fig. 2–16c. From this sketch, the magnitude of $\mathbf{F}_R$ is then found from the Pythagorean theorem; that is,

$$F_R = \sqrt{F_{Rx}^2 + F_{Ry}^2}$$

Also, the direction angle θ, which specifies the orientation of the force, is determined from trigonometry:

$$\theta = \tan^{-1}\left|\frac{F_{Ry}}{F_{Rx}}\right|$$

The above concepts are illustrated numerically in the examples which follow.

IMPORTANT POINTS

- The resultant of several coplanar forces can easily be determined if an x, y coordinate system is established and the forces are resolved along the axes.

- The direction of each force is specified by the angle its line of action makes with one of the axes, or by a sloped triangle.

- The orientation of the x and y axes is arbitrary, and their positive direction can be specified by the Cartesian unit vectors $\mathbf{i}$ and $\mathbf{j}$.

- The x and y components of the *resultant force* are simply the algebraic addition of the components of all the coplanar forces.

- The magnitude of the resultant force is determined from the Pythagorean theorem, and when the components are sketched on the x and y axes, the direction can be determined from trigonometry.

EXAMPLE 2.5

Determine the x and y components of $\mathbf{F}_1$ and $\mathbf{F}_2$ acting on the boom shown in Fig. 2–17a. Express each force as a Cartesian vector.

$F_1 = 200$ N

(a)

Solution

Scalar Notation. By the parallelogram law, $\mathbf{F}_1$ is resolved into x and y components, Fig. 2–17b. The magnitude of each component is determined by trigonometry. Since $\mathbf{F}_{1x}$ acts in the $-x$ direction, and $\mathbf{F}_{1y}$ acts in the $+y$ direction, we have

$$F_{1x} = -200 \sin 30° \text{ N} = -100 \text{ N} = 100 \text{ N} \leftarrow \qquad \textit{Ans.}$$
$$F_{1y} = 200 \cos 30° \text{ N} = 173 \text{ N} = 173 \text{ N}\uparrow \qquad \textit{Ans.}$$

The force $\mathbf{F}_2$ is resolved into its x and y components as shown in Fig. 2–17c. Here the *slope* of the line of action for the force is indicated. From this "slope triangle" we could obtain the angle θ, e.g., $\theta = \tan^{-1}\left(\frac{5}{12}\right)$, and then proceed to determine the magnitudes of the components in the same manner as for $\mathbf{F}_1$. An easier method, however, consists of using proportional parts of similar triangles, i.e.,

$$\frac{F_{2x}}{260 \text{ N}} = \frac{12}{13} \qquad F_{2x} = 260 \text{ N}\left(\frac{12}{13}\right) = 240 \text{ N}$$

(b)

Similarly,

$$F_{2y} = 260 \text{ N}\left(\frac{5}{13}\right) = 100 \text{ N}$$

Notice that the magnitude of the *horizontal component*, F_{2x}, was obtained by multiplying the force magnitude by the ratio of the *horizontal leg* of the slope triangle divided by the hypotenuse; whereas the magnitude of the *vertical component*, F_{2y}, was obtained by multiplying the force magnitude by the ratio of the *vertical leg* divided by the hypotenuse. Hence, using scalar notation,

$$F_{2x} = 240 \text{ N} = 240 \text{ N} \rightarrow \qquad \textit{Ans.}$$
$$F_{2y} = -100 \text{ N} = 100 \text{ N} \downarrow \qquad \textit{Ans.}$$

Cartesian Vector Notation. Having determined the magnitudes and directions of the components of each force, we can express each force as a Cartesian vector.

$$\mathbf{F}_1 = \{-100\mathbf{i} + 173\mathbf{j}\} \text{ N} \qquad \textit{Ans.}$$
$$\mathbf{F}_2 = \{240\mathbf{i} - 100\mathbf{j}\} \text{ N} \qquad \textit{Ans.}$$

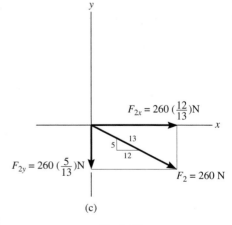

(c)

Fig. 2–17

E X A M P L E 2.6

(a)

(b)

(c)

Fig. 2–18

The link in Fig. 2–18a is subjected to two forces $\mathbf{F}_1$ and $\mathbf{F}_2$. Determine the magnitude and orientation of the resultant force.

Solution I

Scalar Notation. This problem can be solved by using the parallelogram law; however, here we will resolve each force into its x and y components, Fig. 2–18b, and sum these components algebraically. Indicating the "positive" sense of the x and y force components alongside each equation, we have

$$\xrightarrow{+} F_{Rx} = \Sigma F_x; \qquad F_{Rx} = 600 \cos 30° \text{ N} - 400 \sin 45° \text{ N}$$
$$= 236.8 \text{ N} \rightarrow$$

$$+\uparrow F_{Ry} = \Sigma F_y; \qquad F_{Ry} = 600 \sin 30° \text{ N} + 400 \cos 45° \text{ N}$$
$$= 582.8 \text{ N} \uparrow$$

The resultant force, shown in Fig. 2–18c, has a *magnitude* of

$$F_R = \sqrt{(236.8 \text{ N})^2 + (582.8 \text{ N})^2}$$
$$= 629 \text{ N} \qquad\qquad\qquad Ans.$$

From the vector addition, Fig. 2–18c, the direction angle θ is

$$\theta = \tan^{-1}\left(\frac{582.8\text{N}}{236.8\text{N}}\right) = 67.9° \qquad\qquad Ans.$$

Solution II

Cartesian Vector Notation. From Fig. 2–18b, each force is expressed as a Cartesian vector

$$\mathbf{F}_1 = \{600 \cos 30°\mathbf{i} + 600 \sin 30°\mathbf{j}\} \text{ N}$$
$$\mathbf{F}_2 = \{-400 \sin 45°\mathbf{i} + 400 \cos 45°\mathbf{j}\} \text{ N}$$

Thus,

$$\mathbf{F}_R = \mathbf{F}_1 + \mathbf{F}_2 = (600 \cos 30° \text{ N} - 400 \sin 45° \text{ N})\mathbf{i}$$
$$+ (600 \sin 30° \text{ N} + 400 \cos 45° \text{ N})\mathbf{j}$$
$$= \{236.8\mathbf{i} + 582.8\mathbf{j}\} \text{ N}$$

The magnitude and direction of $\mathbf{F}_R$ are determined in the same manner as shown above.

Comparing the two methods of solution, note that use of scalar notation is more efficient since the components can be found *directly*, without first having to express each force as a Cartesian vector before adding the components. Later we will show that Cartesian vector analysis is very beneficial for solving three-dimensional problems.

E X A M P L E **2.7**

The end of the boom O in Fig. 2–19a is subjected to three concurrent and coplanar forces. Determine the magnitude and orientation of the resultant force.

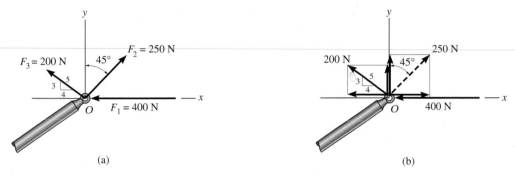

(a)   (b)

Fig. 2–19

Solution

Each force is resolved into its x and y components, Fig. 2–19b. Summing the x components, we have

$$\xrightarrow{+} F_{Rx} = \Sigma F_x; \qquad F_{Rx} = -400 \text{ N} + 250 \sin 45° \text{ N} - 200(\tfrac{4}{5}) \text{ N}$$
$$= -383.2 \text{ N} = 383.2 \text{ N} \leftarrow$$

The negative sign indicates that F_{Rx} acts to the left, i.e., in the negative x direction as noted by the small arrow. Summing the y components yields

$$+\uparrow F_{Ry} = \Sigma F_y; \qquad F_{Ry} = 250 \cos 45° \text{ N} + 200(\tfrac{3}{5}) \text{ N}$$
$$= 296.8 \text{ N}\uparrow$$

The resultant force, shown in Fig. 2–19c, has a *magnitude* of

$$F_R = \sqrt{(-383.2\text{N})^2 + (296.8\text{N})^2}$$
$$= 485 \text{ N} \qquad\qquad \textit{Ans.}$$

From the vector addition in Fig. 2–19c, the direction angle θ is

$$\theta = \tan^{-1}\left(\frac{296.8}{383.2}\right) = 37.8° \qquad\qquad \textit{Ans.}$$

Note how convenient it is to use this method, compared to two applications of the parallelogram law.

PROBLEMS

2-31. Determine the x and y components of the 800-N force.

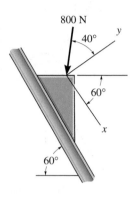

Prob. 2–31

2-33. Determine the magnitude of force **F** so that the resultant $\mathbf{F}_R$ of the three forces is as small as possible.

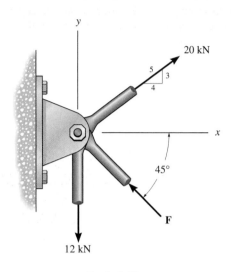

Prob. 2–33

*2-32. Determine the magnitude of the resultant force and its direction, measured clockwise from the positive x axis.

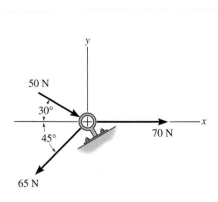

Prob. 2–32

2-34. Determine the magnitude of the resultant force and its direction, measured counterclockwise from the positive x axis.

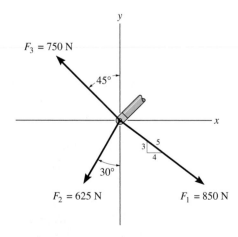

Prob. 2–34

2-35. Three forces act on the bracket. Determine the magnitude and direction θ of $\mathbf{F}_1$ so that the resultant force is directed along the positive x' axis and has a magnitude of 1 kN.

*2-36. If $F_1 = 300$ N and $\theta = 20°$, determine the magnitude and direction, measured counterclockwise from the x' axis, of the resultant force of the three forces acting on the bracket.

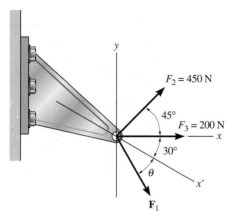

Probs. 2–35/36

2-37. Determine the magnitude and direction θ of $\mathbf{F}_1$ so that the resultant force is directed vertically upward and has a magnitude of 800 N.

2-38. Determine the magnitude and direction measured counterclockwise from the positive x axis of the resultant force of the three forces acting on the ring A. Take $F_1 = 500$ N and $\theta = 20°$.

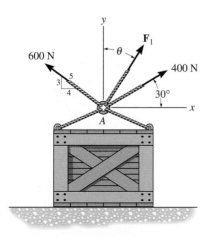

Probs. 2–37/38

2-39. Express $\mathbf{F}_1$ and $\mathbf{F}_2$ as Cartesian vectors.

*2-40. Determine the magnitude of the resultant force and its direction measured counterclockwise from the positive x axis.

Probs. 2–39/40

2-41. Solve Prob. 2–1 by summing the rectangular or x, y components of the forces to obtain the resultant force.

2-42. Solve Prob. 2–22 by summing the rectangular or x, y components of the forces to obtain the resultant force.

2-43. Determine the magnitude and orientation θ of $\mathbf{F}_B$ so that the resultant force is directed along the positive y axis and has a magnitude of 1500 N.

*2-44. Determine the magnitude and orientation, measured counterclockwise from the positive y axis, of the resultant force acting on the bracket, if $F_B = 600$ N and $\theta = 20°$.

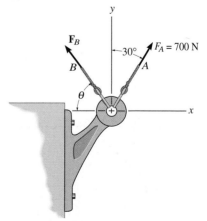

Probs. 2–43/44

2-45. Determine the x and y components of $\mathbf{F}_1$ and $\mathbf{F}_2$.

2-46. Determine the magnitude of the resultant force and its direction, measured counterclockwise from the positive x axis.

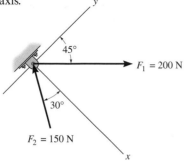

Probs. 2–45/46

2-47. Determine the x and y components of each force acting on the *gusset plate* of the bridge truss. Show that the resultant force is zero.

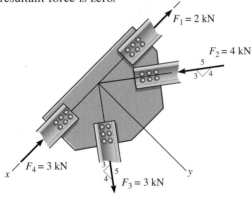

Prob. 2–47

*2-48. If $\theta = 60°$ and $F = 20$ kN, determine the magnitude of the resultant force and its direction measured clockwise from the positive x axis.

Prob. 2–48

2-49. Determine the magnitude and direction θ of $\mathbf{F}_A$ so that the resultant force is directed along the positive x axis and has a magnitude of 1250 N.

2-50. Determine the magnitude and direction, measured counterclockwise from the positive x axis, of the resultant force acting on the ring at O, if $F_A = 750$ N and $\theta = 45°$.

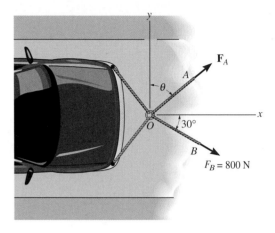

Probs. 2–49/50

2-51. Express each of the three forces acting on the column in Cartesian vector form and compute the magnitude of the resultant force.

Prob. 2–51

*2-52. The three concurrent forces acting on the screw eye produce a resultant force $F_R = 0$. If $F_2 = \frac{2}{3}F_1$ and F_1 is to be 90° from F_2 as shown, determine the required magnitude of F_3 expressed in terms of F_1 and the angle θ.

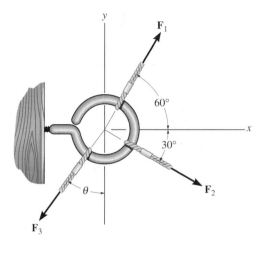

Prob. 2–52

2-53. Determine the magnitude of force **F** so that the resultant F_R of the three forces is as small as possible. What is the minimum magnitude of F_R?

Prob. 2–53

2-54. Express each of the three forces acting on the bracket in Cartesian vector form with respect to the x and y axes. Determine the magnitude and direction θ of F_1 so that the resultant force is directed along the positive x' axis and has a magnitude of $F_R = 600$ N.

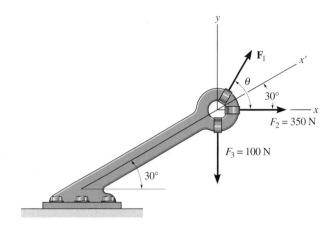

Prob. 2–54

2-55. The three concurrent forces acting on the post produce a resultant force $F_R = 0$. If $F_2 = \frac{1}{2}F_1$, and F_1 is to be 90° from F_2 as shown, determine the required magnitude F_3 expressed in terms of F_1 and the angle θ.

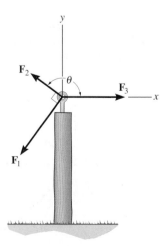

Prob. 2–55

*2-56. Three forces act on the bracket. Determine the magnitude and orientation θ of $\mathbf{F}_2$ so that the resultant force is directed along the positive u axis and has a magnitude of 50 kN.

2-57. If $F_2 = 150$ kN and $\theta = 55°$, determine the magnitude and orientation, measured clockwise from the positive x axis, of the resultant force of the three forces acting on the bracket.

2-58. Determine the magnitude of force $\mathbf{F}$ so that the resultant force of the three forces is as small as possible. What is the magnitude of the resultant force?

Prob. 2–58

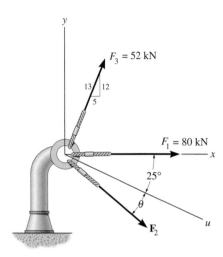

Probs. 2–56/57

2.5 Cartesian Vectors

Right-handed coordinate system.

Fig. 2–20

The operations of vector algebra, when applied to solving problems in *three dimensions*, are greatly simplified if the vectors are first represented in Cartesian vector form. In this section we will present a general method for doing this; then in the next section we will apply this method to solving problems involving the addition of forces. Similar applications will be illustrated for the position and moment vectors given in later sections of the book.

Right-Handed Coordinate System. A right-handed coordinate system will be used for developing the theory of vector algebra that follows. A rectangular or Cartesian coordinate system is said to be *right-handed* provided the thumb of the right hand points in the direction of the positive z axis when the right-hand fingers are curled about this axis and directed from the positive x toward the positive y axis, Fig. 2–20. Furthermore, according to this rule, the z axis for a two-dimensional problem as in Fig. 2–19 would be directed outward, perpendicular to the page.

Rectangular Components of a Vector. A vector **A** may have one, two, or three rectangular components along the *x, y, z* coordinate axes, depending on how the vector is oriented relative to the axes. In general, though, when **A** is directed within an octant of the *x, y, z* frame, Fig. 2–21, then by two successive applications of the parallelogram law, we may resolve the vector into components as $\mathbf{A} = \mathbf{A}' + \mathbf{A}_z$ and then $\mathbf{A}' = \mathbf{A}_x + \mathbf{A}_y$. Combining these equations, **A** is represented by the vector sum of its *three* rectangular components,

$$\mathbf{A} = \mathbf{A}_x + \mathbf{A}_y + \mathbf{A}_z \qquad (2\text{–}2)$$

Unit Vector. The direction of **A** can be specified using a unit vector. This vector is so named since it has a magnitude of 1. If **A** is a vector having a magnitude $A \neq 0$, then the unit vector having the *same direction* as **A** is represented by

$$\mathbf{u}_A = \frac{\mathbf{A}}{A} \qquad (2\text{–}3)$$

So that

$$\mathbf{A} = A\mathbf{u}_A \qquad (2\text{–}4)$$

Since **A** is of a certain type, e.g., a force vector, it is customary to use the proper set of units for its description. The magnitude A also has this same set of units; hence, from Eq. 2–3, the *unit vector will be dimensionless* since the units will cancel out. Equation 2–4 therefore indicates that vector **A** may be expressed in terms of both its magnitude and direction *separately*; i.e., A (a positive scalar) defines the *magnitude* of **A**, and $\mathbf{u}_A$ (a dimensionless vector) defines the *direction* and sense of **A**, Fig. 2–22.

Cartesian Unit Vectors. In three dimensions, the set of Cartesian unit vectors, **i, j, k**, is used to designate the directions of the *x, y, z* axes respectively. As stated in Sec. 2.4, the *sense* (or arrowhead) of these vectors will be described analytically by a plus or minus sign, depending on whether they are pointing along the positive or negative *x, y,* or *z* axes. The positive Cartesian unit vectors are shown in Fig. 2–23.

Fig. 2–21

Fig. 2–22

Fig. 2–23

Fig. 2–24

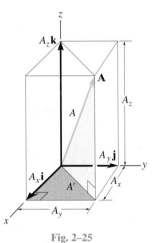

Fig. 2–25

Cartesian Vector Representation. Since the three components of **A** in Eq. 2–2 act in the positive **i**, **j**, and **k** directions, Fig. 2–24, we can write **A** in Cartesian vector form as

$$\mathbf{A} = A_x\mathbf{i} + A_y\mathbf{j} + A_z\mathbf{k} \tag{2–5}$$

There is a distinct advantage to writing vectors in this manner. Note that the *magnitude* and *direction* of each *component vector* are *separated*, and as a result this will simplify the operations of vector algebra, particularly in three dimensions.

Magnitude of a Cartesian Vector. It is always possible to obtain the magnitude of **A** provided it is expressed in Cartesian vector form. As shown in Fig. 2–25, from the colored right triangle, $A = \sqrt{A'^2 + A_z^2}$, and from the shaded right triangle, $A' = \sqrt{A_x^2 + A_y^2}$. Combining these equations yields

$$A = \sqrt{A_x^2 + A_y^2 + A_z^2} \tag{2–6}$$

*Hence, the magnitude of **A** is equal to the positive square root of the sum of the squares of its components.*

Direction of a Cartesian Vector. The *orientation* of **A** is defined by the *coordinate direction angles* α (alpha), β (beta), and γ (gamma), measured between the *tail* of **A** and the *positive x, y, z axes* located at the tail of **A**, Fig. 2–26. Note that regardless of where **A** is directed, each of these angles will be between 0° and 180°.

To determine α, β, and γ, consider the projection of **A** onto the *x, y, z* axes, Fig. 2–27. Referring to the blue colored right triangles shown in each figure, we have

$$\cos\alpha = \frac{A_x}{A} \qquad \cos\beta = \frac{A_y}{A} \qquad \cos\gamma = \frac{A_z}{A} \tag{2–7}$$

These numbers are known as the *direction cosines* of **A**. Once they have been obtained, the coordinate direction angles α, β, γ can then be determined from the inverse cosines.

An easy way of obtaining the direction cosines of **A** is to form a unit vector in the direction of **A**, Eq. 2–3. Provided **A** is expressed in Cartesian vector form, $\mathbf{A} = A_x\mathbf{i} + A_y\mathbf{j} + A_z\mathbf{k}$ (Eq. 2–5), then

Fig. 2–26

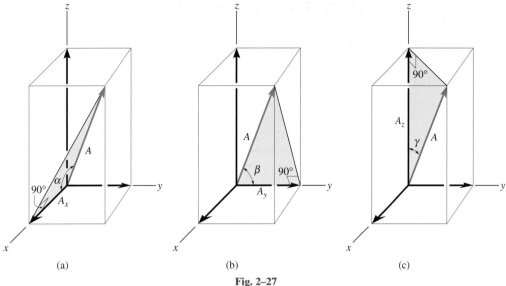

Fig. 2–27

$$\mathbf{u}_A = \frac{\mathbf{A}}{A} = \frac{A_x}{A}\mathbf{i} + \frac{A_y}{A}\mathbf{j} + \frac{A_z}{A}\mathbf{k} \qquad (2\text{–}8)$$

where $A = \sqrt{A_x^2 + A_y^2 + A_z^2}$ (Eq. 2–6). By comparison with Eqs. 2–7, it is seen that *the* $\mathbf{i}$, $\mathbf{j}$, $\mathbf{k}$ *components of* $\mathbf{u}_A$ *represent the direction cosines of* $\mathbf{A}$, i.e.,

$$\mathbf{u}_A = \cos\alpha\,\mathbf{i} + \cos\beta\,\mathbf{j} + \cos\gamma\,\mathbf{k} \qquad (2\text{–}9)$$

Since the magnitude of a vector is equal to the positive square root of the sum of the squares of the magnitudes of its components, and $\mathbf{u}_A$ has a magnitude of 1, then from Eq. 2–9 an important relation between the direction cosines can be formulated as

$$\cos^2\alpha + \cos^2\beta + \cos^2\gamma = 1 \qquad (2\text{–}10)$$

Provided vector $\mathbf{A}$ lies in a known octant, this equation can be used to determine one of the coordinate direction angles if the other two are known.

Finally, if the magnitude and coordinate direction angles of $\mathbf{A}$ are given, $\mathbf{A}$ may be expressed in Cartesian vector form as

$$\begin{aligned}
\mathbf{A} &= A\mathbf{u}_A \\
&= A\cos\alpha\,\mathbf{i} + A\cos\beta\,\mathbf{j} + A\cos\gamma\,\mathbf{k} \\
&= A_x\mathbf{i} + A_y\mathbf{j} + A_z\mathbf{k}
\end{aligned} \qquad (2\text{–}11)$$

2.6 Addition and Subtraction of Cartesian Vectors

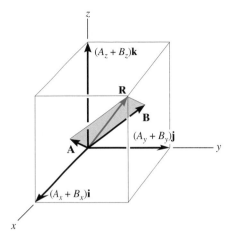

Fig. 2–28

The vector operations of addition and subtraction of two or more vectors are greatly simplified if the vectors are expressed in terms of their Cartesian components. For example, if $\mathbf{A} = A_x\mathbf{i} + A_y\mathbf{j} + A_z\mathbf{k}$ and $\mathbf{B} = B_x\mathbf{i} + B_y\mathbf{j} + B_z\mathbf{k}$, Fig. 2–28, then the resultant vector, $\mathbf{R}$, has components which represent the scalar sums of the $\mathbf{i}, \mathbf{j}, \mathbf{k}$ components of $\mathbf{A}$ and $\mathbf{B}$, i.e.,

$$\mathbf{R} = \mathbf{A} + \mathbf{B} = (A_x + B_x)\mathbf{i} + (A_y + B_y)\mathbf{j} + (A_z + B_z)\mathbf{k}$$

Vector subtraction, being a special case of vector addition, simply requires a scalar subtraction of the respective $\mathbf{i}, \mathbf{j}, \mathbf{k}$ components of either $\mathbf{A}$ or $\mathbf{B}$. For example,

$$\mathbf{R}' = \mathbf{A} - \mathbf{B} = (A_x - B_x)\mathbf{i} + (A_y - B_y)\mathbf{j} + (A_z - B_z)\mathbf{k}$$

Concurrent Force Systems. If the above concept of vector addition is generalized and applied to a system of several concurrent forces, then the force resultant is the vector sum of all the forces in the system and can be written as

$$\mathbf{F}_R = \Sigma\mathbf{F} = \Sigma F_x\mathbf{i} + \Sigma F_y\mathbf{j} + \Sigma F_z\mathbf{k} \qquad (2\text{–}12)$$

Here ΣF_x, ΣF_y, and ΣF_z represent the algebraic sums of the respective x, y, z or $\mathbf{i}, \mathbf{j}, \mathbf{k}$ components of each force in the system.

The examples which follow illustrate numerically the methods used to apply the above theory to the solution of problems involving force as a vector quantity.

The force $\mathbf{F}$ that the tie-down rope exerts on the ground support at O is directed along the rope. Using the local x, y, z axes, the coordinate direction angles α, β, γ can be measured. The cosines of their values form the components of a unit vector $\mathbf{u}$ which acts in the direction of the rope. If the force has a magnitude F, then the force can be written in Cartesian vector form, as $\mathbf{F} = F\mathbf{u} = F\cos\alpha\mathbf{i} + F\cos\beta\mathbf{j} + F\cos\gamma\mathbf{k}$.

IMPORTANT POINTS

- Cartesian vector analysis is often used to solve problems in three dimensions.
- The positive direction of the x, y, z axes are defined by the Cartesian unit vectors $\mathbf{i}$, $\mathbf{j}$, $\mathbf{k}$, respectively.
- The *magnitude* of a Cartesian vector is $A = \sqrt{A_x^2 + A_y^2 + A_z^2}$.
- The *direction* of a Cartesian vector is specified using coordinate direction angles which the tail of the vector makes with the positive x, y, z axes, respectively. The components of the unit vector $\mathbf{u} = \mathbf{A}/A$ represent the direction cosines of α, β, γ. Only two of the angles α, β, γ have to be specified. The third angle is determined from the relationship $\cos^2 \alpha + \cos^2 \beta + \cos^2 \gamma = 1$.
- To find the *resultant* of a concurrent force system, express each force as a Cartesian vector and add the $\mathbf{i}$, $\mathbf{j}$, $\mathbf{k}$ components of all the forces in the system.

EXAMPLE 2.8

Express the force $\mathbf{F}$ shown in Fig. 2–29 as a Cartesian vector.

Solution
Since only two coordinate direction angles are specified, the third angle α must be determined from Eq. 2–10; i.e.,

$$\cos^2 \alpha + \cos^2 \beta + \cos^2 \gamma = 1$$
$$\cos^2 \alpha + \cos^2 60° + \cos^2 45° = 1$$
$$\cos \alpha = \sqrt{1 - (0.5)^2 - (0.707)^2} = \pm 0.5$$

Hence, two possibilities exist, namely,

$$\alpha = \cos^{-1}(0.5) = 60° \quad \text{or} \quad \alpha = \cos^{-1}(-0.5) = 120°$$

By inspection of Fig. 2–29, it is necessary that $\alpha = 60°$, since $\mathbf{F}_x$ is in the $+x$ direction.

Using Eq. 2–11, with $F = 200$ N, we have

$$\mathbf{F} = F \cos \alpha \mathbf{i} + F \cos \beta \mathbf{j} + F \cos \gamma \mathbf{k}$$
$$= (200 \cos 60° \text{ N})\mathbf{i} + (200 \cos 60° \text{ N})\mathbf{j} + (200 \cos 45° \text{ N})\mathbf{k}$$
$$= \{100.0\mathbf{i} + 100.0\mathbf{j} + 141.4\mathbf{k}\}\text{N} \qquad Ans.$$

By applying Eq. 2–6, note that indeed the magnitude of $F = 200$ N.

$$F = \sqrt{F_x^2 + F_y^2 + F_z^2}$$
$$= \sqrt{(100.0)^2 + (100.0)^2 + (141.4)^2} = 200 \text{ N}$$

Fig. 2–29

EXAMPLE 2.9

Determine the magnitude and the coordinate direction angles of the resultant force acting on the ring in Fig. 2–30a.

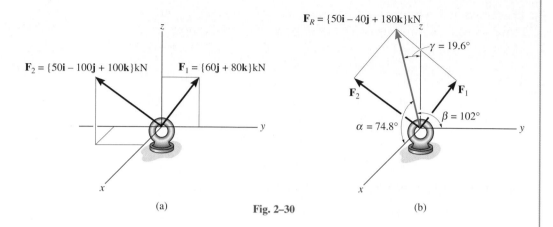

(a)   Fig. 2–30   (b)

Solution

Since each force is represented in Cartesian vector form, the resultant force, shown in Fig. 2–30b, is

$$\mathbf{F}_R = \Sigma\mathbf{F} = \mathbf{F}_1 + \mathbf{F}_2 = \{60\mathbf{j} + 80\mathbf{k}\}\,\text{lb} + \{50\mathbf{i} - 100\mathbf{j} + 100\mathbf{k}\}\,\text{kN}$$
$$= \{50\mathbf{i} - 40\mathbf{j} + 180\mathbf{k}\}\,\text{kN}$$

The magnitude of $\mathbf{F}_R$ is found from Eq. 2–6, i.e.,

$$F_R = \sqrt{(50)^2 + (-40)^2 + (180)^2} = 191.0$$
$$= 191\,\text{kN} \qquad\qquad\qquad\qquad Ans.$$

The coordinate direction angles α, β, γ are determined from the components of the unit vector acting in the direction of $\mathbf{F}_R$.

$$\mathbf{u}_{F_R} = \frac{\mathbf{F}_R}{F_R} = \frac{50}{191.0}\mathbf{i} - \frac{40}{191.0}\mathbf{j} + \frac{180}{191.0}\mathbf{k}$$
$$= 0.2617\mathbf{i} - 0.2094\mathbf{j} + 0.9422\mathbf{k}$$

so that

$$\cos\alpha = 0.2617 \qquad \alpha = 74.8° \qquad\qquad Ans.$$
$$\cos\beta = -0.2094 \qquad \beta = 102° \qquad\qquad Ans.$$
$$\cos\gamma = 0.9422 \qquad \gamma = 19.6° \qquad\qquad Ans.$$

These angles are shown in Fig. 2–30b. In particular, note that $\beta > 90°$ since the $\mathbf{j}$ component of $\mathbf{u}_{F_R}$ is negative.

EXAMPLE 2.10

Express the force $\mathbf{F}_1$, shown in Fig. 2–31a as a Cartesian vector.

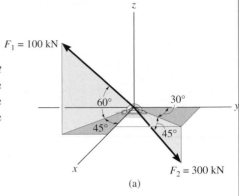

$F_1 = 100$ kN

$F_2 = 300$ kN

(a)

Solution

The angles of 60° and 45° defining the direction of $\mathbf{F}_1$ are *not* coordinate direction angles. The two successive applications of the parallelogram law needed to resolve $\mathbf{F}_1$ into its x, y, z components are shown in Fig. 2–31b. By trigonometry, the magnitudes of the components are

$$F_{1z} = 100 \sin 60° \text{ kN} = 86.6 \text{ kN}$$
$$F' = 100 \cos 60° \text{ kN} = 50 \text{ kN}$$
$$F_{1x} = 50 \cos 45° \text{ kN} = 35.4 \text{ kN}$$
$$F_{1y} = 50 \sin 45° \text{ kN} = 35.4 \text{ kN}$$

Realizing that $\mathbf{F}_{1y}$ has a direction defined by $-\mathbf{j}$, we have

$$\mathbf{F}_1 = \{35.4\mathbf{i} - 35.4\mathbf{j} + 86.6\mathbf{k}\} \text{ kN} \qquad \textit{Ans.}$$

To show that the magnitude of this vector is indeed 100 kN, apply Eq. 2–6,

$$F_1 = \sqrt{F_{1x}^2 + F_{1y}^2 + F_{1z}^2}$$
$$= \sqrt{(35.4)^2 + (-35.4)^2 + (86.6)^2} = 100 \text{ kN}$$

If needed, the coordinate direction angles of $\mathbf{F}_1$ can be determined from the components of the unit vector acting in the direction of $\mathbf{F}_1$. Hence,

$$\mathbf{u}_1 = \frac{\mathbf{F}_1}{F_1} = \frac{F_{1x}}{F_1}\mathbf{i} + \frac{F_{1y}}{F_1}\mathbf{j} + \frac{F_{1z}}{F_1}\mathbf{k}$$
$$= \frac{35.4}{100}\mathbf{i} - \frac{35.4}{100}\mathbf{j} + \frac{86.6}{100}\mathbf{k}$$
$$= 0.354\mathbf{i} - 0.354\mathbf{j} + 0.866\mathbf{k}$$

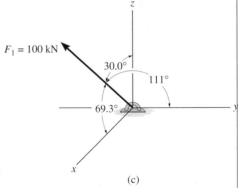

$F_1 = 100$ kN

(c)

Fig. 2–31

so that

$$\alpha_1 = \cos^{-1}(0.354) = 69.3°$$
$$\beta_1 = \cos^{-1}(-0.354) = 111°$$
$$\gamma_1 = \cos^{-1}(0.866) = 30.0°$$

These results are shown in Fig. 2–31c.

Using this same method, show that $\mathbf{F}_2$ in Fig. 2–31a can be written in Cartesian vector form as

$$\mathbf{F}_2 = \{106\mathbf{i} + 184\mathbf{j} - 212\mathbf{k}\} \text{ kN} \qquad \textit{Ans.}$$

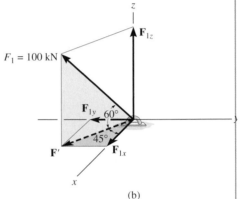

$F_1 = 100$ kN

(b)

E X A M P L E **2.11**

(a)

Fig. 2–32

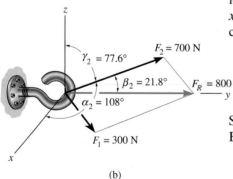

(b)

Fig. 2–32

Two forces act on the hook shown in Fig. 2–32a. Specify the coordinate direction angles of $\mathbf{F}_2$ so that the resultant force $\mathbf{F}_R$ acts along the positive y axis and has a magnitude of 800 N.

Solution

To solve this problem, the resultant force $\mathbf{F}_R$ and its two components, $\mathbf{F}_1$ and $\mathbf{F}_2$, will each be expressed in Cartesian vector form. Then, as shown in Fig. 2–32b, it is necessary that $\mathbf{F}_R = \mathbf{F}_1 + \mathbf{F}_2$.

Applying Eq. 2–11,

$$\mathbf{F}_1 = F_1 \cos \alpha_1 \mathbf{i} + F_1 \cos \beta_1 \mathbf{j} + F_1 \cos \gamma_1 \mathbf{k}$$
$$= 300 \cos 45° \ \mathbf{Ni} + 300 \cos 60° \ \mathbf{Nj} + 300 \cos 120° \ \mathbf{k}$$
$$= \{212.1\mathbf{i} + 150\mathbf{j} - 150\mathbf{k}\} \ \text{N}$$
$$\mathbf{F}_2 = F_{2x}\mathbf{i} + F_{2y}\mathbf{j} + F_{2z}\mathbf{k}$$

Since the resultant force $\mathbf{F}_R$ has a magnitude of 800 N and acts in the $+\mathbf{j}$ direction.

$$\mathbf{F}_R = (800 \ \text{N})(+\mathbf{j}) = \{800\mathbf{j}\} \ \text{N}$$

We require

$$\mathbf{F}_R = \mathbf{F}_1 + \mathbf{F}_2$$
$$800\mathbf{j} = 212.1\mathbf{i} + 150\mathbf{j} - 150\mathbf{k} + F_{2x}\mathbf{i} + F_{2y}\mathbf{j} + F_{2z}\mathbf{k}$$
$$800\mathbf{j} = (212.1 + F_{2x})\mathbf{i} + (150 + F_{2y})\mathbf{j} + (-150 + F_{2z})\mathbf{k}$$

To satisfy this equation, the corresponding $\mathbf{i}$, $\mathbf{j}$, $\mathbf{k}$ components on the left and right sides must be equal. This is equivalent to stating that the x, y, z components of $\mathbf{F}_R$ must be equal to the corresponding x, y, z components of $(\mathbf{F}_1 + \mathbf{F}_2)$. Hence,

$$0 = 212.1 + F_{2x} \qquad F_{2x} = -212.1 \ \text{N}$$
$$800 = 150 + F_{2y} \qquad F_{2y} = 650 \ \text{N}$$
$$0 = -150 + F_{2z} \qquad F_{2z} = 150 \ \text{N}$$

Since the magnitudes of $\mathbf{F}_2$ and its components are known, we can use Eq. 2–11 to determine α_2, β_2, γ_2.

$$-212.1 = 700 \cos \alpha_2; \qquad \alpha_2 = \cos^{-1}\left(\frac{-212.1}{700}\right) = 108° \qquad \textit{Ans.}$$

$$650 = 700 \cos \beta_2; \qquad \beta_2 = \cos^{-1}\left(\frac{650}{700}\right) = 21.8° \qquad \textit{Ans.}$$

$$150 = 700 \cos \gamma_2; \qquad \gamma_2 = \cos^{-1}\left(\frac{150}{700}\right) = 77.6° \qquad \textit{Ans.}$$

These results are shown in Fig. 2–32b.

PROBLEMS

2-59. Determine the magnitude and coordinate direction angles of $\mathbf{F}_1 = \{60\mathbf{i} - 50\mathbf{j} + 40\mathbf{k}\}$ N and $\mathbf{F}_2 = \{-40\mathbf{i} - 85\mathbf{j} + 30\mathbf{k}\}$ N. Sketch each force on an x, y, z reference.

*2-60. The cable at the end of the crane boom exerts a force of 2500 N on the boom as shown. Express $\mathbf{F}$ as a Cartesian vector.

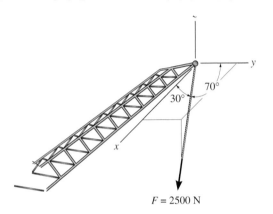

$F = 2500$ N

Prob. 2–60

2-61. Determine the magnitude and coordinate direction angles of the force $\mathbf{F}$ acting on the stake.

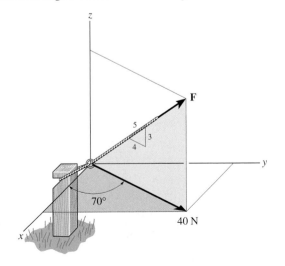

Prob. 2–61

2-62. Determine the magnitude and the coordinate direction angles of the resultant force.

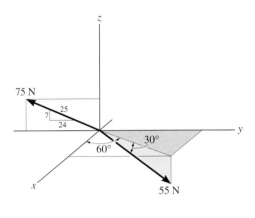

Prob. 2–62

2-63. The stock mounted on the lathe is subjected to a force of 60 N. Determine the coordinate direction angle β and express the force as a Cartesian vector.

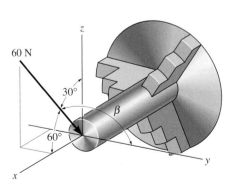

Prob. 2–63

*2-64. Determine the magnitude and coordinate direction angles of the resultant force and sketch this vector on the coordinate system.

2-65. Specify the coordinate direction angles of F_1 and F_2 and express each force as a Cartesian vector.

*2-68. The cables attached to the screw eye are subjected to the three forces shown. Express each force in Cartesian vector form and determine the magnitude and coordinate direction angles of the resultant force.

Probs. 2–64/65

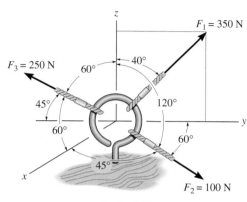

Prob. 2–68

2-66. The mast is subjected to the three forces shown. Determine the coordinate direction angles α_1, β_1, γ_1 of F_1 so that the resultant force acting on the mast is $F_R = \{350\mathbf{i}\}$ N.

2-67. The mast is subjected to the three forces shown. Determine the coordinate direction angles α_1, β_1, γ_1 of F_1 so that the resultant force acting on the mast is zero.

2-69. The beam is subjected to the two forces shown. Express each force in Cartesian vector form and determine the magnitude and coordinate direction angles of the resultant force.

Probs. 2–66/67

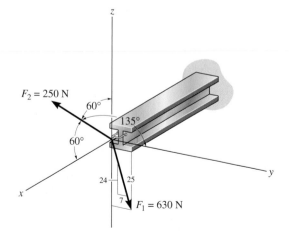

Prob. 2–69

2-70. Determine the magnitude and coordinate direction angles of the resultant force and sketch this vector on the coordinate system.

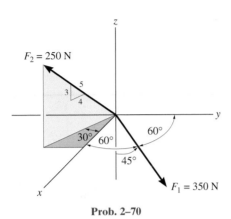

Prob. 2–70

2-71. The two forces $\mathbf{F}_1$ and $\mathbf{F}_2$ acting at A have a resultant force of $\mathbf{F}_R = \{-100\mathbf{k}\}$ N. Determine the magnitude and coordinate direction angles of $\mathbf{F}_2$.

*2-72. Determine the coordinate direction angles of the force $\mathbf{F}_1$ and indicate them on the figure.

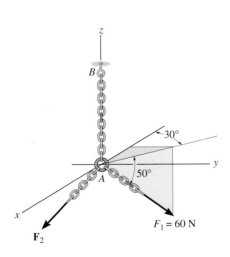

Probs. 2–71/72

2-73. The bracket is subjected to the two forces shown. Express each force in Cartesian vector form and then determine the resultant force $\mathbf{F}_R$. Find the magnitude and coordinate direction angles of the resultant force.

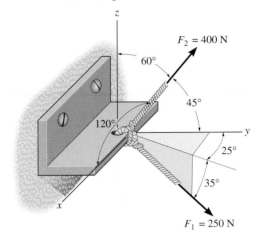

Prob. 2–73

2-74. The pole is subjected to the force $\mathbf{F}$, which has components acting along the x, y, z axes as shown. If the magnitude of $\mathbf{F}$ is 3 kN, and $\beta = 30°$ and $\gamma = 75°$, determine the magnitudes of its three components.

2-75. The pole is subjected to the force $\mathbf{F}$ which has components $F_x = 1.5$ kN and $F_z = 1.25$ kN. If $\beta = 75°$, determine the magnitudes of $\mathbf{F}$ and $\mathbf{F}_y$.

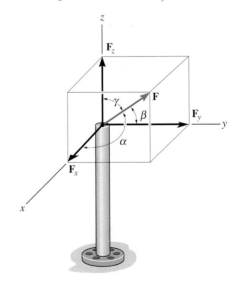

Probs. 2–74/75

*2-76. A force **F** is applied at the top of the tower at A. If it acts in the direction shown such that one of its components lying in the shaded y-z plane has a magnitude of 800 N, determine its magnitude F and coordinate direction angles α, β, γ.

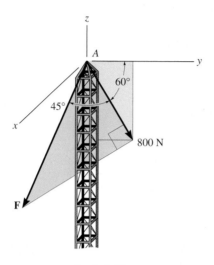

Prob. 2–76

2-77. Three forces act on the hook. If the resultant force $\mathbf{F}_R$ has a magnitude and direction as shown, determine the magnitude and the coordinate direction angles of force $\mathbf{F}_3$.

2-78. Determine the coordinate direction angles of $\mathbf{F}_1$ and $\mathbf{F}_R$.

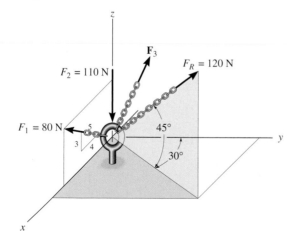

Probs. 2–77/78

2-79. The bolt is subjected to the force **F**, which has components acting along the x, y, z axes as shown. If the magnitude of **F** is 80 N, and $\alpha = 60°$ and $\gamma = 45°$, determine the magnitudes of its components.

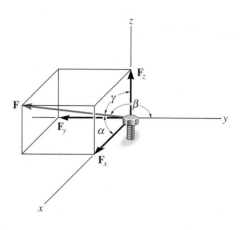

Prob. 2–79

*2-80. Two forces $\mathbf{F}_1$ and $\mathbf{F}_2$ act on the bolt. If the resultant force $\mathbf{F}_R$ has a magnitude of 50 N and coordinate direction angles $\alpha = 110°$ and $\beta = 80°$, as shown, determine the magnitude of $\mathbf{F}_2$ and its coordinate direction angles.

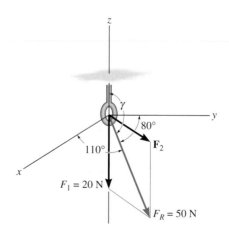

Prob. 2–80

2.7 Position Vectors

In this section we will introduce the concept of a position vector. It will be shown that this vector is of importance in formulating a Cartesian force vector directed between any two points in space. Later, in Chapter 4, we will use it for finding the moment of a force.

***x, y, z* Coordinates.**   Throughout the book we will use a *right-handed* coordinate system to reference the location of points in space. Furthermore, we will use the convention followed in many technical books, and that is to require the positive *z* axis to be directed *upward* (the zenith direction) so that it measures the height of an object or the altitude of a point. The *x, y* axes then lie in the horizontal plane, Fig. 2–33. Points in space are located relative to the origin of coordinates, *O*, by successive measurements along the *x, y, z* axes. For example, in Fig. 2–33 the coordinates of point *A* are obtained by starting at *O* and measuring $x_A = +4$ m along the *x* axis, $y_A = +2$ m along the *y* axis, and $z_A = -6$ m along the *z* axis. Thus, $A(4, 2, -6)$. In a similar manner, measurements along the *x, y, z* axes from *O* to *B* yield the coordinates of *B*, i.e., $B(0,2,0)$. Also notice that $C(6, -1, 4)$.

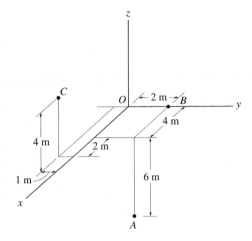

Fig. 2–33

Position Vector.   The *position vector* **r** is defined as a fixed vector which locates a point in space relative to another point. For example, if **r** extends from the origin of coordinates, *O*, to point $P(x, y, z)$, Fig. 2–34a, then **r** can be expressed in Cartesian vector form as

$$\mathbf{r} = x\mathbf{i} + y\mathbf{j} + z\mathbf{k}$$

Note how the head-to-tail vector addition of the three components yields vector **r**, Fig. 2–34b. Starting at the origin *O*, one travels *x* in the $+\mathbf{i}$ direction, then *y* in the $+\mathbf{j}$ direction, and finally *z* in the $+\mathbf{k}$ direction to arrive at point $P(x, y, z)$.

(a)   (b)

Fig. 2–34

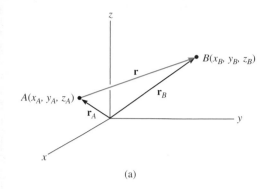

(a)

In the more general case, the position vector may be directed from point *A* to point *B* in space, Fig. 2–35*a*. As noted, this vector is also designated by the symbol **r**. As a matter of convention, however, we will *sometimes* refer to this vector with *two subscripts* to indicate from and to the point where it is directed. Thus, **r** can also be designated as $\mathbf{r}_{AB}$. Also, note that $\mathbf{r}_A$ and $\mathbf{r}_B$ in Fig. 2–35*a* are referenced with only one subscript since they extend from the origin of coordinates.

From Fig. 2–35*a*, by the head-to-tail vector addition, we require

$$\mathbf{r}_A + \mathbf{r} = \mathbf{r}_B$$

Solving for **r** and expressing $\mathbf{r}_A$ and $\mathbf{r}_B$ in Cartesian vector form yields

$$\mathbf{r} = \mathbf{r}_B - \mathbf{r}_A = (x_B\mathbf{i} + y_B\mathbf{j} + z_B\mathbf{k}) - (x_A\mathbf{i} + y_A\mathbf{j} + z_A\mathbf{k})$$

or

$$\mathbf{r} = (x_B - x_A)\mathbf{i} + (y_B - y_A)\mathbf{j} + (z_B - z_A)\mathbf{k} \qquad (2\text{–}13)$$

Thus, the **i**, **j**, **k** *components of the position vector* **r** *may be formed by taking the coordinates of the tail of the vector,* $A(x_A, y_A, z_A)$, *and subtracting them from the corresponding coordinates of the head,* $B(x_B, y_B, z_B)$. *Again note how the head-to-tail addition of these three components yields* **r**, *i.e., going from A to B*, Fig. 2–35*b*, *one first travels* $(x_B - x_A)$ *in the* +**i** *direction, then* $(y_B - y_A)$ *in the* +**j** *direction, and finally* $(z_B - z_A)$ *in the* +**k** *direction.*

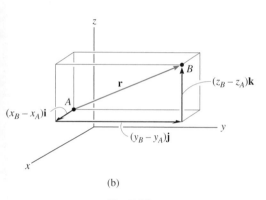

(b)

Fig. 2–35

The length and direction of cable *AB* used to support the stack can be determined by measuring the coordinates of points *A* and *B* using the *x, y, z* axes. The position vector **r** along the cable can then be established. The magnitude *r* represents the length of the cable, and the direction of the cable is defined by α, β, γ, which are determined from the components of the unit vector found from the position vector, $\mathbf{u} = \mathbf{r}/r$.

EXAMPLE 2.12

An elastic rubber band is attached to points A and B as shown in Fig. 2–36a. Determine its length and its direction measured from A toward B.

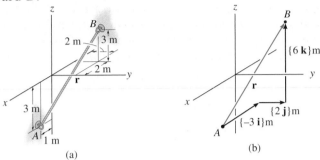

(a)

(b)

Fig. 2–36

Solution

We first establish a position vector from A to B, Fig. 2–36b. In accordance with Eq. 2–13, the coordinates of the tail $A(1 \text{ m}, 0, -3 \text{ m})$ are subtracted from the coordinates of the head $B(-2 \text{ m}, 2 \text{ m}, 3 \text{ m})$, which yields

$$\mathbf{r} = [-2 \text{ m} - 1 \text{ m}]\mathbf{i} + [2 \text{ m} - 0]\mathbf{j} + [3 \text{ m} - (-3 \text{ m})]\mathbf{k}$$
$$= \{-3\mathbf{i} + 2\mathbf{j} + 6\mathbf{k}\} \text{ m}$$

These components of $\mathbf{r}$ can also be determined *directly* by realizing from Fig. 2–36a that they represent the direction and distance one must go along each axis in order to move from A to B, i.e., along the x axis $\{-3\mathbf{i}\}$ m, along the y axis $\{2\mathbf{j}\}$ m, and finally along the z axis $\{6\mathbf{k}\}$ m.

The magnitude of $\mathbf{r}$ represents the length of the rubber band.

$$r = \sqrt{(-3)^2 + (2)^2 + (6)^2} = 7 \text{ m} \qquad Ans.$$

Formulating a unit vector in the direction of $\mathbf{r}$, we have

$$\mathbf{u} = \frac{\mathbf{r}}{r} = \frac{-3}{7}\mathbf{i} + \frac{2}{7}\mathbf{j} + \frac{6}{7}\mathbf{k}$$

The components of this unit vector yield the coordinate direction angles

$$\alpha = \cos^{-1}\left(\frac{-3}{7}\right) = 115° \qquad Ans.$$

$$\beta = \cos^{-1}\left(\frac{2}{7}\right) = 73.4° \qquad Ans.$$

$$\gamma = \cos^{-1}\left(\frac{6}{7}\right) = 31.0° \qquad Ans.$$

These angles are measured from the *positive axes* of a localized coordinate system placed at the tail of $\mathbf{r}$, point A, as shown in Fig. 2–36c.

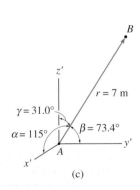

(c)

2.8 Force Vector Directed along a Line

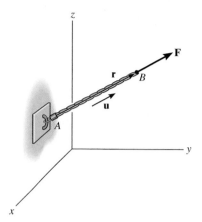

Fig. 2–37

Quite often in three-dimensional statics problems, the direction of a force is specified by two points through which its line of action passes. Such a situation is shown in Fig. 2–37, where the force **F** is directed along the cord *AB*. We can formulate **F** as a Cartesian vector by realizing that it has the *same direction* and *sense* as the position vector **r** directed from point *A* to point *B* on the cord. This common direction is specified by the *unit vector* **u** = **r**/*r*. Hence,

$$\mathbf{F} = F\mathbf{u} = F\left(\frac{\mathbf{r}}{r}\right)$$

Although we have represented **F** symbolically in Fig. 2–37, note that it has *units of force*, unlike **r**, which has units of length.

The force **F** acting along the chain can be represented as a Cartesian vector by first establishing *x, y, z* axes and forming a position vector **r** along the length of the chain, then finding the corresponding unit vector **u** = **r**/*r* that defines the direction of both the chain and the force. Finally, the magnitude of the force is combined with its direction, **F** = *F***u**.

IMPORTANT POINTS

- A position vector locates one point in space relative to another point.

- The easiest way to formulate the components of a position vector is to determine the distance and direction that must be traveled along the *x, y, z* directions—going from the tail to the head of the vector.

- A force **F** acting in the direction of a position vector **r** can be represented in Cartesian form if the unit vector **u** of the position vector is determined and this is multiplied by the magnitude of the force, i.e., **F** = *F***u** = *F*(**r**/*r*).

E X A M P L E 2.13

The man shown in Fig. 2–38a pulls on the cord with a force of 350 N. Represent this force, acting on the support A, as a Cartesian vector and determine its direction.

Solution

Force $\mathbf{F}$ is shown in Fig. 2–38b. The *direction* of this vector, $\mathbf{u}$, is determined from the position vector $\mathbf{r}$, which extends from A to B, Fig. 2–38b. The coordinates of the end points of the cord are $A(0, 0, 7.5$ m$)$ and $B(3$ m, -2 m, 1.5 m$)$. Forming the position vector by subtracting the corresponding x, y, and z coordinates of A from those of B, we have

$$\mathbf{r} = (3\text{ m} - 0)\mathbf{i} + (-2\text{ m} - 0)\mathbf{j} + (1.5\text{ m} - 7.5\text{ m})\mathbf{k}$$

$$= \{3\mathbf{i} - 2\mathbf{j} - 6\mathbf{k}\}\text{ m}$$

This result can also be determined *directly* by noting in Fig. 2–38a, that one must go from A $\{-6\mathbf{k}\}$ m, then $\{-2\mathbf{j}\}$ m, and finally $\{3\mathbf{i}\}$ m to get to B.

The magnitude of $\mathbf{r}$, which represents the *length* of cord AB, is

$$r = \sqrt{(3\text{ m})^2 + (-2\text{ m})^2 + (-6\text{ m})^2} = 7\text{ m}$$

Forming the unit vector that defines the direction and sense of both $\mathbf{r}$ and $\mathbf{F}$ yields

$$\mathbf{u} = \frac{\mathbf{r}}{r} = \frac{3}{7}\mathbf{i} - \frac{2}{7}\mathbf{j} - \frac{6}{7}\mathbf{k}$$

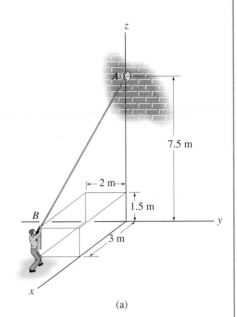

(a)

Since $\mathbf{F}$ has a *magnitude* of 350 N and a *direction* specified by $\mathbf{u}$, then

$$\mathbf{F} = F\mathbf{u} = 350\text{ N}\left(\frac{3}{7}\mathbf{i} - \frac{2}{7}\mathbf{j} - \frac{6}{7}\mathbf{k}\right)$$

$$= \{150\mathbf{i} - 100\mathbf{j} - 300\mathbf{k}\}\text{ N} \qquad Ans.$$

The coordinate direction angles are measured between $\mathbf{r}$ (or $\mathbf{F}$) and the *positive axes* of a localized coordinate system with origin placed at A, Fig. 2–38b. From the components of the unit vector:

$$\alpha = \cos^{-1}\left(\frac{3}{7}\right) = 64.6° \qquad Ans.$$

$$\beta = \cos^{-1}\left(\frac{-2}{7}\right) = 107° \qquad Ans.$$

$$\gamma = \cos^{-1}\left(\frac{-6}{7}\right) = 149° \qquad Ans.$$

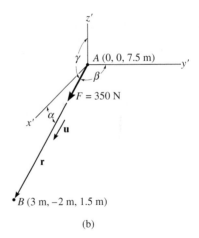

(b)

Fig. 2–38

E X A M P L E   2.14

(a)

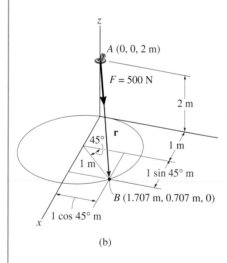

(b)

Fig. 2–39

The circular plate in Fig. 2–39*a* is partially supported by the cable *AB*. If the force of the cable on the hook at *A* is $F = 500$ N, express **F** as a Cartesian vector.

Solution

As shown in Fig. 2–39*b*, **F** has the same direction and sense as the position vector **r**, which extends from *A* to *B*. The coordinates of the end points of the cable are $A(0, 0, 2\text{ m})$ and $B(1.707\text{ m}, 0.707\text{ m}, 0)$, as indicated in the figure. Thus,

$$\mathbf{r} = (1.707\text{ m} - 0)\mathbf{i} + (0.707\text{ m} - 0)\mathbf{j} + (0 - 2\text{ m})\mathbf{k}$$

$$= \{1.707\mathbf{i} + 0.707\mathbf{j} - 2\mathbf{k}\}\text{ m}$$

Note how one can calculate these components *directly* by going from A, $\{-2\mathbf{k}\}$ m along the z axis, then $\{1.707\mathbf{i}\}$ m along the x axis, and finally $\{0.707\mathbf{j}\}$ m along the y axis to get to B.

The magnitude of **r** is

$$r = \sqrt{(1.707)^2 + (0.707)^2 + (-2)^2} = 2.723\text{ m}$$

Thus,

$$\mathbf{u} = \frac{\mathbf{r}}{r} = \frac{1.707}{2.723}\mathbf{i} + \frac{0.707}{2.723}\mathbf{j} - \frac{2}{2.723}\mathbf{k}$$

$$= 0.6269\mathbf{i} + 0.2597\mathbf{j} - 0.7345\mathbf{k}$$

Since $F = 500$ N and **F** has the direction **u**, we have

$$\mathbf{F} = F\mathbf{u} = 500\text{ N}(0.6269\mathbf{i} + 0.2597\mathbf{j} - 0.7345\mathbf{k})$$

$$= \{313\mathbf{i} + 130\mathbf{j} - 367\mathbf{k}\}\text{ N} \qquad\qquad \textit{Ans.}$$

Using these components, notice that indeed the magnitude of **F** is 500 N; i.e.,

$$F = \sqrt{(313)^2 + (130)^2 + (-367)^2} = 500\text{ N}$$

Show that the coordinate direction angle $\gamma = 137°$, and indicate this angle on the figure.

E X A M P L E   2.15

The roof is supported by cables as shown in the photo. If the cables exert forces $F_{AB} = 100$ N and $F_{AC} = 120$ N on the wall hook at A as shown in Fig. 2–40a, determine the magnitude of the resultant force acting at A.

Solution

The resultant force $\mathbf{F}_R$ is shown graphically in Fig. 2–40b. We can express this force as a Cartesian vector by first formulating $\mathbf{F}_{AB}$ and $\mathbf{F}_{AC}$ as Cartesian vectors and then adding their components. The directions of $\mathbf{F}_{AB}$ and $\mathbf{F}_{AC}$ are specified by forming unit vectors $\mathbf{u}_{AB}$ and $\mathbf{u}_{AC}$ along the cables. These unit vectors are obtained from the associated position vectors $\mathbf{r}_{AB}$ and $\mathbf{r}_{AC}$. With reference to Fig. 2–40b, for $\mathbf{F}_{AB}$ we have

$$\mathbf{r}_{AB} = (4 \text{ m} - 0)\mathbf{i} + (0 - 0)\mathbf{j} + (0 - 4 \text{ m})\mathbf{k}$$

$$= \{4\mathbf{i} - 4\mathbf{k}\} \text{ m}$$

$$r_{AB} = \sqrt{(4)^2 + (-4)^2} = 5.66 \text{ m}$$

$$\mathbf{F}_{AB} = 100 \text{ N}\left(\frac{\mathbf{r}_{AB}}{r_{AB}}\right) = 100 \text{ N}\left(\frac{4}{5.66}\mathbf{i} - \frac{4}{5.66}\mathbf{k}\right)$$

$$\mathbf{F}_{AB} = \{70.7\mathbf{i} - 70.7\mathbf{k}\} \text{ N}$$

For $\mathbf{F}_{AC}$ we have

$$\mathbf{r}_{AC} = (4 \text{ m} - 0)\mathbf{i} + (2 \text{ m} - 0)\mathbf{j} + (0 - 4 \text{ m})\mathbf{k}$$

$$= \{4\mathbf{i} + 2\mathbf{j} - 4\mathbf{k}\} \text{ m}$$

$$r_{AC} = \sqrt{(4)^2 + (2)^2 + (-4)^2} = 6 \text{ m}$$

$$\mathbf{F}_{AC} = 120 \text{ N}\left(\frac{\mathbf{r}_{AC}}{r_{AC}}\right) = 120 \text{ N}\left(\frac{4}{6}\mathbf{i} + \frac{2}{6}\mathbf{j} - \frac{4}{6}\mathbf{k}\right)$$

$$= \{80\mathbf{i} + 40\mathbf{j} - 80\mathbf{k}\} \text{ N}$$

The resultant force is therefore

$$\mathbf{F}_R = \mathbf{F}_{AB} + \mathbf{F}_{AC} = \{70.7\mathbf{i} - 70.7\mathbf{k}\} \text{ N} + \{80\mathbf{i} + 40\mathbf{j} - 80\mathbf{k}\} \text{ N}$$

$$= \{150.7\mathbf{i} + 40\mathbf{j} - 150.7\mathbf{k}\} \text{ N}$$

The magnitude of $\mathbf{F}_R$ is thus

$$F_R = \sqrt{(150.7)^2 + (40)^2 + (-150.7)^2}$$

$$= 217 \text{ N} \qquad \qquad Ans.$$

(a)

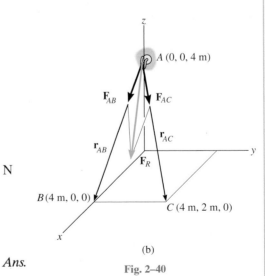

(b)

Fig. 2–40

PROBLEMS

2-81. If $r_1 = \{3i - 4j + 3k\}$ m, $r_2 = \{4i - 5k\}$ m, $r_3 = \{3i - 2j + 5k\}$ m, determine the magnitude and direction of $r = 2r_1 - r_2 + 3r_3$.

2-82. Represent the position vector r acting from point $A(3\text{ m}, 5\text{ m}, 6\text{ m})$ to point $B(5\text{ m}, -2\text{ m}, 1\text{ m})$ in Cartesian vector form. Determine its coordinate direction angles and find the distance between points A and B.

2-83. A position vector extends from the origin to point $A(2\text{ m}, 3\text{ m}, 6\text{ m})$. Determine the angles α, β, γ which the tail of the vector makes with the x, y, z axes, respectively.

*2-84. Express the position vector r in Cartesian vector form; then determine its magnitude and coordinate direction angles.

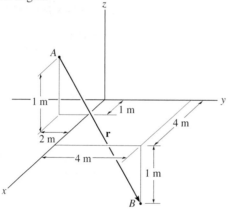

Prob. 2–84

2-85. Express the position vector r in Cartesian vector form; then determine its magnitude and coordinate direction angles.

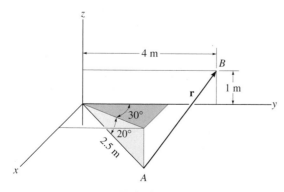

Prob. 2–85

2-86. Express force F as a Cartesian vector; then determine its coordinate direction angles.

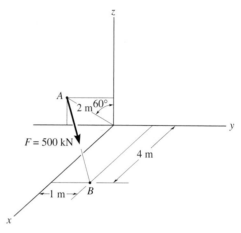

Prob. 2–86

2-87. Determine the length of member AB of the truss by first establishing a Cartesian position vector from A to B and then determining its magnitude.

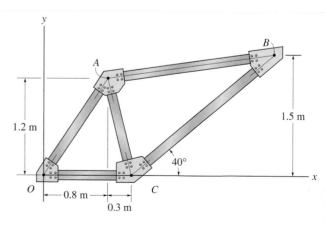

Prob. 2–87

*2-88. At a given instant, the position of a plane at A and a train at B are measured relative to a radar antenna at O. Determine the distance d between A and B at this instant. To solve the problem, formulate a position vector, directed from A to B, and then determine its magnitude.

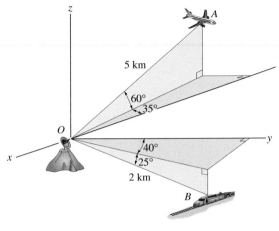

Prob. 2–88

2-90. Determine the length of the crankshaft AB by first formulating a Cartesian position vector from A to B and then determining its magnitude.

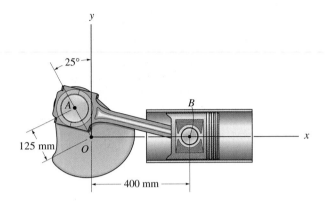

Prob. 2–90

2-89. The hinged plate is supported by the cord AB. If the force in the cord is $F = 340$ N, express this force, directed from A toward B, as a Cartesian vector. What is the length of the cord?

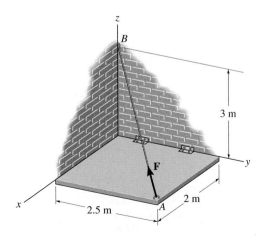

Prob. 2–89

2-91. Determine the lengths of wires AD, BD, and CD. The ring at D is midway between A and B.

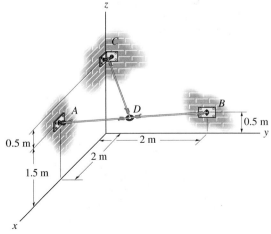

Prob. 2–91

*2-92. Express force **F** as a Cartesian vector; then determine its coordinate direction angles.

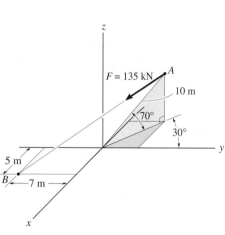

Prob. 2–92

2-93. Express force **F** as a Cartesian vector; then determine its coordinate direction angles.

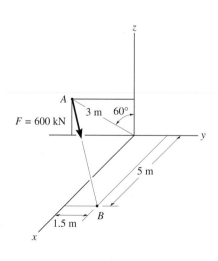

Prob. 2–93

2-94. Determine the magnitude and coordinate direction angles of the resultant force acting at point A.

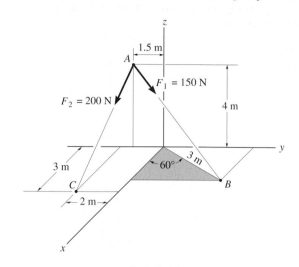

Prob. 2–94

2-95. The door is held opened by means of two chains. If the tension in AB and CD is $F_A = 300$ N and $F_C = 250$ N, respectively, express each of these forces in Cartesian vector form.

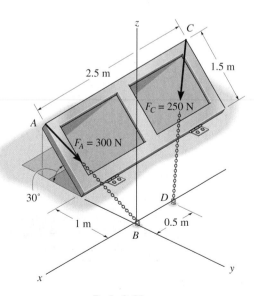

Prob. 2–95

*2-96. The two mooring cables exert forces on the stern of a ship as shown. Represent each force as as Cartesian vector and determine the magnitude and direction of the resultant.

Prob. 2–96

2-98. The guy wires are used to support the telephone pole. Represent the force in each wire in Cartesian vector form.

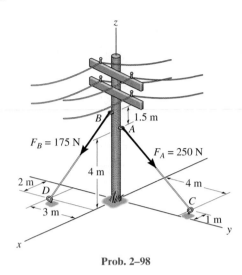

Prob. 2–98

2-97. Two tractors pull on the tree with the forces shown. Represent each force as a Cartesian vector and then determine the magnitude and coordinate direction angles of the resultant force.

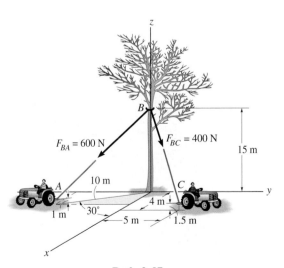

Prob. 2–97

2-99. Express each of the forces in Cartesian vector form and determine the magnitude and coordinate direction angles of the resultant force.

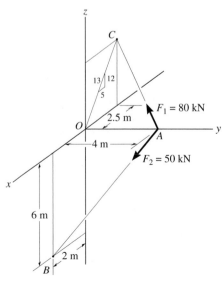

Prob. 2–99

*2-100. The cable attached to the tractor at B exerts a force of 3500 N on the framework. Express this force as a Cartesian vector.

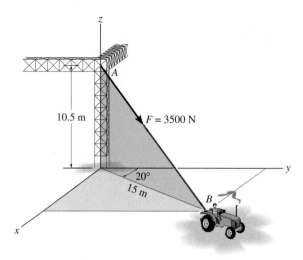

Prob. 2–100

2-101. The load at A creates a force of 300 N in wire AB. Express this force as a Cartesian vector acting on A and directed toward B as shown.

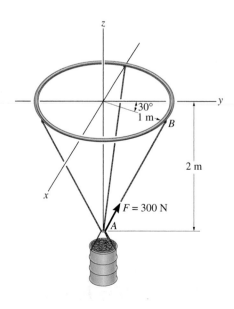

Prob. 2–101

2-102. The pipe is supported at its ends by a cord AB. If the cord exerts a force of $F = 120$ N on the pipe at A, express this force as a Cartesian vector.

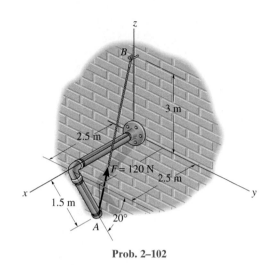

Prob. 2–102

2-103. The cord exerts a force of $\mathbf{F} = \{12\mathbf{i} + 9\mathbf{j} - 8\mathbf{k}\}$ kN on the hook. If the cord is 4 m long, determine the location x, y of the point of attachment B, and the height z of the hook.

*2-104. The cord exerts a force of $F = 30$ kN on the hook. If the cord is 4 m long, $z = 2$ m, and the x component of the force is $F_x = 25$ kN, determine the location x, y of the point of attachment B of the cord to the ground.

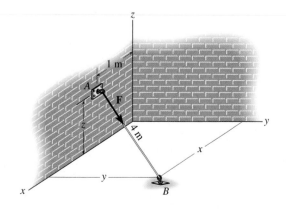

Probs. 2–103/104

2-105. Each of the four forces acting at E has a magnitude of 28 kN. Express each force as a Cartesian vector and determine the resultant force.

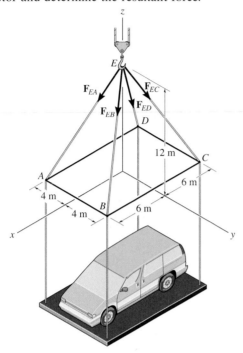

Prob. 2–105

2-106. The tower is held in place by three cables. If the force of each cable acting on the tower is shown, determine the magnitude and coordinate direction angles α, β, γ of the resultant force. Take $x = 20$ m, $y = 15$ m.

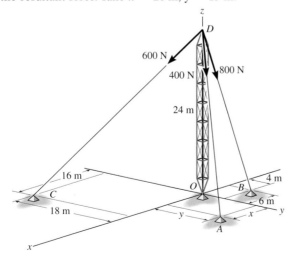

Prob. 2–106

2-107. The cable, attached to the shear-leg derrick, exerts a force on the derrick of $F = 3500$ N. Express this force as a Cartesian vector.

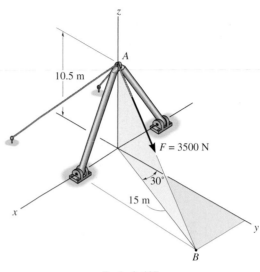

Prob. 2–107

*2-108. The window is held open by chain AB. Determine the length of the chain, and express the 200-N force acting at A along the chain as a Cartesian vector and determine its coordinate direction angles.

Prob. 2–108

2.9 Dot Product

Fig. 2–41

Occasionally in statics one has to find the angle between two lines or the components of a force parallel and perpendicular to a line. In two dimensions, these problems can readily be solved by trigonometry since the geometry is easy to visualize. In three dimensions, however, this is often difficult, and consequently vector methods should be employed for the solution. The dot product defines a particular method for "multiplying" two vectors and is used to solve the above-mentioned problems.

The *dot product* of vectors **A** and **B**, written **A · B**, and read "**A** dot **B**," is defined as the product of the magnitudes of **A** and **B** and the cosine of the angle θ between their tails, Fig. 2–41. Expressed in equation form,

$$\mathbf{A} \cdot \mathbf{B} = AB \cos \theta \qquad\qquad (2\text{–}14)$$

where $0° \leq \theta \leq 180°$. The dot product is often referred to as the *scalar product* of vectors since the result is a *scalar* and not a vector.

Laws of Operation

1. Commutative law:

$$\mathbf{A} \cdot \mathbf{B} = \mathbf{B} \cdot \mathbf{A}$$

2. Multiplication by a scalar:

$$a(\mathbf{A} \cdot \mathbf{B}) = (a\mathbf{A}) \cdot \mathbf{B} = \mathbf{A} \cdot (a\mathbf{B}) = (\mathbf{A} \cdot \mathbf{B})a$$

3. Distributive law:

$$\mathbf{A} \cdot (\mathbf{B} + \mathbf{D}) = (\mathbf{A} \cdot \mathbf{B}) + (\mathbf{A} \cdot \mathbf{D})$$

It is easy to prove the first and second laws by using Eq. 2–14. The proof of the distributive law is left as an exercise (see Prob. 2–109).

Cartesian Vector Formulation. Equation 2–14 may be used to find the dot product for each of the Cartesian unit vectors. For example, $\mathbf{i} \cdot \mathbf{i} = (1)(1) \cos 0° = 1$ and $\mathbf{i} \cdot \mathbf{j} = (1)(1) \cos 90° = 0$. In a similar manner,

$$\mathbf{i} \cdot \mathbf{i} = 1 \qquad \mathbf{j} \cdot \mathbf{j} = 1 \qquad \mathbf{k} \cdot \mathbf{k} = 1$$
$$\mathbf{i} \cdot \mathbf{j} = 0 \qquad \mathbf{i} \cdot \mathbf{k} = 0 \qquad \mathbf{k} \cdot \mathbf{j} = 0$$

These results should not be memorized; rather, it should be clearly understood how each is obtained.

Consider now the dot product of two general vectors **A** and **B** which are expressed in Cartesian vector form. We have

$$\begin{aligned}
\mathbf{A} \cdot \mathbf{B} &= (A_x\mathbf{i} + A_y\mathbf{j} + A_z\mathbf{k}) \cdot (B_x\mathbf{i} + B_y\mathbf{j} + B_z\mathbf{k}) \\
&= A_xB_x(\mathbf{i} \cdot \mathbf{i}) + A_xB_y(\mathbf{i} \cdot \mathbf{j}) + A_xB_z(\mathbf{i} \cdot \mathbf{k}) \\
&\quad + A_yB_x(\mathbf{j} \cdot \mathbf{i}) + A_yB_y(\mathbf{j} \cdot \mathbf{j}) + A_yB_z(\mathbf{j} \cdot \mathbf{k}) \\
&\quad + A_zB_x(\mathbf{k} \cdot \mathbf{i}) + A_zB_y(\mathbf{k} \cdot \mathbf{j}) + A_zB_z(\mathbf{k} \cdot \mathbf{k})
\end{aligned}$$

Carrying out the dot-product operations, the final result becomes

$$\mathbf{A} \cdot \mathbf{B} = A_x B_x + A_y B_y + A_z B_z \qquad (2\text{--}15)$$

Thus, to determine the dot product of two Cartesian vectors, multiply their corresponding x, y, z components and sum their products algebraically. Since the result is a scalar, be careful *not* to include any unit vectors in the final result.

Applications. The dot product has two important applications in mechanics.

1. *The angle formed between two vectors or intersecting lines.* The angle θ between the tails of vectors $\mathbf{A}$ and $\mathbf{B}$ in Fig. 2–41 can be determined from Eq. 2–14 and written as

$$\theta = \cos^{-1}\!\left(\frac{\mathbf{A} \cdot \mathbf{B}}{AB}\right) \qquad 0° \leq \theta \leq 180°$$

   Here $\mathbf{A} \cdot \mathbf{B}$ is found from Eq. 2–15. In particular, notice that if $\mathbf{A} \cdot \mathbf{B} = 0, \theta = \cos^{-1} 0 = 90°$, so that $\mathbf{A}$ will be *perpendicular* to $\mathbf{B}$.

2. *The components of a vector parallel and perpendicular to a line.* The component of vector $\mathbf{A}$ parallel to or collinear with the line aa' in Fig. 2–42 is defined by $\mathbf{A}_\parallel$, where $A_\parallel = A\cos\theta$. This component is sometimes referred to as the *projection* of $\mathbf{A}$ onto the line, since a right angle is formed in the construction. If the *direction* of the line is specified by the unit vector $\mathbf{u}$, then, since $u = 1$, we can determine $A_\parallel$ directly from the dot product (Eq. 2–14); i.e.,

$$A_\parallel = A\cos\theta = \mathbf{A} \cdot \mathbf{u}$$

$$\mathbf{A}_\parallel = A\cos\theta\,\mathbf{u}$$

Fig. 2–42

   Hence, the scalar projection of $\mathbf{A}$ along a line is determined from the dot product of $\mathbf{A}$ and the unit vector $\mathbf{u}$ which defines the direction of the line. Notice that if this result is positive, then $\mathbf{A}_\parallel$ has a directional sense which is the same as $\mathbf{u}$, whereas if $A_\parallel$ is a negative scalar, then $\mathbf{A}_\parallel$ has the opposite sense of direction to $\mathbf{u}$. The component $\mathbf{A}_\parallel$ represented as a *vector* is therefore

$$\mathbf{A}_\parallel = A\cos\theta\,\mathbf{u} = (\mathbf{A} \cdot \mathbf{u})\mathbf{u}$$

   The component of $\mathbf{A}$ which is *perpendicular* to line aa' can also be obtained, Fig. 2–42. Since $\mathbf{A} = \mathbf{A}_\parallel + \mathbf{A}_\perp$, then $\mathbf{A}_\perp = \mathbf{A} - \mathbf{A}_\parallel$. There are two possible ways of obtaining $A_\perp$. One way would be to determine θ from the dot product, $\theta = \cos^{-1}(\mathbf{A} \cdot \mathbf{u}/A)$, then $A_\perp = A\sin\theta$. Alternatively, if $A_\parallel$ is known, then by the Pythagorean theorem we can also write $A_\perp = \sqrt{A^2 - A_\parallel^2}$.

The angle θ which is made between the rope and the connecting beam A can be determined by using the dot product. Simply formulate position vectors or unit vectors along the beam, $\mathbf{u}_A = \mathbf{r}_A/r_A$, and along the rope, $\mathbf{u}_r = \mathbf{r}_r/r_r$. Since θ is defined between the tails of these vectors we can solve for θ using $\theta = \cos^{-1}(\mathbf{r}_A \cdot \mathbf{r}_r/r_A r_r) = \cos^{-1} \mathbf{u}_A \cdot \mathbf{u}_r$.

If the rope exerts a force $\mathbf{F}$ on the joint, the projection of this force along beam A can be determined by first defining the *direction of the beam* using the unit vector $\mathbf{u}_A = \mathbf{r}_A/r_A$ and then formulating the force as a Cartesian vector $\mathbf{F} = F(\mathbf{r}_r/r_r) = F\mathbf{u}_r$. Applying the dot product, the projection is $F_\parallel = \mathbf{F} \cdot \mathbf{u}_A$.

IMPORTANT POINTS

- The dot product is used to determine the angle between two vectors or the projection of a vector in a specified direction.
- If the vectors $\mathbf{A}$ and $\mathbf{B}$ are expressed in Cartesian form, the dot product is determined by multiplying the respective x, y, z scalar components together and algebraically adding the results, i.e., $\mathbf{A} \cdot \mathbf{B} = A_x B_x + A_y B_y + A_z B_z$.
- From the definition of the dot product, the angle formed between the tails of vectors $\mathbf{A}$ and $\mathbf{B}$ is $\theta = \cos^{-1}(\mathbf{A} \cdot \mathbf{B}/AB)$.
- The magnitude of the projection of vector $\mathbf{A}$ along a line whose direction is specified by $\mathbf{u}$ is determined from the dot product $A_\parallel = \mathbf{A} \cdot \mathbf{u}$.

EXAMPLE 2.16

The frame shown in Fig. 2–43a is subjected to a horizontal force $\mathbf{F} = \{300\mathbf{j}\}$ N. Determine the magnitude of the components of this force parallel and perpendicular to member AB.

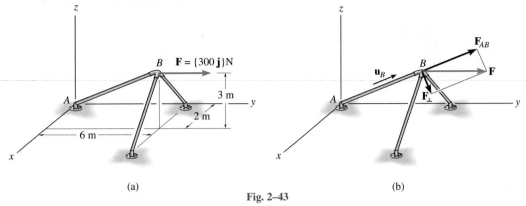

(a)   (b)

Fig. 2–43

Solution

The magnitude of the component of $\mathbf{F}$ along AB is equal to the dot product of $\mathbf{F}$ and the unit vector $\mathbf{u}_B$, which defines the direction of AB, Fig. 2–43b. Since

$$\mathbf{u}_B = \frac{\mathbf{r}_B}{r_B} = \frac{2\mathbf{i} + 6\mathbf{j} + 3\mathbf{k}}{\sqrt{(2)^2 + (6)^2 + (3)^2}} = 0.286\mathbf{i} + 0.857\mathbf{j} + 0.429\mathbf{k}$$

then

$$F_{AB} = F\cos\theta = \mathbf{F} \cdot \mathbf{u}_B = (300\mathbf{j}) \cdot (0.286\mathbf{i} + 0.857\mathbf{j} + 0.429\mathbf{k})$$
$$= (0)(0.286) + (300)(0.857) + (0)(0.429)$$
$$= 257.1 \text{ N} \hspace{3cm} \textit{Ans.}$$

Since the result is a positive scalar, $\mathbf{F}_{AB}$ has the same sense of direction as $\mathbf{u}_B$, Fig. 2–43b.

Expressing $\mathbf{F}_{AB}$ in Cartesian vector form, we have

$$F_{AB} = F_{AB}\mathbf{u}_B = (257.1 \text{ N})(0.286\mathbf{i} + 0.857\mathbf{j} + 0.429\mathbf{k})$$
$$= \{73.5\mathbf{i} + 220\mathbf{j} + 110\mathbf{k}\} \text{ N} \hspace{2cm} \textit{Ans.}$$

The perpendicular component, Fig. 2–43b, is therefore

$$\mathbf{F}_{\perp} = \mathbf{F} - \mathbf{F}_{AB} = 300\mathbf{j} - (73.5\mathbf{i} + 220\mathbf{j} + 110\mathbf{k})$$
$$= \{-73.5\mathbf{i} + 80\mathbf{j} - 110\mathbf{k}\} \text{ N}$$

Its magnitude can be determined either from this vector or from the Pythagorean theorem, Fig. 2–43b:

$$F_{\perp} = \sqrt{F^2 - F^2_{AB}}$$
$$= \sqrt{(300 \text{ N})^2 - (257.1 \text{ N})^2}$$
$$= 155 \text{ N} \hspace{3cm} \textit{Ans.}$$

EXAMPLE 2.17

The pipe in Fig. 2–44*a* is subjected to the force of $F = 800$ N. Determine the angle θ between **F** and the pipe segment *BA*, and the magnitudes of the components of **F**, which are parallel and perpendicular to *BA*.

Solution

Angle θ. First we will establish position vectors from *B* to *A* and *B* to *C*. Then we will determine the angle θ between the tails of these two vectors.

$$\mathbf{r}_{BA} = \{-2\mathbf{i} - 2\mathbf{j} + 1\mathbf{k}\}\ \dot{\mathrm{m}}$$
$$\mathbf{r}_{BC} = \{-3\mathbf{j} + 1\mathbf{k}\}\ \dot{\mathrm{m}}$$

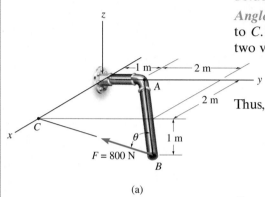

Thus,

$$\cos\theta = \frac{\mathbf{r}_{BA} \cdot \mathbf{r}_{BC}}{\mathbf{r}_{BA}\mathbf{r}_{BC}} = \frac{(-2)(0) + (-2)(-3) + (1)(1)}{3\sqrt{10}}$$
$$= 0.7379$$
$$\theta = 42.5° \qquad\qquad Ans.$$

Components of F. The force **F** is resolved into components as shown in Fig. 2–44*b*. Since $F_{BA} = \mathbf{F} \cdot \mathbf{u}_{BA}$, we must first formulate the unit vector along *BA* and force **F** as Cartesian vectors.

$$\mathbf{u}_{BA} = \frac{\mathbf{r}_{BA}}{r_{BA}} = \frac{(-2\mathbf{i} - 2\mathbf{j} + 1\mathbf{k})}{3} = -\frac{2}{3}\mathbf{i} - \frac{2}{3}\mathbf{j} + \frac{1}{3}\mathbf{k}$$

$$\mathbf{F} = 800\ \mathrm{N}\left(\frac{\mathbf{r}_{BC}}{r_{BC}}\right) = 800\left(\frac{23\mathbf{j} + 1\mathbf{k}}{\sqrt{10}}\right) = -758.9\mathbf{j} + 253.0\mathbf{k}$$

Thus,

$$F_{BA} = \mathbf{F} \cdot \mathbf{u}_{BA} = (-758.9\mathbf{j} + 253.0\mathbf{k}) \cdot \left(-\frac{2}{3}\mathbf{i} - \frac{2}{3}\mathbf{j} + \frac{1}{3}\mathbf{k}\right)$$
$$= 0 + 506.0 + 84.3$$
$$= 590\ \mathrm{N} \qquad\qquad Ans.$$

Since θ was calculated in Fig. 2–44*b*, this same result can also be obtained directly from trigonometry.

$$F_{BA} = 800\cos 42.5°\ \mathrm{N} = 590\ \mathrm{N} \qquad\qquad Ans.$$

The perpendicular component can be obtained by trigonometry,

$$F_\perp = F\sin\theta$$
$$= 800\cos 42.5°\ \mathrm{N}$$
$$= 540\ \mathrm{N} \qquad\qquad Ans.$$

Or, by the Pythagorean theorem,

$$F_\perp = \sqrt{F^2 - F_{BA}^2} = \sqrt{(800)^2 - (590)^2}$$
$$= 540\ \mathrm{N} \qquad\qquad Ans.$$

Fig. 2–44

PROBLEMS

2-109. Given the three vectors **A, B,** and **D,** show that $\mathbf{A} \cdot (\mathbf{B} + \mathbf{D}) = (\mathbf{A} \cdot \mathbf{B}) + (\mathbf{A} \cdot \mathbf{D})$.

2-110. Determine the angle θ between the tails of the two vectors.

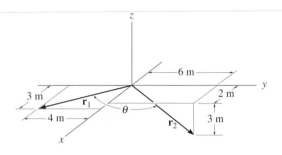

Prob. 2–110

2-111. Determine the angle θ between the tails of the two vectors.

***2-112.** Determine the magnitude of the projected component of $\mathbf{r}_1$ along $\mathbf{r}_2$, and the projection of $\mathbf{r}_2$ along $\mathbf{r}_1$.

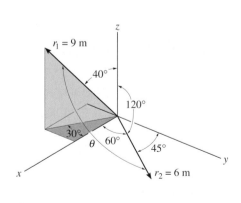

Probs. 2–111/112

2-113. Determine the angle θ between the y axis of the pole and the wire AB.

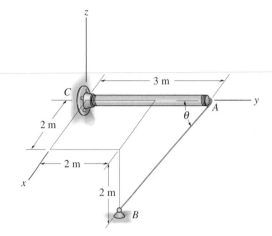

Prob. 2–113

2-114. The force $\mathbf{F} = \{25\mathbf{i} - 50\mathbf{j} + 10\mathbf{k}\}$ N acts at the end A of the pipe assembly. Determine the magnitude of the components $\mathbf{F}_1$ and $\mathbf{F}_2$ which act along the axis of AB and perpendicular to it.

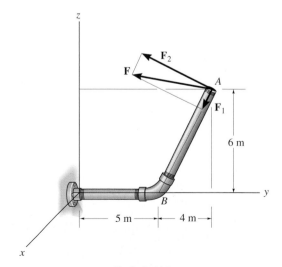

Prob. 2–114

2-115. Determine the angle θ between the sides of the triangular plate.

*2-116. Determine the length of side BC of the triangular plate. Solve the problem by finding the magnitude of $\mathbf{r}_{BC}$; then check the result by first finding θ, r_{AB}, and r_{AC} and then use the cosine law.

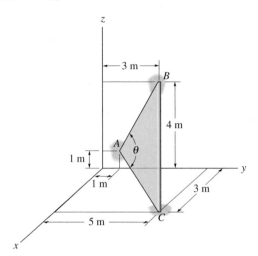

Probs. 2–115/116

2-117. Determine the components of $\mathbf{F}$ that act along rod AC and perpendicular to it. Point B is located at the midpoint of the rod.

2-118. Determine the components of $\mathbf{F}$ that act along rod AC and perpendicular to it. Point B is located 3 m along the rod from end C.

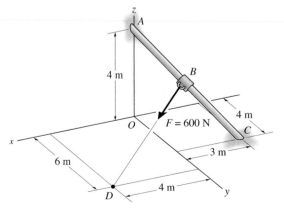

Probs. 2–117/118

2-119. The clamp is used on a jig. If the vertical force acting on the bolt is $\mathbf{F} = \{-500\mathbf{k}\}$ N, determine the magnitudes of the components $\mathbf{F}_1$ and $\mathbf{F}_2$ which act along the OA axis and perpendicular to it.

$$\mathbf{F} = \{-500 \ \mathbf{k}\}N$$

Prob. 2–119

*2-120. Determine the projection of the force $\mathbf{F}$ along the pole.

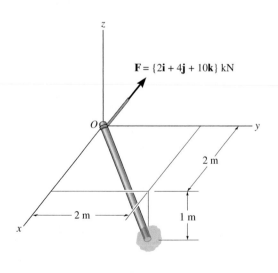

Prob. 2–120

2-121. Determine the projected component of the 80-N force acting along the axis AB of the pipe.

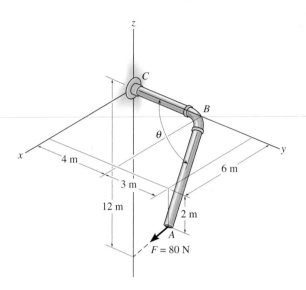

Prob. 2–121

2-122. Cable OA is used to support column OB. Determine the angle θ it makes with beam OC.

2-123. Cable OA is used to support column OB. Determine the angle ϕ it makes with beam OD.

Probs. 2–122/123

*2-124. The force **F** acts at the end A of the pipe assembly. Determine the magnitudes of the components $\mathbf{F}_1$ and $\mathbf{F}_2$ which act along the axis of AB and perpendicular to it.

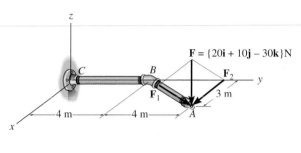

Prob. 2–124

2-125. Two cables exert forces on the pipe. Determine the magnitude of the projected component of $\mathbf{F}_1$ along the line of action of $\mathbf{F}_2$.

2-126. Determine the angle θ between the two cables attached to the pipe.

Probs. 2–125/126

2-127. Determine the angle θ between cables AB and AC.

*2-128. If $\mathbf{F}$ has a magnitude of 55 N, determine the magnitude of its projected component acting along the x axis and along cable AC.

2-130. The cables each exert a force of 400 N on the post. Determine the magnitude of the projected component of $\mathbf{F}_1$ along the line of action of $\mathbf{F}_2$.

2-131. Determine the angle θ between the two cables attached to the post.

Probs. 2–127/128

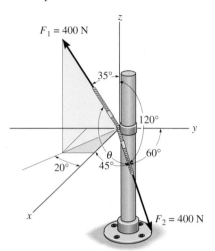

Probs. 2–130/131

2-129. Determine the angle θ between the edges of the sheet-metal bracket.

*2-132. Determine the angles θ and ϕ made between the axes OA of the flag pole and AB and AC, respectively, of each cable.

Prob. 2–129

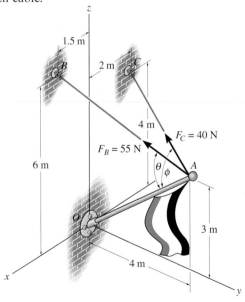

Prob. 2–132

CHAPTER REVIEW

- **Parallelogram Law.** Two vectors add according to the parallelogram law. The *components* form the sides of the parallelogram and the *resultant* is the diagonal. To obtain the components or the resultant, show how the vectors add by the tip-to-tail addition using the triangle rule, and then use the law of sines and the law of cosines to calculate their values.

- **Cartesian Vectors.** A vector can be resolved into its Cartesian components along the x, y, z axes so that $\mathbf{F} = F_x\mathbf{i} + F_y\mathbf{j} + F_z\mathbf{k}$.

  The magnitude of $\mathbf{F}$ is determined from $F = \sqrt{F_x^2 + F_y^2 + F_z^2}$ and the coordinate direction angles α, β, γ are determined by formulating a unit vector in the direction of $\mathbf{F}$, that is $\mathbf{u} = (F_x/F)\mathbf{i} + (F_y/F)\mathbf{j} + (F_z/F)\mathbf{k}$. The components of $\mathbf{u}$ represent $\cos\alpha$, $\cos\beta$, $\cos\gamma$. These three angles are related by $\cos^2\alpha + \cos^2\beta + \cos^2\gamma = 1$, so that only two of the three angles are independent of one another.

- **Force and Position Vectors.** A position vector is directed between two points. It can be formulated by finding the distance and the direction one has to travel along the x, y, z axes from one point (the tail) to the other point (the tip). If the line of action of a force passes through these two points, then it acts in the same direction $\mathbf{u}$ as the position vector. The force can be expressed as a Cartesian vector using $\mathbf{F} = F\mathbf{u} = F(\mathbf{r}/r)$.

- **Dot Product.** The dot product between two vectors $\mathbf{A}$ and $\mathbf{B}$ is defined by $\mathbf{A} \cdot \mathbf{B} = AB\cos\theta$. If $\mathbf{A}$ and $\mathbf{B}$ are expressed as Cartesian vectors, then $\mathbf{A} \cdot \mathbf{B} = A_xB_x + A_yB_y + A_zB_z$. In statics the dot product is used to determine the angle between the tails of the vectors, $\theta = \cos^{-1}(\mathbf{A} \cdot \mathbf{B}/AB)$. It is also used to determine the projected component of a vector $\mathbf{A}$ onto an axis defined by its unit vector $\mathbf{u}$, so that $A = A\cos\theta = \mathbf{A} \cdot \mathbf{u}$.

REVIEW PROBLEMS

2-133. Determine the magnitude and coordinate direction angles of $\mathbf{F}_3$ so that the resultant of the three forces acts along the positive y axis and has a magnitude of 600 N.

2-134. Determine the magnitude and coordinate direction angles of $\mathbf{F}_3$ so that the resultant of the three forces is zero.

*2-136. The force $\mathbf{F}$ has a magnitude of 80 N and acts at the midpoint C of the thin rod. Express the force as a Cartesian vector.

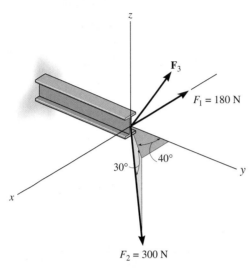

$F_1 = 180$ N
$40°$
$30°$
$F_2 = 300$ N

Probs. 2–133/134

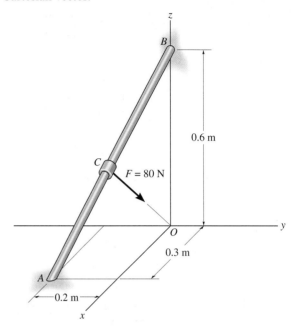

$F = 80$ N
0.6 m
0.3 m
0.2 m

Prob. 2–136

2-135. Determine the design angle θ ($\theta < 90°$) between the two struts so that the 500-N horizontal force has a component of 600-N directed from A toward C. What is the component of force acting along member BA?

2-137. Two forces $\mathbf{F}_1$ and $\mathbf{F}_2$ act on the hook. If their lines of action are at an angle θ apart and the magnitude of each force is $F_1 = F_2 = F$, determine the magnitude of the resultant force $\mathbf{F}_R$ and the angle between $\mathbf{F}_R$ and $\mathbf{F}_1$.

$20°$
θ
500 N

Prob. 2–135

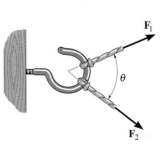

$\mathbf{F}_1$
θ
$\mathbf{F}_2$

Prob. 2–137

2-138. Determine the angles θ and ϕ between the wire segments.

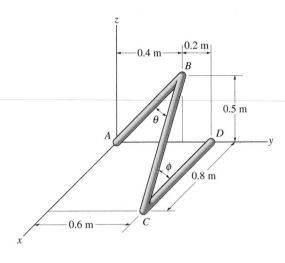

Prob. 2–138

2-139. Determine the magnitudes of the projected components of the force $\mathbf{F} = \{60\mathbf{i} + 12\mathbf{j} - 40\mathbf{k}\}$ N in the direction of the cables AB and AC.

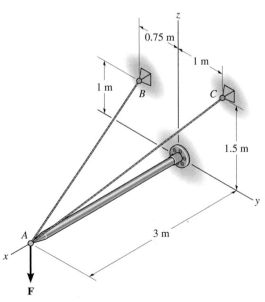

Prob. 2–139

*2-140. Determine the magnitude of the projected component of the 100-N force acting along the axis BC of the pipe.

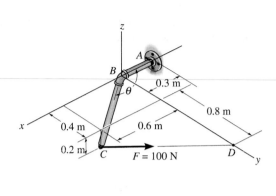

Prob. 2–140

2-141. The boat is to be pulled onto the shore using two ropes. If the resultant force is to be 400 N, directed along the keel aa, as shown, determine the magnitudes of forces $\mathbf{T}$ and $\mathbf{P}$ acting in each rope and the angle θ so that $\mathbf{P}$ is a *minimum*. $\mathbf{T}$ acts at 30° from the keel as shown.

Prob. 2–141

Whenever cables are used for hoisting loads, they must be selected so that they do not fail when they are placed at their points of attachment. In this chapter, we will show how to calculate cable loadings for such cases.

Equilibrium of a Particle

- To introduce the concept of the free-body diagram for a particle.
- To show how to solve particle equilibrium problems using the equations of equilibrium.

3.1   Condition for the Equilibrium of a Particle

A particle is in *equilibrium* provided it is at rest if originally at rest or has a constant velocity if originally in motion. Most often, however, the term "equilibrium" or, more specifically, "static equilibrium" is used to describe an object at rest. To maintain equilibrium, it is *necessary* to satisfy Newton's first law of motion, which requires the *resultant force* acting on a particle to be equal to *zero*. This condition may be stated mathematically as

$$\Sigma \mathbf{F} = \mathbf{0} \qquad\qquad (3\text{--}1)$$

where $\Sigma \mathbf{F}$ is the vector *sum of all the forces* acting on the particle.

Not only is Eq. 3–1 a necessary condition for equilibrium, it is also a *sufficient* condition. This follows from Newton's second law of motion, which can be written as $\Sigma \mathbf{F} = m\mathbf{a}$. Since the force system satisfies Eq. 3–1, then $m\mathbf{a} = \mathbf{0}$, and therefore the particle's acceleration $\mathbf{a} = \mathbf{0}$. Consequently the particle indeed moves with constant velocity or remains at rest.

3.2 The Free-Body Diagram

To apply the equation of equilibrium, we must account for *all* the known and unknown forces ($\Sigma\mathbf{F}$) which act *on* the particle. The best way to do this is to draw the particle's *free-body diagram*. This diagram is simply a sketch which shows the particle "free" from its surroundings with *all* the forces that act *on* it.

Before presenting a formal procedure as to how to draw a free-body diagram, we will first consider two types of connections often encountered in particle equilibrium problems.

Springs. If a *linear elastic spring* is used for support, the length of the spring will change in direct proportion to the force acting on it. A characteristic that defines the "elasticity" of a spring is the *spring constant* or *stiffness k*. The magnitude of force exerted on a linear elastic spring which has a stiffness k and is deformed (elongated or compressed) a distance s, measured from its *unloaded* position, is

$$F = ks \qquad (3\text{–}2)$$

Here s is determined from the difference in the spring's deformed length l and its undeformed length l_0, i.e., $s = l - l_0$. If s is positive, $\mathbf{F}$ "pulls" on the spring; whereas if s is negative, $\mathbf{F}$ must "push" on it. For example, the spring shown in Fig. 3–1 has an undeformed length $l_0 = 0.4$ m and stiffness $k = 500$ N/m. To stretch it so that $l = 0.6$ m, a force $F = ks = (500 \text{ N/m})(0.6 \text{ m} - 0.4 \text{ m}) = 100$ N is needed. Likewise, to compress it to a length $l = 0.2$ m, a force $F = ks = (500 \text{ N/m})(0.2 \text{ m} - 0.4 \text{ m}) = -100$ N is required, Fig. 3–1.

Fig. 3–1

Cables and Pulleys. Throughout this book, except in Sec. 7.4, all cables (or cords) are assumed to have negligible weight and they cannot stretch. Also, a cable can support *only* a tension or "pulling" force, and this force always acts in the direction of the cable. In Chapter 5 it will be shown that the tension force developed in a *continuous cable* which passes over a frictionless pulley must have a *constant* magnitude to keep the cable in equilibrium. Hence, for any angle θ, shown in Fig. 3–2, the cable is subjected to a constant tension T throughout its length.

Cable is in tension

Fig. 3–2

PROCEDURE FOR DRAWING A FREE-BODY DIAGRAM

Since we must account for *all the forces acting on the particle* when applying the equations of equilibrium, the importance of first drawing a free-body diagram cannot be overemphasized. To construct a free-body diagram, the following three steps are necessary.

Draw Outlined Shape. Imagine the particle to be *isolated* or cut "free" from its surroundings by drawing its outlined shape.

Show All Forces. Indicate on this sketch *all* the forces that act *on the particle*. These forces can be *active forces*, which tend to set the particle in motion, or they can be *reactive forces* which are the result of the constraints or supports that tend to prevent motion. To account for all these forces, it may help to trace around the particle's boundary, carefully noting each force acting on it.

Identify Each Force. The forces that are *known* should be labeled with their proper magnitudes and directions. Letters are used to represent the magnitudes and directions of forces that are unknown.

The bucket is held in equilibrium by the cable, and instinctively we know that the force in the cable must equal the weight of the bucket. By drawing a free-body diagram of the bucket we can understand why this is so. This diagram shows that there are only two forces *acting on the bucket*, namely, its weight **W** and the force **T** of the cable. For equilibrium, the resultant of these forces must be equal to zero, and so $T = W$. The important point is that by *isolating the bucket* the unknown cable force **T** becomes "exposed" and must be considered as a requirement for equilibrium.

Consider the spool having a weight W which is suspended from the crane boom. If we wish to obtain the forces in cables AB and AC then we can consider the free-body diagram of the ring at A since these forces act on the ring. Here the cables AD exert a resultant force of **W** on the ring and the condition of equilibrium is used to obtain $\mathbf{T}_B$ and $\mathbf{T}_C$.

E X A M P L E 3.1

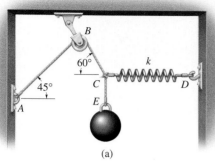

(a)

The sphere in Fig. 3–3a has a mass of 6 kg and is supported as shown. Draw a free-body diagram of the sphere, the cord *CE*, and the knot at *C*.

Solution

Sphere. By inspection, there are only two forces acting on the sphere, namely, its weight and the force of cord *CE*. The sphere has a weight of 6 kg $(9.81 \text{ m/s}^2) = 58.9$ N. The free-body diagram is shown in Fig. 3–3b.

Cord CE. When the cord *CE* is isolated from its surroundings, its free-body diagram shows only two forces acting on it, namely, the force of the sphere and the force of the knot, Fig. 3–3c. Notice that $\mathbf{F}_{CE}$ shown here is equal but opposite to that shown in Fig. 3–3b, a consequence of Newton's third law. Also, $\mathbf{F}_{CE}$ and $\mathbf{F}_{EC}$ pull on the cord and keep it in tension so that it doesn't collapse. For equilibrium, $F_{CE} = F_{EC}$.

Knot. The knot at *C* is subjected to three forces, Fig. 3–3d. They are caused by the cords *CBA* and *CE* and the spring *CD*. As required the free-body diagram shows all these forces labeled with their magnitudes and directions. It is important to recognize that the weight of the sphere does not directly act on the knot. Instead, the cord *CE* subjects the knot to this force.

$\mathbf{F}_{CE}$ (Force of cord *CE* acting on sphere)

58.9N (Weight or gravity acting on sphere)

(b)

$\mathbf{F}_{EC}$ (Force of knot acting on cord *CE*)

$\mathbf{F}_{CE}$ (Force of sphere acting on cord *CE*)

(c)

$\mathbf{F}_{CBA}$ (Force of cord *CBA* acting on knot)

60° C

$\mathbf{F}_{CD}$ (Force of spring acting on knot)

$\mathbf{F}_{CE}$ (Force of cord *CE* acting on knot)

(d)

Fig. 3–3

3.3 Coplanar Force Systems

If a particle is subjected to a system of coplanar forces that lie in the x–y plane, Fig. 3–4, then each force can be resolved into its $\mathbf{i}$ and $\mathbf{j}$ components. For equilibrium, Eq. 3–1 can be written as

$$\Sigma \mathbf{F} = \mathbf{0}$$
$$\Sigma F_x \mathbf{i} + \Sigma F_y \mathbf{j} = \mathbf{0}$$

For this vector equation to be satisfied, both the x and y components must be equal to zero. Hence,

$$\Sigma F_x = 0$$
$$\Sigma F_y = 0 \tag{3-3}$$

These *scalar equations of equilibrium* require that the *algebraic sum* of the x and y components of all the forces acting on the particle be equal to zero. As a result, Eqs. 3–3 can be solved for at most two unknowns, generally represented as angles and magnitudes of forces shown on the particle's free-body diagram.

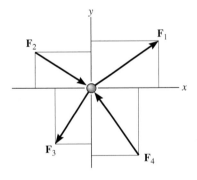

Fig. 3–4

Scalar Notation. Since each of the two equilibrium equations requires the resolution of vector components along a specified x or y axis, we will use scalar notation to represent the components when applying these equations. When doing this, the sense of direction for each component is accounted for by an *algebraic sign* which corresponds to the arrowhead direction of the component along each axis. If a force has an *unknown magnitude*, then the arrowhead sense of the force on the free-body diagram can be *assumed*. Since the magnitude of a force is *always positive*, then if the *solution* yields a *negative scalar*, this indicates that the sense of the force acts in the opposite direction.

The chains exert three forces on the ring at A. The ring will not move, or will move with constant velocity, provided the summation of these forces along the x and along the y axis on the free-body diagram is zero. If one of the three forces is known, the magnitudes of the other two forces can be obtained from the two equations of equilibrium.

Fig. 3–5

For example, consider the free-body diagram of the particle subjected to the two forces shown in Fig. 3–5. Here it is *assumed* that the *unknown force* **F** acts to the right to maintain equilibrium. Applying the equation of equilibrium along the x axis, we have

$$\xrightarrow{+} \Sigma F_x = 0; \qquad +F + 10\,\text{N} = 0$$

Both terms are "positive" since both forces act in the positive x direction. When this equation is solved, $F = -10\,\text{N}$. Here the *negative sign* indicates that **F** must act to the left to hold the particle in equilibrium, Fig. 3–5. Notice that if the $+x$ axis in Fig. 3–5 was directed to the left, both terms in the above equation would be negative, but again, after solving, $F = -10\,\text{N}$, indicating again **F** would be directed to the left.

PROCEDURE FOR ANALYSIS

Coplanar force equilibrium problems for a particle can be solved using the following procedure.

Free-Body Diagram.

- Establish the x, y axes in any suitable orientation.
- Label all the known and unknown force magnitudes and directions on the diagram.
- The sense of a force having an unknown magnitude can be assumed.

Equations of Equilibrium.

- Apply the equations of equilibrium $\Sigma F_x = 0$ and $\Sigma F_y = 0$.
- Components are positive if they are directed along a positive axis, and negative if they are directed along a negative axis.
- If more than two unknowns exist and the problem involves a spring, apply $F = ks$ to relate the spring force to the deformation s of the spring.
- If the solution yields a negative result, this indicates the sense of the force is the reverse of that shown on the free-body diagram.

EXAMPLE 3.2

Determine the tension in cables AB and AD for equilibrium of the 250-kg engine shown in Fig. 3–6a.

Solution

Free-Body Diagram. To solve this problem, we will investigate the equilibrium of the ring at A because this "particle" is subjected to the forces of both cables AB and AD. First, however, note that the engine has a weight $(250 \text{ kg})(9.81 \text{ m/s}^2) = 2.452 \text{ kN}$ which is supported by cable CA. Therefore, as shown in Fig. 3–6b, there are three concurrent forces *acting on the ring*. The forces $\mathbf{T}_B$ and $\mathbf{T}_D$ have unknown magnitudes but known directions, and cable AC exerts a downward force on A equal to 2.452 kN.

(a)

Fig. 3–6

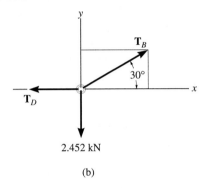

2.452 kN

(b)

Equations of Equilibrium. The two unknown magnitudes T_B and T_D can be obtained from the two scalar equations of equilibrium, $\Sigma F_x = 0$ and $\Sigma F_y = 0$. To apply these equations, the x, y axes are established on the free-body diagram and $\mathbf{T}_B$ must be resolved into its x and y components. Thus,

$$\xrightarrow{+} \Sigma F_x = 0; \qquad T_B \cos 30° - T_D = 0 \qquad (1)$$
$$+\uparrow \Sigma F_y = 0; \qquad T_B \sin 30° - 2.452 \text{ kN} = 0 \qquad (2)$$

Solving Eq. 2 for T_B and substituting into Eq. 1 to obtain T_D yields

$$T_B = 4.90 \text{ kN} \qquad \qquad Ans.$$
$$T_D = 4.25 \text{ kN} \qquad \qquad Ans.$$

The accuracy of these results, of course, depends on the accuracy of the data, i.e., measurements of geometry and loads. For most engineering work involving a problem such as this, the data as measured to three significant figures would be sufficient. Also, note that here we have neglected the weights of the cables, a reasonable assumption since they would be small in comparison with the weight of the engine.

EXAMPLE 3.3

(a)

(b)

(c)

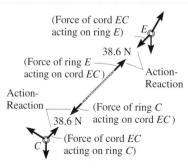

(d)

Fig. 3–7

If the sack at A in Fig. 3–7a has a weight of 20 N ($\approx$ 2 kg), determine the weight of the sack at B and the force in each cord needed to hold the system in the equilibrium position shown.

Solution
Since the weight of A is known, the unknown tension in the two cords EG and EC can be determined by investigating the equilibrium of the ring at E. Why?

Free-Body Diagram. There are three forces acting on E, as shown in Fig. 3–7b.

Equations of Equilibrium. Establishing the x, y axes and resolving each force onto its x and y components using trigonometry, we have

$$\xrightarrow{+} \Sigma F_x = 0; \qquad T_{EG} \sin 30° - T_{EC} \cos 45° = 0 \qquad (1)$$
$$+\uparrow \Sigma F_y = 0; \qquad T_{EG} \cos 30° - T_{EC} \sin 45° - 20\text{ N} = 0 \qquad (2)$$

Solving Eq. 1 for T_{EG} in terms of T_{EC} and substituting the result into Eq. 2 allows a solution for T_{EC}. One then obtains T_{EG} from Eq. 1. The results are

$$T_{EC} = 38.6\text{ N} \qquad \qquad Ans.$$
$$T_{EG} = 54.6\text{ N} \qquad \qquad Ans.$$

Using the calculated result for T_{EC}, the equilibrium of the ring at C can now be investigated to determine the tension in CD and the weight of B.

Free-Body Diagram. As shown in Fig. 3–7c, $T_{EC} = 38.6$ N "pulls" on C. The reason for this becomes clear when one draws the free-body diagram of cord CE and applies both equilibrium and the principle of action, equal but opposite force reaction (Newton's third law), Fig. 3–7d.

Equations of Equilibrium. Establishing the x, y axes and noting the components of $\mathbf{T}_{CD}$ are proportional to the slope of the cord as defined by the 3–4–5 triangle, we have

$$\xrightarrow{+} \Sigma F_x = 0; \qquad 38.6 \cos 45°\text{ N} - (\tfrac{4}{5})T_{CD} = 0 \qquad (3)$$
$$\xrightarrow{+} \Sigma F_y = 0; \qquad (\tfrac{3}{5})T_{CD} + 38.6 \sin 45°\text{ N} - W_B = 0 \qquad (4)$$

Solving Eq. 3 and substituting the result into Eq. 4 yields

$$T_{CD} = 34.1\text{ N} \qquad \qquad Ans.$$
$$W_B = 47.8\text{ N} \qquad \qquad Ans.$$

EXAMPLE 3.4

Determine the required length of cord AC in Fig. 3–8a so that the 8-kg lamp is suspended in the position shown. The *undeformed* length of spring AB is $l'_{AB} = 0.4$ m, and the spring has a stiffness of $k_{AB} = 300$ N/m.

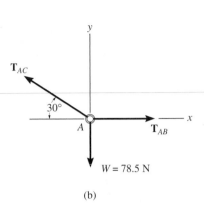

(a)

(b)

Fig. 3–8

Solution

If the force in spring AB is known, the stretch of the spring can be found using $F = ks$. From the problem geometry, it is then possible to calculate the required length of AC.

Free-Body Diagram. The lamp has a weight $W = 8(9.81) = 78.5$ N. The free-body diagram of the ring at A is shown in Fig. 3–8b.

Equations of Equilibrium. Using the x, y axes,

$$\xrightarrow{+} \Sigma F_x = 0; \qquad T_{AB} - T_{AC} \cos 30° = 0$$
$$+\uparrow \Sigma F_y = 0; \qquad T_{AC} \sin 30° - 78.5 \text{ N} = 0$$

Solving, we obtain

$$T_{AC} = 157.0 \text{ N}$$
$$T_{AB} = 136.0 \text{ N}$$

The stretch of spring AB is therefore

$$T_{AB} = k_{AB}s_{AB}; \qquad 136.0 \text{ N} = 300 \text{ N/m}(s_{AB})$$
$$s_{AB} = 0.453 \text{ m}$$

so the stretched length is

$$l_{AB} = l'_{AB} + s_{AB}$$
$$l_{AB} = 0.4 \text{ m} + 0.453 \text{ m} = 0.853 \text{ m}$$

The horizontal distance from C to B, Fig. 3–8a, requires

$$2 \text{ m} = l_{AC} \cos 30° + 0.853 \text{ m}$$
$$l_{AC} = 1.32 \text{ m} \qquad\qquad Ans.$$

PROBLEMS

3-1. Determine the magnitudes of F_1 and F_2 so that particle P is in equilibrium.

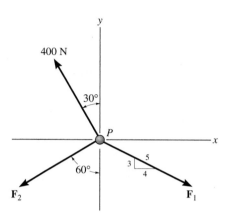

Prob. 3–1

3-3. Determine the magnitude and angle θ of F_1 so that particle P is in equilibrium.

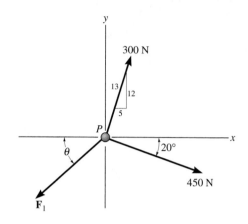

Prob. 3–3

3-2. Determine the magnitude and direction θ of F so that the particle is in equilibrium.

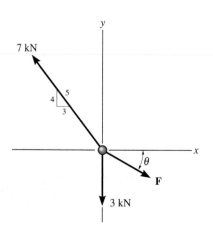

Prob. 3–2

*3-4. Determine the magnitude and angle θ of F so that the particle is in equilibrium.

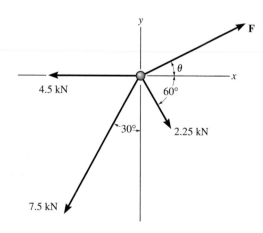

Prob. 3–4

3-5. The members of a truss are pin-connected at joint O. Determine the magnitudes of   and   for equilibrium. Set $\theta = 60°$.

3-6. The members of a truss are pin-connected at joint O. Determine the magnitude of $\mathbf{F}_1$ and its angle θ for equilibrium. Set $F_2 = 6$ kN.

*3-8. Determine the force in cables AB and AC necessary to support the 12-kg traffic light.

Prob. 3–8

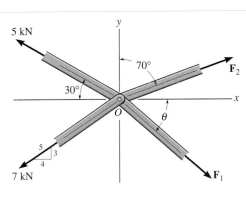

Probs. 3–5/6

3-7. The device shown is used to straighten the frames of wrecked autos. Determine the tension of each segment of the chain, i.e., AB and BC, if the force which the hydraulic cylinder DB exerts on point B is 3.50 kN, as shown.

3-9. Cords AB and AC can each sustain a maximum tension of 8000 N. If the drum has a weight of 9000 N (≈ 900 kg), determine the smallest angle θ at which they can be attached to the drum.

Prob. 3–7

Prob. 3–9

3-10. The 500-N ($\approx$ 50-kg) crate is hoisted using the ropes AB and AC. Each rope can withstand a maximum tension of 2500 N before it breaks. If AB always remains horizontal, determine the smallest angle θ to which the crate can be hoisted.

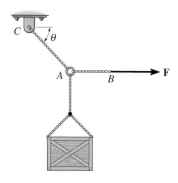

Prob. 3–10

3-11. Two electrically charged pith balls, each having a mass of 0.2 g, are suspended from light threads of equal length. Determine the resultant horizontal force of repulsion, F, acting on each ball if the measured distance between them is $r = 200$ mm.

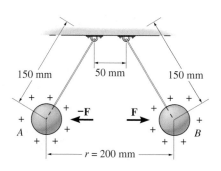

Prob. 3–11

*3-12. The concrete pipe elbow has a weight of 2000 N ($\approx$ 200 kg) and the center of gravity is located at point G. Determine the force in the cables AB and CD needed to support it.

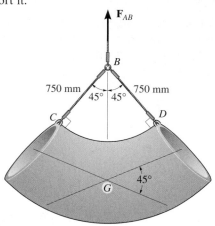

Prob. 3–12

3-13. Determine the stretch in each spring for equilibrium of the 2-kg block. The springs are shown in the equilibrium position.

3-14. The unstretched length of spring AB is 2 m. If the block is held in the equilibrium position shown, determine the mass of the block at D.

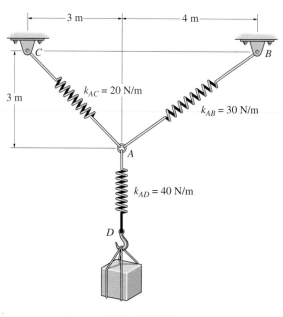

Probs. 3–13/14

3-15. The spring *ABC* has a stiffness of 500 N/m and an unstretched length of 6 m. Determine the horizontal force **F** applied to the cord which is attached to the *small* pulley *B* so that the displacement of the pulley from the wall is *d* = 1.5 m.

*3-16. The spring *ABC* has a stiffness of 500 N/m and an unstretched length of 6 m. Determine the displacement *d* of the cord from the wall when a force *F* = 175 N is applied to the cord.

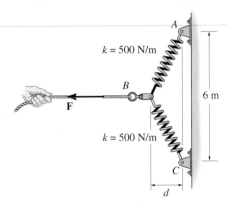

Probs. 3–15/16

3-17. Determine the maximum weight of the flowerpot that can be supported without exceeding a cable tension of 250 N in either cable *AB* or *AC*.

Prob. 3–17

3-18. The motor at *B* winds up the cord attached to the 325-N (≈ 32.5kg) crate with a constant speed. Determine the force in cord *CD* supporting the pulley and the angle θ for equilibrium. Neglect the size of the pulley at *C*.

3-19. The cords *BCA* and *CD* can each support a maximum load of 500 N (≈ 50 kg). Determine the maximum weight of the crate that can be hoisted at constant velocity, and the angle θ for equilibrium.

Probs. 3–18/19

*3-20. Determine the forces in cables *AC* and *AB* needed to hold the 20-kg ball *D* in equilibrium. Take *F* = 300 N and *d* = 1 m.

3-21. The ball *D* has a mass of 20 kg. If a force of *F* = 100 N is applied horizontally to the ring at *A*, determine the largest dimension *d* so that the force in cable *AC* is zero.

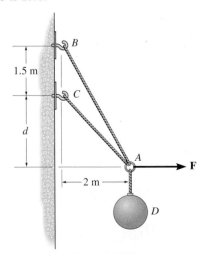

Probs. 3–20/21

3-22. The block has a weight of 20 N (≈ 2 kg) and is being hoisted at uniform velocity. Determine the angle θ for equilibrium and the required force in each cord.

3-23. Determine the maximum weight W of the block that can be suspended in the position shown if each cord can support a maximum tension of 80 N. Also, what is the angle θ for equilibrium?

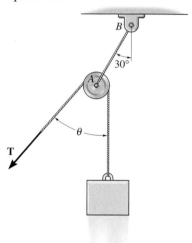

Probs. 3–22/23

*3-24. Determine the magnitude and direction θ of the equilibrium force F_{AB} exerted along link AB by the tractive apparatus shown. The suspended mass is 10 kg. Neglect the size of the pulley at A.

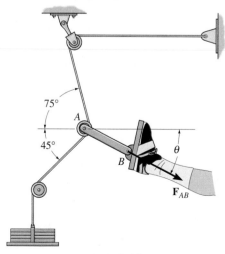

Prob. 3–24

3-25. Blocks D and F weigh 5 N (≈ 0.5 kg) each and block E weighs 8 N (≈ 0.8 kg). Determine the sag s for equilibrium. Neglect the size of the pulleys.

3-26. If blocks D and F weigh 5 N (≈ 0.5 kg) each, determine the weight of block E if the sag s = 0.3 m. Neglect the size of the pulleys.

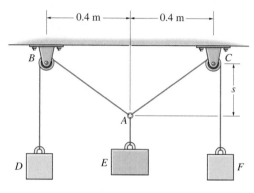

Probs. 3–25/26

3-27. The lift sling is used to hoist a container having a mass of 500 kg. Determine the force in each of the cables AB and AC as a function of θ. If the maximum tension allowed in each cable is 5 kN, determine the shortest lengths of cables AB and AC that can be used for the lift. The center of gravity of the container is located at G.

Prob. 3–27

*3-28. The load has a mass of 15 kg and is lifted by the pulley system shown. Determine the force **F** in the cord as a function of the angle θ. Plot the function of force F versus the angle θ for $0 \le \theta \le 90°$.

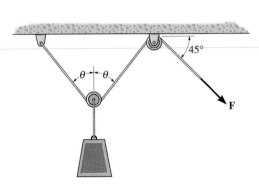

Prob. 3–28

3-30. The 400-N ($\approx$ 40-kg) uniform tank is suspended by means of a 3-m-long cable, which is attached to the sides of the tank and passes over the small pulley located at O. If the cable can be attached at either points A and B, or C and D, determine which attachment produces the least amount of tension in the cable. What is this tension?

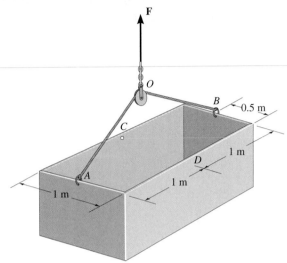

Prob. 3–30

3-29. The picture has a weight of 50 N ($\approx$ 5 kg) and is to be hung over the smooth pin B. If a string is attached to the frame at points A and C, and the maximum force the string can support is 75 N, determine the shortest string that can be safely used.

■ **3-31.** A vertical force $P = 10$ N is applied to the ends of the 0.2 m cord AB and spring AC. If the spring has an unstretched length of 0.2 m, determine the angle θ for equilibrium. Take $k = 150$ N/m.

*3-32. Determine the unstretched length of spring AC if a force $P = 80$ N causes the angle $\theta = 60°$ for equilibrium. Cord AB is 0.2 m long. Take $k = 500$ N/m.

Prob. 3–29

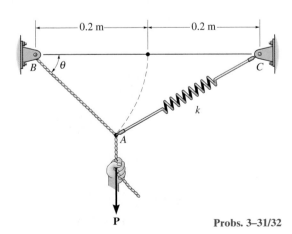

Probs. 3–31/32

■ **3-33.** A "scale" is constructed with a 1.2-m-long cord and the 50-N (≈ 5-kg) block *D*. The cord is fixed to a pin at *A* and passes over two *small* pulleys. Determine the weight of the suspended block *B* if the system is in equilibrium when *s* = 0.45 m.

Prob. 3–33

■ **3-34.** A car is to be towed using the rope arrangement shown. The towing force required is 3000 N. Determine the minimum length *l* of rope *AB* so that the tension in either rope *AB* or *AC* does not exceed 3750 N. *Hint:* Use the equilibrium condition at point *A* to determine the required angle θ for attachment, then determine *l* using trigonometry applied to triangle *ABC*.

Prob. 3–34

■ **3-35.** The spring has a stiffness of *k* = 800 N/m and an unstretched length of 200 mm. Determine the force in cables *BC* and *BD* when the spring is held in the position shown.

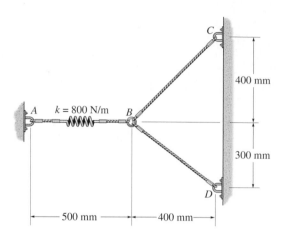

Prob. 3–35

3-36. The sling *BAC* is used to lift the 1000-N (≈ 100-kg) load with constant velocity. Determine the force in the sling and plot its value *T* (ordinate) as a function of its orientation θ, where 0 ≤ θ ≤ 90°.

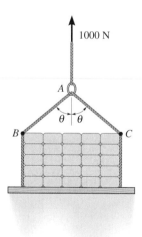

Prob. 3–36

■ **3-37.** The 45 kN (≈ 4.5 kg) lamp fixture is suspended from two springs, each having an unstretched length of 1.2 m and stiffness of $k = 75$ N/m. Determine the angle θ for equilibrium.

3-39. A 4-kg sphere rests on the smooth parabolic surface. Determine the normal force it exerts on the surface and the mass m_B of block B needed to hold it in the equilibrium position shown.

Prob. 3–37

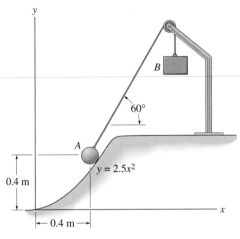

Prob. 3–39

3-38. The pail and its contents have a mass of 60 kg. If the cable is 15 m long, determine the distance y of the pulley for equilibrium. Neglect the size of the pulley at A.

*3-40.** The 30-kg pipe is supported at A by a system of five cords. Determine the force in each cord for equilibrium.

Prob. 3–38

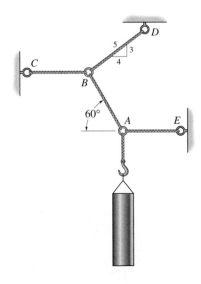

Prob. 3–40

3.4 Three-Dimensional Force Systems

For particle equilibrium we require

$$\Sigma \mathbf{F} = \mathbf{0} \qquad (3\text{–}4)$$

If the forces are resolved into their respective $\mathbf{i}$, $\mathbf{j}$, $\mathbf{k}$ components, Fig. 3–9, then we have

$$\Sigma F_x \mathbf{i} + \Sigma F_y \mathbf{j} + \Sigma F_z \mathbf{k} = \mathbf{0}$$

To ensure equilibrium, we must therefore require that the following three scalar component equations be satisfied:

$$\begin{aligned} \Sigma F_x &= 0 \\ \Sigma F_y &= 0 \\ \Sigma F_z &= 0 \end{aligned} \qquad (3\text{–}5)$$

These equations represent the *algebraic sums* of the x, y, z force components acting on the particle. Using them we can solve for at most three unknowns, generally represented as angles or magnitudes of forces shown on the particle's free-body diagram.

Fig. 3–9

The ring at A is subjected to the force from the hook as well as forces from each of the three chains. If the electromagnet and its load has a weight W, then the hook force will be $\mathbf{W}$, and the three scalar equations of equilibrium can be applied to the free-body diagram of the ring in order to determine the chain forces, $\mathbf{F}_B$, $\mathbf{F}_C$ and $\mathbf{F}_D$.

PROCEDURE FOR ANALYSIS

Three-dimensional force equilibrium problems for a particle can be solved using the following procedure.

Free-Body Diagram.

- Establish the x, y, z axes in any suitable orientation.
- Label all the known and unknown force magnitudes and directions on the diagram.
- The sense of a force having an unknown magnitude can be assumed.

Equations of Equilibrium.

- Use the scalar equations of equilibrium, $\Sigma F_x = 0$, $\Sigma F_y = 0$, $\Sigma F_z = 0$, in cases where it is easy to resolve each force into its x, y, z components.
- If the three-dimensional geometry appears difficult, then first express each force as a Cartesian vector, substitute these vectors into $\Sigma \mathbf{F} = \mathbf{0}$, and then set the $\mathbf{i}$, $\mathbf{j}$, $\mathbf{k}$ components equal to zero.
- If the solution yields a negative result, this indicates the sense of the force is the reverse of that shown on the free-body diagram.

EXAMPLE 3.5

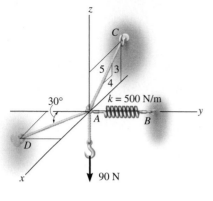

(a)

A 90-N load is suspended from the hook shown in Fig. 3–10a. The load is supported by two cables and a spring having a stiffness $k = 500$ N/m. Determine the force in the cables and the stretch of the spring for equilibrium. Cable AD lies in the x–y plane and cable AC lies in the x–z plane.

Solution

The stretch of the spring can be determined once the force in the spring is determined.

Free-Body Diagram. The connection at A is chosen for the equilibrium analysis since the cable forces are concurrent at this point. The free-body diagram is shown in Fig. 3–10b.

Equations of Equilibrium. By inspection, each force can easily be resolved into its x, y, z components, and therefore the three scalar equations of equilibrium can be directly applied. Considering components directed along the positive axes as "positive," we have

$$\Sigma F_x = 0; \qquad\qquad F_D \sin 30° - \tfrac{4}{5}F_C = 0 \qquad (1)$$

$$\Sigma F_y = 0; \qquad\qquad -F_D \cos 30° + F_B = 0 \qquad (2)$$

$$\Sigma F_z = 0; \qquad\qquad \tfrac{3}{5}F_C - 90 \text{ N} = 0 \qquad (3)$$

(b)

Fig. 3–10

Solving Eq. 3 for F_C, then Eq. 1 for F_D, and finally Eq. 2 for F_B, yields

$$F_C = 150 \text{ N} \qquad\qquad \textit{Ans.}$$
$$F_D = 240 \text{ N} \qquad\qquad \textit{Ans.}$$
$$F_B = 208 \text{ N} \qquad\qquad \textit{Ans.}$$

The stretch of the spring is therefore

$$F_B = ks_{AB}$$
$$208 \text{ N} = 500 \text{ N/m } (s_{AB})$$
$$s_{AB} = 0.416 \text{ m} \qquad\qquad \textit{Ans.}$$

EXAMPLE 3.6

Determine the magnitude and coordinate direction angles of force **F** in Fig. 3–11*a* that are required for equilibrium of particle *O*.

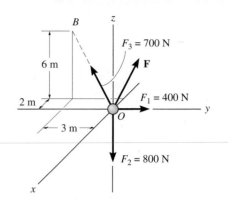

(a)

Solution

Free-Body Diagram. Four forces act on particle *O*, Fig. 3–11*b*.

Equations of Equilibrium. Each of the forces can be expressed in Cartesian vector form, and the equations of equilibrium can be applied to determine the *x*, *y*, *z* components of **F**. Noting that the coordinates of *B* are $B(-2 \text{ m}, -3 \text{ m}, 6 \text{ m})$, we have

$$\mathbf{F}_1 = \{400\mathbf{j}\} \text{ N}$$

$$\mathbf{F}_2 = \{-800\mathbf{k}\} \text{ N}$$

$$\mathbf{F}_3 = F_3\left(\frac{\mathbf{r}_B}{r_B}\right) = 700 \text{ N}\left[\frac{-2\mathbf{i} - 3\mathbf{j} + 6\mathbf{k}}{\sqrt{(-2)^2 + (-3)^2 + (6)^2}}\right]$$

$$= \{-200\mathbf{i} - 300\mathbf{j} + 600\mathbf{k}\} \text{ N}$$

$$\mathbf{F} = F_x\mathbf{i} + F_y\mathbf{j} + F_z\mathbf{k}$$

For equilibrium

$$\Sigma\mathbf{F} = \mathbf{0}; \qquad \mathbf{F}_1 + \mathbf{F}_2 + \mathbf{F}_3 + \mathbf{F} = \mathbf{0}$$

$$400\mathbf{j} - 800\mathbf{k} - 200\mathbf{i} - 300\mathbf{j} + 600\mathbf{k} + F_x\mathbf{i} + F_y\mathbf{j} + F_z\mathbf{k} = \mathbf{0}$$

Equating the respective **i**, **j**, **k** components to zero, we have

$$\Sigma F_x = 0; \qquad -200 + F_x = 0 \qquad F_x = 200 \text{ N}$$

$$\Sigma F_y = 0; \qquad 400 - 300 + F_y = 0 \qquad F_y = -100 \text{ N}$$

$$\Sigma F_z = 0; \qquad -800 + 600 + F_z = 0 \qquad F_z = 200 \text{ N}$$

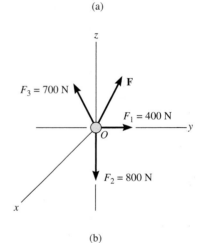

(b)

Thus,

$$\mathbf{F} = \{200\mathbf{i} - 100\mathbf{j} + 200\mathbf{k}\} \text{ N}$$

$$F = \sqrt{(200)^2 + (-100)^2 + (200)^2} = 300 \text{ N} \qquad \textit{Ans.}$$

$$\mathbf{u}_F = \frac{\mathbf{F}}{F} = \frac{200}{300}\mathbf{i} - \frac{100}{300}\mathbf{j} + \frac{200}{300}\mathbf{k}$$

$$\alpha = \cos^{-1}\left(\frac{200}{300}\right) = 48.2° \qquad \textit{Ans.}$$

$$\beta = \cos^{-1}\left(\frac{-100}{300}\right) = 109° \qquad \textit{Ans.}$$

$$\gamma = \cos^{-1}\left(\frac{200}{300}\right) = 48.2° \qquad \textit{Ans.}$$

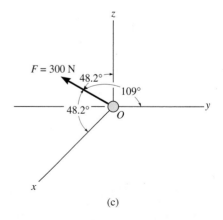

(c)

Fig. 3–11

The magnitude and correct direction of **F** are shown in Fig. 3–11*c*.

E X A M P L E 3.7

(a)

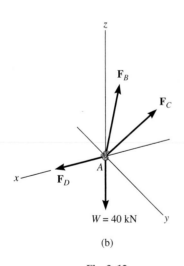

(b)

Fig. 3–12

Determine the force developed in each cable used to support the 40-kN (≈ 4-tonne) crate shown in Fig. 3–12*a*.

Solution

Free-Body Diagram. As shown in Fig. 3–12*b*, the free-body diagram of point *A* is considered in order to "expose" the three unknown forces in the cables.

Equations of Equilibrium. First we will express each force in Cartesian vector form. Since the coordinates of points *B* and *C* are $B(-3 \text{ m}, -4 \text{ m}, 8 \text{ m})$ and $C(-3 \text{ m}, 4 \text{ m}, 8 \text{ m})$, we have

$$\mathbf{F}_B = F_B\left[\frac{-3\mathbf{i} - 4\mathbf{j} + 8\mathbf{k}}{\sqrt{(-3)^2 + (-4)^2 + (8)^2}}\right]$$

$$= -0.318F_B\mathbf{i} - 0.424F_B\mathbf{j} + 0.848F_B\mathbf{k}$$

$$\mathbf{F}_C = F_C\left[\frac{-3\mathbf{i} + 4\mathbf{j} + 8\mathbf{k}}{\sqrt{(-3)^2 + (4)^2 + (8)^2}}\right]$$

$$= -0.318F_C\mathbf{i} + 0.424F_C\mathbf{j} + 0.484F_C\mathbf{k}$$

$$\mathbf{F}_D = F_D\mathbf{i}$$

$$\mathbf{W} = \{-40\mathbf{k}\} \text{ kN}$$

Equilibrium requires

$$\Sigma\mathbf{F} = \mathbf{0}; \qquad \mathbf{F}_B + \mathbf{F}_C + \mathbf{F}_D + \mathbf{W} = \mathbf{0}$$

$$-0.318F_B\mathbf{i} - 0.424F_B\mathbf{j} + 0.848F_B\mathbf{k} - 0.318F_C\mathbf{i} + 0.424F_C\mathbf{j}$$

$$+0.848F_C\mathbf{k} + F_D\mathbf{i} - 40\mathbf{k} = \mathbf{0}$$

Equating the respective **i**, **j**, **k** components to zero yields

$$\Sigma F_x = 0; \qquad -0.318F_B - 0.318F_C + F_D = 0 \qquad (1)$$

$$\Sigma F_y = 0; \qquad -0.424F_B + 0.424F_C = 0 \qquad (2)$$

$$\Sigma F_z = 0; \qquad 0.848F_B + 0.848F_C - 40 = 0 \qquad (3)$$

Equation 2 states that $F_B = F_C$. Thus, solving Eq. 3 for F_B and F_C and substituting the result into Eq. 1 to obtain F_D, we have

$$F_B = F_C = 23.6 \text{ kN} \qquad \qquad Ans.$$

$$F_D = 15.0 \text{ kN} \qquad \qquad Ans.$$

E X A M P L E **3.8**

The 100-kg crate shown in Fig. 3–13a is supported by three cords, one of which is connected to a spring. Determine the tension in cords AC and AD and the stretch of the spring.

Solution

Free-Body Diagram. The force in each of the cords can be determined by investigating the equilibrium of point A. The free-body diagram is shown in Fig. 3–13b. The weight of the crate is $W = 100(9.81) = 981$ N.

(a)

Equations of Equilibrium. Each vector on the free-body diagram is first expressed in Cartesian vector form. Using Eq. 2–11 for $\mathbf{F}_C$ and noting point $D(-1$ m, 2 m, 2 m) for $\mathbf{F}_D$, we have

$$\mathbf{F}_B = F_B\mathbf{i}$$

$$\mathbf{F}_C = F_C \cos 120°\mathbf{i} + F_C \cos 135°\mathbf{j} + F_C \cos 60°\mathbf{k}$$

$$= -0.5F_C\mathbf{i} - 0.707F_C\mathbf{j} + 0.5F_C\mathbf{k}$$

$$\mathbf{F}_D = F_D\left[\frac{-1\mathbf{i} + 2\mathbf{j} + 2\mathbf{k}}{\sqrt{(-1)^2 + (2)^2 + (2)^2}}\right]$$

$$= -0.333F_D\mathbf{i} + 0.667F_D\mathbf{j} + 0.667F_D\mathbf{k}$$

$$\mathbf{W} = \{-981\mathbf{k}\} \text{ N}$$

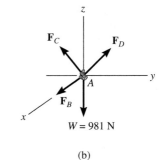

(b)

Fig. 3–13

Equilibrium requires

$$\Sigma\mathbf{F} = \mathbf{0};\qquad \mathbf{F}_B + \mathbf{F}_C + \mathbf{F}_D + \mathbf{W} = \mathbf{0}$$

$$F_B\mathbf{i} - 0.5F_C\mathbf{i} - 0.707F_C\mathbf{j} + 0.5F_C\mathbf{k} - 0.333F_D\mathbf{i} + 0.667F_D\mathbf{j}$$

$$+ 0.667F_D\mathbf{k} - 981\mathbf{k} = \mathbf{0}$$

Equating the respective $\mathbf{i}$, $\mathbf{j}$, $\mathbf{k}$ components to zero,

$$\Sigma F_x = 0;\qquad F_B - 0.5F_C - 0.333F_D = 0 \tag{1}$$

$$\Sigma F_y = 0;\qquad -0.707F_C + 0.667F_D = 0 \tag{2}$$

$$\Sigma F_z = 0;\qquad 0.5F_C + 0.667F_D - 981 = 0 \tag{3}$$

Solving Eq. 2 for F_D in terms of F_C and substituting into Eq. 3 yields F_C. F_D is determined from Eq. 2. Finally, substituting the results into Eq. 1 gives F_B. Hence,

$$F_C = 813 \text{ N} \qquad\qquad Ans.$$

$$F_D = 862 \text{ N} \qquad\qquad Ans.$$

$$F_B = 693.7 \text{ N}$$

The stretch of the spring is therefore

$$F = ks;\qquad\qquad 693.7 = 1500s$$

$$s = 0.462 \text{ m} \qquad\qquad Ans.$$

PROBLEMS

3-41. Determine the magnitude and direction of F_1 required to keep the concurrent force system in equilibrium.

3-43. Determine the magnitudes of F_1, F_2, and F_3 for equilibrium of the particle.

Prob. 3–41

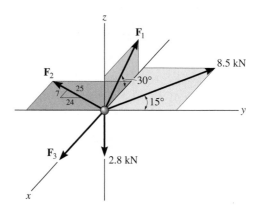

Prob. 3–43

3-42. Determine the magnitudes of F_1, F_2, and F_3 for equilibrium of the particle.

*■ 3-44. Determine the magnitude and direction of the force **P** required to keep the concurrent force system in equilibrium.

Prob. 3–42

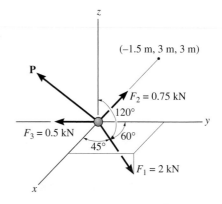

Prob. 3–44

3-45. The three cables are used to support the 800-N lamp. Determine the force developed in each cable for equilibrium.

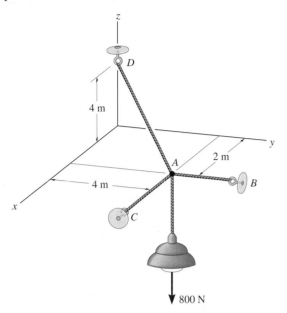

800 N

Prob. 3–45

3-46. If cable AB is subjected to a tension of 700 N, determine the tension in cables AC and AD and the magnitude of the vertical force **F**.

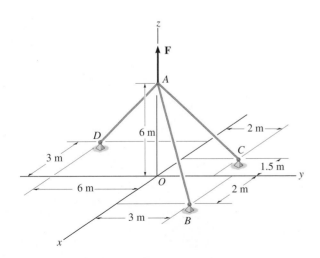

Prob. 3–46

3-47. Determine the stretch in each of the two springs required to hold the 20-kg crate in the equilibrium position shown. Each spring has an unstretched length of 2 m and a stiffness of $k = 300$ N/m.

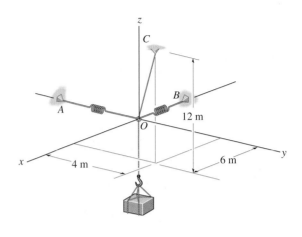

Prob. 3–47

*3-48. If the bucket and its contents have a total weight of 200 ($\approx$ 20 kg), determine the force in the supporting cables DA, DB, and DC.

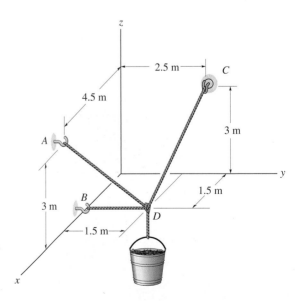

Prob. 3–48

■ **3-49.** The 2500-N crate is to be hoisted with constant velocity from the hold of a ship using the cable arrangement shown. Determine the tension in each of the three cables for equilibrium.

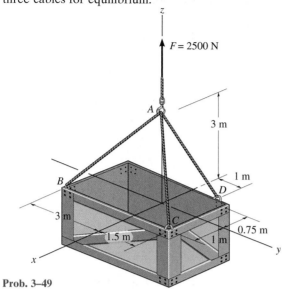

Prob. 3–49

■ **3-50.** The lamp has a mass of 15 kg and is supported by a pole AO and cables AB and AC. If the force in the pole acts along its axis, determine the forces in AO, AB, and AC for equilibrium.

3-51. Cables AB and AC can sustain a maximum tension of 500 N, and the pole can support a maximum compression of 300 N. Determine the maximum weight of the lamp that can be supported in the position shown. The force in the pole acts along the axis of the pole.

Probs. 3–50/51

*3-52.** Determine the tension in cables AB, AC, and AD, required to hold the 300-N ($\approx$ 30-kg) crate in equilibrium.

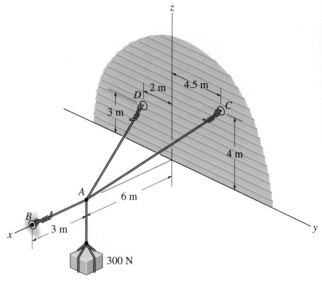

Prob. 3–52

3-53. The boom supports a bucket and contents, which have a total mass of 300 kg. Determine the forces developed in struts AD and AE and the tension in cable AB for equilibrium. The force in each strut acts along its axis.

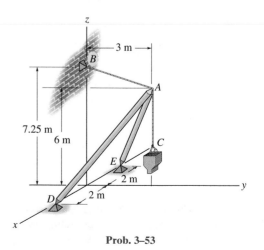

Prob. 3–53

3-54. Determine the force in each of the three cables needed to lift the tractor which has a mass of 8 Mg.

Prob. 3–54

3-55. Determine the force acting along the axis of each of the three struts needed to support the 500-kg block.

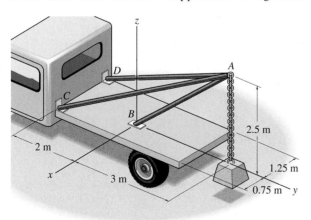

Prob. 3–55

*3-56. The 50-kg pot is supported from A by the three cables. Determine the force acting in each cable for equilibrium. Take $d = 2.5$ m.

3-57. Determine the height d of cable AB so that the force in cables AD and AC is one-half as great as the force in cable AB. What is the force in each cable for this case? The flowerpot has a mass of 50 kg.

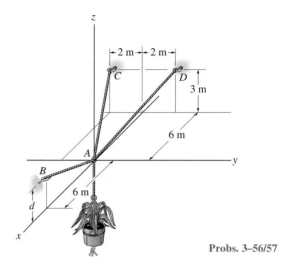

Probs. 3–56/57

3-58. The 400-N ($\approx$ 40-kg) chandelier is supported by three wires as shown. Determine the force in each wire for equilibrium.

3-59. If each wire can sustain a maximum tension of 600 N before it fails, determine the greatest weight of the chandelier the wires will support in the position shown.

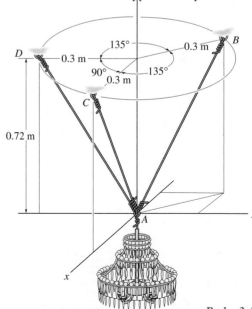

Probs. 3–58/59

*3-60. Three cables are used to support a 900-N ($\approx$ 90-kg) ring. Determine the tension in each cable for equilibrium.

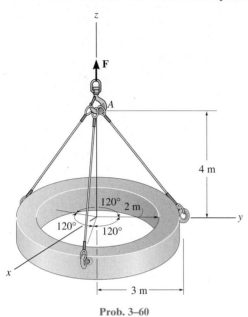

Prob. 3–60

3-61. The 800-N ($\approx$ 80-kg) cylinder is supported by three chains as shown. Determine the force in each chain for equilibrium. Take $d = 1$ m.

Prob. 3–61

3-62. A small peg P rests on a spring that is contained inside the smooth pipe. When the spring is compressed so that $s = 0.15$ m, the spring exerts an upward force of 60 N on the peg. Determine the point of attachment $A(x, y, 0)$ of cord PA so that the tension in cords PB and PC equals 30 N and 50 N, respectively.

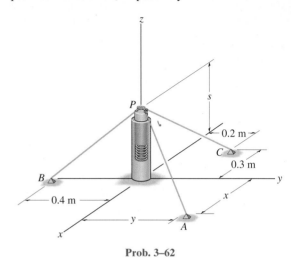

Prob. 3–62

3-63. Determine the force in each cable needed to support the 17500-N ($\approx$ 1.75-tonne) platform. Set $d = 2$ m.

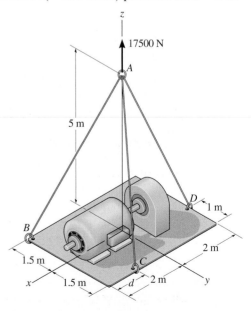

Prob. 3–63

■*3-64. The 400-N ball is suspended from the horizontal ring using three springs each having an unstretched length of 0.5 m and stiffness of 1000 N/m. Determine the vertical distance h from the ring to point A for equilibrium.

3-65. Determine the tension developed in cables OD and OB and the strut OC, required to support the 50-kg crate. The spring OA has an unstretched length of 0.8 m and a stiffness $k_{OA} = 1.2$ kN/m. The force in the strut acts along the axis of the strut.

Prob. 3–64

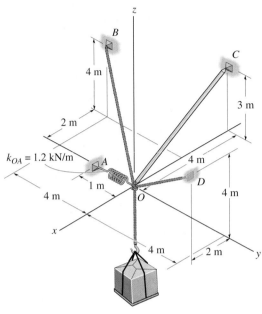

Prob. 3–65

CHAPTER REVIEW

- **Equilibrium.** When a particle is at rest or moves with constant velocity, it is in equilibrium. This requires that all the forces acting on the particle form a zero force resultant. In order to account for all the forces, it is necessary to draw a free-body diagram. This diagram is an outlined shape of the particle that shows all the forces, listed with their known or unknown magnitudes and directions.

- **Two Dimensions.** The two scalar equations of force equilibrium $\Sigma F_x = 0$ and $\Sigma F_y = 0$ can be applied when referenced from an established x, y coordinate system. If the solution for a force magnitude yields a negative scalar, then the force acts in the opposite direction to that shown on the free-body diagram. If the problem involves a linear elastic spring then the stretch or compression s of the spring can be related to the force applied to it using $F = ks$.

- **Three Dimensions.** Since three-dimensional geometry can be difficult to visualize, the equilibrium equation $\Sigma \mathbf{F} = \mathbf{0}$ should be applied using a Cartesian vector analysis. This requires first expressing each force on the free-body diagram as a Cartesian vector. When the forces are summed and set equal to zero, then the $\mathbf{i}$, $\mathbf{j}$, and $\mathbf{k}$ components are also zero, so that $\Sigma F_x = 0$, $\Sigma F_y = 0$ and $\Sigma F_z = 0$.

REVIEW PROBLEMS

3-66. The pipe is held in place by the vice. If the bolt exerts a force of 250 N on the pipe in the direction shown, determine the forces F_A and F_B that the smooth contacts at A and B exert on the pipe.

Prob. 3–66

3-67. When y is zero, the springs sustain a force of 300 Nb. Determine the magnitude of the applied vertical forces $\mathbf{F}$ and $-\mathbf{F}$ required to pull point A away from point B a distance of $y = 0.2$ m. The ends of cords CAD and CBD are attached to rings at C and D.

*3-68. When y is zero, the springs are each stretched 0.15 m. Determine the distance y if a force of $F = 300$ N is applied to points A and B as shown. The ends of cords CAD and CBD are attached to rings at C and D.

3-69. Romeo tries to reach Juliet by climbing with constant velocity up a rope which is knotted at point A. Any of the three segments of the rope can sustain a maximum force of 2 kN before it breaks. Determine if Romeo, who has a mass of 65 kg, can climb the rope, and if so, can he along with his Juliet, who has a mass of 60 kg, climb down with constant velocity?

Prob. 3–69

■ **3-70.** Determine the magnitudes of forces $\mathbf{F}_1$, $\mathbf{F}_2$, and $\mathbf{F}_3$ necessary to hold the force $\mathbf{F} = \{-9\mathbf{i} - 8\mathbf{j} - 5\mathbf{k}\}$ kN in equilibrium.

Prob. 3–67/68

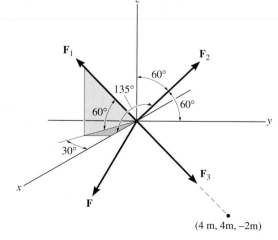

Prob. 3–70

3-71. The man attempts to pull the log at C by using the three ropes. Determine the direction θ in which he should pull on his rope with a force of 400 N, so that he exerts a maximum force on the log. What is the force on the log for this case? Also, determine the direction in which he should pull in order to maximize the force in the rope attached to B. What is this maximum force?

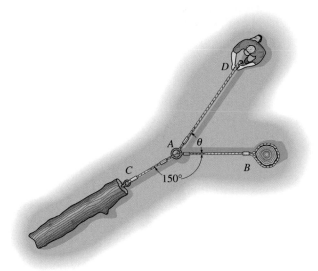

Prob. 3–71

*■3-72. The ring of negligible size is subjected to a vertical force of 200 N. Determine the required length l of cord AC such that the tension acting in AC is 160 N. Also, what is the force acting in cord AB? *Hint:* Use the equilibrium condition to determine the required angle θ for attachment, then determine l using trigonometry applied to $\triangle ABC$.

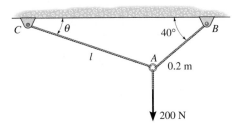

Prob. 3–72

3-73. Determine the maximum weight of the engine that can be supported without exceeding a tension of 900 N in chain AB and 960 N in chain AC.

Prob. 3–73

3-74. Determine the force in each cable needed to support the 500-N ($\approx$ 50-kg) load.

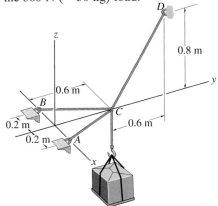

Prob. 3–74

3-75. The joint of a space frame is subjected to four member forces. Member OA lies in the $x - y$ plane and member OB lies in the $y - z$ plane. Determine the forces acting in each of the members required for equilibrium of the joint.

Prob. 3–75

This utility pole is subjected to many forces, caused by the cables and the weight of the transformer. In some cases, it is important to be able to simplify this system to a single resultant force and specify where this resultant acts on the pole.

Force System Resultants

- To discuss the concept of the moment of a force and show how to calculate it in two and three dimensions.
- To provide a method for finding the moment of a force about a specified axis.
- To define the moment of a couple.
- To present methods for determining the resultants of nonconcurrent force systems.
- To indicate how to reduce a simple distributed loading to a resultant force having a specified location.

4.1 Moment of a Force—Scalar Formulation

The *moment* of a force about a point or axis provides a measure of the tendency of the force to cause a body to rotate about the point or axis. For example, consider the horizontal force $\mathbf{F}_x$, which acts perpendicular to the handle of the wrench and is located a distance d_y from point O, Fig. 4–1a. It is seen that this force tends to cause the pipe to turn about the z axis. The larger the force or the distance d_y, the greater the turning effect. This tendency for rotation caused by $\mathbf{F}_x$ is sometimes called a *torque*, but most often it is called the *moment of a force* or simply the *moment* $(\mathbf{M}_O)_z$. Note that the *moment axis* (z) is perpendicular to the shaded plane (x–y) which contains both $\mathbf{F}_x$ and d_y and that this axis intersects the plane at point O.

(a)

Fig. 4–1

(b)

(c)

Fig. 4–1

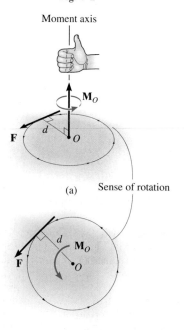

Moment axis

(a)   Sense of rotation

(b)

Fig. 4–2

Now consider applying the force $\mathbf{F}_z$ to the wrench, Fig. 4–1b. This force will *not* rotate the pipe about the z axis. Instead, it tends to rotate it about the x axis. Keep in mind that although it may not be possible to actually "rotate" or turn the pipe in this manner, $\mathbf{F}_z$ still creates the *tendency* for rotation and so the moment $(\mathbf{M}_O)_x$ is produced. As before, the force and distance d_y lie in the shaded plane (y–z) which is perpendicular to the moment axis (x). Lastly, if a force $\mathbf{F}_y$ is applied to the wrench, Fig. 4–1c, no moment is produced about point O. This results in a lack of turning since the line of action of the force passes through O and therefore no tendency for rotation is possible.

We will now generalize the above discussion and consider the force $\mathbf{F}$ and point O which lie in a shaded plane as shown in Fig. 4–2a. The moment $\mathbf{M}_O$ about point O, or about an axis passing through O and perpendicular to the plane, is a *vector quantity* since it has a specified magnitude and direction.

Magnitude.   The magnitude of M_O is

$$M_O = Fd \qquad (4\text{–}1)$$

where d is referred to as the *moment arm* or perpendicular distance from the axis at point O to the line of action of the force. Units of moment magnitude consist of force times distance, e.g., $N \cdot m$.

Direction.   The direction of $\mathbf{M}_O$ will be specified by using the "right-hand rule." To do this, the fingers of the right hand are curled such that they follow the sense of rotation, which would occur if the force could rotate about point O, Fig. 4–2a. The *thumb* then *points* along the *moment axis* so that it gives the direction and sense of the moment vector, which is *upward* and *perpendicular* to the shaded plane containing $\mathbf{F}$ and d.

In three dimensions, $\mathbf{M}_O$ is illustrated by a vector arrow with a curl on it to *distinguish* it from a force vector, Fig. 4–2a. Many problems in mechanics, however, involve coplanar force systems that may be conveniently viewed in two dimensions. For example, a two-dimensional view of Fig. 4–2a is given in Fig. 4–2b. Here $\mathbf{M}_O$ is simply represented by the (counterclockwise) curl, which indicates the action of $\mathbf{F}$. The arrowhead on this curl is used to show the *sense of rotation* caused by $\mathbf{F}$. Using the right-hand rule, however, realize that the direction and sense of the moment vector in Fig. 4–2b are specified by the thumb, which points *out* of the page since the fingers follow the curl. In particular, notice that *this curl or sense of rotation can always be determined by observing in which direction the force would "orbit" about point O* (counterclockwise in Fig. 4–2b). In two dimensions we will often refer to finding the moment of a force "about a point" (O). Keep in mind, however, that the moment *always acts about an axis* which is perpendicular to the plane containing $\mathbf{F}$ and d, and this axis intersects the plane at the point (O), Fig. 4–2a.

Resultant Moment of a System of Coplanar Forces. If a system of forces lies in an x–y plane, then the moment produced by each force about point O will be directed along the z axis, Fig. 4–3. Consequently, the resultant moment $\mathbf{M}_{R_O}$ of the system can be determined by simply adding the moments of all forces *algebraically* since all the moment vectors are collinear. We can write this vector sum symbolically as

$$\zeta + M_{R_O} = \Sigma Fd \qquad (4\text{–}2)$$

Here the counterclockwise curl written alongside the equation indicates that, by the scalar sign convention, the moment of any force will be positive if it is directed along the $+z$ axis, whereas a negative moment is directed along the $-z$ axis.

The following examples illustrate numerical application of Eqs. 4–1 and 4–2.

Fig. 4–3

By pushing down on the pry bar the load on the ground at A can be lifted. The turning effect, caused by the applied force, is due to the moment about A. To produce this moment with minimum effort we instinctively know that the force should be applied to the *end* of the bar; however, the *direction* in which this force is applied is also important. This is because moment is the product of the force and the moment arm. Notice that when the force is at an angle $\theta < 90°$, then the moment arm distance is *shorter* than when the force is applied perpendicular to the bar $\theta = 90°$, i.e., $d' < d$. Hence the greatest moment is produced when the force is farthest from point A and applied perpendicular to the axis of the bar so as to maximize the moment arm.

The moment of a force does not always cause a rotation. For example, the force $\mathbf{F}$ tends to rotate the beam clockwise about its support at A with a moment $M_A = Fd_A$. The actual rotation would occur if the support at B were removed. In the same manner. $\mathbf{F}$ creates a tendency to rotate the beam counterclockwise about B with a moment $M_B = Fd_B$. Here the support at A prevents the rotation.

EXAMPLE 4.1

For each case illustrated in Fig. 4–4, determine the moment of the force about point O.

Solution *(Scalar Analysis)*
The line of action of each force is extended as a dashed line in order to establish the moment arm d. Also illustrated is the tendency of rotation of the member as caused by the force. Furthermore, the orbit of the force is shown as a colored curl. Thus,

Fig. 4–4a $M_O = (100\ \text{N})(2\ \text{m}) = 200\ \text{N} \cdot \text{m} \downarrow$ *Ans.*

Fig. 4–4b $M_O = (50\ \text{N})(0.75\ \text{m}) = 37.5\ \text{N} \cdot \text{m} \downarrow$ *Ans.*

Fig. 4–4c $M_O = (40\ \text{N})(4\ \text{m} + 2\cos 30°\ \text{m}) = 229\ \text{N} \cdot \text{m} \downarrow$ *Ans.*

Fig. 4–4d $M_O = (60\ \text{N})(1 \sin 45°\ \text{m}) = 42.4\ \text{N} \cdot \text{m} \uparrow$ *Ans.*

Fig. 4–4e $M_O = (7\ \text{kN})(4\ \text{m} - 1\ \text{m}) = 21.0\ \text{kN} \cdot \text{m} \uparrow$ *Ans.*

(a)

(b)

(c)

(d)

(e)

Fig. 4–4

E X A M P L E   **4.2**

Determine the moments of the 800-N force acting on the frame in Fig. 4–5 about points A, B, C, and D.

Solution (Scalar Analysis)
In general, $M = Fd$, where d is the moment arm or *perpendicular distance* from the *point* on the moment axis to the *line of action* of the force. Hence,

$M_A = 800 \text{ N}(2.5 \text{ m}) = 2000 \text{ N} \cdot \text{m} \downarrow$ *Ans.*

$M_B = 800 \text{ N}(1.5 \text{ m}) = 1200 \text{ N} \cdot \text{m} \downarrow$ *Ans.*

$M_C = 800 \text{ N}(0) = 0$ (line of action of **F** passes through C)   *Ans.*

$M_D = 800 \text{ N}(0.5 \text{ m}) = 400 \text{ N} \cdot \text{m} \uparrow$ *Ans.*

The curls indicate the sense of rotation of the moment, which is defined by the direction the force orbits about each point.

Fig. 4–5

E X A M P L E   **4.3**

Determine the resultant moment of the four forces acting on the rod shown in Fig. 4–6 about point O.

Solution
Assuming that positive moments act in the $+\mathbf{k}$ direction, i.e., counterclockwise, we have

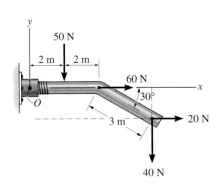

$\curvearrowleft + M_{R_O} = \Sigma Fd;$

$\quad M_{R_O} = -50 \text{ N}(2\text{m}) + 60 \text{ N}(0) + 20 \text{ N}(3 \sin 30° \text{ m})$

$\qquad\quad -40 \text{ N}(4 \text{ m} + 3 \cos 30° \text{ m})$

$\quad M_{R_O} = -334 \text{ N} \cdot \text{m} = 334 \text{ N} \cdot \text{m} \downarrow$ *Ans.*

For this calculation, note how the moment-arm distances for the 20-N and 40-N forces are established from the extended (dashed) lines of action of each of these forces.

Fig. 4–6

4.2 Cross Product

The moment of a force will be formulated using Cartesian vectors in the next section. Before doing this, however, it is first necessary to expand our knowledge of vector algebra and introduce the cross-product method of vector multiplication.

The *cross product* of two vectors **A** and **B** yields the vector **C**, which is written

$$C = A \times B$$

and is read "**C** equals **A** cross **B**."

Magnitude. The *magnitude* of **C** is defined as the product of the magnitudes of **A** and **B** and the sine of the angle θ between their tails $(0° \leq \theta \leq 180°)$. Thus, $C = AB \sin \theta$.

Direction. Vector **C** has a *direction* that is perpendicular to the plane containing **A** and **B** such that **C** is specified by the right-hand rule; i.e., curling the fingers of the right hand from vector **A** (cross) to vector **B**, the thumb then points in the direction of **C**, as shown in Fig. 4–7.

Knowing both the magnitude and direction of **C**, we can write

$$C = A \times B = (AB \sin \theta)u_C \qquad (4\text{–}3)$$

where the scalar $AB \sin \theta$ defines the *magnitude* of **C** and the unit vector u_C defines the *direction* of **C**. The terms of Eq. 4–3 are illustrated graphically in Fig. 4–8.

$C = A \times B$

Fig. 4–7

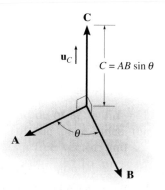

$C = AB \sin \theta$

Fig. 4–8

Laws of Operation.

1. The commutative law is *not* valid; i.e.,

$$\mathbf{A} \times \mathbf{B} \neq \mathbf{B} \times \mathbf{A}$$

Rather,

$$\mathbf{A} \times \mathbf{B} = -\mathbf{B} \times \mathbf{A}$$

This is shown in Fig. 4–9 by using the right-hand rule. The cross product $\mathbf{B} \times \mathbf{A}$ yields a vector that acts in the opposite direction to $\mathbf{C}$; i.e., $\mathbf{B} \times \mathbf{A} = -\mathbf{C}$.

2. Multiplication by a scalar:

$$a(\mathbf{A} \times \mathbf{B}) = (a\mathbf{A}) \times \mathbf{B} = \mathbf{A} \times (a\mathbf{B}) = (\mathbf{A} \times \mathbf{B})a$$

This property is easily shown since the magnitude of the resultant vector ($|a| AB \sin \theta$) and its direction are the same in each case.

3. The distributive law:

$$\mathbf{A} \times (\mathbf{B} + \mathbf{D}) = (\mathbf{A} \times \mathbf{B}) + (\mathbf{A} \times \mathbf{D})$$

The proof of this identity is left as an exercise (see Prob. 4–1). It is important to note that *proper order* of the cross products must be maintained, since they are not commutative.

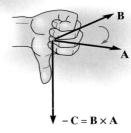

Fig. 4–9

Cartesian Vector Formulation.   Equation 4–3 may be used to find the cross product of a pair of Cartesian unit vectors. For example, to find $\mathbf{i} \times \mathbf{j}$, the *magnitude* of the resultant vector is $(i)(j)(\sin 90°) = (1)(1)(1) = 1$, and its *direction* is determined using the right-hand rule. As shown in Fig. 4–10, the resultant vector points in the $+\mathbf{k}$ direction. Thus, $\mathbf{i} \times \mathbf{j} = (1)\mathbf{k}$. In a similar manner,

$$\mathbf{i} \times \mathbf{j} = \mathbf{k} \quad \mathbf{i} \times \mathbf{k} = -\mathbf{j} \quad \mathbf{i} \times \mathbf{i} = 0$$
$$\mathbf{j} \times \mathbf{k} = \mathbf{i} \quad \mathbf{j} \times \mathbf{i} = -\mathbf{k} \quad \mathbf{j} \times \mathbf{j} = 0$$
$$\mathbf{k} \times \mathbf{i} = \mathbf{j} \quad \mathbf{k} \times \mathbf{j} = -\mathbf{i} \quad \mathbf{k} \times \mathbf{k} = 0$$

These results should *not* be memorized; rather, it should be clearly understood how each is obtained by using the right-hand rule and the definition of the cross product. A simple scheme shown in Fig. 4–11 is helpful for obtaining the same results when the need arises. If the circle is constructed as shown, then "crossing" two unit vectors in a *counterclockwise* fashion around the circle yields the *positive* third unit vector; e.g., $\mathbf{k} \times \mathbf{i} = \mathbf{j}$. Moving *clockwise*, a *negative* unit vector is obtained; e.g., $\mathbf{i} \times \mathbf{k} = -\mathbf{j}$.

Fig. 4–10

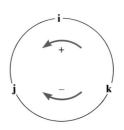

Fig. 4–11

Consider now the cross product of two general vectors **A** and **B** which are expressed in Cartesian vector form. We have

$$\mathbf{A} \times \mathbf{B} = (A_x\mathbf{i} + A_y\mathbf{j} + A_z\mathbf{k}) \times (B_x\mathbf{i} + B_y\mathbf{j} + B_z\mathbf{k})$$
$$= A_xB_x(\mathbf{i} \times \mathbf{i}) + A_xB_y(\mathbf{i} \times \mathbf{j}) + A_xB_z(\mathbf{i} \times \mathbf{k})$$
$$+ A_yB_x(\mathbf{j} \times \mathbf{i}) + A_yB_y(\mathbf{j} \times \mathbf{j}) + A_yB_z(\mathbf{j} \times \mathbf{k})$$
$$+ A_zB_x(\mathbf{k} \times \mathbf{i}) + A_zB_y(\mathbf{k} \times \mathbf{j}) + A_zB_z(\mathbf{k} \times \mathbf{k})$$

Carrying out the cross-product operations and combining terms yields

$$\mathbf{A} \times \mathbf{B} = (A_yB_z - A_zB_y)\mathbf{i} - (A_xB_z - A_zB_x)\mathbf{j} + (A_xB_y - A_yB_x)\mathbf{k} \qquad (4\text{–}4)$$

This equation may also be written in a more compact determinant form as

$$\mathbf{A} \times \mathbf{B} = \begin{vmatrix} \mathbf{i} & \mathbf{j} & \mathbf{k} \\ A_x & A_y & A_z \\ B_x & B_y & B_z \end{vmatrix} \qquad (4\text{–}5)$$

Thus, to find the cross product of any two Cartesian vectors **A** and **B**, it is necessary to expand a determinant whose first row of elements consists of the unit vectors **i**, **j**, and **k** and whose second and third rows represent the *x, y, z* components of the two vectors **A** and **B**, respectively.*

---

*A determinant having three rows and three columns can be expanded using three minors, each of which is multiplied by one of the three terms in the first row. There are four elements in each minor, e.g.,

$$\begin{vmatrix} A_{11} & A_{12} \\ A_{21} & A_{22} \end{vmatrix}$$

By *definition*, this notation represents the terms $(A_{11}A_{22} - A_{12}A_{21})$, which is simply the product of the two elements of the arrow slanting downward to the right $(A_{11}A_{22})$ *minus* the product of the two elements intersected by the arrow slanting downward to the left $(A_{12}A_{21})$. For a 3×3 determinant, such as Eq. 4–5, the three minors can be generated in accordance with the following scheme:

For element **i**: $\begin{vmatrix} \mathbf{i} & \mathbf{j} & \mathbf{k} \\ A_x & A_y & A_z \\ B_x & B_y & B_z \end{vmatrix} = \mathbf{i}(A_yB_z - A_zB_y)$

For element **j**: $\begin{vmatrix} \mathbf{i} & \mathbf{j} & \mathbf{k} \\ A_x & A_y & A_z \\ B_x & B_y & B_z \end{vmatrix} = -\mathbf{j}(A_xB_z - A_zB_x)$

For element **k**: $\begin{vmatrix} \mathbf{i} & \mathbf{j} & \mathbf{k} \\ A_x & A_y & A_z \\ B_x & B_y & B_z \end{vmatrix} = \mathbf{k}(A_xB_y - A_yB_x)$

Adding the results and noting that the **j** element *must include the minus sign* yields the expanded form of **A** $\times$ **B** given by Eq. 4–4.

4.3 Moment of a Force—Vector Formulation

The moment of a force $\mathbf{F}$ about point O, or actually about the moment axis passing through O and perpendicular to the plane containing O and $\mathbf{F}$, Fig. 4–12a, can be expressed using the vector cross product, namely,

$$\mathbf{M}_O = \mathbf{r} \times \mathbf{F} \qquad (4\text{--}6)$$

Here $\mathbf{r}$ represents a position vector drawn *from O* to *any point* lying on the line of action of $\mathbf{F}$. We will now show that indeed the moment $\mathbf{M}_O$, when determined by this cross product, has the proper magnitude and direction.

Magnitude. The magnitude of the cross product is defined from Eq. 4–3 as $M_O = rF \sin\theta$, where the angle θ is measured between the *tails* of $\mathbf{r}$ and $\mathbf{F}$. To establish this angle, $\mathbf{r}$ must be treated as a sliding vector so that θ can be constructed properly, Fig. 4–12b. Since the moment arm $d = r \sin\theta$, then

$$M_O = rF \sin\theta = F(r\sin\theta) = Fd$$

which agrees with Eq. 4–1.

Direction. The direction and sense of $\mathbf{M}_O$ in Eq. 4–6 are determined by the right-hand rule as it applies to the cross product. Thus, extending $\mathbf{r}$ to the dashed position and curling the right-hand fingers from $\mathbf{r}$ toward $\mathbf{F}$, "$\mathbf{r}$ cross $\mathbf{F}$," the thumb is directed upward or perpendicular to the plane containing $\mathbf{r}$ and $\mathbf{F}$ and this is in the *same direction* as $\mathbf{M}_O$, the moment of the force about point O, Fig. 4–12b. Note that the "curl" of the fingers, like the curl around the moment vector, indicates the sense of rotation caused by the force. Since the cross product is not commutative, it is important that the *proper order* of $\mathbf{r}$ and $\mathbf{F}$ be maintained in Eq. 4–6.

(a)   (b)

Fig. 4–12

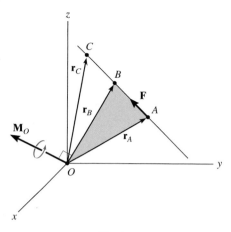

Fig. 4–13

Principle of Transmissibility. Consider the force **F** applied at point A in Fig. 4–13. The moment created by **F** about O is $\mathbf{M}_O = \mathbf{r}_A \times \mathbf{F}$; however, it was shown that "**r**" can extend from O to *any point* on the line of action of **F**. Consequently, **F** may be applied at point B or C, and the same moment $\mathbf{M}_O = \mathbf{r}_B \times \mathbf{F} = \mathbf{r}_C \times \mathbf{F}$ will be computed. As a result, **F** has the properties of a *sliding vector* and can therefore act at *any point along its line of action and still create the same moment about point O*. We refer to this as the *principle of transmissibility*, and we will discuss this property further in Sec. 4.7.

Cartesian Vector Formulation. If we establish x, y, z coordinate axes, then the position vector **r** and force **F** can be expressed as Cartesian vectors, Fig. 4–14. Applying Eq. 4–5 we have

$$\mathbf{M}_O = \mathbf{r} \times \mathbf{F} = \begin{vmatrix} \mathbf{i} & \mathbf{j} & \mathbf{k} \\ r_x & r_y & r_z \\ F_x & F_y & F_z \end{vmatrix} \qquad (4\text{–}7)$$

where

r_x, r_y, r_z represent the x, y, z components of the position vector drawn from point O to *any point* on the line of action of the force

F_x, F_y, F_z represent the x, y, z components of the force vector

If the determinant is expanded, then like Eq. 4–4 we have

$$\mathbf{M}_O = (r_y F_z - r_z F_y)\mathbf{i} - (r_x F_z - r_z F_x)\mathbf{j} + (r_x F_y - r_y F_x)\mathbf{k} \quad (4\text{–}8)$$

The physical meaning of these three moment components becomes evident by studying Fig. 4–14a. For example, the **i** component of $\mathbf{M}_O$ is

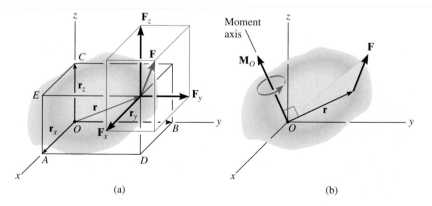

(a)   (b)

Fig. 4–14

determined from the moments of $\mathbf{F}_x$, $\mathbf{F}_y$, and $\mathbf{F}_z$ about the x axis. In particular, note that $\mathbf{F}_x$ does *not* create a moment or tendency to cause turning about the x axis since this force is *parallel* to the x axis. The line of action of $\mathbf{F}_y$ passes through point E, and so the magnitude of the moment of $\mathbf{F}_y$ about point A on the x axis is $r_z F_y$. By the right-hand rule this component acts in the negative $\mathbf{i}$ direction. Likewise, $\mathbf{F}_z$ contributes a moment component of $r_y F_z \mathbf{i}$. Thus, $(M_O)_x = (r_y F_z - r_z F_y)$ as shown in Eq. 4–8. As an exercise, establish the $\mathbf{j}$ and $\mathbf{k}$ components of $\mathbf{M}_O$ in this manner and show that indeed the expanded form of the determinant, Eq. 4–8, represents the moment of $\mathbf{F}$ about point O. Once $\mathbf{M}_O$ is determined, realize that it will always be *perpendicular* to the shaded plane containing vectors $\mathbf{r}$ and $\mathbf{F}$, Fig. 4–14b.

It will be shown in Example 4.4 that the computation of the moment using the cross product has a distinct advantage over the scalar formulation when solving problems in *three dimensions*. This is because it is generally easier to establish the position vector $\mathbf{r}$ to the force, rather than determining the moment-arm distance d that must be directed *perpendicular* to the line of action of the force.

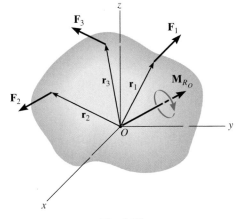

Resultant Moment of a System of Forces. If a body is acted upon by a system of forces, Fig. 4–15, the resultant moment of the forces about point O can be determined by vector addition resulting from successive applications of Eq. 4–6. This resultant can be written symbolically as

$$\mathbf{M}_{R_o} = \Sigma(\mathbf{r} \times \mathbf{F}) \qquad (4\text{–}9)$$

and is shown in Fig. 4–15.

Fig. 4–15

If we pull on cable BC with a force $\mathbf{F}$ at *any point along the cable*, the moment of this force about the base of the utility pole at A will always be the *same*. This is a consequence of the principle of transmissibility. Note that the moment arm, or perpendicular distance from A to the cable, is r_d, and so $M_A = r_d F$. In three dimensions this distance is often difficult to determine, and so we can use the vector cross product to obtain the moment in a more direct manner. For example, $\mathbf{M}_A = \mathbf{r}_{AB} \times \mathbf{F} = \mathbf{r}_{AC} \times \mathbf{F}$. As required, both of these vectors are directed from point A to a point on the line of action of the force.

E X A M P L E **4.4**

(a)

(b)

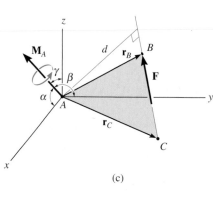

(c)

Fig. 4–16

The pole in Fig. 4–16a is subjected to a 60-N force that is directed from C to B. Determine the magnitude of the moment created by this force about the support at A.

Solution *(Vector Analysis)*
As shown in Fig. 4–16b, either one of two position vectors can be used for the solution, since $\mathbf{M}_A = \mathbf{r}_B \times \mathbf{F}$ or $\mathbf{M}_A = \mathbf{r}_C \times \mathbf{F}$. The position vectors are represented as

$$\mathbf{r}_B = \{1\mathbf{i} + 3\mathbf{j} + 2\mathbf{k}\} \text{ m} \quad \text{and} \quad \mathbf{r}_C = \{3\mathbf{i} + 4\mathbf{j}\} \text{ m}$$

The force has a magnitude of 60 N and a direction specified by the unit vector $\mathbf{u}_F$, directed from C to B. Thus,

$$\mathbf{F} = (60 \text{ N})\mathbf{u}_F = (60 \text{ N})\left[\frac{(1-3)\mathbf{i} + (3-4)\mathbf{j} + (2-0)\mathbf{k}}{\sqrt{(-2)^2 + (-1)^2 + (2)^2}}\right]$$
$$= \{-40\mathbf{i} - 20\mathbf{j} + 40\mathbf{k}\} \text{ N}$$

Substituting into the determinant formulation, Eq. 4–7, and following the scheme for determinant expansion as stated in the footnote on page 120, we have

$$\mathbf{M}_A = \mathbf{r}_B \times \mathbf{F} = \begin{vmatrix} \mathbf{i} & \mathbf{j} & \mathbf{k} \\ 1 & 3 & 2 \\ -40 & -20 & 40 \end{vmatrix}$$
$$= [3(40)-2(-20)]\mathbf{i}-[1(40)-2(-40)]\mathbf{j} + [1(-20)-3(-40)]\mathbf{k}$$

or

$$\mathbf{M}_A = \mathbf{r}_C \times \mathbf{F} = \begin{vmatrix} \mathbf{i} & \mathbf{j} & \mathbf{k} \\ 3 & 4 & 0 \\ -40 & -20 & 40 \end{vmatrix}$$
$$= [4(40)-0(-20)]\mathbf{i}-[3(40)-0(-40)]\mathbf{j} + [3(-20)-4(-40)]\mathbf{k}$$

In both cases,

$$\mathbf{M}_A = \{160\mathbf{i} - 120\mathbf{j} + 100\mathbf{k}] \text{ N} \cdot \text{m}$$

The *magnitude* of $\mathbf{M}_A$ is therefore

$$M_A = \sqrt{(160)^2 + (-120)^2 + (100)^2} = 224 \text{ N} \cdot \text{m} \quad \textit{Ans.}$$

As expected, $\mathbf{M}_A$ acts perpendicular to the shaded plane containing vectors $\mathbf{F}$, $\mathbf{r}_B$, and $\mathbf{r}_C$, Fig. 4–16c. (How would you find its coordinate direction angles $\alpha = 44.3°$, $\beta = 122°$, $\gamma = 63.4°$?) Had this problem been worked using a scalar approach, where $M_A = Fd$, notice the difficulty that can arise in obtaining the moment arm d.

EXAMPLE 4.5

Three forces act on the rod shown in Fig. 4–17a. Determine the resultant moment they create about the flange at O and determine the coordinate direction angles of the moment axis.

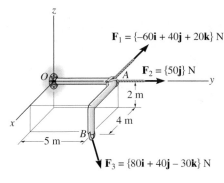

$F_1 = \{-60i + 40j + 20k\}$ N

$F_2 = \{50j\}$ N

(a)

Solution

Position vectors are directed from point O to each force as shown in Fig. 4–17b. These vectors are

$$r_A = \{5j\} \text{ m}$$
$$r_B = \{4i + 5j - 2k\} \text{ m}$$

The resultant moment about O is therefore

$$M_{R_O} = \Sigma(r \times F)$$
$$= r_A \times F_1 + r_A \times F_2 + r_B \times F_3$$
$$= \begin{vmatrix} i & j & k \\ 0 & 5 & 0 \\ -60 & 40 & 20 \end{vmatrix} + \begin{vmatrix} i & j & k \\ 0 & 5 & 0 \\ 0 & 50 & 0 \end{vmatrix} + \begin{vmatrix} i & j & k \\ 4 & 5 & -2 \\ 80 & 40 & -30 \end{vmatrix}$$
$$= [5(20) - 40(0)]i - [0j] + [0(40) - (-60)(5)]k + [0i - 0j + 0k]$$
$$+ [5(-30) - (40)(-2)]i - [4(-30) - 80(-2)]j + [4(40) - 80(5)]k$$
$$= \{30i - 40j + 60k\} \text{ N} \cdot \text{m} \qquad \textit{Ans.}$$

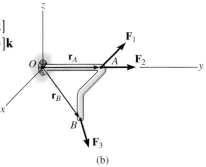

(b)

The moment axis is directed along the line of action of M_{R_O}. Since the magnitude of this moment is

$$M_{R_O} = \sqrt{(30)^2 + (-40)^2 + (60)^2} = 78.10 \text{ N} \cdot \text{m}$$

the unit vector which defines the direction of the moment axis is

$$u = \frac{M_{R_O}}{M_{R_O}} = \frac{30i - 40j + 60k}{78.10} = 0.3841i - 0.5121j + 0.7682k$$

Therefore, the coordinate direction angles of the moment axis are

$$\cos \alpha = 0.3841; \qquad \alpha = 67.4° \qquad \textit{Ans.}$$
$$\cos \beta = -0.5121; \qquad \beta = 121° \qquad \textit{Ans.}$$
$$\cos \gamma = 0.7682; \qquad \gamma = 39.8° \qquad \textit{Ans.}$$

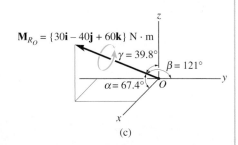

$M_{R_O} = \{30i - 40j + 60k\}$ N · m

(c)

Fig. 4–17

These results are shown in Fig. 4–17c. Realize that the three forces tend to cause the rod to rotate about this axis in the manner shown by the curl indicated on the moment vector.

4.4 Principle of Moments

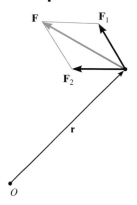

Fig. 4–18

A concept often used in mechanics is the *principle of moments*, which is sometimes referred to as *Varignon's theorem* since it was originally developed by the French mathematician Varignon (1654–1722). It states that *the moment of a force about a point is equal to the sum of the moments of the force's components about the point.* The proof follows directly from the distributive law of the vector cross product. To show this, consider the force $\mathbf{F}$ and two of its rectangular components, where $\mathbf{F} = \mathbf{F}_1 + \mathbf{F}_2$, Fig. 4–18. We have

$$\mathbf{M}_O = \mathbf{r} \times \mathbf{F}_1 + \mathbf{r} \times \mathbf{F}_2 = \mathbf{r} \times (\mathbf{F}_1 + \mathbf{F}_2) = \mathbf{r} \times \mathbf{F}$$

This concept has important applications to the solution of problems and proofs of theorems that follow, since it is often easier to determine the moments of a force's components rather than the moment of the force itself.

The guy cable exerts a force $\mathbf{F}$ on the pole and this creates a moment about the base at A of $M_A = Fd$. If the force is replaced by its two components $\mathbf{F}_x$ and $\mathbf{F}_y$ at point B where the cable acts on the pole, then the sum of the moments of these two components about A will yield the *same* resultant moment. For the calculation $\mathbf{F}_y$ will create zero moment about A and so $M_A = F_x h$. This is an application of the *principle of moments*. In addition we can apply the *principle of transmissibility* and slide the force to where its line of action intersects the ground at C. In this case $\mathbf{F}_x$ will create zero moment about A, and so $M_A = F_y b$.

IMPORTANT POINTS

- The moment of a force indicates the tendency of a body to turn about an axis passing through a specific point O.

- Using the right-hand rule, the sense of rotation is indicated by the fingers, and the thumb is directed along the moment axis, or line of action of the moment.

- The magnitude of the moment is determined from $M_O = Fd$, where d is the perpendicular or shortest distance from point O to the line of action of the force $\mathbf{F}$.

- In three dimensions use the vector cross product to determine the moment, i.e., $\mathbf{M}_O = \mathbf{r} \times \mathbf{F}$. Remember that $\mathbf{r}$ is directed *from* point O *to any point* on the line of action of $\mathbf{F}$.

- The principle of moments states that the moment of a force about a point is equal to the sum of the moments of the force's components about the point. This is a very convenient method to use in two dimensions.

EXAMPLE 4.6

A 200-N force acts on the bracket shown in Fig. 4–19a. Determine the moment of the force about point A.

(a)

Fig. 4–19

Solution I

The moment arm d can be found by trigonometry, using the construction shown in Fig. 4–19b. From the right triangle BCD,

$$CB = d = 100 \cos 45° = 70.71 \text{ mm} = 0.070\ 71 \text{ m}$$

Thus,

$$M_A = Fd = 200 \text{ N}(0.070\ 71 \text{ m}) = 14.1 \text{ N} \cdot \text{m} \curvearrowleft$$

According to the right-hand rule, $\mathbf{M}_A$ is directed in the $+\mathbf{k}$ direction since the force tends to rotate or orbit *counterclockwise* about point A. Hence, reporting the moment as a Cartesian vector, we have

$$\mathbf{M}_A = \{14.1\mathbf{k}\} \text{ N} \cdot \text{m} \qquad\qquad Ans.$$

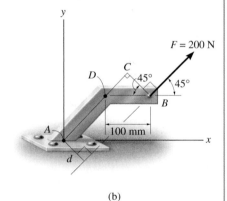

(b)

Solution II

The 200-N force may be resolved into x and y components, as shown in Fig. 4–19c. In accordance with the principle of moments, the moment of $\mathbf{F}$ computed about point A is equivalent to the sum of the moments produced by the two force components. Assuming counterclockwise rotation as positive, i.e., in the $+\mathbf{k}$ direction, we can apply Eq. 4–2 ($M_A = \Sigma Fd$), in which case

$$\curvearrowleft + M_A = (200 \sin 45° \text{ N})(0.20 \text{ m}) - (200 \cos 45° \text{ N})(0.10 \text{ m})$$
$$= 14.1 \text{ N} \cdot \text{m} \curvearrowleft$$

Thus

$$\mathbf{M}_A = \{14.1\mathbf{k}\} \text{ N} \cdot \text{m} \qquad\qquad Ans.$$

(c)

By comparison, it is seen that Solution II provides a more *convenient method* for analysis than Solution I since the moment arm for each component force is easier to establish.

EXAMPLE 4.7

(a)

The force **F** acts at the end of the angle bracket shown in Fig. 4–20a. Determine the moment of the force about point O.

Solution I (Scalar Analysis)
The force is resolved into its x and y components as shown in Fig. 4–20b, and the moments of the components are computed about point O. Taking positive moments as counterclockwise, i.e., in the $+\mathbf{k}$ direction, we have

$$\zeta + M_O = 400 \sin 30° \text{ N}(0.2 \text{ m}) - 400 \cos 30° \text{ N}(0.4 \text{ m})$$
$$= -98.6 \text{ N} \cdot \text{m} = 98.6 \text{ N} \cdot \text{m} \downarrow$$

or

$$\mathbf{M}_O = \{-98.6\mathbf{k}\} \text{ N} \cdot \text{m} \qquad \qquad Ans.$$

Solution II (Vector Analysis)
Using a Cartesian vector approach, the force and position vectors shown in Fig. 4–20c can be represented as

$$\mathbf{r} = \{0.4\mathbf{i} - 0.2\mathbf{j}\} \text{ m}$$
$$\mathbf{F} = \{400 \sin 30°\mathbf{i} - 400 \cos 30°\mathbf{j}\} \text{ N}$$
$$= \{200.0\mathbf{i} - 346.4\mathbf{j}\} \text{ N}$$

The moment is therefore

$$\mathbf{M}_O = \mathbf{r} \times \mathbf{F} = \begin{vmatrix} \mathbf{i} & \mathbf{j} & \mathbf{k} \\ 0.4 & -0.2 & 0 \\ 200.0 & -346.4 & 0 \end{vmatrix}$$
$$= 0\mathbf{i} - 0\mathbf{j} + [0.4(-346.4) - (-0.2)(200.0)]\mathbf{k}$$
$$= \{-98.6\mathbf{k}\} \text{ N} \cdot \text{m} \qquad \qquad Ans.$$

By comparison, it is seen that the scalar analysis (Solution I) provides a more *convenient method* for analysis than Solution II since the direction of the moment and the moment arm for each component force are easy to establish. Hence, this method is generally recommended for solving problems displayed in two dimensions. On the other hand, Cartesian vector analysis is generally recommended only for solving three-dimensional problems, where the moment arms and force components are often more difficult to determine.

(b)

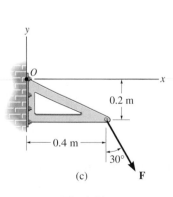

(c)

Fig. 4–20

PROBLEMS

4-1. If **A**, **B**, and **D** are given vectors, prove the distributive law for the vector cross product, i.e., $\mathbf{A} \times (\mathbf{B} + \mathbf{D}) = (\mathbf{A} \times \mathbf{B}) + (\mathbf{A} \times \mathbf{D})$.

4-2. Prove the triple scalar product identity $\mathbf{A} \cdot (\mathbf{B} \times \mathbf{C}) = (\mathbf{A} \times \mathbf{B}) \cdot \mathbf{C}$.

4-3. Given the three nonzero vectors **A**, **B**, and **C**, show that if $\mathbf{A} \cdot (\mathbf{B} \times \mathbf{C}) = 0$, the three vectors *must* lie in the same plane.

*4-4. Determine the magnitude and directional sense of the moment of the force at A about point O.

4-5. Determine the magnitude and directional sense of the moment of the force at A about point P.

*4-8. Determine the magnitude and directional sense of the resultant moment of the forces about point O.

4-9. Determine the magnitude and directional sense of the resultant moment of the forces about point P.

Probs. 4–8/9

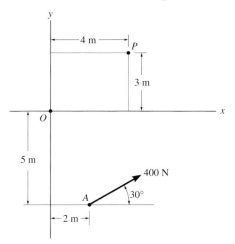

Probs. 4–4/5

4-6. Determine the magnitude and directional sense of the moment of the force at A about point O.

4-7. Determine the magnitude and directional sense of the moment of the force at A about point P.

4-10. The wrench is used to loosen the bolt. Determine the moment of each force about the bolt's axis passing through point O.

Probs. 4–6/7

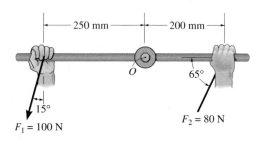

Prob. 4–10

4-11. Determine the magnitude and directional sense of the resultant moment of the forces about point O.

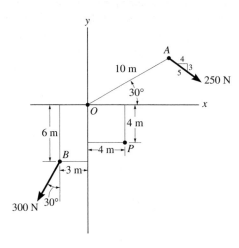

Prob. 4–11

4-14. Determine the moment of each force about the bolt located at A. Take $F_B = 200$ N, $F_C = 250$ N.

4-15. If $F_B = 150$ N and $F_C = 225$ N, determine the resultant moment about the bolt located at A.

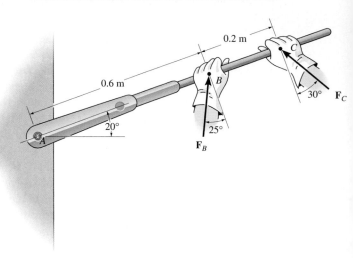

Probs. 4–14/15

*4-12.** Determine the moment about point A of each of the three forces acting on the beam.

4-13. Determine the moment about point B of each of the three forces acting on the beam.

*4-16.** The power pole supports the three lines, each line exerting a vertical force on the pole due to its weight as shown. Determine the resultant moment at the base D due to all of these forces. If it is possible for wind or ice to snap the lines, determine which line(s) when removed create(s) a condition for the greatest moment about the base. What is this resultant moment?

Probs. 4–12/13

Prob. 4–16

4-17. A force of 80 N acts on the handle of the paper cutter at A. Determine the moment created by this force about the hinge at O, if $\theta = 60°$. At what angle θ should the force be applied so that the moment it creates about point O is a maximum (clockwise)? What is this maximum moment?

Prob. 4–17

4-19. The hub of the wheel can be attached to the axle either with negative offset (left) or with positive offset (right). If the tire is subjected to both a normal and radial load as shown, determine the resultant moment of these loads about the axle, point O for both cases.

Prob. 4–19

4-18. Determine the direction $\theta(0° \leq \theta \leq 180°)$ of the force $F = 200$ N so that it produces (a) the maximum moment about point A and (b) the minimum moment about point A. Compute the moment in each case.

*4-20.** The boom has a length of 9 m, a weight of 4000 N, and mass center at G. If the maximum moment that can be developed by the motor at A is $M = 30(10^3)$ N · m, determine the maximum load W, having a mass center at G', that can be lifted. Take $\theta = 30°$.

Prob. 4–18

Prob. 4–20

4-21. The tool at A is used to hold a power lawnmower blade stationary while the nut is being loosened with the wrench. If a force of 50 N is applied to the wrench at B in the direction shown, determine the moment it creates about the nut at C. What is the magnitude of force **F** at A so that it creates the opposite moment about C?

Prob. 4–21

4-22. Determine the moment of each of the three forces about point A. Solve the problem first by using each force as a whole, and then by using the principle of moments.

$F_1 = 250$ N $\quad 30°$

$F_2 = 300$ N

$60°$

2 m | 3 m

4 m

4 | 5
3

$F_3 = 500$ N

Prob. 4–22

4-23. As part of an acrobatic stunt, a man supports a girl who has a weight of 500 N ($\approx$ 50 kg) and is seated on a chair on top of the pole. If her center of gravity is at G, and if the maximum counterclockwise moment the man can exert on the pole at A is 350 N · m, determine the maximum angle of tilt, θ, which will not allow the girl to fall, i.e., so her clockwise moment about A does not exceed 350 N · m.

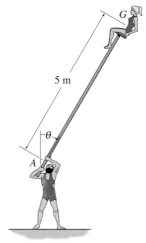

G

5 m

θ

A

Prob. 4–23

*4-24. The two boys push on the gate with forces of $F_A = 120$ N and $F_B = 200$ N as shown. Determine the moment of each force about C. Which way will the gate rotate, clockwise or counterclockwise? Neglect the thickness of the gate.

4-25. Two boys push on the gate as shown. If the boy at B exerts a force of $F_B = 120$ N, determine the magnitude of the force F_A the boy at A must exert in order to prevent the gate from turning. Neglect the thickness of the gate.

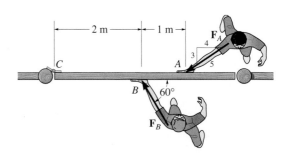

2 m | 1 m

F_A

4
3
5

C | A

B | $60°$

F_B

Probs. 4–24/25

4-26. The towline exerts a force of $P = 4$ kN at the end of the 20-m-long crane boom. If $\theta = 30°$, determine the placement x of the hook at A so that this force creates a maximum moment about point O. What is this moment?

4-27. The towline exerts a force of $P = 4$ kN at the end of the 20-m-long crane boom. If $x = 25$ m, determine the position θ of the boom so that this force creates a maximum moment about point O. What is this moment?

Probs. 4–26/27

*4-28. Determine the direction θ for $0° \leq \theta \leq 180°$ of the force $\mathbf{F}$ so that $\mathbf{F}$ produces (a) the maximum moment about point A and (b) the minimum moment about point A. Calculate the moment in each case.

4-29. Determine the moment of the force $\mathbf{F}$ about point A as a function of θ. Plot the results of M (ordinate) θ (abscissa) for $0° \leq \theta \leq 180°$.

Probs. 4–28/29

4-30. The total hip replacement is subjected to a force of $F = 120$ N. Determine the moment of this force about the neck at A and at the stem B.

Prob. 4–30

4-31. The crane can be adjusted for any angle $0° \leq \theta \leq 90°$ and any extension $0 \leq x \leq 5$ m. For a suspended mass of 120 kg, determine the moment developed at A as a function of x and θ. What values of both x and θ develop the maximum possible moment at A? Compute this moment. Neglect the size of the pulley at B.

Prob. 4–31

*4-32. Determine the angle θ at which the 500-N force must act at A so that the moment of this force about point B is equal to zero.

Prob. 4–32

4-33. Segments of drill pipe D for an oil well are tightened a prescribed amount by using a set of tongs T, which grip the pipe, and a hydraulic cylinder (not shown) to regulate the force **F** applied to the tongs. This force acts along the cable which passes around the small pulley P. If the cable is originally perpendicular to the tongs as shown, determine the magnitude of force **F** which must be applied so that the moment about the pipe is $M = 2000$ N · m. In order to maintain this same moment what magnitude of **F** is required when the tongs rotate 30° to the dashed position? *Note:* The angle DAP is not 90° in this position.

4-34. Determine the moment of the force at A about point O. Express the result as a Cartesian vector.

4-35. Determine the moment of the force at A about point P. Express the result as a Cartesian vector.

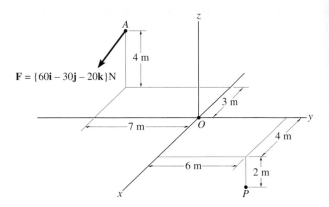

Probs. 4–34/35

*4-36. Determine the moment of the force **F** at A about point O. Express the result as a Cartesian vector.

4-37. Determine the moment of the force **F** at A about point P. Express the result as a Cartesian vector.

Probs. 4–36/37

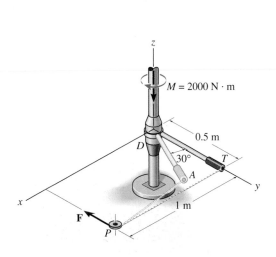

Prob. 4–33

4-38. The curved rod lies in the *x-y* plane and has a radius of 3 m. If a force of $F = 80$ N acts at its end as shown, determine the moment of this force about point O.

4-39. The curved rod lies in the *x-y* plane and has a radius of 3 m. If a force of $F = 80$ N acts at its end as shown, determine the moment of this force about point B.

4-41. The curved rod has a radius of 0.5 m. If a force of 60 N acts at its end as shown, determine the moment of this force about point C.

Prob. 4–41

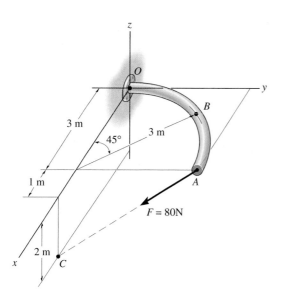

Probs. 4–38/39

*****4-40.** The force $\mathbf{F} = \{600\mathbf{i} + 300\mathbf{j} - 600\mathbf{k}\}$ N acts at the end of the beam. Determine the moment of the force about point A.

4-42. A force $\mathbf{F}$ having a magnitude of $F = 100$ N acts along the diagonal of the parallelepiped. Determine the moment of $\mathbf{F}$ about point A, using $\mathbf{M}_A = \mathbf{r}_B \times \mathbf{F}$ and $\mathbf{M}_A = \mathbf{r}_C \times \mathbf{F}$.

Prob. 4–40

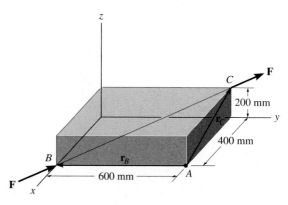

Prob. 4–42

4-43. Determine the smallest force F that must be applied along the rope in order to cause the curved rod, which has a radius of 0.5 m, to fail at the support C. This requires a moment of $M = 80$ N · m to be developed at C.

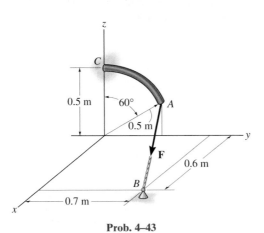

Prob. 4–43

4-46. Strut AB of the 1-m-diameter hatch door exerts a force of 450 N on point B. Determine the moment of this force about point O.

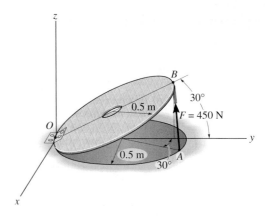

Prob. 4–46

*4-44.** The pipe assembly is subjected to the 80-N force. Determine the moment of this force about point A.

4-45. The pipe assembly is subjected to the 80-N force. Determine the moment of this force about point B.

Probs. 4–44/45

4-47. Using Cartesian vector analysis, determine the resultant moment of the three forces about the base of the column at A. Take $\mathbf{F}_1 = \{400\mathbf{i} + 300\mathbf{j} + 120\mathbf{k}\}$ N.

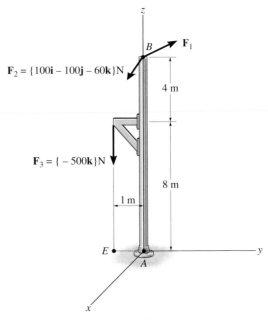

Prob. 4–47

*4-48. A force of $\mathbf{F} = \{6\mathbf{i} - 2\mathbf{j} + 1\mathbf{k}\}$ kN produces a moment of $\mathbf{M}_O = \{4\mathbf{i} + 5\mathbf{j} - 14\mathbf{k}\}$ kN·m about the origin of coordinates, point O. If the force acts at a point having an x coordinate of $x = 1$ m, determine the y and z coordinates.

■ 4-50. Using a ring collar the 75-N force can act in the vertical plane at various angles θ. Determine the magnitude of the moment it produces about point A, plot the result of M (ordinate) versus θ (abscissa) for $0° \leq \theta \leq 180°$, and specify the angles that give the maximum and minimum moment.

Prob. 4–48

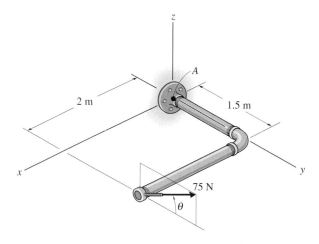

Prob. 4–50

4-49. The force $\mathbf{F} = \{6\mathbf{i} + 8\mathbf{j} + 10\mathbf{k}\}$ N creates a moment about point O of $\mathbf{M}_O = \{-14\mathbf{i} + 8\mathbf{j} + 2\mathbf{k}\}$ N·m. If the force passes through a point having an x coordinate of 1 m, determine the y and z coordinates of the point. Also, realizing that $M_O = Fd$, determine the perpendicular distance d from point O to the line of action of $\mathbf{F}$.

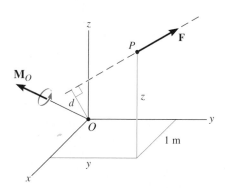

Prob. 4–49

4.5 Moment of a Force about a Specified Axis

Recall that when the moment of a force is computed about a point, the moment and its axis are *always* perpendicular to the plane containing the force and the moment arm. In some problems it is important to find the *component* of this moment along a *specified axis* that passes through the point. To solve this problem either a scalar or vector analysis can be used.

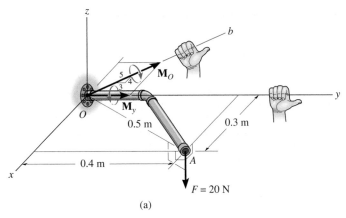

(a)

Fig. 4–21

Scalar Analysis. As a numerical example of this problem, consider the pipe assembly shown in Fig. 4–21a, which lies in the horizontal plane and is subjected to the vertical force of $F = 20$ N applied at point A. The moment of this force about point O has a *magnitude* of $M_O = (20 \text{ N})(0.5 \text{ m}) = 10 \text{ N} \cdot \text{m}$, and a *direction* defined by the right-hand rule, as shown in Fig. 4–21a. This moment tends to turn the pipe about the Ob axis. For practical reasons, however, it may be necessary to determine the *component* of $\mathbf{M}_O$ about the y axis, $\mathbf{M}_y$, since this component tends to unscrew the pipe from the flange at O. From Fig. 4–21a, $\mathbf{M}_y$ has a magnitude of $M_y = \frac{3}{5}(10 \text{ N} \cdot \text{m}) = 6 \text{ N} \cdot \text{m}$ and a sense of direction shown by the vector resolution. Rather than performing this *two-step* process of first finding the moment of the force about point O and then resolving the moment along the y axis, it is also possible to solve this problem *directly*. To do so, it is necessary to determine the perpendicular or moment-arm distance from the line of action of $\mathbf{F}$ to the y axis. From Fig. 4–21a this distance is 0.3 m. Thus the *magnitude* of the moment of the force about the y axis is again $M_y = 0.3(20 \text{ N}) = 6 \text{ N} \cdot \text{m}$, and the *direction* is determined by the right-hand rule as shown.

In general, then, *if the line of action of a force $\mathbf{F}$ is perpendicular to any specified axis aa*, the magnitude of the moment of $\mathbf{F}$ about the axis can be determined from the equation

$$M_a = Fd_a \qquad\qquad (4\text{--}10)$$

Here d_a is the *perpendicular or shortest distance* from the force line of action to the axis. The direction is determined from the thumb of the right hand when the fingers are curled in accordance with the direction of rotation as produced by the force. In particular, realize that a *force will not contribute a moment about a specified axis if the force line of action is parallel to the axis or its line of action passes through the axis.*

If a horizontal force $\mathbf{F}$ is applied to the handle of the flex-headed wrench, it tends to turn the socket at A about the z axis. This effect is caused by the moment of $\mathbf{F}$ about the z axis. The *maximum moment* is determined when the wrench is in the horizontal plane so that full leverage from the handle can be achieved, i.e., $(M_z)_{max} = Fd$. If the handle is not in the horizontal position, then the moment about the z axis is determined from $M_z = Fd'$, where d' is the perpendicular distance from the force line of action to the axis. We can also determine this moment by first finding the moment of $\mathbf{F}$ about A, $M_A = Fd$, then finding the projection or component of this moment along z, i.e., $M_z = M_A \cos\theta$.

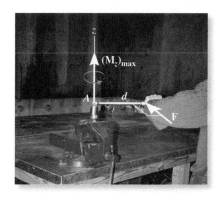

Vector Analysis. The previous two-step solution of first finding the moment of the force about a point on the axis and then finding the projected component of the moment about the axis can also be performed using a vector analysis, Fig. 4–21b. Here the moment about point O is first determined from $\mathbf{M}_O = \mathbf{r}_A \times \mathbf{F} = (0.3\mathbf{i} + 0.4\mathbf{j}) \times (-20\mathbf{k}) = \{-8\mathbf{i} + 6\mathbf{j}\}\text{N}\cdot\text{m}$. The component or projection of this moment along the y axis is then determined from the dot product (Sec. 2.9). Since the unit vector for this axis (or line) is $\mathbf{u}_a = \mathbf{j}$, then $M_y = \mathbf{M}_O \cdot \mathbf{u}_a = (-8\mathbf{i} + 6\mathbf{j}) \cdot \mathbf{j} = 6\,\text{N}\cdot\text{m}$. This result, of course, is to be expected, since it represents the $\mathbf{j}$ component of $\mathbf{M}_O$.

(b)

Fig. 4–21

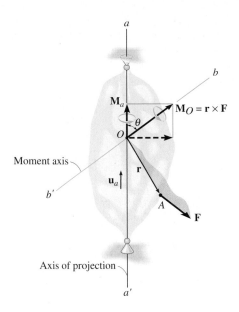

Fig. 4–22

A vector analysis such as this is particularly advantageous for finding the moment of a force about an axis when the force components or the appropriate moment arms are difficult to determine. For this reason, the above two-step process will now be generalized and applied to a body of arbitrary shape. To do so, consider the body in Fig. 4–22, which is subjected to the force **F** acting at point A. Here we wish to determine the effect of **F** in tending to rotate the body about the aa' axis. This tendency for rotation is measured by the moment component **M**$_a$. To determine **M**$_a$ we first compute the moment of **F** about any *arbitrary point* O that lies on the aa' axis. In this case, **M**$_O$ is expressed by the cross product **M**$_O$ = **r** × **F**, where **r** is directed from O to A. Here **M**$_O$ acts along the moment axis bb', and so the component or projection of **M**$_O$ onto the aa' axis is then **M**$_a$. The *magnitude* of **M**$_a$ is determined by the dot product, $M_a = M_O \cos\theta = \mathbf{M}_O \cdot \mathbf{u}_a$ where **u**$_a$ is a unit vector that defines the direction of the aa' axis. Combining these two steps as a general expression, we have $M_a = (\mathbf{r} \times \mathbf{F}) \cdot \mathbf{u}_a$. Since the dot product is commutative, we can also write

$$M_a = \mathbf{u}_a \cdot (\mathbf{r} \times \mathbf{F})$$

In vector algebra, this combination of dot and cross product yielding the scalar M_a is called the *triple scalar product*. Provided x, y, z axes are established and the Cartesian components of each of the vectors can be determined, then the triple scalar product may be written in determinant form as

$$M_a = (u_{a_x}\mathbf{i} + u_{a_y}\mathbf{j} + u_{a_z}\mathbf{k}) \cdot \begin{vmatrix} \mathbf{i} & \mathbf{j} & \mathbf{k} \\ r_x & r_y & r_z \\ F_x & F_y & F_z \end{vmatrix}$$

or simply

$$M_a = \mathbf{u}_a \cdot (\mathbf{r} \times \mathbf{F}) = \begin{vmatrix} u_{a_x} & u_{a_y} & u_{a_z} \\ r_x & r_y & r_z \\ F_x & F_y & F_z \end{vmatrix} \qquad (4\text{–}11)$$

where

$u_{a_x}, u_{a_y}, u_{a_z}$	represent the x, y, z components of the unit vector defining the direction of the aa' axis
r_x, r_y, r_z	represent the x, y, z components of the position vector drawn from *any point* O on the aa' axis to *any point* A on the line of action of the force
F_x, F_y, F_z	represent the x, y, z components of the force vector.

When M_a is evaluated from Eq. 4–11, it will yield a positive or negative scalar. The sign of this scalar indicates the sense of direction of $\mathbf{M}_a$ along the aa' axis. If it is positive, then $\mathbf{M}_a$ will have the same sense as $\mathbf{u}_a$, whereas if it is negative, then $\mathbf{M}_a$ will act opposite to $\mathbf{u}_a$.

Once M_a is determined, we can then express $\mathbf{M}_a$ as a Cartesian vector, namely,

$$\mathbf{M}_a = M_a\mathbf{u}_a = [\mathbf{u}_a \cdot (\mathbf{r} \times \mathbf{F})]\mathbf{u}_a \qquad (4\text{–}12)$$

Finally, if the resultant moment of a series of forces is to be computed about the aa' axis, then the moment components of each force are added together *algebraically*, since each component lies along the same axis. Thus the magnitude of $\mathbf{M}_a$ is

$$M_a = \Sigma[\mathbf{u}_a \cdot (\mathbf{r} \times \mathbf{F})] = \mathbf{u}_a \cdot \Sigma(\mathbf{r} \times \mathbf{F})$$

The examples which follow illustrate a numerical application of the above concepts.

Wind blowing on the face of this traffic sign creates a resultant force $\mathbf{F}$ that tends to tip the sign over due to the moment $\mathbf{M}_A$ created about the $a - a$ axis. The moment of $\mathbf{F}$ about a point A that lies on the axis is $\mathbf{M}_A = \mathbf{r} \times \mathbf{F}$. The projection of this moment along the axis, whose direction is defined by the unit vector $\mathbf{u}_a$, is $M_a = \mathbf{u}_a \cdot (\mathbf{r} \times \mathbf{F})$. Had this moment been calculated using scalar methods, then the perpendicular distance from the force line of action to the $a - a$ axis would have to be determined, which in this case would be a more difficult task.

IMPORTANT POINTS

- The moment of a force about a specified axis can be determined provided the perpendicular distance d_a from *both* the force line of action and the axis can be determined. $M_a = Fd_a$.

- If vector analysis is used, $M_a = \mathbf{u}_a \cdot (\mathbf{r} \times \mathbf{F})$, where $\mathbf{u}_a$ defines the direction of the axis and $\mathbf{r}$ is directed from *any point* on the axis to *any point* on the line of action of the force.

- If M_a is calculated as a negative scalar, then the sense of direction of $\mathbf{M}_a$ is opposite to $\mathbf{u}_a$.

- The moment $\mathbf{M}_a$ expressed as a Cartesian vector is determined from $\mathbf{M}_a = M_a\mathbf{u}_a$.

E X A M P L E 4.8

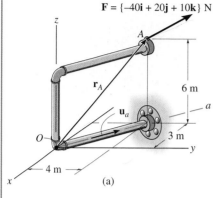

F = {−40i + 20j + 10k} N

(a)

80 N · m

120 N · m

(b)

10 N 40 N

20 N

6 m

3 m

4 m

(c)

Fig. 4–23

The force $\mathbf{F} = \{-40\mathbf{i} + 20\mathbf{j} + 10\mathbf{k}\}$ N acts at point A shown in Fig. 4–23a. Determine the moments of this force about the x and a axes.

Solution I *(Vector Analysis)*

We can solve this problem by using the position vector $\mathbf{r}_A$. Why? Since $\mathbf{r}_A = \{-3\mathbf{i} + 4\mathbf{j} + 6\mathbf{k}\}$ m and $\mathbf{u}_x = \mathbf{i}$, then applying Eq. 4–11,

$$M_x = \mathbf{i} \cdot (\mathbf{r}_A \times \mathbf{F}) = \begin{vmatrix} 1 & 0 & 0 \\ -3 & 4 & 6 \\ -40 & 20 & 10 \end{vmatrix}$$

$$= 1[4(10)-6(20)]-0[(-3)(10)-6(-40)]+0[(-3)(20)-4(-40)]$$

$$= -80 \text{ N} \cdot \text{m} \qquad \qquad \textit{Ans.}$$

The negative sign indicates that the sense of $\mathbf{M}_x$ is opposite to $\mathbf{i}$.

We can compute M_a also using $\mathbf{r}_A$ because $\mathbf{r}_A$ extends from a point on the a axis to the force. Also, $\mathbf{u}_a = -\frac{3}{5}\mathbf{i} + \frac{4}{5}\mathbf{j}$. Thus,

$$M_a = \mathbf{u}_a \cdot (\mathbf{r}_A \times \mathbf{F}) = \begin{vmatrix} -\frac{3}{5} & \frac{4}{5} & 0 \\ -3 & 4 & 6 \\ -40 & 20 & 10 \end{vmatrix}$$

$$= -\tfrac{3}{5}[4(10)-6(20)]-\tfrac{4}{5}[(-3)(10)-6(-40)]+0[(-3)(20)-4(-40)]$$

$$= -120 \text{ N} \cdot \text{m} \qquad \qquad \textit{Ans.}$$

What does the negative sign indicate?

The moment components are shown in Fig. 4–23b.

Solution II *(Scalar Analysis)*

Since the force components and moment arms are easy to determine for computing M_x, a scalar analysis can be used to solve this problem. Referring to Fig. 4–23c, only the 10-N and 20-N forces contribute moments about the x axis. (The line of action of the 40-N force is parallel to this axis and hence its moment about the x axis is zero.) Using the right-hand rule, the algebraic sum of the moment components about the x axis is therefore

$$M_x = (10 \text{ N})(4 \text{ m}) - (20 \text{ N})(6 \text{ m}) = -80 \text{ N} \cdot \text{m} \qquad \textit{Ans.}$$

Although not required here, note also that

$$M_y = (10 \text{ N})(3 \text{ m}) - (40 \text{ N})(6 \text{ m}) = -210 \text{ N} \cdot \text{m}$$

$$M_z = (40 \text{ N})(4 \text{ m}) - (20 \text{ N})(3 \text{ m}) = 100 \text{ N} \cdot \text{m}$$

If we were to determine M_a by this scalar method, it would require much more effort since the force components of 40 N and 20 N are *not perpendicular* to the direction of the a axis. The vector analysis yields a more direct solution.

EXAMPLE 4.9

The rod shown in Fig. 4–24a is supported by two brackets at A and B. Determine the moment $\mathbf{M}_{AB}$ produced by $\mathbf{F} = \{-600\mathbf{i}+200\mathbf{j}-300\mathbf{k}\}$ N, which tends to rotate the rod about the AB axis.

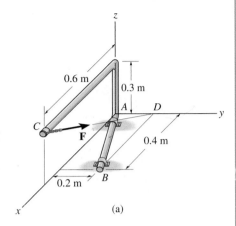

0.6 m

0.3 m

C

A D

y

$\mathbf{F}$

0.4 m

0.2 m B

x

(a)

Solution

A vector analysis using $M_{AB} = \mathbf{u}_B \cdot (\mathbf{r} \times \mathbf{F})$ will be considered for the solution since the moment arm or perpendicular distance from the line of action of $\mathbf{F}$ to the AB axis is difficult to determine. Each of the terms in the equation will now be identified.

Unit vector $\mathbf{u}_B$ defines the direction of the AB axis of the rod, Fig. 4–24b, where

$$\mathbf{u}_B = \frac{\mathbf{r}_B}{r_B} = \frac{0.4\mathbf{i} + 0.2\mathbf{j}}{\sqrt{(0.4)^2 + (0.2)^2}} = 0.894\mathbf{i} + 0.447\mathbf{j}$$

Vector $\mathbf{r}$ is directed from *any point* on the AB axis to *any point* on the line of action of the force. For example, position vectors $\mathbf{r}_C$ and $\mathbf{r}_D$ are suitable, Fig. 4–24b. (Although not shown, $\mathbf{r}_{BC}$ or $\mathbf{r}_{BD}$ can also be used.) For simplicity, we choose $\mathbf{r}_D$, where

$$\mathbf{r}_D = \{0.2\mathbf{j}\} \text{ m}$$

The force is

$$\mathbf{F} = \{-600\mathbf{i} + 200\mathbf{j} - 300\mathbf{k}\} \text{ N}$$

Substituting these vectors into the determinant form and expanding, we have

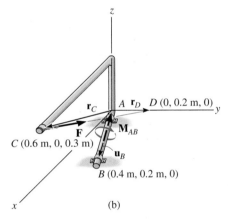

z

$\mathbf{r}_C$

A $\mathbf{r}_D$ $D\,(0, 0.2\text{ m}, 0)$ y

$\mathbf{F}$ $\mathbf{M}_{AB}$

$C\,(0.6\text{ m}, 0, 0.3\text{ m})$ $\mathbf{u}_B$

$B\,(0.4\text{ m}, 0.2\text{ m}, 0)$

x

(b)

Fig. 4–24

$$M_{AB} = \mathbf{u}_B \cdot (\mathbf{r}_D \times \mathbf{F}) = \begin{vmatrix} 0.894 & 0.447 & 0 \\ 0 & 0.2 & 0 \\ -600 & 200 & -300 \end{vmatrix}$$

$$= 0.894[0.2(-300) - 0(200)] - 0.447[0(-300) - 0(-600)] + 0[0(200) - 0.2(-600)]$$

$$= -53.67 \text{ N} \cdot \text{m}$$

The negative sign indicates that the sense of $\mathbf{M}_{AB}$ is opposite to that of $\mathbf{u}_B$.

Expressing $\mathbf{M}_{AB}$ as a Cartesian vector yields

$$\mathbf{M}_{AB} = M_{AB}\mathbf{u}_B = (-53.67 \text{ N} \cdot \text{m})(0.894\mathbf{i} + 0.447\mathbf{j})$$
$$= \{-48.0\mathbf{i} - 24.0\mathbf{j}\} \text{ N} \cdot \text{m} \qquad \textit{Ans.}$$

The result is shown in Fig. 4–24b.

Note that if axis AB is defined using a unit vector directed from B toward A, then in the above formulation $-\mathbf{u}_B$ would have to be used. This would lead to $M_{AB} = +53.67 \text{ N} \cdot \text{m}$. Consequently, $\mathbf{M}_{AB} = M_{AB}(-\mathbf{u}_B)$, and the above result would again be determined.

PROBLEMS

4-51. Determine the moment of the force **F** about the *Oa* axis. Express the result as a Cartesian vector.

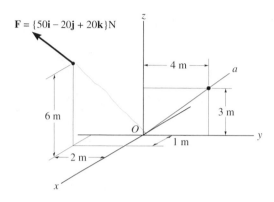

Prob. 4–51

4-53. Determine the resultant moment of the two forces about the *Oa* axis. Express the result as a Cartesian vector.

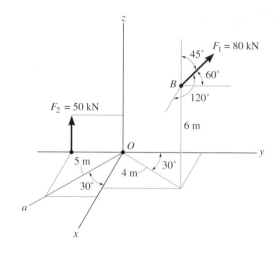

Prob. 4–53

4-54. Determine the magnitude of the moment of each of the three forces about the axis *AB*. Solve the problem (a) using a Cartesian vector approach and (b) using a scalar approach.

*4-52. Determine the moment of the force **F** about the *aa* axis. Express the result as a Cartesian vector.

Prob. 4–52

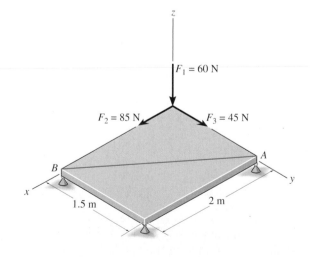

Prob. 4–54

4-55. The chain AB exerts a force of 100 N on the door at B. Determine the magnitude of the moment of this force along the hinged axis x of the door.

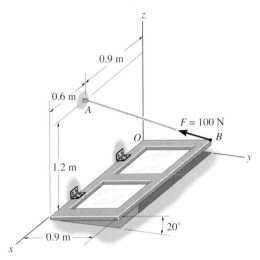

0.9 m

0.6 m

A

$F = 100$ N

O

B

y

1.2 m

20°

0.9 m

x

Prob. 4–55

4-57. The cutting tool on the lathe exerts a force $\mathbf{F}$ on the shaft in the direction shown. Determine the moment of this force about the y axis of the shaft.

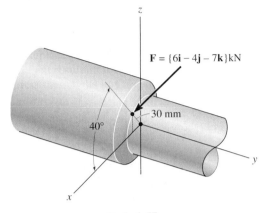

$\mathbf{F} = \{6\mathbf{i} - 4\mathbf{j} - 7\mathbf{k}\}$kN

30 mm

40°

x

y

z

Prob. 4–57

*4-56. The force of $F = 30$ N acts on the bracket as shown. Determine the moment of the force about the $a-a$ axis of the pipe. Also, determine the coordinate direction angles of F in order to produce the maximum moment about the $a-a$ axis. What is this moment?

z

$F = 30$ N

45°

60°

60°

y

x

50 mm

100 mm

100 mm

a

a

Prob. 4–56

4-58. The hood of the automobile is supported by the strut AB, which exerts a force of $F = 100$ N on the hood. Determine the moment of this force about the hinged axis y.

z

B

$\mathbf{F}$

1.2 m

A

0.6 m

0.6 m

1.2 m

y

x

Prob. 4–58

4-59. Determine the magnitude of the moments of the force **F** about the x, y, and z axes. Solve the problem (a) using a Cartesian vector approach and (b) using a scalar approach.

***4-60.** Determine the moment of the force **F** about an axis extending between A and C. Express the result as a Cartesian vector.

F = {4i + 12j – 3k} kN

Probs. 4–59/60

4-61. The lug and box wrenches are used in combination to remove the lug nut from the wheel hub. If the applied force on the end of the box wrench is **F** = {4i – 12j + 2k} N, determine the magnitude of the moment of this force about the x axis which is effective in unscrewing the lug nut.

Prob. 4–61

4-62. A 70-N ($\approx$ 7-kg) force acts vertically on the "Z" bracket. Determine the magnitude of the moment of this force about the bolt axis (z axis).

Prob. 4–62

4-63. Determine the magnitude of the moment of the force **F** = {50i – 20j – 80k} N about the base line CA of the tripod.

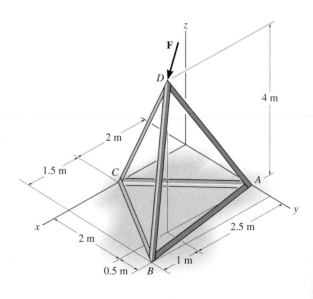

Prob. 4–63

*4-64. The flex-headed ratchet wrench is subjected to a force of $P = 80$ N, applied perpendicular to the handle as shown. Determine the moment or torque this imparts along the vertical axis of the bolt at A.

4-65. If a torque or moment of 8 N · m is required to loosen the bolt at A, determine the force P that must be applied perpendicular to the handle of the flex-headed ratchet wrench.

4-67. A horizontal force of _____ is applied perpendicular to the handle of the pipe wrench. Determine the moment that this force exerts along the axis OA (z axis) of the pipe assembly. Both the wrench and pipe assembly. $OABC$, lie in the y-z plane. *Suggestion:* Use a scalar analysis.

Prob. 4–67

Probs. 4–64/65

4-66. The A-frame is being hoisted into an upright position by the vertical force of $F = 800$ N. Determine the moment of this force about the y axis when the frame is in the position shown.

*4-68. Determine the magnitude of the horizontal force $\mathbf{F} = -F\mathbf{i}$ acting on the handle of the wrench so that this force produces a component of moment along the OA axis (z axis) of the pipe assembly of $\mathbf{M}_z = \{4\mathbf{k}\}$ N · m. Both the wrench and the pipe assembly, $OABC$, lie in the y-z plane. *Suggestion:* Use a scalar analysis.

Prob. 4–66

Prob. 4–68

4.6 Moment of a Couple

F

d

−F

Fig. 4–25

A *couple* is defined as two parallel forces that have the same magnitude, have opposite directions, and are separated by a perpendicular distance d, Fig. 4–25. Since the resultant force is zero, the only effect of a couple is to produce a rotation or tendency of rotation in a specified direction.

The moment produced by a couple is called a *couple moment*. We can determine its value by finding the sum of the moments of both couple forces about *any* arbitrary point. For example, in Fig. 4–26, position vectors $\mathbf{r}_A$ and $\mathbf{r}_B$ are directed from point O to points A and B lying on the line of action of $-\mathbf{F}$ and $\mathbf{F}$. The couple moment computed about O is therefore

$$\mathbf{M} = \mathbf{r}_A \times (-\mathbf{F}) + \mathbf{r}_B \times \mathbf{F}$$

Rather than sum the moments of both forces to determine the couple moment, it is simpler to take moments about a point lying on the line of action of one of the forces. If point A is chosen, then the moment of $-\mathbf{F}$ about A is zero, and we have

$$\mathbf{M} = \mathbf{r} \times \mathbf{F} \qquad (4\text{--}13)$$

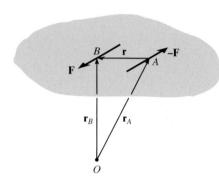

Fig. 4–26

The fact that we obtain the *same result* in both cases can be demonstrated by noting that in the first case we can write $\mathbf{M} = (\mathbf{r}_B - \mathbf{r}_A) \times \mathbf{F}$; and by the triangle rule of vector addition, $\mathbf{r}_A + \mathbf{r} = \mathbf{r}_B$ or $\mathbf{r} = \mathbf{r}_B - \mathbf{r}_A$, so that upon substitution we obtain Eq. 4–13. This result indicates that a couple moment is a *free vector*, i.e., it can act at *any point* since $\mathbf{M}$ depends *only* upon the position vector $\mathbf{r}$ directed *between* the forces and *not* the position vectors $\mathbf{r}_A$ and $\mathbf{r}_B$, directed from the arbitrary point O to the forces. This concept is therefore unlike the moment of a force, which requires a definite point (or axis) about which moments are determined.

Scalar Formulation. The moment of a couple, $\mathbf{M}$, Fig. 4–27, is defined as having a *magnitude* of

$$\boxed{M = Fd} \qquad (4\text{--}14)$$

where F is the magnitude of one of the forces and d is the perpendicular distance or moment arm between the forces. The *direction* and sense of the couple moment are determined by the right-hand rule, where the thumb indicates the direction when the fingers are curled with the sense of rotation caused by the two forces. In all cases, $\mathbf{M}$ acts perpendicular to the plane containing these forces.

M

−F

d

F

Fig. 4–27

Vector Formulation. The moment of a couple can also be expressed by the vector cross product using Eq. 4–13, i.e.,

$$\boxed{\mathbf{M} = \mathbf{r} \times \mathbf{F}} \qquad (4\text{--}15)$$

Application of this equation is easily remembered if one thinks of taking the moments of both forces about a point lying on the line of action of one of the forces. For example, if moments are taken about point A in Fig. 4–26, the moment of $-\mathbf{F}$ is *zero* about this point, and the moment or **F** is defined from Eq. 4–15. Therefore, in the formulation **r** is crossed with the force **F** to which it is directed.

Equivalent Couples. Two couples are said to be equivalent if they produce the same moment. Since the moment produced by a couple is always perpendicular to the plane containing the couple forces, it is therefore necessary that the forces of equal couples lie either in the same plane or in planes that are *parallel* to one another. In this way, the direction of each couple moment will be the same, that is, perpendicular to the parallel planes.

Resultant Couple Moment. Since couple moments are free vectors, they may be applied at any point P on a body and added vectorially. For example, the two couples acting on different planes of the body in Fig. 4–28a may be replaced by their corresponding couple moments $\mathbf{M}_1$ and $\mathbf{M}_2$, Fig. 4–28b, and then these free vectors may be moved to the *arbitrary point P* and added to obtain the resultant couple moment $\mathbf{M}_R = \mathbf{M}_1 + \mathbf{M}_2$, shown in Fig. 4–28c.

If more than two couple moments act on the body, we may generalize this concept and write the vector resultant as

$$\mathbf{M}_R = \Sigma(\mathbf{r} \times \mathbf{F}) \tag{4–16}$$

These concepts are illustrated numerically in the examples which follow. In general, problems projected in two dimensions should be solved using a scalar analysis since the moment arms and force components are easy to compute.

(a)

(b)

(c)

Fig. 4–28

The frictional forces of the floor on the blades of the concrete finishing machine create a couple moment $\mathbf{M}_c$ on the machine that tends to turn it. An equal but opposite couple moment must be applied by the hands of the operator to prevent the turning. Here the couple moment, $M_c = Fd$, is applied on the handle, although it could be applied at any other point on the machine.

A moment of 12 N · m is needed to turn the shaft connected to the center of the wheel. To do this it is efficient to apply a couple since this effect produces a pure rotation. The couple forces can be made as small as possible by placing the hands on the *rim* of the wheel, where the spacing is 0.4 m. In this case 12 N · m = F(0.4 m), F = 30 N. An equivalent couple moment of 12 N · m can be produced if one grips the wheel within the inner hub, although here much larger forces are needed. If the distance between the hands becomes 0.3 m, then 12 N · m = F'(0.3), F' = 40 N. Also, realize that if the wheel was connected to the shaft at a point other than at its center, the wheel would still turn when the forces are applied since the 12-N · m couple moment is a *free vector*.

IMPORTANT POINTS

- A couple moment is produced by two noncollinear forces that are equal but opposite. Its effect is to produce pure rotation, or tendency for rotation in a specified direction.

- A couple moment is a free vector, and as a result it causes the same effect of rotation on a body regardless of where the couple moment is applied to the body.

- The moment of the two couple forces can be computed about *any point*. For convenience, this point is often chosen on the line of action of one of the forces in order to eliminate the moment of this force about the point.

- In three dimensions the couple moment is often determined using the vector formulation, $\mathbf{M} = \mathbf{r} \times \mathbf{F}$, where $\mathbf{r}$ is directed from *any point* on the line of action of one of the forces to *any point* on the line of action of the other force $\mathbf{F}$.

- A resultant couple moment is simply the vector sum of all the couple moments of the system.

E X A M P L E 4.10

A couple acts on the gear teeth as shown in Fig. 4–29*a*. Replace it by an equivalent couple having a pair of forces that act through points *A* and *B*.

(a)

Fig. 4–29

(b)

(c)

Solution (*Scalar Analysis*)

The couple has a magnitude of $M = Fd = 40(0.6) = 24\ \text{N} \cdot \text{m}$ and a direction that is out of the page since the forces tend to rotate counterclockwise. $\mathbf{M}$ is a free vector, and so it can be placed at any point on the gear, Fig. 4–29*b*. To preserve the counterclockwise rotation of $\mathbf{M}$, *vertical* forces acting through points *A* and *B* must be directed as shown in Fig. 4–29*c*. The magnitude of each force is

$$M = Fd \qquad 24\ \text{N} \cdot \text{m} = F(0.2\ \text{m})$$
$$F = 120\ \text{N} \qquad\qquad\qquad Ans.$$

E X A M P L E **4.11**

Determine the moment of the couple acting on the member shown in Fig. 4–30a.

(a)   (b)

Solution (Scalar Analysis)

Here it is somewhat difficult to determine the perpendicular distance between the forces and compute the couple moment as $M = Fd$. Instead, we can resolve each force into its horizontal and vertical components, $F_x = \frac{4}{5}(150 \text{ kN}) = 120 \text{ kN}$ and $F_y = \frac{3}{5}(150 \text{ kN}) = 90 \text{ kN}$, Fig. 4–30b, and then use the principle of moments. The couple moment can be determined about *any point*. For example, if point D is chosen, we have for all four forces,

$\zeta + M = 120 \text{ kN } (0 \text{ m}) - 90 \text{ kN } (2 \text{ m}) + 90 \text{ kN } (5 \text{ m}) + 120 \text{ kN } (1 \text{ m})$

$= 390 \text{ kN} \cdot \text{m} \;\uparrow$ *Ans.*

It is easier, however, to determine the moments about point A or B in order to *eliminate* the moment of the forces acting at the moment point. For point A, Fig. 4-30b, we have

$\zeta + M = 90 \text{ kN } (3 \text{ m}) + 120 \text{ kN } (1 \text{ m})$

$= 390 \text{ kN} \cdot \text{m} \;\uparrow$ *Ans.*

Show that one obtains this same result if moments are summed about point B. Notice also that the couple in Fig. 4–30a can be replaced by *two* couples in Fig. 4–30b. Using $M = Fd$, one couple has a moment of $M_1 = 90 \text{ kN } (3 \text{ m}) = 270 \text{ kN} \cdot \text{m}$ and the other has a moment of $M_2 = 120 \text{ kN } (1 \text{ m}) = 120 \text{ kN} \cdot \text{m}$. By the right-hand rule, both couple moments are counterclockwise and are therefore directed out of the page. Since these couples are free vectors, they can be moved to any point and added, which yields $M = 270 \text{ kN} \cdot \text{m} + 120 \text{ kN} \cdot \text{m} = 390 \text{ kN} \cdot \text{m} \;\uparrow$, the same result determined above. **M** is a free vector and can therefore act at any point on the member, Fig. 4–30c. Also, realize that the external effect, such as the support reactions on the member, will be the *same* if the member supports the couple, Fig. 4–30a, or the couple moment, Fig. 4–30c.

(c)

Fig. 4–30

E X A M P L E **4.12**

Determine the couple moment acting on the pipe shown in Fig. 4–31a. Segment AB is directed 30° below the x-y plane.

(a)

(d)

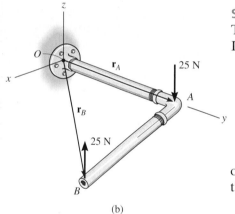

(b)

Solution I *(Vector Analysis)*

The moment of the two couple forces can be found about *any point*. If point O is considered, Fig. 4–31b, we have

$$\mathbf{M} = \mathbf{r}_A \times (-25\mathbf{k}) + \mathbf{r}_B \times (25\mathbf{k})$$
$$= (8\mathbf{j}) \times (-25\mathbf{k}) + (6\cos 30°\mathbf{i} + 8\mathbf{j} - 6\sin 30°\mathbf{k}) \times (25\mathbf{k})$$
$$= -200\mathbf{i} - 129.9\mathbf{j} + 200\mathbf{i}$$
$$= \{-130\mathbf{j}\} \text{ N} \cdot \text{cm} \qquad\qquad Ans.$$

It is *easier* to take moments of the couple forces about a point lying on the line of action of one of the forces, e.g., point A, Fig. 4–31c. In this case the moment of the force A is zero, so that

$$\mathbf{M} = \mathbf{r}_{AB} \times (25\mathbf{k})$$
$$= (6\cos 30°\mathbf{i} - 6\sin 30°\mathbf{k}) \times (25\mathbf{k})$$
$$= \{-130\mathbf{j}\} \text{ N} \cdot \text{cm} \qquad\qquad Ans.$$

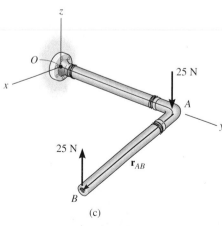

(c)

Fig. 4–31

Solution II *(Scalar Analysis)*

Although this problem is shown in three dimensions, the geometry is simple enough to use the scalar equation $M = Fd$. The perpendicular distance between the lines of action of the forces is $d = 6\cos 30° = 5.20$ cm, Fig. 4–31d. Hence, taking moments of the forces about either point A or B yields

$$M = Fd = 25 \text{ N} (5.20 \text{ cm}) = 129.9 \text{ N} \cdot \text{cm}$$

Applying the right-hand rule, $\mathbf{M}$ acts in the $-\mathbf{j}$ direction. Thus,

$$\mathbf{M} = \{-130\mathbf{j}\} \text{ N} \cdot \text{cm} \qquad\qquad Ans.$$

EXAMPLE 4.13

Replace the two couples acting on the pipe column in Fig. 4–32a by a resultant couple moment.

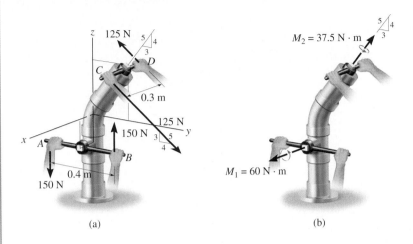

Fig. 4–32

Solution *(Vector Analysis)*

The couple moment $\mathbf{M}_1$, developed by the forces at A and B, can easily be determined from a scalar formulation.

$$M_1 = Fd = 150 \text{ N}(0.4 \text{ m}) = 60 \text{ N} \cdot \text{m}$$

By the right-hand rule, $\mathbf{M}_1$ acts in the $+\mathbf{i}$ direction, Fig. 4–32b. Hence,

$$\mathbf{M}_1 = \{60\mathbf{i}\} \text{ N} \cdot \text{m}$$

Vector analysis will be used to determine $\mathbf{M}_2$, caused by forces at C and D. If moments are computed about point D, Fig. 4–32a, $\mathbf{M}_2 = \mathbf{r}_{DC} \times \mathbf{F}_C$, then

$$\begin{aligned}
\mathbf{M}_2 = \mathbf{r}_{DC} \times \mathbf{F}_C &= (0.3\mathbf{i}) \times [125(\tfrac{4}{5})\mathbf{j} - 125(\tfrac{3}{5})\mathbf{k}] \\
&= (0.3\mathbf{i}) \times [100\mathbf{j} - 75\mathbf{k}] = 30(\mathbf{i} \times \mathbf{j}) - 22.5(\mathbf{i} \times \mathbf{k}) \\
&= \{22.5\mathbf{j} + 30\mathbf{k}\} \text{ N} \cdot \text{m}
\end{aligned}$$

Try to establish $\mathbf{M}_2$ by using a scalar formulation, Fig. 4–32b.

Since $\mathbf{M}_1$ and $\mathbf{M}_2$ are free vectors, they may be moved to some arbitrary point P and added vectorially, Fig. 4–32c. The resultant couple moment becomes

$$\mathbf{M}_R = \mathbf{M}_1 + \mathbf{M}_2 = \{60\mathbf{i} + 22.5\mathbf{j} + 30\mathbf{k}\} \text{ N} \cdot \text{m} \qquad \textit{Ans.}$$

PROBLEMS

4-69. Determine the magnitude and sense of the couple moment.

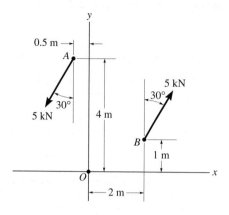

Prob. 4–69

4-71. Determine the magnitude and sense of the couple moment. Each force has a magnitude of $F = 8$ kN.

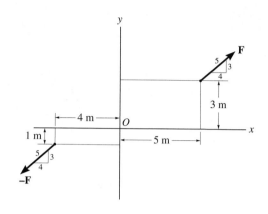

Prob. 4–71

4-70. If the couple moment has a magnitude of 220 N · m, determine the magnitude F of the couple forces.

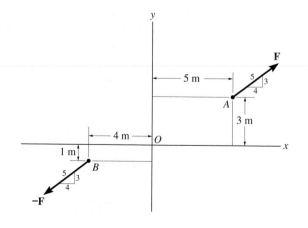

Prob. 4–70

*4-72. If the couple moment has a magnitude of 300 N · m, determine the magnitude F of the couple forces.

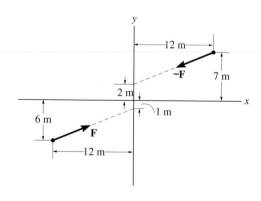

Prob. 4–72

4-73. A twist of 4 N · m is applied to the handle of the screwdriver. Resolve this couple moment into a pair of couple forces **F** exerted on the handle and **P** exerted on the blade.

Prob. 4–73

4-75. A device called a rolamite is used in various ways to replace slipping motion with rolling motion. If the belt, which wraps between the rollers, is subjected to a tension of 15 N, determine the reactive forces *N* of the top and bottom plates on the rollers so that the resultant couple acting on the rollers is equal to zero.

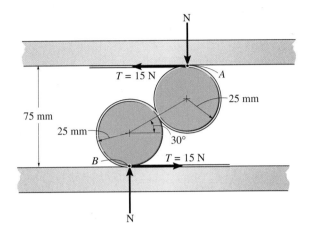

Prob. 4–75

4-74. The resultant couple moment created by the two couples acting on the disk is $M_R = \{10\mathbf{k}\}$ kN · cm. Determine the magnitude of force **T**.

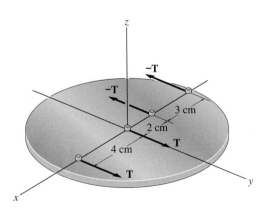

Prob. 4–74

*4-76. The caster wheel is subjected to the two couples. Determine the forces **F** that the bearings create on the shaft so that the resultant couple moment on the caster is zero.

Prob. 4–76

4-77. When the engine of the plane is running, the vertical reaction that the ground exerts on the wheel at A is measured as 2900 N. When the engine is turned off, however, the vertical reactions at A and B are 2600 N each. The difference in readings at A is caused by a couple acting on the propeller when the engine is running. This couple tends to overturn the plane counterclockwise, which is opposite to the propeller's clockwise rotation. Determine the magnitude of this couple and the magnitude of the reaction force exerted at B when the engine is running.

4-79. Express the moment of the couple acting on the pipe assembly in Cartesian vector form. Solve the problem (a) using Eq. 4–13, and (b) summing the moment of each force about point O. Take F = {25k} N.

*4-80. If the couple moment acting on the pipe has a magnitude of 400 N·m, determine the magnitude F of the vertical force applied to each wrench.

Probs. 4–79/80

Prob. 4–77

4-78. Two couples act on the beam. Determine the magnitude of **F** so that the resultant couple moment is 450 N · m, counterclockwise. Where on the beam does the resultant couple moment act?

4-81. The ends of the triangular plate are subjected to three couples. Determine the plate dimension d so that the resultant couple is 350 N·m clockwise.

Prob. 4–78

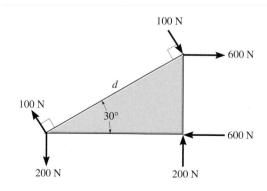

Prob. 4–81

4-82. Two couples act on the beam as shown. Determine the magnitude of **F** so that the resultant couple moment is 100 N · m counterclockwise. Where on the beam does the resultant couple act?

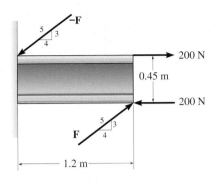

Prob. 4–82

4-83. Two couples act on the frame. If the resultant couple moment is to be zero, determine the distance d between the 80-N couple forces.

*4-84. Two couples act on the frame. If $d = 0.4$ m, determine the resultant couple moment. Compute the result by resolving each force into x and y components and (a) finding the moment of each couple (Eq. 4–13) and (b) summing the moments of all the force components about point A.

4-85. Two couples act on the frame. If $d = 0.4$ m, determine the resultant couple moment. Compute the result by resolving each force into x and y components and (a) finding the moment of each couple (Eq. 4–13) and (b) summing the moments of all the force components about point B.

Probs. 4–83/84/85

4-86. Determine the couple moment. Express the result as a Cartesian vector.

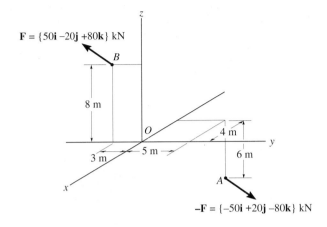

Prob. 4–86

4-87. Determine the couple moment. Express the result as a Cartesian vector. Each force has a magnitude of $F = 120$ kN.

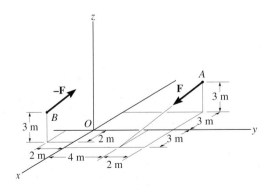

Prob. 4–87

*4–88. The gear reducer is subjected to the four couple moments. Determine the magnitude of the resultant couple moment and its coordinate direction angles.

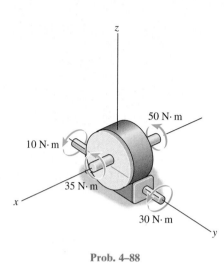

Prob. 4–88

4–89. The main beam along the wing of an airplane is swept back at an angle of 25°. From load calculations it is determined that the beam is subjected to couple moments $M_x = 20\ \text{kN} \cdot \text{m}$ and $M_y = 30\ \text{kN} \cdot \text{m}$. Determine the resultant couple moments created about the x' and y' axes. The axes all lie in the same horizontal plane.

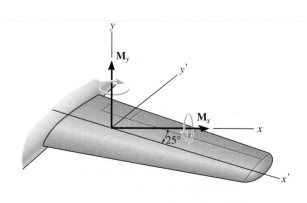

Prob. 4–89

4–90. If $\mathbf{F} = \{100\mathbf{k}\}$ N, determine the couple moment that acts on the assembly. Express the result as a Cartesian vector. Member BA lies in the x-y plane.

4–91. If the magnitude of the resultant couple moment is $15\ \text{N} \cdot \text{m}$, determine the magnitude F of the forces applied to the wrenches.

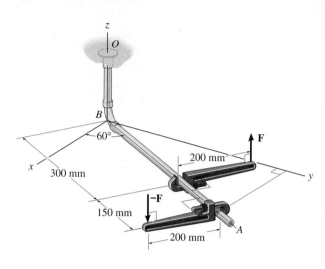

Probs. 4–90/91

*4–92. The gear reducer is subjected to the couple moments shown. Determine the resultant couple moment and specify its magnitude and coordinate direction angles.

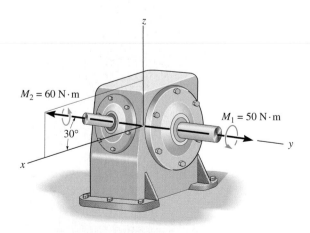

Prob. 4–92

4-93. The gear reducer is subject to the couple moments shown. Determine the resultant couple moment and specify its magnitude and coordinate direction angles.

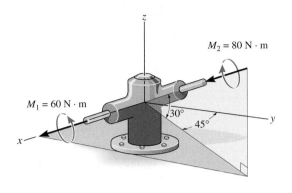

Prob. 4–93

4-94. The meshed gears are subjected to the couple moments shown. Determine the magnitude of the resultant couple moment and specify its coordinate direction angles.

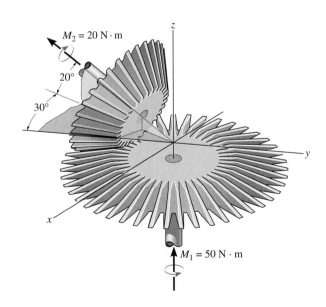

Prob. 4–94

4-95. A couple acts on each of the handles of the minidual valve. Determine the magnitude and coordinate direction angles of the resultant couple moment.

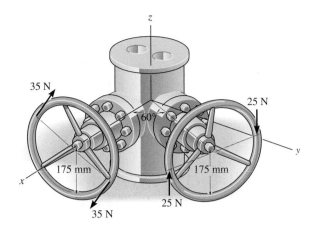

Prob. 4–95

*4-96. Determine the resultant couple moment of the two couples that act on the pipe assembly. The distance from A to B is $d = 400$ mm. Express the result as a Cartesian vector.

4-97. Determine the distance d between A and B so that the resultant couple moment has a magnitude of $M_R = 20$ N $\cdot$ m.

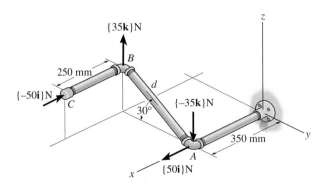

Probs. 4–96/97

4.7 Equivalent System

A force has the effect of both translating and rotating a body, and the amount by which it does so depends upon where and how the force is applied. In the next section we will discuss the method used to *simplify* a system of forces and couple moments acting on a body to a single resultant force and couple moment acting at a specified point O. To do this, however, it is necessary that the force and couple moment system produce the *same* "external" effects of translation and rotation of the body as their resultants. When this occurs these two sets of loadings are said to be *equivalent*.

In this section we wish to show how to maintain this equivalency when a single force is applied to a specific point on a body and when it is located at another point O. Two cases for the location of point O will now be considered.

Point O Is On the Line of Action of the Force. Consider the body shown in Fig. 4–33a, which is subjected to the force $\mathbf{F}$ applied to point A. In order to apply the force to point O without altering the external effects on the body, we will first apply equal but opposite forces $\mathbf{F}$ and $-\mathbf{F}$ at O, as shown in Fig. 4–33b. The two forces indicated by the slash across them can be canceled, leaving the force at point O as required, Fig. 4–33c. By using this construction procedure, an *equivalent system* has been maintained between each of the diagrams, as shown by the equal signs. Note, however, that the force has simply been "transmitted" along its line of action, from point A, Fig. 4–33a, to point O, Fig. 4–33c. In other words, the force can be considered as a *sliding vector* since it can act at any point O along its line of action. In Sec. 4.3 we referred to this concept as the *principle of transmissibility*. It is important to realize that only the *external effects*, such as the body's motion or the forces needed to support the body if it is stationary, remain *unchanged* after $\mathbf{F}$ is moved. Certainly the *internal effects* depend on where $\mathbf{F}$ is located. For example, when $\mathbf{F}$ acts at A, the internal forces in the body have a high intensity around A; whereas movement of $\mathbf{F}$ away from this point will cause these internal forces to decrease.

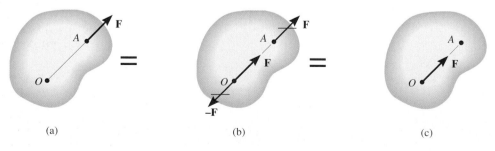

(a)

(b)

(c)

Fig. 4–33

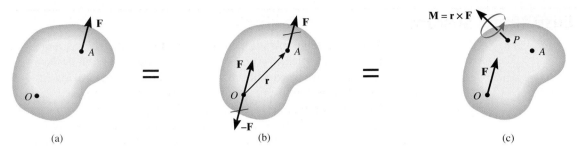

Fig. 4–34

Point *O* Is Not On the Line of Action of the Force.

This case is shown in Fig. 4–34*a*, where **F** is to be moved to point *O* without altering the external effects on the body. Following the same procedure as before, we first apply equal but opposite forces **F** and −**F** at point *O*, Fig. 4–34*b*. Here the two forces indicated by a slash across them form a couple which has a moment that is perpendicular to **F** and is defined by the cross product **M** = **r** × **F**. Since the couple moment is a *free vector*, it may be applied at *any point P* on the body as shown in Fig. 4–34*c*. In addition to this couple moment, **F** now acts at point *O* as required.

To summarize these concepts, when the point on the body is *on the line of action of the force*, simply transmit or slide the force along its line of action to the point. When the point is not on the line of action of the force, then move the force to the point and add a couple moment anywhere to the body. This couple moment is found by taking the moment of the force about the point. When these rules are carried out, equivalent external effects will be produced.

Consider the effects on the hand when a stick of negligible weight supports a force **F** at its end. When the force is applied horizontally, the same force is felt at the grip, regardless of where it is applied along its line of action. This is a consequence of the principle of transmissibility.

When the force is applied vertically it causes both a downward force **F** to be felt at the grip and a clockwise couple moment or twist of $M = Fd$. These same effects are felt if **F** is applied at the grip and **M** is applied anywhere on the stick. In both cases the systems are equivalent.

4.8 Resultants of a Force and Couple System

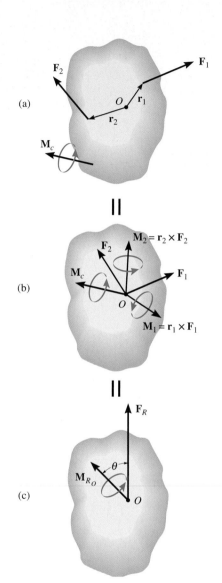

Fig. 4–35

When a rigid body is subjected to a *system* of forces and couple moments, it is often simpler to study the external effects on the body by *replacing* the system by an equivalent single resultant force acting at a specified point O and a resultant couple moment. To show how to determine these resultants we will consider the rigid body in Fig. 4–35*a* and use the concepts discussed in the previous section. Since point O is not on the line of action of the forces, an equivalent effect is produced if the forces are moved to point O *and* the corresponding couple moments $\mathbf{M}_1 = \mathbf{r}_1 \times \mathbf{F}_1$ and $\mathbf{M}_2 = \mathbf{r}_2 \times \mathbf{F}_2$ are applied to the body. Furthermore, the couple moment $\mathbf{M}_c$ is simply moved to point O since it is a free vector. These results are shown in Fig. 4–35*b*. By vector addition, the resultant force is $\mathbf{F}_R = \mathbf{F}_1 + \mathbf{F}_2$, and the resultant couple moment is $\mathbf{M}_{R_O} = \mathbf{M}_c + \mathbf{M}_1 + \mathbf{M}_2$, Fig. 4–35*c*. Since equivalency is maintained between the diagrams in Fig. 4–35, each force and couple system will cause the *same external effects*, i.e., the same translation and rotation of the body. Note that both the magnitude and direction of $\mathbf{F}_R$ are independent of the location of point O; however, $\mathbf{M}_{R_O}$ depends upon this location since the moments $\mathbf{M}_1$ and $\mathbf{M}_2$ are determined using the position vectors $\mathbf{r}_1$ and $\mathbf{r}_2$. Also note that $\mathbf{M}_{R_O}$ is a free vector and can act at *any point* on the body, although point O is generally chosen as its point of application.

The above method of simplifying any force and couple moment system to a resultant force acting at point O and a resultant couple moment can be generalized and represented by application of the following two equations.

$$
\begin{aligned}
\mathbf{F}_R &= \Sigma \mathbf{F} \\
\mathbf{M}_{R_O} &= \Sigma \mathbf{M}_c + \Sigma \mathbf{M}_O
\end{aligned}
\tag{4–17}
$$

The first equation states that the resultant force of the system is equivalent to the sum of all the forces; and the second equation states that the resultant couple moment of the system is equivalent to the sum of all the couple moments $\Sigma \mathbf{M}_c$, plus the moments about point O of all the forces $\Sigma \mathbf{M}_O$. If the force system lies in the $x-y$ plane and any couple moments are perpendicular to this plane, that is along the z axis, then the above equations reduce to the following three scalar equations.

$$
\begin{aligned}
F_{R_x} &= \Sigma F_x \\
F_{R_y} &= \Sigma F_y \\
M_{R_O} &= \Sigma M_c + \Sigma M_O
\end{aligned}
\tag{4–18}
$$

Note that the resultant force $\mathbf{F}_R$ is equivalent to the vector sum of its two components $\mathbf{F}_{R_x}$ and $\mathbf{F}_{R_y}$.

If the two forces acting on the stick are replaced by an equivalent resultant force and couple moment at point A, or by the equivalent resultant force and couple moment at point B, then in each case the hand must provide the same resistance to translation and rotation in order to keep the stick in the horizontal position. In other words, the external effects on the stick are the *same* in each case.

PROCEDURE FOR ANALYSIS

The following points should be kept in mind when applying Eqs. 4–17 or 4–18.

• Establish the coordinate axes with the origin located at the point O and the axes having a selected orientation.

Force Summation.

• If the force system is *coplanar*, resolve each force into its x and y components. If a component is directed along the positive x or y axis, it represents a positive scalar; whereas if it is directed along the negative x or y axis, it is a negative scalar.

• In three dimensions, represent each force as a Cartesian vector before summing the forces.

Moment Summation.

• When determining the moments of a *coplanar* force system about point O, it is generally advantageous to use the principle of moments, i.e., determine the moments of the components of each force rather than the moment of the force itself.

• In three dimensions use the vector cross product to determine the moment of each force about the point. Here the position vectors extend from point O to any point on the line of action of each force.

EXAMPLE 4.14

Replace the forces acting on the brace shown in Fig. 4–36a by an equivalent resultant force and couple moment acting at point A.

Fig. 4–36

Solution *(Scalar Analysis)*

The principle of moments will be applied to the 400-N force, whereby the moments of its two rectangular components will be considered.

Force Summation. The resultant force has x and y components of

$$\xrightarrow{+} F_{R_x} = \Sigma F_x; \qquad F_{R_x} = -100 \text{ N} - 400 \cos 45°\text{N} = -382.8 \text{ N} = 382.8 \text{ N} \leftarrow$$
$$+\uparrow F_{R_y} = \Sigma F_y; \qquad F_{R_y} = -600 \text{ N} - 400 \sin 45°\text{N} = -882.8 \text{ N} = 882.8 \text{ N}\downarrow$$

As shown in Fig. 4–36b, $\mathbf{F}_R$ has a magnitude of

$$F_R = \sqrt{(F_{R_x})^2 + (F_{R_y})^2} = \sqrt{(382.8)^2 + (882.8)^2} = 962 \text{ N} \quad \textit{Ans.}$$

and a direction of

$$\theta = \tan^{-1}\left(\frac{F_{R_y}}{F_{R_x}}\right) = \tan^{-1}\left(\frac{882.8}{382.8}\right) = 66.6° \qquad \textit{Ans.}$$

Moment Summation. The resultant couple moment $\mathbf{M}_{R_A}$ is determined by summing the moments of the forces about point A. Assuming that positive moments act counterclockwise, i.e., in the $+\mathbf{k}$ direction, we have

$$\zeta+M_{R_A} = \Sigma M_A;$$
$$M_{R_A} = 100 \text{ N}(0) - 600 \text{ N}(0.4 \text{ m}) - (400 \sin 45°\text{N})(0.8 \text{ m})$$
$$- (400 \cos 45°\text{N})(0.3 \text{ m})$$
$$= -551 \text{ N} \cdot \text{m} = 551 \text{ N} \cdot \text{m}\,\zeta \qquad \textit{Ans.}$$

In conclusion, when $\mathbf{M}_{R_A}$ and $\mathbf{F}_R$ act on the brace at point A, Fig. 4–36b, they will produce the *same* external effect or reactions at the supports as that produced by the force system in Fig. 4–36a.

EXAMPLE 4.15

A structural member is subjected to a couple moment **M** and forces **F**$_1$ and **F**$_2$ as shown in Fig. 4–37a. Replace this system by an equivalent resultant force and couple moment acting at its base, point O.

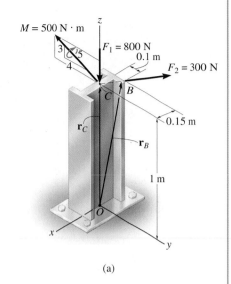

(a)

Solution *(Vector Analysis)*

The three-dimensional aspects of the problem can be simplified by using a Cartesian vector analysis. Expressing the forces and couple moment as Cartesian vectors, we have

$$\mathbf{F}_1 = \{-800\mathbf{k}\} \text{ N}$$

$$\mathbf{F}_2 = (300 \text{ N})\mathbf{u}_{CB} = (300 \text{ N})\left(\frac{\mathbf{r}_{CB}}{r_{CB}}\right)$$

$$= 300\left[\frac{-0.15\mathbf{i} + 0.1\mathbf{j}}{\sqrt{(-0.15)^2 + (0.1)^2}}\right] = \{-249.6\mathbf{i} + 166.4\mathbf{j}\} \text{ N}$$

$$\mathbf{M} = -500(\tfrac{4}{5})\mathbf{j} + 500(\tfrac{3}{5})\mathbf{k} = \{-400\mathbf{j} + 300\mathbf{k}\} \text{ N} \cdot \text{m}$$

Force Summation.

$$\mathbf{F}_R = \Sigma\mathbf{F}; \qquad \mathbf{F}_R = \mathbf{F}_1 + \mathbf{F}_2 = -800\mathbf{k} - 249.6\mathbf{i} + 166.4\mathbf{j}$$

$$= \{-249.6\mathbf{i} + 166.4\mathbf{j} - 800\mathbf{k}\} \text{ N} \qquad\qquad \textit{Ans.}$$

Moment Summation.

(b)

$$\mathbf{M}_{R_O} = \Sigma\mathbf{M}_C + \Sigma\mathbf{M}_O$$
$$\mathbf{M}_{R_O} = \mathbf{M} + \mathbf{r}_C \times \mathbf{F}_1 + \mathbf{r}_B \times \mathbf{F}_2$$

$$\mathbf{M}_{R_O} = (-400\mathbf{j} + 300\mathbf{k}) + (1\mathbf{k}) \times (-800\mathbf{k}) + \begin{vmatrix} \mathbf{i} & \mathbf{j} & \mathbf{k} \\ -0.15 & 0.1 & 1 \\ -249.6 & 166.4 & 0 \end{vmatrix}$$

$$= (-400\mathbf{j} + 300\mathbf{k}) + (\mathbf{0}) + (-166.4\mathbf{i} - 249.6\mathbf{j})$$

$$= \{-166\mathbf{i} - 650\mathbf{j} + 300\mathbf{k}\} \text{ N} \cdot \text{m} \qquad\qquad \textit{Ans.}$$

The results are shown in Fig. 4–37b.

Fig. 4–37

4.9 Further Reduction of a Force and Couple System

Simplification to a Single Resultant Force. Consider now a special case for which the system of forces and couple moments acting on a rigid body, Fig. 4–38a, reduces at point O to a resultant force $\mathbf{F}_R = \Sigma\mathbf{F}$ and resultant couple moment $\mathbf{M}_{R_O} = \Sigma\mathbf{M}_O$, which are *perpendicular* to one another, Fig. 4–38b. Whenever this occurs, we can further simplify the force and couple moment system by moving $\mathbf{F}_R$ to another point P, located either on or off the body so that no resultant couple moment has to be applied to the body, Fig. 4–38c. In other words, if the force and couple moment system in Fig. 4–38a is reduced to a resultant system at point P, only the force resultant will have to be applied to the body, Fig. 4–38c.

The location of point P, measured from point O, can always be determined provided $\mathbf{F}_R$ and $\mathbf{M}_{R_O}$ are known, Fig. 4–38b. As shown in Fig. 4–38c, P must lie on the bb axis, which is perpendicular to both the line of action of $\mathbf{F}_R$ and the aa axis. This point is chosen such that the distance d satisfies the scalar equation $M_{R_O} = F_R d$ or $d = M_{R_O}/F_R$. With $\mathbf{F}_R$ so located, it will produce the same external effects on the body as the force and couple moment system in Fig. 4–38a, or the force and couple moment resultants in Fig. 4–38b.

If a system of forces is either concurrent, coplanar, or parallel, it can always be reduced, as in the above case, to a single resultant force $\mathbf{F}_R$ acting through. This is because in each of these cases $\mathbf{F}_R$ and $\mathbf{M}_{R_O}$ will always be perpendicular to each other when the force system is simplified at *any* point O.

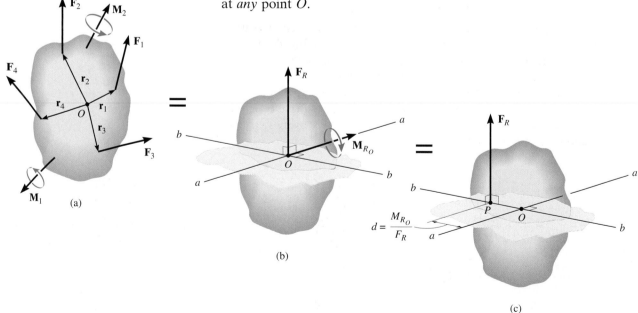

Fig. 4–38

Concurrent Force Systems. A concurrent force system has been treated in detail in Chapter 2. Obviously, all the forces act at a point for which there is no resultant couple moment, so the point P is automatically specified, Fig. 4–39.

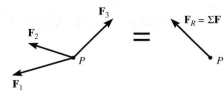

Fig. 4–39

Coplanar Force Systems. Coplanar force systems, which may include couple moments directed perpendicular to the plane of the forces as shown in Fig. 4–40a, can be reduced to a single resultant force, because when each force in the system is moved to any point O in the x–y plane, it produces a couple moment that is *perpendicular* to the plane, i.e., in the $\pm\mathbf{k}$ direction. The resultant moment $\mathbf{M}_{R_O} = \Sigma\mathbf{M} + \Sigma(\mathbf{r} \times \mathbf{F})$ is thus perpendicular to the resultant force $\mathbf{F}_R$. Fig. 4–40b; and so $\mathbf{F}_R$ can be positioned a distance d from O so as to create this same moment $\mathbf{M}_{R_O}$ about O, Fig. 4–40c.

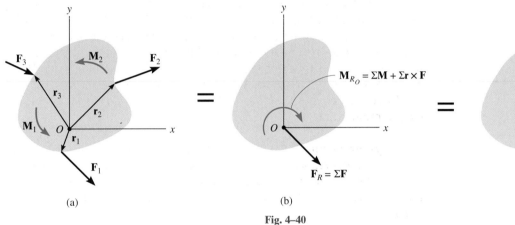

(a)   (b)   (c)

Fig. 4–40

Parallel Force Systems. Parallel force systems, which can include couple moments that are perpendicular to the forces, as shown in Fig. 4–41a, can be reduced to a single resultant force because when each force is moved to any point O in the x–y plane, it produces a couple moment that has components only about the x and y axes. The resultant moment $\mathbf{M}_{R_O} = \Sigma\mathbf{M}_O + \Sigma(\mathbf{r} \times \mathbf{F})$ is thus perpendicular to the resultant force $\mathbf{F}_R$, Fig. 4–41b; and so $\mathbf{F}_R$ can be moved to a point a distance d away so that it produces the same moment about O.

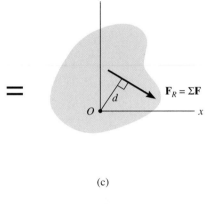

(a)   (b)   (c)

Fig. 4–41

The three parallel forces acting on the stick can be replaced by a single resultant force F_R acting at a distance d from the grip. To be equivalent we require the resultant force to equal the sum of the forces, $F_R = F_1 + F_2 + F_3$, and to find the distance d the moment of the resultant force about the grip must be equal to the moment of all the forces about the grip, $F_R d = F_1 d_1 + F_2 d_2 + F_3 d_3$.

PROCEDURE FOR ANALYSIS

The technique used to reduce a coplanar or parallel force system to a single resultant force follows a similar procedure outlined in the previous section.

- Establish the x, y, z, axes and locate the resultant force $\mathbf{F}_R$ an arbitrary distance away from the origin of the coordinates.

Force Summation.

- The resultant force is equal to the sum of all the forces in the system.

- For a coplanar force system, resolve each force into its x and y components. Positive components are directed along the positive x and y axes, and negative components are directed along the negative x and y axes.

Moment Summation.

- The moment of the resultant force about point O is equal to the sum of all the couple moments in the system plus the moments about point O of all the forces in the system.

- This moment condition is used to find the location of the resultant force from point O.

Reduction to a Wrench. In the general case, the force and couple moment system acting on a body, Fig. 4–35a, will reduce to a single resultant force $\mathbf{F}_R$ and couple moment $\mathbf{M}_{R_O}$ at O which are *not* perpendicular. Instead, $\mathbf{F}_R$ will act at an angle θ from $\mathbf{M}_{R_O}$, Fig. 4–35c. As shown in Fig. 4–42a, however, $\mathbf{M}_{R_O}$ may be resolved into two components: one perpendicular, $\mathbf{M}_\perp$, and the other parallel $\mathbf{M}_\parallel$, to the line of action of $\mathbf{F}_R$. As in the previous discussion, the perpendicular component $\mathbf{M}_\perp$ may be *eliminated* by moving $\mathbf{F}_R$ to point P, as shown in Fig. 4–42b. This point lies on axis bb, which is perpendicular to both $\mathbf{M}_{R_O}$ and $\mathbf{F}_R$. In order to maintain an equivalency of loading, the distance from O to P is $d = M_\perp/F_R$. Furthermore, when $\mathbf{F}_R$ is applied at P, the moment of $\mathbf{F}_R$ tending to cause rotation of the body *about O* is in the *same direction* as $\mathbf{M}_\perp$, Fig. 4–42a. Finally, since $\mathbf{M}_\parallel$ is a free vector, it may be moved to P so that it is collinear with $\mathbf{F}_R$, Fig. 4–42c. This combination of a collinear force and couple moment is called a *wrench* or *screw*. The *axis of the wrench* has the same line of action as the force. Hence, the wrench tends to cause both a translation along and a rotation about this axis. Comparing Fig. 4–42a to Fig. 4–42c, it is seen that a general force and couple moment system acting on a body can be reduced to a wrench. The axis of the wrench and the point through which this axis passes can always be determined.

Fig. 4–35, (Repeated)

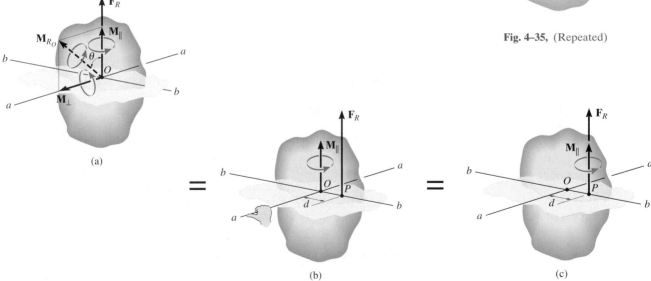

Fig. 4–42

EXAMPLE 4.16

The beam AE in Fig. 4–43a is subjected to a system of coplanar forces. Determine the magnitude, direction, and location on the beam of a resultant force which is equivalent to the given system of forces measured from E.

(a)   (b)

Fig. 4–43

Solution

The origin of coordinates is located at point E as shown in Fig. 4–43a.

Force Summation. Resolving the 500-N force into x and y components and summing the force components yields

$$\xrightarrow{+} F_{R_x} = \Sigma F_x; \quad F_{R_x} = 500 \cos 60° \text{ N} + 100 \text{ N} = 350.0 \text{ N} \rightarrow$$

$$+\uparrow F_{R_y} = \Sigma F_y; \quad F_{R_y} = -500 \sin 60° \text{ N} + 200 \text{ N} = -233.0 \text{ N}$$

$$= 233.0 \text{ N} \downarrow$$

The magnitude and direction of the resultant force are established from the vector addition shown in Fig. 4–43b. We have

$$F_R = \sqrt{(350.0)^2 + (233.0)^2} = 420.5 \text{ N} \qquad \text{Ans.}$$

$$\theta = \tan^{-1}\left(\frac{233.0}{350.0}\right) = 33.7° \searrow_\theta \qquad \text{Ans.}$$

Moment Summation. Moments will be summed about point E. Hence, from Figs. 4–43a and 4–43b, we require the moments of the components of $\mathbf{F}_R$ (or the moment of $\mathbf{F}_R$) about point E to equal the moments of the force system about E. Assuming positive moments are counterclockwise, we have

$$\zeta + M_{R_E} = \Sigma M_E$$

$$233.0 \text{ N}(d) + 350.0 \text{ N}(0) = (500 \sin 60° \text{ N})(4 \text{ m}) + (500 \cos 60° \text{ N})(0)$$
$$- (100 \text{ N})(0.5 \text{ m}) - (200 \text{ N})(2.5 \text{ m})$$

$$d = \frac{1182.1}{233.0} = 5.07 \text{ m} \qquad \text{Ans.}$$

Note that using a clockwise sign convention would yield this same result. Since d is *positive*, $\mathbf{F}_R$ acts to the left of E as shown. Try to solve this problem by summing moments about point A and show $d' = 0.927$ m, measured to the right of A.

EXAMPLE 4.17

The jib crane shown in Fig. 4–44a is subjected to three coplanar forces. Replace this loading by an equivalent resultant force and specify where the resultant's line of action intersects the column AB and boom BC.

Solution

Force Summation. Resolving the 2.50 kN force into x and y components and summing the force components yields

$$\xrightarrow{+} F_{R_x} = \Sigma F_x; \quad F_{R_x} = -2.50 \text{ kN}(\tfrac{3}{5}) - 1.75 \text{ kN} = -3.25 \text{ kN} = 3.25 \text{ kN} \leftarrow$$
$$+\uparrow F_{R_y} = \Sigma F_y; \quad F_{R_y} = -2.50 \text{ kN}(\tfrac{4}{5}) - 0.60 \text{ kN} = -2.60 \text{ kN} = 2.60 \text{ kN} \downarrow$$

As shown by the vector addition in Fig. 4–44b,

$$F_R = \sqrt{(3.25)^2 + (2.60)^2} = 4.16 \text{ kN} \qquad \textit{Ans.}$$

$$\theta = \tan^{-1}\left(\frac{2.60}{3.25}\right) = 38.7° \quad \textit{Ans.}$$

(a)

Moment Summation. Moments will be summed about the arbitrary point A. Assuming the line of action of $\mathbf{F}_R$ intersects AB, Fig. 4–44b, we require the moment of the components of $\mathbf{F}_R$ in Fig. 4–44b about A to equal the moments of the force system in Fig. 4–44a about A; i.e.,

$$\downdownarrows + M_{R_A} = \Sigma M_A; \quad 3.25 \text{ kN } (y) + 2.60 \text{ kN } (0)$$
$$= 1.75 \text{ kN}(1 \text{ m}) - 0.60 \text{ kN}(0.6 \text{ m}) + 2.50 \text{ kN}(\tfrac{3}{5})(2.2 \text{ m}) - 2.50 \text{ kN}(\tfrac{4}{5})(1.6 \text{ m})$$
$$y = 0.458 \text{ m} \qquad \textit{Ans.}$$

By the principle of transmissibility, $\mathbf{F}_R$ can also be treated as intersecting BC, Fig. 4–44b, in which case we have

$$\downdownarrows + M_{R_A} = \Sigma M_A; \quad 3.25 \text{ kN } (2.2 \text{ m}) - 2.60 \text{ kN } (x)$$
$$= 1.75 \text{ kN}(1 \text{ m}) - 0.60 \text{ kN}(0.6 \text{ m}) + 2.50 \text{ kN}(\tfrac{3}{5})(2.2 \text{ m}) - 2.50 \text{ kN}(\tfrac{4}{5})(1.6 \text{ m})$$
$$x = 2.177 \text{ m} \qquad \textit{Ans.}$$

We can also solve for these positions by assuming $\mathbf{F}_R$ acts at the arbitrary point (x, y) on its line of action, Fig. 4–44b. Summing moments about point A yields

(b)

Fig. 4–44

$$\downdownarrows + M_{R_A} = \Sigma M_A; \quad 3.25 \text{ kN } (y) - 2.60 \text{ kN } (x)$$
$$= 1.75 \text{ kN}(1 \text{ m}) - 0.60 \text{ kN}(0.6 \text{ m}) + 2.50 \text{ kN}(\tfrac{3}{5})(2.2 \text{ m}) - 2.50 \text{ kN}(\tfrac{4}{5})(1.6 \text{ m})$$
$$3.25y - 2.60x = 1.49$$

which is the equation of the colored dashed line in Fig. 4–44b. To find the points of intersection with the crane along AB, set $x = 0$, then $y = 0.458$ m, and along BC set $y = 2.2$ m, then $x = 2.177$ m.

E X A M P L E **4.18**

The slab in Fig. 4–45a is subjected to four parallel forces. Determine the magnitude and direction of a resultant force equivalent to the given force system and locate its point of application on the slab.

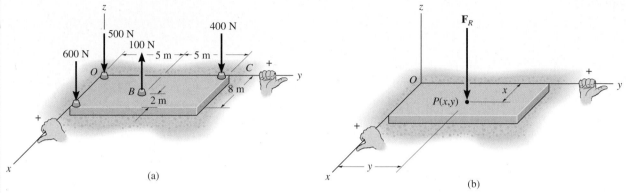

Fig. 4–45

Solution *(Scalar Analysis)*

Force Summation. From Fig. 4–45a, the resultant force is

$$+\uparrow F_R = \Sigma F; \quad F_R = -600 \text{ N} + 100 \text{ N} - 400 \text{ N} - 500 \text{ N}$$
$$= -1400 \text{ N} = 1400 \text{ N} \downarrow \qquad \textit{Ans.}$$

Moment Summation. We require the moment about the x axis of the resultant force, Fig. 4–45b, to be equal to the sum of the moments about the x axis of all the forces in the system, Fig. 4–45a. The moment arms are determined from the y coordinates since these coordinates represent the *perpendicular distances* from the x axis to the lines of action of the forces. Using the right-hand rule, where positive moments act in the $+\mathbf{i}$ direction, we have

$$M_{R_x} = \Sigma M_x;$$
$$-(1400 \text{ N})y = 600 \text{ N}(0) + 100 \text{ N}(5 \text{ m}) - 400 \text{ N}(10 \text{ m}) + 500 \text{ N}(0)$$
$$-1400y = -3500 \qquad y = 2.50 \text{ m} \qquad \textit{Ans.}$$

In a similar manner, assuming that positive moments act in the $+\mathbf{j}$ direction, a moment equation can be written about the y axis using moment arms defined by the x coordinates of each force.

$$M_{R_y} = \Sigma M_y;$$
$$(1400 \text{ N})x = 600 \text{ N}(8 \text{ m}) - 100 \text{ N}(6 \text{ m}) + 400 \text{ N}(0) + 500 \text{ N}(0)$$
$$1400x = 4200 \qquad x = 3.00 \text{ m} \qquad \textit{Ans.}$$

Hence, a force of $F_R = 1400$ N placed at point $P(3.00 \text{ m}, 2.50 \text{ m})$ on the slab, Fig. 4–45b, is equivalent to the parallel force system acting on the slab in Fig. 4–45a.

E X A M P L E 4.19

Three parallel bolting forces act on the rim of the circular cover plate in Fig. 4–46a. Determine the magnitude and direction of a resultant force equivalent to the given force system and locate its point of application, P, on the cover plate.

Solution (*Vector Analysis*)

Force Summation. From Fig. 4–46a, the force resultant $\mathbf{F}_R$ is

$$\mathbf{F}_R = \Sigma\mathbf{F}; \qquad \mathbf{F}_R = -300\mathbf{k} - 200\mathbf{k} - 150\mathbf{k}$$

$$= \{-650\mathbf{k}\}\ \text{N} \qquad\qquad\qquad Ans.$$

Moment Summation. Choosing point O as a reference for computing moments and assuming that $\mathbf{F}_R$ acts at a point $P(x, y)$, Fig. 4–46b, we require

$$\mathbf{M}_{R_O} = \Sigma\mathbf{M}_O;$$
$$\mathbf{r} \times \mathbf{F}_R = \mathbf{r}_A \times (-300\mathbf{k}) + \mathbf{r}_B \times (-200\mathbf{k}) + \mathbf{r}_C \times (-150\mathbf{k})$$

$$(x\mathbf{i} + y\mathbf{j}) \times (-650\mathbf{k}) = (0.8\mathbf{i}) \times (-300\mathbf{k}) + (-0.8\mathbf{j}) \times (-200\mathbf{k})$$
$$+ (-0.8\sin 45°\mathbf{i} + 0.8\cos 45°\mathbf{j}) \times (-150\mathbf{k})$$

$$650x\mathbf{j} - 650y\mathbf{i} = 240\mathbf{j} + 160\mathbf{i} - 84.85\mathbf{j} - 84.85\mathbf{i}$$

Equating the corresponding $\mathbf{j}$ and $\mathbf{i}$ components yields

$$650x = 240 - 84.85 \qquad\qquad (1)$$
$$-650y = 160 - 84.85 \qquad\qquad (2)$$

Solving these equations, we obtain the coordinates of point P,

$$x = 0.239\ \text{m} \qquad y = -0.116\ \text{m} \qquad Ans.$$

The negative sign indicates that it was wrong to have assumed a $+y$ position for $\mathbf{F}_R$ as shown in Fig. 4–46b.

It is also possible to establish Eqs. 1 and 2 directly by summing moments about the y and x axes. Using the right-hand rule we have

$$M_{R_y} = \Sigma M_y; \qquad 650x = 300\ \text{N}\ (0.8\ \text{m}) - 150\ \text{N}\ (0.8\sin 45°\ \text{m})$$
$$M_{R_x} = \Sigma M_x; \qquad -650x = 200\ \text{N}\ (0.8\ \text{m}) - 150\ \text{N}\ (0.8\cos 45°\ \text{m})$$

(a)

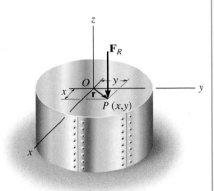

(b)

Fig. 4–46

PROBLEMS

4-98. Replace the force at A by an equivalent force and couple moment at point O.

4-99. Replace the force at A by an equivalent force and couple moment at point P.

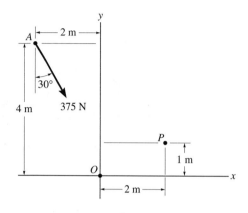

Probs. 4–98/99

*4-100. Replace the force and couple moment system by an equivalent force and couple moment acting at point O.

4-101. Replace the force and couple moment system by an equivalent force and couple moment acting at point P.

Probs. 4–100/101

4-102. Replace the force system by an equivalent force and couple moment at point O.

4-103. Replace the force system by an equivalent force and couple moment at point P.

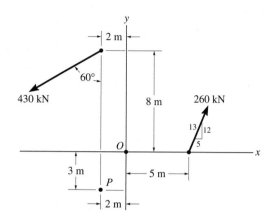

Probs. 4–102/103

*4-104. Replace the force and couple system by an equivalent force and couple moment acting at point O.

4-105. Replace the force and couple system by an equivalent force and couple moment acting at point P.

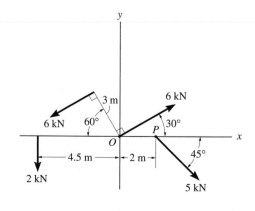

Probs. 4–104/105

4-106. Replace the force and couple system by an equivalent force and couple moment at point O.

4-107. Replace the force and couple system by an equivalent force and couple moment at point P.

4-110. Replace the force system acting on the beam by an equivalent force and couple moment at point A.

4-111. Replace the force system acting on the beam by an equivalent force and couple moment at point B.

Probs. 4–106/107

Probs. 4–110/111

*4-108.** Replace the force system by a single force resultant and specify its point of application, measured along the x axis from point O.

4-109. Replace the force system by a single force resultant and specify its point of application, measured along the x axis from point P.

*4-112.** Replace the three forces acting on the shaft by a single resultant force. Specify where the force acts, measured from end A.

4-113. Replace the three forces acting on the shaft by a single resultant force. Specify where the force acts, measured from end B.

Probs. 4–108/109

Probs. 4–112/113

4-114. Replace the loading on the frame by a single resultant force. Specify where its line of action intersects member AB, measured from A.

Prob. 4–114

4-117. Determine the magnitudes of $\mathbf{F}_1$ and $\mathbf{F}_2$ and the direction of $\mathbf{F}_1$ so that the loading creates a zero resultant force and couple moment on the wheel.

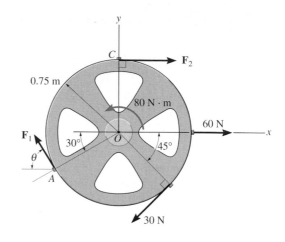

Prob. 4–117

4-115. Replace the loading acting on the beam by a single resultant force. Specify where the force acts, measured from end A.

*4-116. Replace the loading acting on the beam by a single resultant force. Specify where the force acts, measured from B.

4-118. The weights of the various components of the truck are shown. Replace this system of forces by an equivalent resultant force and couple moment acting at point A.

4-119. The weights of the various components of the truck are shown. Replace this system of forces by an equivalent resultant force and specify its location measured from point A.

Probs. 4–115/116

Probs. 4–118/119

*4-120. Replace the loading on the frame by a single resultant force. Specify where its line of action intersects member *AB*, measured from *A*.

4-121. Replace the loading on the frame by a single resultant force. Specify where its line of action intersects member *CD*, measured from end *C*.

Probs. 4–120/121

4-122. Replace the force system acting on the frame by an equivalent resultant force and specify where the resultant's line of action intersects member *AB*, measured from point *A*.

4-123. Replace the force system acting on the frame by an equivalent resultant force and specify where the resultant's line of action intersects member *BC*, measured from point *B*.

*4-124. Replace the force system acting on the frame by an equivalent resultant force and couple moment acting at point *A*.

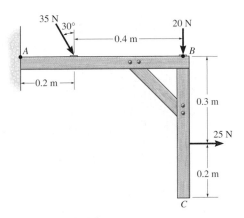

Probs. 4–122/123/124

4-125. Replace the force and couple-moment system by an equivalent resultant force and couple moment at point *O*. Express the results in Cartesian vector form.

4-126. Replace the force and couple-moment system by an equivalent resultant force and couple moment at point *P*. Express the results in Cartesian vector form.

4-127. Replace the force and couple-moment system by an equivalent resultant force and couple moment at point *Q*. Express the results in Cartesian vector form.

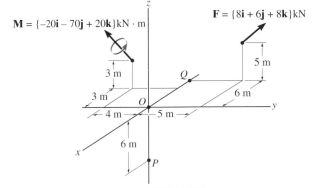

Probs. 4–125/126/127

*4-128. The belt passing over the pulley is subjected to forces F_1 and F_2, each having a magnitude of 40 N. F_1 acts in the $-\mathbf{k}$ direction. Replace these forces by an equivalent force and couple moment at point *A*. Express the result in Cartesian vector form. Set $\theta = 0°$ so that F_2 acts in the $-\mathbf{j}$ direction.

4-129. The belt passing over the pulley is subjected to two forces F_1 and F_2, each having a magnitude of 40 N. F_1 acts in the $-\mathbf{k}$ direction. Replace these forces by an equivalent force and couple moment at point *A*. Express the result in Cartesian vector form. Take $\theta = 45°$.

Probs. 4–128/129

4-130. Replace the force system by an equivalent force and couple moment at point A.

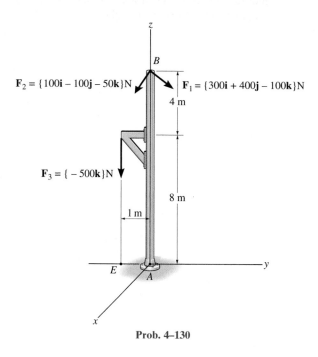

$F_2 = \{100i - 100j - 50k\}N$

$F_1 = \{300i + 400j - 100k\}N$

4 m

$F_3 = \{-500k\}N$

8 m

1 m

B

z

E

A

y

x

Prob. 4–130

4-131. The slab is to be hoisted using the three slings shown. Replace the system of forces acting on slings by an equivalent force and couple moment at point O. The force F_1, is vertical.

$F_3 = 4$ kN

$F_2 = 5$ kN

45°

60°

60°

O

$F_1 = 6$ kN

45°

30°

6 m

2 m

2 m

z

x

y

Prob. 4–131

*4-132. A biomechanical model of the lumber region of the human trunk is shown. The forces acting in the four muscle groups consist of $F_R = 35$ N for the rectus, $F_O = 45$ N for the oblique, $F_L = 23$ N for the lumbar latissimus dorsi, and $F_E = 32$ N for the erector spinae. These loadings are symmetric with respect to the y-z plane. Replace this system of parallel forces by an equivalent force and couple moment acting at the spine, point O. Express the results in Cartesian vector form.

z

F_R

F_O

F_R

F_E

F_E

F_L

F_O

F_L

O

75 mm

15 mm

45 mm

50 mm

40 mm

30 mm

x

y

Prob. 4–132

4-133. The building slab is subjected to four parallel column loadings. Determine the equivalent resultant force and specify its location (x, y) on the slab. Take $F_1 = 30$ kN, $F_2 = 40$ kN.

4-134. The building slab is subjected to four parallel column loadings. Determine the equivalent resultant force and specify its location (x, y) on the slab. Take $F_1 = 20$ kN, $F_2 = 50$ kN.

z

20 kN

50 kN

F_1

F_2

x

3 m

8 m

2 m

4 m

6 m

y

Probs. 4–133/134

4-135. Replace the two wrenches and the force, acting on the pipe assembly, by an equivalent resultant force and couple moment at point O.

Prob. 4–135

4-137. Replace the three forces acting on the plate by a wrench. Specify the magnitude of the force and couple moment for the wrench and the point $P(x, y)$ where its line of action intersects the plate.

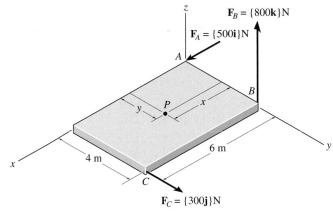

Prob. 4–137

*4-136. The three forces acting on the block each have a magnitude of 10 N. Replace this system by a wrench and specify the point where the wrench intersects the z axis, measured from point O.

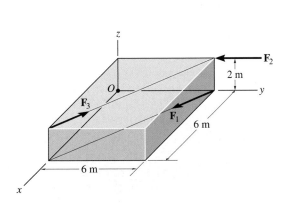

Prob. 4–136

4-138. Replace the three forces acting on the plate by a wrench. Specify the magnitude of the force and couple moment for the wrench and the point $P(y, z)$ where its line of action intersects the plate.

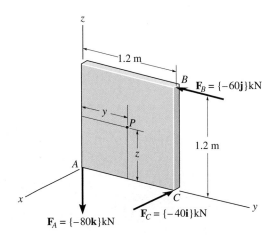

Prob. 4–138

4.10 Reduction of a Simple Distributed Loading

(a)

(b)

(c)

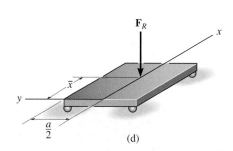

(d)

Fig. 4–47

In many situations a very large surface area of a body may be subjected to *distributed loadings* such as those caused by wind, fluids, or simply the weight of material supported over the body's surface. The *intensity* of these loadings at each point on the surface is defined as the *pressure p* (force per unit area), which can be measured in unit of pascals (Pa), where $1 \text{ Pa} = 1 \text{ N/m}^2$.

In this section we will consider the most common case of a distributed pressure loading, which is *uniform* along one axis of a flat rectangular body upon which the loading is applied.* An example of such a loading is shown in Fig. 4–47a. The direction of the intensity of the pressure load is indicated by arrows shown on the *load-intensity diagram*. The entire loading on the plate is therefore a system of parallel forces, infinite in number and each acting on a separate differential area of the plate. Here the *loading function*, $p = p(x)$ Pa, is only a function of x since the pressure is uniform along the y axis. If we multiply $p = p(x)$ by the *width* a m of the plate, we obtain $w = [p(x) \text{ N/m}^2]a \text{ m} = w(x) \text{ N/m}$. This loading function, shown in Fig. 4–47b, is a measure of load distribution along the line $y = 0$ which is in the plane of symmetry of the loading, Fig. 4–47a. As noted, it is measured as a force per unit length, rather than a force per unit area. Consequently, the load-intensity diagram for $w = w(x)$ can be represented by a system of *coplanar* parallel forces, shown in two dimensions in Fig. 4–47b. Using the methods of Sec. 4.9, this system of forces can be simplified to a single resultant force $\mathbf{F}_R$ and its location $\bar{x}$ can be specified, Fig. 4–47c.

Magnitude of Resultant Force. From Eq. 4–17 ($F_R = \Sigma F$), the magnitude of $\mathbf{F}_R$ is equivalent to the sum of all the forces in the system. In this case integration must be used since there is an infinite number of parallel forces $d\mathbf{F}$ acting along the plate, Fig. 4–47b. Since $d\mathbf{F}$ is acting on an element of length dx and $w(x)$ is a force per unit length, then at the location x, $dF = w(x) \, dx = dA$. In other words, the magnitude of $d\mathbf{F}$ is determined from the colored differential *area dA* under the loading curve. For the entire plate length,

$$+\downarrow F_R = \Sigma F; \qquad \boxed{F_R = \int_L w(x) \, dx = \int_A dA = A} \qquad (4\text{--}19)$$

Hence, the magnitude of the resultant force is equal to the total area A under the loading diagram $w = w(x)$, Fig. 4–47c.

---

*The more general case of a nonuniform surface loading acting on a body is considered in Sec. 9.5.

Location of Resultant Force.   Applying Eq. 4–17 ($M_{R_O} = \Sigma M_O$), the location $\overline{x}$ of the line of action of $\mathbf{F}_R$ can be determined by equating the moments of the force resultant and the force distribution about point O (the y axis). Since $d\mathbf{F}$ produces a moment of $x\, dF = x\, w(x)\, dx$ about O, Fig. 4-47b, then for the entire plate, Fig. 4-47c,

$$\uparrow + M_{R_O} = \Sigma M_O; \qquad \overline{x} F_R = \int_L x\, w(x)\, dx$$

Solving for $\overline{x}$, using Eq. 4–19, we can write

$$\overline{x} = \frac{\displaystyle\int_L x\, w(x)\, dx}{\displaystyle\int_L w(x)\, dx} = \frac{\displaystyle\int_A x\, dA}{\displaystyle\int_A dA} \qquad (4\text{–}20)$$

This equation represents the x coordinate for the geometric center or *centroid* of the *area* under the distributed-loading diagram $w(x)$. *Therefore, the resultant force has a line of action which passes through the centroid C (geometric center) of the area defined by the distributed-loading diagram* $w(x)$, Fig. 4–47c.

Once $\overline{x}$ is determined, $\mathbf{F}_R$ by symmetry passes through point $(\overline{x}, 0)$ on the surface of the plate, Fig. 4–47d. If we now consider the three-dimensional pressure loading $p(x)$, Fig. 4–47a, we can therefore conclude that *the resultant force has a magnitude equal to the volume under the distributed-loading curve* $p = p(x)$ *and a line of action which passes through the centroid (geometric center) of this volume.* Detailed treatment of the integration techniques for computing the centroids of volumes or areas is given in Chapter 9. In many cases, however, the distributed-loading diagram is in the shape of a rectangle, triangle, or some other simple geometric form. The centroids for such common shapes do not have to be determined from Eq. 4–20; rather, they can be obtained directly from the tabulation given on the inside back cover.

The beam supporting this stack of lumber is subjected to a *uniform* distributed loading, and so the load-intensity diagram has a rectangular shape. If the load intensity is w_0, then the resultant force is determined from the area of the rectangle, $F_R = w_0 b$. The line of action of this force passes through the centroid or center of this area, $\overline{x} = a + b/2$. This resultant is equivalent to the distributed load, and so both loadings produce the same "external" effects or support reactions on the beam.

IMPORTANT POINTS

- Distributed loadings are defined by using a loading function $w = w(x)$ that indicates the intensity of the loading along the length of the member. This intensity is measured in N/m.

- The external effects caused by a coplanar distributed load acting on a body can be represented by a single resultant force.

- The resultant force is equivalent to the *area* under the distributed loading diagram, and has a line of action that passes through the *centroid* or geometric center of this area.

E X A M P L E 4.20

(a)

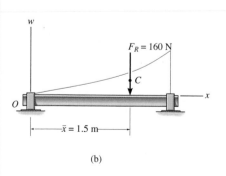

(b)

Fig. 4–48

Determine the magnitude and location of the equivalent resultant force acting on the shaft in Fig. 4–48a.

Solution

Since $w = w(x)$ is given, this problem will be solved by integration. The colored differential area element $dA = w\,dx = 60x^2\,dx$. Applying Eq. 4–19, by summing these elements from $x = 0$ to $x = 2$ m, we obtain the resultant force $\mathbf{F}_R$.

$$F_R = \Sigma F;$$

$$F_R = \int_A dA = \int_0^2 60x^2\,dx = 60\left[\frac{x^3}{3}\right]_0^2 = 60\left[\frac{2^3}{3} - \frac{0^3}{3}\right]$$

$$= 160 \text{ N} \qquad\qquad Ans.$$

Since the element of area dA is located an arbitrary distance x from O, the location $\bar{x}$ of $\mathbf{F}_R$ *measured from* O, Fig. 4–48b, is determined from Eq. 4–20.

$$\bar{x} = \frac{\int_A x\,dA}{\int_A dA} = \frac{\int_0^2 x(60x^2)dx}{160} = \frac{60\left[\frac{x^4}{4}\right]_0^2}{160} = \frac{60\left[\frac{2^4}{4} - \frac{0^4}{4}\right]}{160}$$

$$= 1.5 \text{ m} \qquad\qquad Ans.$$

These results may be checked by using the table on the inside back cover, where it is shown that for an exparabolic area of length a, height b, and shape shown in Fig. 4–48a,

$$A = \frac{ab}{3} = \frac{2\text{ m}(240\text{ N/m})}{3} = 160 \text{ N and } \bar{x} = \frac{3}{4}a = \frac{3}{4}(2\text{ m}) = 1.5 \text{ m}$$

EXAMPLE 4.21

A distributed loading of $p = 800x$ Pa acts over the top surface of the beam shown in Fig. 4–49a. Determine the magnitude and location of the equivalent resultant force.

7200 Pa

$p = 800x$ Pa

x

p

y

x

9 m

0.2 m

(a)

Fig. 4–49

Solution

The loading function $p = 800x$ Pa indicates that the load intensity varies uniformly from $p = 0$ at $x = 0$ to $p = 7200$ Pa at $x = 9$ m. Since the intensity is uniform along the width of the beam (the y axis), the loading may be viewed in two dimensions as shown in Fig. 4–49b. Here

w

$w = 160x$ N/m

1440 N/m

x

x

9 m

(b)

$$w = (800x \text{ N/m}^2)(0.2 \text{ m})$$
$$= (160x) \text{ N/m}$$

At $x = 9$ m, note that $w = 1440$ N/m. Although we may again apply Eqs. 4–19 and 4–20 as in Example 4.20, it is simpler to use the table on the inside back cover.

The magnitude of the resultant force is equivalent to the area under the triangle.

$$F_R = \tfrac{1}{2}(9 \text{ m})(1440 \text{ N/m}) = 6480 \text{ N} = 6.48 \text{ kN} \qquad Ans.$$

The line of action of $\mathbf{F}_R$ passes through the *centroid C* of the triangle. Hence,

$$\bar{x} = 9 \text{ m} - \tfrac{1}{3}(9 \text{ m}) = 6 \text{ m} \qquad Ans.$$

$F_R = 6.48$ kN

$\bar{x} = 6$ m

3 m

C

(c)

The results are shown in Fig. 4–49c.

We may also view the resultant $\mathbf{F}_R$ as *acting* through the *centroid* of the *volume* of the loading diagram $p = p(x)$ in Fig. 4–49a. Hence $\mathbf{F}_R$ intersects the x–y plane at the point (6 m, 0). Furthermore, the magnitude of $\mathbf{F}_R$ is equal to the volume under the loading diagram; i.e.,

$$F_R = V = \tfrac{1}{2}(7200 \text{ N/m}^2)(9 \text{ m})(0.2 \text{ m}) = 6.48 \text{ kN} \qquad Ans.$$

E X A M P L E 4.22

The granular material exerts the distributed loading on the beam as shown in Fig. 4–50a. Determine the magnitude and location of the equivalent resultant of this load.

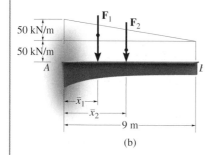

(a)

Solution

The area of the loading diagram is a *trapezoid*, and therefore the solution can be obtained directly from the area and centroid formulas for a trapezoid listed on the inside back cover. Since these formulas are not easily remembered, instead we will solve this problem by using "composite" areas. In this regard, we can divide the trapezoidal loading into a rectangular and triangular loading as shown in Fig. 4–50b. The magnitude of the force represented by each of these loadings is equal to its associated *area*,

$$F_1 = \tfrac{1}{2}(9 \text{ m})(50 \text{ kN/m}) = 225 \text{ kN}$$

$$F_2 = \tfrac{1}{2}(9 \text{ m})(50 \text{ kN/m}) = 450 \text{ kN}$$

The lines of action of these parallel forces act through the *centroid* of their associated areas and therefore intersect the beam at

$$\bar{x}_1 = \tfrac{1}{3}(9 \text{ m}) = 3 \text{ m}$$

$$\bar{x}_2 = \tfrac{1}{2}(9 \text{ m}) = 4.5 \text{ m}$$

(b)

The two parallel forces $\mathbf{F}_1$ and $\mathbf{F}_2$ can be reduced to a single resultant $\mathbf{F}_R$. The magnitude of $\mathbf{F}_R$ is

$$+\downarrow F_R = \Sigma F; \qquad F_R = \ 225 + 450 \ = 675 \text{ kN} \qquad \textit{Ans.}$$

With reference to point A, Fig. 4–50b and 4–50c, we can find the location of $\mathbf{F}_R$. We require

$$\zeta + M_{R_A} = \Sigma M_A; \qquad \bar{x}(675) = 3(225) + 4.5(450)$$

$$\bar{x} = 4 \text{ m} \qquad \textit{Ans.}$$

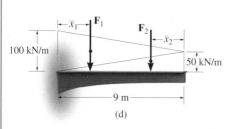

(c)

Note: The trapezoidal area in Fig. 4–50a can also be divided into two triangular areas as shown in Fig. 4–50d. In this case

$$F_1 = \tfrac{1}{2}(9 \text{ m})(100 \text{ kN/m}) = 450 \text{ kN}$$

$$F_2 = \tfrac{1}{2}(9 \text{ m})(50 \text{ kN/m}) = 225 \text{ kN}$$

and

$$\bar{x}_1 = \tfrac{1}{3}(9 \text{ m}) = 3 \text{ m}$$

$$\bar{x}_2 = \tfrac{1}{3}(9 \text{ m}) = 3 \text{ m}$$

(d)

Fig. 4–50

Using these results, show that again $F_R = 675 \text{ kN}$ and $\bar{x} = 4 \text{ m}$.

PROBLEMS

4-139. The loading on the bookshelf is distributed as shown. Determine the magnitude of the equivalent resultant location, measured from point O.

30 N/m

50 N/m

A

O

1 m

1.5 m

0.5 m

Prob. 4–139

*4-140. The masonry support creates the loading distribution acting on the end of the beam. Simplify this load to a single resultant force and specify its location measured from point O.

0.3 m

O

1 kN/m

2.5 kN/m

Prob. 4–140

4-141. Replace the loading by an equivalent force and couple moment acting at point O.

4-142. Replace the loading by a single resultant force, and specify the location of the force on the beam measured from point O.

6 kN/m

15 kN

500 kN · m

O

7.5 m

4.5 m

Probs. 4–141/142

4-143. The column is used to support the floor which exerts a force of 30000 N on the top of the column. The effect of soil pressure along its side is distributed as shown. Replace this loading by an equivalent resultant force and specify where it acts along the column, measured from its base A.

30000 N

800 N/m

3 m

2000 N/m

A

Prob. 4–143

*4-144. Replace the loading by an equivalent force and couple moment acting at point O.

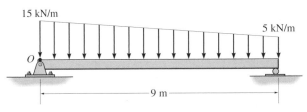

15 kN/m

5 kN/m

O

9 m

Prob. 4–144

4-145. Replace the distributed loading by an equivalent resultant force, and specify its location on the beam, measured from the pin at C.

C

A

30°

B

8000 N/m

5 m

5 m

Prob. 4–145

4-146. Replace the loading by an equivalent force and couple moment acting at point O.

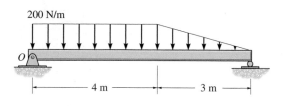

200 N/m

4 m   3 m

Prob. 4–146

4-147. The bricks on top of the beam and the supports at the bottom create the distributed loading shown in the second figure. Determine the required intensity w and dimension d of the right support so that the resultant force and couple moment about point A of the system are both zero.

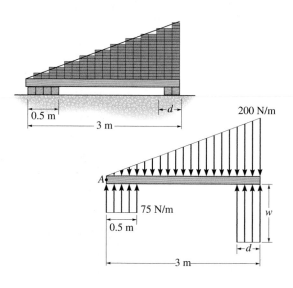

0.5 m
3 m

200 N/m
A
75 N/m
w
0.5 m
d
3 m

Prob. 4–147

*4-148. Replace the distributed loading by an equivalent resultant force and specify its location, measured from point A.

800 N/m
200 N/m
A
B
2 m   3 m

Prob. 4–148

4-149. The distribution of soil loading on the bottom of a building slab is shown. Replace this loading by an equivalent resultant force and specify its location, measured from point O.

O
500 N/m
1000 N/m
3000 N/m
4 m   3 m

Prob. 4–149

4-150. The beam is subjected to the distributed loading. Determine the length b of the uniform load and its position a on the beam such that the resultant force and couple moment acting on the beam are zero.

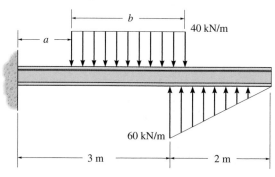

b
40 kN/m
a
60 kN/m
3 m   2 m

Prob. 4–150

4-151. Replace the loading by an equivalent resultant force and specify its location on the beam, measured from point B.

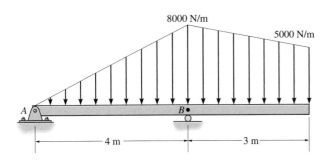

8000 N/m

5000 N/m

A

B

4 m

3 m

Prob. 4–151

*4-152. Replace the distributed loading by an equivalent resultant force and specify where its line of action intersects member AB, measured from A.

4-153. Replace the distributed loading by an equivalent resultant force and specify where its line of action intersects member BC, measured from C.

200 N/m

100 N/m

B

C

6 m

5 m

200 N/m

A

Probs. 4–152/153

4-154. Replace the loading by an equivalent resultant force and couple moment acting at point O.

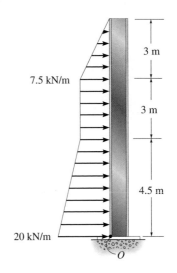

3 m

7.5 kN/m

3 m

4.5 m

20 kN/m

O

Prob. 4–154

4-155. Wet concrete exerts a pressure distribution along the wall of the form. Determine the resultant force of this distribution and specify the height h where the bracing strut should be placed so that it lies through the line of action of the resultant force. The wall has a width of 5 m.

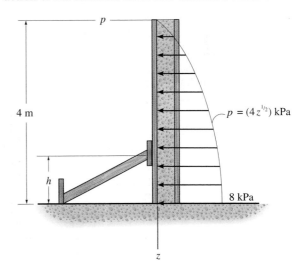

p

4 m

$p = (4z^{1/2})$ kPa

h

8 kPa

z

Prob. 4–155

*4-156. Wind has blown sand over a platform such that the intensity of the load can be approximated by the function $w = (0.5x^3)$ N/m. Simplify this distributed loading to an equivalent resultant force and specify the magnitude and location of the force, measured from A.

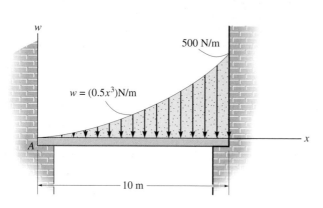

Prob. 4–156

4-157. Replace the loading by an equivalent force and couple moment acting at point O.

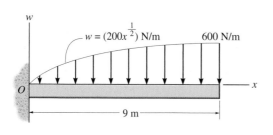

Prob. 4–157

4-158. The lifting force along the wing of a jet aircraft consists of a uniform distribution along AB, and a semiparabolic distribution along BC with origin at B. Replace this loading by a single resultant force and specify its location measured from point A.

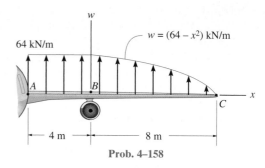

Prob. 4–158

4-159. Determine the magnitude of the equivalent resultant force of the distributed load and specify its location on the beam measured from point A.

Prob. 4–159

*4-160. Determine the equivalent resultant force of the distributed loading and its location, measured from point A. Evaluate the integrals using Simpson's rule.

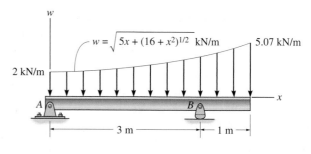

Prob. 4–160

CHAPTER REVIEW

- *Moment of a Force.* A force produces a turning effect about a point O that does not lie on its line of action. In scalar form, the moment *magnitude* is $M_O = Fd$, where d is the moment arm or perpendicular distance from point O to the line of action of the force. The *direction* of the moment is defined using the right-hand rule. Rather than finding d, it is normally easier to resolve the force into its x and y components, determine the moment of each component about the point, and then sum the results. Since three-dimensional geometry is generally more difficult to visualize, the vector cross product can be used to determine the moment, $\mathbf{M}_O = \mathbf{r} \times \mathbf{F}$, where $\mathbf{r}$ is a position vector that extends from point O to any point on the line of action of $\mathbf{F}$.

- *Moment about a Specified Axis.* If the moment of a force is to be determined about an arbitrary axis, then the projection of the moment onto the axis must be obtained. Provided the distance d_a that is perpendicular to *both* the line of action of the force and the axis can be determined, then the moment of the force about the axis is simply $M_a = F\, d_a$. If this distance d_a cannot be found, then the vector triple product should be used, where $M_a = \mathbf{u}_a \cdot \mathbf{r} \times \mathbf{F}$. Here $\mathbf{u}_a$ is the unit vector that specifies the direction of the axis and $\mathbf{r}$ is a position vector that is directed from any point on the axis to any point on the line of action of the force.

- *Couple Moment.* A couple consists of two equal but opposite forces that act a perpendicular distance d apart. Couples tend to produce a rotation without translation. The moment of the couple is determined from $M = Fd$, and its direction is established using the right-hand rule. If the vector cross product is used to determine the moment of the couple then $\mathbf{M} = \mathbf{r} \times \mathbf{F}$. Here $\mathbf{r}$ extends from any point on the line of action of one of the forces to any point on the line of action of the force $\mathbf{F}$ used in the cross product.

- *Reduction of a Force and Couple System.* Any system of forces and couples can be reduced to a single resultant force and resultant couple moment acting at a point. The resultant force is the sum of all the forces in the system, and the resultant couple moment is equal to the sum of all the forces and couple moments about the point. Further simplification to a single resultant force is possible provided the force system is *concurrent, coplanar,* or *parallel.* For this case, to find the location of the resultant force from a point, it is necessary to equate the moment of the resultant force about the point to the moment of the forces and couples in the system about the same point. Doing this for any *other type* of force system would yield a *wrench*, which consists of the resultant force and a resultant collinear couple moment.

- *Distributed Loading.* A simple distributed loading can be replaced by a *resultant force*, which is equivalent to the *area* under the loading curve. This resultant has a line of action that passes through the *centroid* or geometric center of the area or volume under the loading diagram.

REVIEW PROBLEMS

4-161. Determine the coordinate direction angles α, β, γ of **F**, which is applied to the end A of the pipe assembly, so that the moment of **F** about O is zero.

4-162. Determine the moment of the force **F** about point O. The force has coordinate direction angles of $\alpha = 60°$, $\beta = 120°$, $\gamma = 45°$. Express the result as a Cartesian vector.

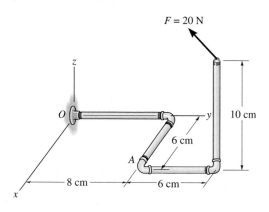

Probs. 4–161/162

4-163. If it takes a force of $F = 500$ N to pull the nail out, determine the smallest vertical force **P** that must be applied to the handle of the crowbar. *Hint:* This requires the moment of **F** about point A to be equal to the moment of **P** about A. Why?

Prob. 4–163

*__4-164.** Determine the moment of the force $\mathbf{F}_c$ about the door hinge at A. Express the result as a Cartesian vector.

4-165. Determine the magnitude of the moment of the force $\mathbf{F}_c$ about the hinged axis aa of the door.

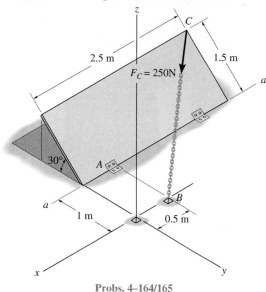

Probs. 4–164/165

4-166. Determine the resultant couple moment of the two couples that act on the assembly. Member OB lies in the x-z plane.

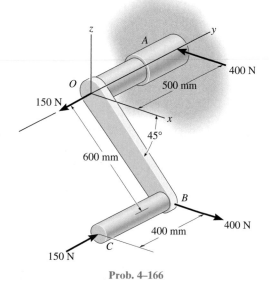

Prob. 4–166

4-167. Replace the force **F** having a magnitude of $F = 50$ kN and acting at point A by an equivalent force and couple moment at point C.

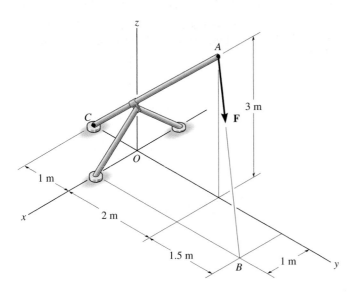

Prob. 4–167

***4-168.** The horizontal 30-N force acts on the handle of the wrench. What is the magnitude of the moment of this force about the z axis?

Prob. 4–168

4-169. The horizontal 30-N force acts on the handle of the wrench. Determine the moment of this force about point O. Specify the coordinate direction angles α, β, γ of the moment axis.

Prob. 4–169

4-170. The forces and couple moments that are exerted on the toe and heel plates of a snow ski are $\mathbf{F}_t = \{-50\mathbf{i}+80\mathbf{j}-158\mathbf{k}\}$ N, $\mathbf{M}_t = \{-6\mathbf{i}+4\mathbf{j}+2\mathbf{k}\}$ N·m, and $\mathbf{F}_h = \{-20\mathbf{i}+60\mathbf{j}-250\mathbf{k}\}$ N, $\mathbf{M}_h = \{-20\mathbf{i}+8\mathbf{j}+3\mathbf{k}\}$ N·m, respectively. Replace this system by an equivalent force and couple moment acting at point P. Express the results in Cartesian vector form.

Prob. 4–170

The tower crane is subjected to its weight and the load it supports.
In order to calculate the support reactions for the crane, it is necessary to
apply the principles of equilibrium.

Equilibrium of a Rigid Body

- To develop the equations of equilibrium for a rigid body.
- To introduce the concept of the free-body diagram for a rigid body.
- To show how to solve rigid body equilibrium problems using the equations of equilibrium.

5.1 Conditions for Rigid-Body Equilibrium

In this section we will develop both the necessary and sufficient conditions required for equilibrium of a rigid body. To do this, consider the rigid body in Fig. 5–1a, which is fixed in the x, y, z reference and is either at rest or moves with the reference at constant velocity. A free-body diagram of the arbitrary ith particle of the body is shown in Fig. 5–1b. There are two types of forces which act on it. The resultant *internal force*, $\mathbf{f}_i$, is caused by interactions with adjacent particles. The resultant *external force* $\mathbf{F}_i$ represents, for example, the effects of gravitational, electrical, magnetic, or contact forces between the ith particle and adjacent bodies or particles *not* included within the body. If the particle is in equilibrium, then applying Newton's first law we have

$$\mathbf{F}_i + \mathbf{f}_i = \mathbf{0}$$

When the equation of equilibrium is applied to each of the other particles of the body, similar equations will result. If all these equations are added together *vectorially*, we obtain

$$\Sigma\mathbf{F}_i + \Sigma\mathbf{f}_i = \mathbf{0}$$

The summation of the internal forces will equal zero since the internal forces between particles within the body will occur in equal

(a)

(b)

Fig. 5–1

(a)

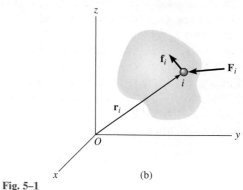

(b)

Fig. 5–1

but opposite collinear pairs, Newton's third law. Consequently, only the sum of the *external forces* will remain; and therefore, letting $\Sigma \mathbf{F}_i = \Sigma \mathbf{F}$, the above equation can be written as

$$\Sigma \mathbf{F} = \mathbf{0}$$

Let us now consider the moments of the forces acting on the ith particle about the arbitrary point O, Fig. 5–1b. Using the above particle equilibrium equation and the distributive law of the vector cross product we have

$$\mathbf{r}_i \times (\mathbf{F}_i + \mathbf{f}_i) = \mathbf{r}_i \times \mathbf{F}_i + \mathbf{r}_i \times \mathbf{f}_i = \mathbf{0}$$

Similar equations can be written for the other particles of the body, and adding them together vectorially, we obtain

$$\Sigma \mathbf{r}_i \times \mathbf{F}_i + \Sigma \mathbf{r}_i \times \mathbf{f}_i = \mathbf{0}$$

The second term is zero since, as stated above, the internal forces occur in equal but opposite collinear pairs, and therefore the resultant moment of each pair of forces about point O is zero. Hence, using the notation $\Sigma \mathbf{M}_O = \Sigma \mathbf{r}_i \times \mathbf{F}_i$, we have

$$\Sigma \mathbf{M}_O = \mathbf{0}$$

Hence the two *equations of equilibrium* for a rigid body can be summarized as follows:

$$\Sigma \mathbf{F} = \mathbf{0}$$
$$\Sigma \mathbf{M}_O = \mathbf{0}$$

(5–1)

These equations require that a rigid body will remain in equilibrium provided the sum of all the *external forces* acting on the body is equal to zero and the sum of the moments of the external forces about a point is equal to zero. The fact that these conditions are *necessary* for equilibrium has now been proven. They are also *sufficient* for maintaining equilibrium. To show this, let us assume that the body is in equilibrium and the force system acting on the body satisfies Eqs. 5–1. Suppose that an *additional force* $\mathbf{F}'$ is applied to the body. As a result, the equilibrium equations become

$$\Sigma \mathbf{F} + \mathbf{F}' = \mathbf{0}$$
$$\Sigma \mathbf{M}_O + \mathbf{M}'_O = \mathbf{0}$$

where $\mathbf{M}'_O$ is the moment of $\mathbf{F}'$ about O. Since $\Sigma \mathbf{F} = \mathbf{0}$ and $\Sigma \mathbf{M}_O = \mathbf{0}$, then we require $\mathbf{F}' = \mathbf{0}$ (also $\mathbf{M}'_O = \mathbf{0}$). Consequently, the additional force $\mathbf{F}'$ is not required, and indeed Eqs. 5–1 are also sufficient conditions for maintaining equilibrium.

Many types of engineering problems involve symmetric loadings and can be solved by projecting all the forces acting on a body onto a single plane. Hence, in the next section, the equilibrium of a body subjected to a *coplanar* or *two-dimensional force system* will be considered. Ordinarily the geometry of such problems is not very complex, so a scalar solution is suitable for analysis. The more general discussion of rigid bodies subjected to *three-dimensional force systems* is given in the latter part of this chapter. It will be seen that many of these types of problems can best be solved by using vector analysis.

Equilibrium in Two Dimensions

5.2   Free-Body Diagrams

Successful application of the equations of equilibrium requires a complete specification of *all* the known and unknown external forces that act *on* the body. The best way to account for these forces is to draw the body's free-body diagram. This diagram is a sketch of the outlined shape of the body, which represents it as being *isolated* or "free" from its surroundings, i.e., a "free body." On this sketch it is necessary to show *all* the forces and couple moments that the surroundings exert *on the body* so that these effects can be accounted for when the equations of equilibrium are applied. For this reason, *a thorough understanding of how to draw a free-body diagram is of primary importance for solving problems in mechanics.*

TABLE 5–1 • Supports for Rigid Bodies Subjected to Two-Dimensional Force Systems

Types of Connection	Reaction	Number of Unknowns
(1) cable		One unknown. The reaction is a tension force which acts away from the member in the direction of the cable.
(2) weightless link	or	One unknown. The reaction is a force which acts along the axis of the link.
(3) roller		One unknown. The reaction is a force which acts perpendicular to the surface at the point of contact.
(4) roller or pin in confined smooth slot	or	One unknown. The reaction is a force which acts perpendicular to the slot.
(5) rocker		One unknown. The reaction is a force which acts perpendicular to the surface at the point of contact.
(6) smooth contacting surface		One unknown. The reaction is a force which acts perpendicular to the surface at the point of contact.
(7) member pin connected to collar on smooth rod	or	One unknown. The reaction is a force which acts perpendicular to the rod.

continued

·TABLE 5–1 • *Continued*

Types of Connection	Reaction	Number of Unknowns
(8) θ smooth pin or hinge	F_y F_x or F ϕ	Two unknowns. The reactions are two components of force, or the magnitude and direction ϕ of the resultant force. Note that ϕ and θ are not necessarily equal [usually not, unless the rod shown is a link as in (2)].
(9) member fixed connected to collar on smooth rod	F M	Two unknowns. The reactions are the couple moment and the force which acts perpendicular to the rod.
(10) fixed support	F_y F_x M or F ϕ M	Three unknowns. The reactions are the couple moment and the two force components, or the couple moment and the magnitude and direction ϕ of the resultant force.

Support Reactions. Before presenting a formal procedure as to how to draw a free-body diagram, we will first consider the various types of reactions that occur at supports and points of support between bodies subjected to coplanar force systems. *As a general rule, if a support prevents the translation of a body in a given direction, then a force is developed on the body in that direction. Likewise, if rotation is prevented, a couple moment is exerted on the body.*

For example, let us consider three ways in which a horizontal member, such as a beam, is supported at its end. One method consists of a *roller* or cylinder, Fig. 5–2*a*. Since this support only prevents the beam from *translating* in the vertical direction, the roller can only exert a *force* on the beam in this direction, Fig. 5–2*b*.

The beam can be supported in a more restrictive manner by using a *pin* as shown in Fig. 5–3*a*. The pin passes through a hole in the beam and two leaves which are fixed to the ground. Here the pin can prevent *translation* of the beam in *any direction* ϕ, Fig. 5–3*b*, and so the pin must exert a *force* **F** on the beam in this direction. For purposes of analysis, it is generally easier to represent this resultant force **F** by its two components F_x and F_y, Fig. 5–3*c*. If F_x and F_y are known, then F and ϕ can be calculated.

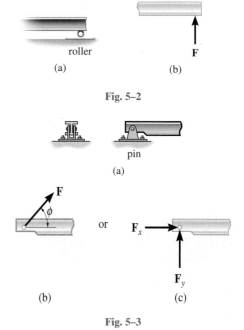

roller

(a)

F

(b)

Fig. 5–2

pin

(a)

F ϕ or F_x

F_y

(b)

(c)

Fig. 5–3

fixed support

(a)

M

F_x

F_y

(b)

Fig. 5–4

The most restrictive way to support the beam would be to use a *fixed support* as shown in Fig. 5–4*a*. This support will prevent both *translation and rotation* of the beam, and so to do this a *force and couple moment* must be developed on the beam at its point of connection, Fig. 5–4*b*. As in the case of the pin, the force is usually represented by its components F_x and F_y.

Table 5–1 lists other common types of supports for bodies subjected to coplanar force systems. (In all cases the angle θ is assumed to be known.) Carefully study each of the symbols used to represent these supports and the types of reactions they exert on their contacting members. Although concentrated forces and couple moments are shown in this table, they actually represent the *resultants* of small *distributed surface loads* that exist between each support and its contacting member. It is these *resultants* which will be determined from the equations of equilibrium.

Typical examples of actual supports that are referenced to Table 5–1 are shown in the following sequence of photos.

The cable exerts a force on the bracket in the direction of the cable. (1)

The rocker support for this bridge girder allows horizontal movement so the bridge is free to expand and contract due to temperature. (5)

This concrete girder rests on the ledge that is assumed to act as a smooth contacting surface. (6)

This utility building is pin supported at the top of the column. (8)

The floor beams of this building are welded together and thus form fixed connections. (10)

External and Internal Forces.   Since a rigid body is a composition of particles, both *external* and *internal* loadings may act on it. It is important to realize, however, that if the free-body diagram for the body is drawn, the forces that are *internal* to the body are *not represented* on the free-body diagram. As discussed in Sec. 5.1, these forces always occur in equal but opposite collinear pairs, and therefore their *net effect* on the body is zero.

In some problems, a free-body diagram for a "system" of connected bodies may be used for an analysis. An example would be the free-body diagram of an entire automobile (system) composed of its many parts. Obviously, the connecting forces between its parts would represent *internal forces* which would *not* be included on the free-body diagram of the automobile. To summarize, internal forces act between particles which are contained within the boundary of the free-body diagram. Particles or bodies outside this boundary exert external forces on the system, and these alone must be shown on the free-body diagram.

Weight and the Center of Gravity.   When a body is subjected to a gravitational field, then each of its particles has a specified weight. For the entire body it is appropriate to consider these gravitational forces to be represented as a *system of parallel forces* acting on all the particles contained within the boundary of the body. It was shown in Sec. 4.9 that such a system can be reduced to a single resultant force acting through a specified point. We refer to this force resultant as the *weight* **W** of the body and to the location of its point of application as the *center of gravity*. The methods used for its calculation will be developed in Chapter 9.

In the examples and problems that follow, if the weight of the body is important for the analysis, this force will then be reported in the problem statement. Also, when the body is *uniform* or made of homogeneous material, the center of gravity will be located at the body's *geometric center* or *centroid*; however, if the body is nonhomogeneous or has an unusual shape, then the location of its center of gravity will be given.

Idealized Models.   In order to perform a correct force analysis of any object, it is important to consider a corresponding analytical or idealized model that gives results that approximate as closely as possible the actual situation. To do this, careful choices have to be made so that selection of the type of supports, the material behavior, and the object's dimensions can be justified. This way the engineer can feel confident that any design or analysis will yield results which can be trusted. In complex cases this process may require developing several different models of the object that must be analyzed, but in any case, this selection process requires both skill and experience.

(a)

(b)

Fig. 5–5

To illustrate what is required to develop a proper model, we will now consider a few cases. As shown in Fig. 5–5a, the steel beam is to be used to support the roof joists of a building. For a force analysis it is reasonable to assume the material is rigid since only very small deflections will occur when the beam is loaded. A bolted connection at A will allow for any slight rotation that occurs when the load is applied, and so a *pin* can be considered for this support. At B a *roller* can be considered since the support offers no resistance to horizontal movement here. Building code requirements are used to specify the roof loading which results in a calculation of the joist loads $\mathbf{F}$. These forces will be larger than any actual loading on the beam since they account for extreme loading cases and for dynamic or vibrational effects. The weight of the beam is generally neglected when it is small compared to the load the beam supports. The idealized model of the beam is shown with average dimensions a, b, c, and d in Fig. 5–5b.

As a second case, consider the lift boom in Fig. 5–6a. By inspection, it is supported by a pin at A and by the hydraulic cylinder BC, which can be approximated as a weightless link. The material can be assumed rigid, and with its density known, the weight of the boom and the location of its center of gravity G are determined. When a design loading $\mathbf{P}$ is specified, the idealized model shown in Fig. 5–6b can be used for a force analysis. Average dimensions (not shown) are used to specify the location of the loads and the supports.

Idealized models of specific objects will be given in some of the examples throughout the text. It should be realized, however, that each case represents the reduction of a practical situation using simplifying assumptions like the ones illustrated here.

(a)

(b)

Fig. 5–6

PROCEDURE FOR DRAWING A FREE-BODY DIAGRAM

To construct a free-body diagram for a rigid body or group of bodies considered as a single system, the following steps should be performed:

Draw Outlined Shape. Imagine the body to be *isolated* or cut "free" from its constraints and connections and draw (sketch) its outlined shape.

Show All Forces and Couple Moments. Identify all the external forces and couple moments that act on the body. Those generally encountered are due to (1) applied loadings, (2) reactions occurring at the supports or at points of contact with other bodies (see Table 5–1), and (3) the weight of the body. To account for all these effects, it may help to trace over the boundary, carefully noting each force or couple moment acting on it.

Identify Each Loading and Give Dimensions. The forces and couple moments that are known should be labeled with their proper magnitudes and directions. Letters are used to represent the magnitudes and direction angles of forces and couple moments that are *unknown*. Establish an *x, y* coordinate system so that these unknowns, A_x, B_y, etc., can be identified. Indicate the dimensions of the body necessary for calculating the moments of forces.

IMPORTANT POINTS

- No equilibrium problem should be solved without *first* drawing the free-body diagram, so as to account for all the forces and couple moments that act on the body.

- If a support *prevents translation* of a body in a particular direction, then the support exerts a *force* on the body in that direction.

- If *rotation is prevented*, then the support exerts a *couple moment* on the body.

- Study Table 5–1.

- Internal forces are never shown on the free-body diagram since they occur in equal but opposite collinear pairs and therefore cancel out.

- The weight of a body is an external force, and its effect is shown as a single resultant force acting through the body's center of gravity *G*.

- *Couple moments* can be placed anywhere on the free-body diagram since they are *free vectors*. Forces can act at any point along their lines of action since they are *sliding vectors*.

EXAMPLE 5.1

Draw the free-body diagram of the uniform beam shown in Fig. 5–7a. The beam has a mass of 100 kg.

(a)

Solution

The free-body diagram of the beam is shown in Fig. 5–7b. Since the support at A is a fixed wall, there are three reactions acting *on the beam* at A, denoted as $\mathbf{A}_x$, $\mathbf{A}_y$, and $\mathbf{M}_A$ drawn in an arbitrary direction. The magnitudes of these vectors are *unknown*, and their sense has been *assumed*. The weight of the beam, $W = 100(9.81) = 981$ N, acts through the beam's center of gravity G, which is 3 m from A since the beam is uniform.

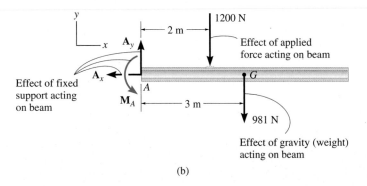

(b)

Fig. 5–7

E X A M P L E   5.2

Draw the free-body diagram of the foot lever shown in Fig. 5–8a. The operator applies a vertical force to the pedal so that the spring is stretched 40 mm and the force in the short link at *B* is 100 N.

(a)

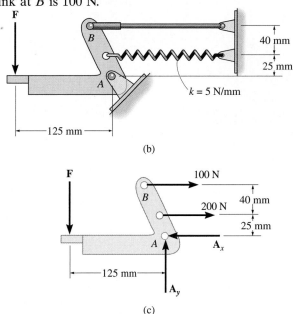

(b)

(c)

Fig. 5–8

Solution

By inspection, the lever is loosely bolted to the frame at *A*. The rod at *B* is pinned at its ends and acts as a "short link." After making the proper measurements, the idealized model of the lever is shown in Fig. 5–8b. From this the free-body diagram must be drawn. As shown in Fig. 5–8c, the pin support at *A* exerts force components $\mathbf{A}_x$ and $\mathbf{A}_y$ on the lever, each force has a known line of action but unknown magnitude. The link at *B* exerts a force of 100 N, acting in the direction of the link. In addition the spring also exerts a horizontal force on the lever. If the stiffness is measured and found to be $k = 5$ N/mm, then since the stretch $s = 40$ mm, using Eq. 3–2, $F_s = ks = 5$ N/mm (40 mm) = 200 N. Finally, the operator's shoe applies a vertical force of $\mathbf{F}$ on the pedal. The dimensions of the lever are also shown on the free-body diagram, since this information will be useful when computing the moments of the forces. As usual, the senses of the unknown forces at *A* have been assumed. The correct senses will become apparent after solving the equilibrium equations.

EXAMPLE **5.3**

Two smooth pipes, each having a mass of 300 kg, are supported by the forks of the tractor in Fig. 5–9a. Draw the free-body diagrams for each pipe and both pipes together.

(a)

B

0.35 m

A

0.35 m

30°

(b)

Effect of *B* acting on *A*

R

A

30°

Effect of sloped
blade acting on *A*

30°

30°

T

2943 N

Effect of gravity
(weight) acting on *A*

F

Effect of sloped
fork acting on *A*

(c)

B

30°

30°

R

2943 N

P

(d)

Fig. 5–9

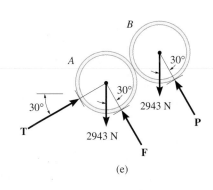

B

A

30°

30°

30°

T

2943 N

2943 N

F

P

(e)

Solution

The idealized model from which we must draw the free-body diagrams is shown in Fig. 5–9b. Here the pipes are identified, the dimensions have been added, and the physical situation reduced to its simplest form.

The free-body diagram for pipe *A* is shown in Fig. 5–9c. Its weight is *W* = 300(9.81) = 2943 N. Assuming all contacting surfaces are *smooth*, the reactive forces **T**, **F**, **R** act in a direction *normal* to the tangent at their surfaces of contact.

The free-body diagram of pipe *B* is shown in Fig. 5–9d. Can you identify each of the three forces acting *on this pipe*? In particular, note that **R**, representing the force of *A* on *B*, Fig. 5–9d, is equal and opposite to **R** representing the force of *B* on *A*, Fig. 5–9c. This is a consequence of Newton's third law of motion.

The free-body diagram of both pipes combined ("system") is shown in Fig. 5–9e. Here the contact force **R**, which acts between *A* and *B*, is considered as an *internal* force and hence is not shown on the free-body diagram. That is, it represents a pair of equal but opposite collinear forces which cancel each other.

EXAMPLE 5.4

Draw the free-body diagram of the unloaded platform that is suspended off the edge of the oil rig shown in Fig. 5–10a. The platform has a mass of 200 kg.

(a)

(b)

Fig. 5–10

(c)

Solution

The idealized model of the platform will be considered in two dimensions because by observation the loading and the dimensions are all symmetrical about a vertical plane passing through its center, Fig. 5–10b. Here the connection at A is assumed to be a pin, and the cable supports the platform at B. The direction of the cable and average dimensions of the platform are listed, and the center of gravity G has been determined. It is from this model that we must proceed to draw the free-body diagram, which is shown in Fig. 5–10c. The platform's weight is $200(9.81) = 1962$ N. The force components $\mathbf{A}_x$ and $\mathbf{A}_y$ along with the cable force $\mathbf{T}$ represent the reactions that both pins and both cables exert on the platform, Fig. 5–10a. Consequently, after the solution for these reactions, half their magnitude is developed at A and half is developed at B.

EXAMPLE **5.5**

The free-body diagram of each object in Fig. 5–11 is drawn. Carefully study each solution and identify what each loading represents, as was done in Fig. 5–7b.

Solution

(a)

(b)

(c)

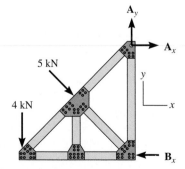

Note: Internal forces of one member on another are equal but opposite collinear forces which are not to be included here since they cancel out.

(d)

Fig. 5–11

PROBLEMS

5-1. Draw the free-body diagram of the 50-kg paper roll which has a center of mass at G and rests on the smooth blade of the paper hauler. Explain the significance of each force acting on the diagram. (See Fig. 5–7b.)

Prob. 5–1

5-2. Draw the free-body diagram of the hand punch, which is pinned at A and bears down on the smooth surface at B.

Prob. 5–2

5-3. Draw the free-body diagram of the dumpster D of the truck, which has a weight of 25 kN ($\approx$ 2.5 tonne) and a center of gravity at G. It is supported by a pin at A and a pin-connected hydraulic cylinder BC (short link) Explain the significance of each force on the diagram. (See Fig. 5–7b.)

Prob. 5–3

*5-4. Draw the free-body diagram of the jib crane AB, which is pin-connected at A and supported by member (link) BC.

Prob. 5–4

5-5. Draw the free-body diagram of the truss that is supported by the cable AB and pin C. Explain the significance of each force acting on the diagram. (See Fig. 5–7b.)

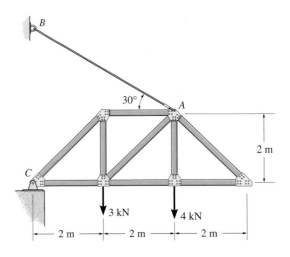

Prob. 5–5

5-6. Draw the free-body diagram of the crane boom AB which has a weight of 2600 N and center of gravity at G. The boom is supported by a pin at A and cable BC. The load of 5000 N is suspended from a cable attached at B. Explain the significance of each force acting on the diagram. (See Fig. 5–7b.)

Prob. 5–6

5-7. Draw the free-body diagram of the beam, which is pin-supported at A and rests on the smooth incline at B.

Prob. 5–7

*5-8. Draw the free-body diagram of member ABC which is supported by a smooth collar at A, roller at B, and short link CD. Explain the significance of each force acting on the diagram. (See Fig. 5–7b.)

Prob. 5–8

5-9. Draw the free-body diagram of the uniform bar, which has a mass of 100 kg and a center of mass at G. The supports A, B, and C are smooth.

Prob. 5–9

5-10. Draw the free-body diagram of the beam, which is pin-connected at A and rocker-supported at B.

Prob. 5–10

5.3 Equations of Equilibrium

In Sec. 5.1 we developed the two equations which are both necessary and sufficient for the equilibrium of a rigid body, namely, $\Sigma\mathbf{F} = \mathbf{0}$ and $\Sigma\mathbf{M}_O = \mathbf{0}$. When the body is subjected to a system of forces, which all lie in the x–y plane, then the forces can be resolved into their x and y components. Consequently, the conditions for equilibrium in two dimensions are

$$\Sigma F_x = 0$$
$$\Sigma F_y = 0 \qquad (5\text{--}2)$$
$$\Sigma M_O = 0$$

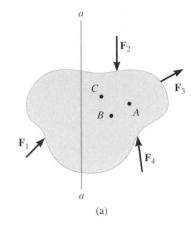

(a)

Here ΣF_x and ΣF_y represent, respectively, the algebraic sums of the x and y components of all the forces acting on the body, and ΣM_O represents the algebraic sum of the couple moments and the moments of all the force components about an axis perpendicular to the x–y plane and passing through the arbitrary point O, which may lie either on or off the body.

Alternative Sets of Equilibrium Equations. Although Eqs. 5–2 are *most often* used for solving coplanar equilibrium problems, two *alternative* sets of three independent equilibrium equations may also be used. One such set is

$$\Sigma F_a = 0$$
$$\Sigma M_A = 0 \qquad (5\text{--}3)$$
$$\Sigma M_B = 0$$

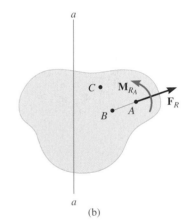

(b)

When using these equations it is required that a line passing through points A and B is *not perpendicular* to the a axis. To prove that Eqs. 5–3 provide the *conditions* for equilibrium, consider the free-body diagram of an arbitrarily shaped body shown in Fig. 5–12a. Using the methods of Sec. 4.8, all the forces on the free-body diagram may be replaced by an equivalent resultant force $\mathbf{F}_R = \Sigma\mathbf{F}$, acting at point A, and a resultant couple moment $\mathbf{M}_{R_A} = \Sigma\mathbf{M}_A$, Fig. 5–12$b$. If $\Sigma M_A = 0$ is satisfied, it is necessary that $\mathbf{M}_{R_A} = \mathbf{0}$. Furthermore, in order that $\mathbf{F}_R$ satisfy $\Sigma F_a = 0$, it must have *no component* along the a axis, and therefore its line of action must be perpendicular to the a axis, Fig. 5–12c. Finally, if it is required that $\Sigma M_B = 0$, where B does not lie on the line of action of $\mathbf{F}_R$, then $\mathbf{F}_R = \mathbf{0}$. Since $\Sigma\mathbf{F} = \mathbf{0}$ and $\Sigma\mathbf{M}_A = \mathbf{0}$, indeed the body in Fig. 5–12a must be in equilibrium.

(c)

Fig. 5–12

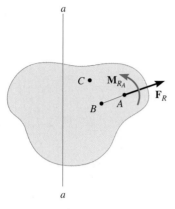

Fig. 5–13

A second alternative set of equilibrium equations is

$$\Sigma M_A = 0$$
$$\Sigma M_B = 0 \qquad\qquad (5\text{–}4)$$
$$\Sigma M_C = 0$$

Here it is necessary that points A, B, and C do not lie on the same line. To prove that these equations, when satisfied, ensure equilibrium, consider the free-body diagram in Fig. 5–13. If $\Sigma M_A = 0$ is to be satisfied, then $\mathbf{M}_{R_A} = \mathbf{0}$. $\Sigma M_B = 0$ is satisfied if the line of action of $\mathbf{F}_R$ passes through point B as shown. Finally, if we require $\Sigma M_C = 0$, where C does not lie on line AB, it is necessary that $\mathbf{F}_R = \mathbf{0}$, and the body in Fig. 5–12a must then be in equilibrium.

PROCEDURE FOR ANALYSIS

Coplanar force equilibrium problems for a rigid body can be solved using the following procedure.

Free-Body Diagram.

- Establish the x, y coordinate axes in any suitable orientation.
- Draw an outlined shape of the body.
- Show all the forces and couple moments acting on the body.
- Label all the loadings and specify their directions relative to the x, y axes. The sense of a force or couple moment having an *unknown* magnitude but known line of action can be *assumed*.
- Indicate the dimensions of the body necessary for computing the moments of forces.

Equations of Equilibrium.

- Apply the moment equation of equilibrium, $\Sigma M_O = 0$, about a point (O) that lies at the intersection of the lines of action of two unknown forces. In this way, the moments of these unknowns are zero about O, and a *direct solution* for the third unknown can be determined.
- When applying the force equilibrium equations, $\Sigma F_x = 0$ and $\Sigma F_y = 0$, orient the x and y axes along lines that will provide the simplest resolution of the forces into their x and y components.
- If the solution of the equilibrium equations yields a negative scalar for a force or couple moment magnitude, this indicates that the sense is opposite to that which was assumed on the free-body diagram.

E X A M P L E 5.6

Determine the horizontal and vertical components of reaction for the beam loaded as shown in Fig. 5–14a. Neglect the weight of the beam in the calculations.

(a)

(b)

Fig. 5–14

Solution

Free-Body Diagram.   Can you identify each of the forces shown on the free-body diagram of the beam, Fig. 5–14b? For simplicity, the 600-N force is represented by its x and y components as shown. Also, note that a 200-N force acts on the beam at B and is independent of the force components $\mathbf{B}_x$ and $\mathbf{B}_y$, which represent the effect of the pin on the beam.

Equations of Equilibrium.   Summing forces in the x direction yields

$$\xrightarrow{+}\ \Sigma F_x = 0; \qquad 600\cos 45°\,\text{N} - B_x = 0$$
$$B_x = 424\ \text{N} \qquad\qquad\qquad Ans.$$

   A direct solution for $\mathbf{A}_y$ can be obtained by applying the moment equation $\Sigma M_B = 0$ about point B. For the calculation, it should be apparent that forces 200 N, $\mathbf{B}_x$, and $\mathbf{B}_y$ all create zero moment about B. Assuming counterclockwise rotation about B to be positive (in the $+\mathbf{k}$ direction), Fig. 5–14b, we have

$$\zeta + \Sigma M_B = 0; \qquad 100\ \text{N}(2\ \text{m}) + (600\sin 45°\,\text{N})(5\ \text{m})$$
$$- (600\cos 45°\,\text{N})(0.2\ \text{m}) - A_y(7\ \text{m}) = 0$$
$$A_y = 319\ \text{N} \qquad\qquad Ans.$$

   Summing forces in the y direction, using this result, gives

$$+\uparrow \Sigma F_y = 0; \qquad 319\ \text{N} - 600\sin 45°\,\text{N} - 100\ \text{N} - 200\ \text{N} + B_y = 0$$
$$B_y = 405\ \text{N} \qquad\qquad Ans.$$

We can check this result by summing moments about point A.

$$\zeta + \Sigma M_A = 0; \qquad -(600\sin 45°\,\text{N})(2\ \text{m}) - (600\cos 45°\,\text{N})(0.2\ \text{m})$$
$$-(100\ \text{N})(5\ \text{m}) - (200\ \text{N})(7\ \text{m}) + B_y(7\ \text{m}) = 0$$
$$B_y = 405\ \text{N} \qquad\qquad Ans.$$

E X A M P L E   **5.7**

The cord shown in Fig. 5–15a supports a force of 500 N and wraps over the frictionless pulley. Determine the tension in the cord at C and the horizontal and vertical components of reaction at pin A.

(a)

Fig. 5–15

(b)

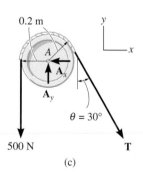

(c)

Solution

Free-Body Diagrams. The free-body diagrams of the cord and pulley are shown in Fig. 5–15b. Note that the principle of action, equal but opposite reaction must be carefully observed when drawing each of these diagrams: the cord exerts an unknown load distribution p along part of the pulley's surface, whereas the pulley exerts an equal but opposite effect on the cord. For the solution, however, it is simpler to *combine* the free-body diagrams of the pulley and the contacting portion of the cord, so that the distributed load becomes *internal* to the system and is therefore eliminated from the analysis, Fig. 5–15c.

Equations of Equilibrium. Summing moments about point A to eliminate $\mathbf{A}_x$ and $\mathbf{A}_y$, Fig. 5–15c, we have

$$\zeta + \Sigma M_A = 0; \qquad 500 \text{ N}(0.2 \text{ m}) - T(0.2 \text{ m}) = 0$$
$$T = 500 \text{ N} \qquad\qquad Ans.$$

It is seen that the tension remains *constant* as the cord passes over the pulley. (This of course is true for *any angle* θ at which the cord is directed and for *any radius* r of the pulley.) Using the result for T, a force summation is applied to determine the components of reaction at pin A.

$$\xrightarrow{+} \Sigma F_x = 0; \qquad -A_x + 500 \sin 30° \text{ N} = 0$$
$$A_x = 250 \text{ N} \qquad\qquad Ans.$$
$$+\uparrow \Sigma F_y = 0; \qquad A_y - 500 \text{ N} - 500 \cos 30° \text{ N} = 0$$
$$A_y = 933 \text{ N} \qquad\qquad Ans.$$

EXAMPLE 5.8

The link shown in Fig. 5–16a is pin-connected at A and rests against a smooth support at B. Compute the horizontal and vertical components of reaction at the pin A.

(a)

(b)

Fig. 5–16

Solution

Free-Body Diagram. As shown in Fig. 5–16b, the reaction N_B is perpendicular to the link at B. Also, horizontal and vertical components of reaction are represented at A.

Equations of Equilibrium. Summing moments about A, we obtain a direct solution for N_B,

$$\zeta + \Sigma M_A = 0; \quad -90 \text{ N} \cdot \text{m} - 60 \text{ N}(1 \text{ m}) + N_B(0.75 \text{ m}) = 0$$
$$N_B = 200 \text{ N}$$

Using this result,

$$\xrightarrow{+} \Sigma F_x = 0; \quad A_x - 200 \sin 30° \text{ N} = 0$$
$$A_x = 100 \text{ N} \qquad \qquad Ans.$$

$$+\uparrow \Sigma F_y = 0; \quad A_y - 200 \cos 30° \text{ N} - 60 \text{ N} = 0$$
$$A_y = 233 \text{ N} \qquad \qquad Ans.$$

E X A M P L E **5.9**

(a)

(b)

Fig. 5–17

The box wrench in Fig. 5–17a is used to tighten the bolt at A. If the wrench does not turn when the load is applied to the handle, determine the torque or moment applied to the bolt and the force of the wrench on the bolt.

Solution

Free-Body Diagram. The free-body diagram for the wrench is shown in Fig. 5–17b. Since the bolt acts as a "fixed support," it exerts force components A_x and A_y and a torque M_A on the wrench at A.

Equations of Equilibrium.

$$\xrightarrow{+} \Sigma F_x = 0; \qquad A_x - 52(\tfrac{5}{13})\text{N} + 30 \cos 60° \,\text{N} = 0$$

$$A_x = 5.00 \,\text{N} \qquad\qquad Ans.$$

$$+\uparrow \Sigma F_y = 0; \qquad A_y - 52(\tfrac{12}{13})\text{N} - 30 \sin 60° \,\text{N} = 0$$

$$A_y = 74.0 \,\text{N} \qquad\qquad Ans.$$

$$\zeta + \Sigma M_A = 0; \qquad M_A - 52(\tfrac{12}{13})\text{N} \,(0.3 \text{ m}) - (30 \sin 60° \,\text{N})(0.7 \text{ m}) = 0$$

$$M_A = 32.6 \,\text{N} \cdot \text{m} \qquad\qquad Ans.$$

Point A was chosen for summing moments because the lines of action of the *unknown* forces A_x and A_y pass through this point, and therefore these forces were not included in the moment summation. Realize, however, that M_A must be *included* in this moment summation. This couple moment is a free vector and represents the twisting resistance of the bolt on the wrench. By Newton's third law, the wrench exerts an equal but opposite moment or torque on the bolt. Furthermore, the resultant force on the wrench is

$$F_A = \sqrt{(5.00)^2 + (74.0)^2} = 74.1 \,\text{N} \qquad\qquad Ans.$$

Because the force components A_x and A_y were calculated as positive quantities, their directional sense is shown correctly on the free-body diagram in Fig. 5–17b. Hence

$$\theta = \tan^{-1}\frac{74.0 \,\text{N}}{5.00 \,\text{N}} = 86.1° \,\measuredangle$$

Realize that $\mathbf{F}_A$ acts in the opposite direction on the bolt. Why?

Although only *three* independent equilibrium equations can be written for a rigid body, it is a good practice to *check* the calculations using a fourth equilibrium equation. For example, the above computations may be verified in part by summing moments about point C:

$$\zeta + \Sigma M_C = 0; \, 52(\tfrac{12}{13})\text{N} \,(0.4 \text{ m}) + 32.6 \,\text{N} \cdot \text{m} - 74.0 \,\text{N}(0.7 \text{ m}) = 0$$

$$19.2 \,\text{N} \cdot \text{m} + 32.6 \,\text{N} \cdot \text{m} - 51.8 \,\text{N} \cdot \text{m} = 0$$

EXAMPLE 5.10

Placement of concrete from the truck is accomplished using the chute shown in the photos, Fig. 5–18a. Determine the force that the hydraulic cylinder and the truck frame exert on the chute to hold it in the position shown. The chute and wet concrete contained along its length have a uniform weight of 560 N/m.

Solution

The idealized model of the chute is shown in Fig. 5–18b. Here the dimensions are given, and it is assumed the chute is pin connected to the frame at A and the hydraulic cylinder BC acts as a short link.

(b)

(c)

Fig. 5–18

(a)

Free-Body Diagram. Since the chute has a length of 4 m, the total supported weight is (560 N/m)(4 m) = 2240 N, which is assumed to act at its midpoint, G. The hydraulic cylinder exerts a horizontal force $\mathbf{F}_{BC}$ on the chute, Fig. 5–18c.

Equations of Equilibrium. A direct solution for $\mathbf{F}_{BC}$ is possible by summing moments about the pin at A. To do this we will use the principle of moments and resolve the weight into components parallel and perpendicular to the chute. We have,

$\zeta + \Sigma M_A = 0;$
$-F_{BC}(0.5 \text{ m}) + 2240 \cos 30° \text{ N } (2 \text{ m}) + 2240 \sin 30° \text{ N } (0.0625 \text{ m}) = 0$
$$F_{BC} = 7900 \text{ N} \qquad \qquad Ans.$$

Summing forces to obtain A_x and A_y, we obtain

$\xrightarrow{+} \Sigma F_x = 0; \qquad -A_x + 7900 \text{ N} = 0$
$$A_x = 7900 \text{ N} \qquad \qquad Ans.$$

$+\uparrow \Sigma F_y = 0; \qquad A_y - 2240 \text{ N} = 0$
$$A_y = 2240 \text{ N} \qquad \qquad Ans.$$

To verify this solution we can sum moments about point B.

$\zeta + \Sigma M_B = 0; \qquad -7900 \text{ N } (0.5 \text{ m}) + 2240 \text{ N } (1 \cos 30° \text{ m}) +$
$2240 \cos 30° \text{ N } (1 \text{ m}) + 2240 \text{ N } \sin 30° (0.0625 \text{ m}) = 0$

E X A M P L E 5.11

The uniform smooth rod shown in Fig. 5–19a is subjected to a force and couple moment. If the rod is supported at A by a smooth wall and at B and C either at the top or bottom by rollers, determine the reactions at these supports. Neglect the weight of the rod.

(a)

Solution

Free-Body Diagram. As shown in Fig. 5–19b, all the support reactions act normal to the surface of contact since the contacting surfaces are smooth. The reactions at B and C are shown acting in the positive y′ direction. This assumes that only the rollers located on the bottom of the rod are used for support.

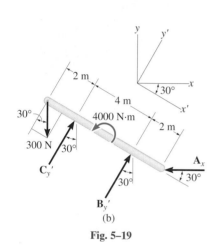

(b)

Fig. 5–19

Equations of Equilibrium. Using the x, y coordinate system in Fig. 5–19b, we have

$$\xrightarrow{+}\ \Sigma F_x = 0; \qquad C_{y'}\sin 30° + B_{y'}\sin 30° - A_x = 0 \qquad (1)$$

$$+\uparrow\Sigma F_y = 0; \qquad -300\ \text{N} + C_{y'}\cos 30° + B_{y'}\cos 30° = 0 \qquad (2)$$

$$\downdownarrows + \Sigma M_A = 0; \qquad -B_{y'}(2\ \text{m}) + 4000\ \text{N}\cdot\text{m} - C_{y'}(6\ \text{m})$$
$$+ (300\cos 30°\,\text{N})(8\ \text{m}) = 0 \qquad (3)$$

When writing the moment equation, it should be noticed that the line of action of the force component 300 sin 30° N passes through point A, and therefore this force is not included in the moment equation.

Solving Eqs. 2 and 3 simultaneously, we obtain

$$B_{y'} = -1000.0\ \text{N} = -1\ \text{kN} \qquad\qquad Ans.$$
$$C_{y'} = 1346.4\ \text{N} = 1.35\ \text{kN} \qquad\qquad Ans.$$

Since $B_{y'}$ is a negative scalar, the sense of $\mathbf{B}_{y'}$ is opposite to that shown on the free-body diagram in Fig. 5–19b. Therefore, the top roller at B serves as the support rather than the bottom one. Retaining the negative sign for $B_{y'}$ (Why?) and substituting the results into Eq. 1, we obtain

$$1346.4\sin 30°\,\text{N} - 1000.0\sin 30°\,\text{N} - A_x = 0$$

$$A_x = 173\ \text{N} \qquad\qquad Ans.$$

EXAMPLE 5.12

The uniform truck ramp shown in Fig. 5–20a has a weight of 1600 N ($\approx$ 160 kg) and is pinned to the body of the truck at each end and held in the position shown by the two side cables. Determine the tension in the cables.

(a)

Solution

The idealized model of the ramp, which indicates all necessary dimensions and supports, is shown in Fig. 5–20b. Here the center of gravity is located at the midpoint since the ramp is approximately uniform.

Free-Body Diagram. Working from the idealized model, the ramp's free-body diagram is shown in Fig. 5–20c.

Equations of Equilibrium. Summing moments about point A will yield a direct solution for the cable tension. Using the principle of moments, there are several ways of determining the moment of **T** about A. If we use x and y components, with **T** applied at B, we have

$$\zeta+\Sigma M_A = 0; \quad -T \cos 20° \,(2 \sin 30° \text{ m}) + T \sin 20° \,(2 \cos 30° \text{ m})$$
$$+ 1600 \text{ N} \,(1.5 \cos 30° \text{ m}) = 0$$
$$T = 5985 \text{ N}$$

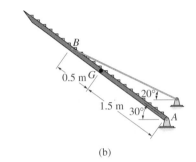

(b)

By the principle of transmissibility, we can locate **T** at C, even though this point is not on the ramp, Fig. 5–20c. In this case the horizontal component of **T** does not create a moment about A. First we must determine d using the sine law.

$$\frac{d}{\sin 10°} = \frac{2 \text{ m}}{\sin 20°}; \quad d = 1.0154 \text{ m}$$

$$\zeta+\Sigma M_A = 0; \quad -T \sin 20° \,(1.0154 \text{ m}) + 1600 \text{ N} \,(1.5 \cos 30° \text{ m}) = 0$$

$$T = 5985 \text{ N}$$

The simplest way to compute the moment of **T** about A is to resolve it into components parallel and perpendicular to the ramp at B. Then the moment of the parallel component is zero about A, so that

$$\zeta+\Sigma M_A = 0; \quad -T \sin 10° \,(2 \text{ m}) + 1600 \text{ N} \,(1.5 \cos 30° \text{ m}) = 0$$

$$T = 5985 \text{ N}$$

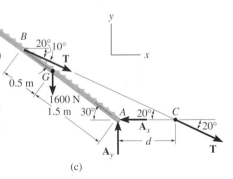

(c)

Since there are two cables supporting the ramp,

$$T' = \frac{T}{2} = 2992.5 \text{ N} \qquad \textit{Ans.}$$

Fig. 5–20

As an exercise, show that $A_x = 5624$ N and $A_y = 3647$ N.

5.4 Two- and Three-Force Members

The solution to some equilibrium problems can be simplified if one is able to recognize members that are subjected to only two or three forces.

Two-Force Members. When a member is subject to *no couple moments* and forces are applied at only two points on a member, the member is called a *two-force member.* An example is shown in Fig. 5–21a. The forces at A and B are summed to obtain their respective *resultants* $\mathbf{F}_A$ and $\mathbf{F}_B$, Fig. 5–21b. These two forces will maintain *translational or force equilibrium* ($\Sigma\mathbf{F} = \mathbf{0}$) provided $\mathbf{F}_A$ is of equal magnitude and opposite direction to $\mathbf{F}_B$. Furthermore, *rotational or moment equilibrium* ($\Sigma\mathbf{M}_O = \mathbf{0}$) is satisfied if $\mathbf{F}_A$ is *collinear* with $\mathbf{F}_B$. As a result, the line of action of both forces is known since it always passes through A and B. Hence, only the force magnitude must be determined or stated. Other examples of two-force members held in equilibrium are shown in Fig. 5–22.

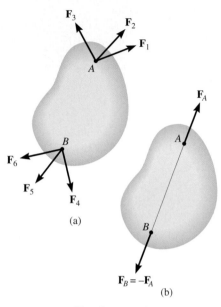

(a)

(b)

Two-force member

Fig. 5–21

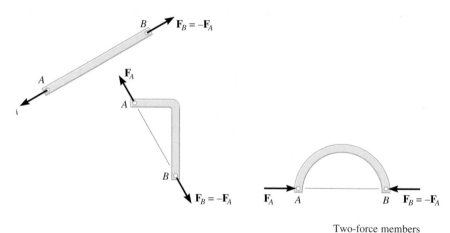

Two-force members

Fig. 5–22

Three-Force Members. If a member is subjected to only three forces, then it is necessary that the forces be either *concurrent* or *parallel* for the member to be in equilibrium. To show the concurrency requirement, consider the body in Fig. 5–23a and suppose that any two of the three forces acting on the body have lines of action that intersect at point O. To satisfy moment equilibrium about O, i.e., $\Sigma M_O = 0$, the third force must also pass through O, which then makes the force system *concurrent*. If two of the three forces are parallel, Fig. 5–23b, the point of concurrency, O, is considered to be at "infinity," and the third force must be parallel to the other two forces to intersect at this "point."

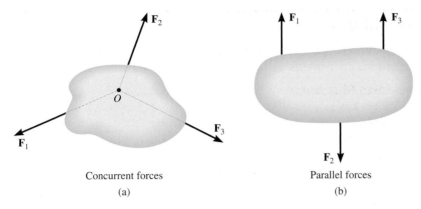

Concurrent forces

(a)

Parallel forces

(b)

Three-force members

Fig. 5–23

Many mechanical elements act as two- or three-force members, and the ability to recognize them in a problem will considerably simplify an equilibrium analysis.

- The bucket link AB on the back-hoe is a typical example of a two-force member since it is pin connected at its ends and, provided its weight is neglected, no other force acts on this member.

- The hydraulic cylinder BC is pin connected at its ends. It is a two-force member. The boom ABD is subjected to the weight of the suspended motor at D, the force of the hydraulic cylinder at B, and the force of the pin at A. If the boom's weight is neglected, it is a three-force member.

- The dump bed of the truck operates by extending the telescopic hydraulic cylinder AB. If the weight of AB is neglected, we can classify it as a two-force member since it is pin connected at its end points.

E X A M P L E 5.13

(a)

(b)

(c)

Fig. 5–24

The lever ABC is pin-supported at A and connected to a short link BD as shown in Fig. 5–24a. If the weight of the members is negligible, determine the force of the pin on the lever at A.

Solution

Free-Body Diagrams. As shown by the free-body diagram, Fig. 5–24b, the short link BD is a *two-force member,* so the *resultant forces* at pins D and B must be equal, opposite, and collinear. Although the magnitude of the force is unknown, the line of action is known since it passes through B and D.

Lever ABC is a *three-force member,* and therefore, in order to satisfy moment equilibrium, the three nonparallel forces acting on it must be concurrent at O, Fig. 5–24c. In particular, note that the force F on the lever at B is equal but opposite to the force F acting at B on the link. Why? The distance CO must be 0.5 m since the lines of action of **F** and the 400-N force are known.

Equations of Equilibrium. By requiring the force system to be concurrent at O, since $\Sigma M_O = 0$, the angle θ which defines the line of action of $\mathbf{F}_A$ can be determined from trigonometry,

$$\theta = \tan^{-1}\left(\frac{0.7}{0.4}\right) = 60.3° \quad \angle\theta \qquad Ans.$$

Using the x, y axes and applying the force equilibrium equations, we can obtain F_A and F.

$$\xrightarrow{+} \Sigma F_x = 0; \qquad F_A \cos 60.3° - F \cos 45° + 400 \text{ N} = 0$$
$$+\uparrow \Sigma F_y = 0; \qquad F_A \sin 60.3° - F \sin 45° = 0$$

Solving, we get

$$F_A = 1.07 \text{ kN} \qquad\qquad Ans.$$
$$F = 1.32 \text{ kN}$$

Note: We can also solve this problem by representing the force at A by its two components $\mathbf{A}_x$ and $\mathbf{A}_y$ and applying $\Sigma M_A = 0, \Sigma F_x = 0, \Sigma F_y = 0$ to the lever. Once A_x and A_y are determined, how would you find F_A and θ?

PROBLEMS

5-11. Determine the reactions at the supports in Prob. 5–1.

*5-12. Determine the magnitude of the resultant force acting at pin A of the handpunch in Prob. 5–2.

5-13. Determine the reactions at the supports for the truss in Prob. 5–5.

5-14. Determine the reactions on the boom in Prob. 5–6.

5-15. Determine the support reactions on the beam in Prob. 5–7.

*5-16. Determine the reactions on the member at A, B, and C in Prob. 5–8.

5-17. Determine the reactions at the points of contact at A, B, and C of the bar in Prob. 5–9.

5-18. Determine the reactions at the pin A and at the roller at B of the beam in Prob. 5–10.

5-19. Determine the magnitude of the reactions on the beam at A and B. Neglect the thickness of the beam.

Prob. 5–19

*5-20. Determine the reactions at the supports A and B of the frame.

Prob. 5–20

5-21. When holding the 20-N ($\approx$ 2-kg) stone in equilibrium, the humerus H, assumed to be smooth, exerts normal forces $\mathbf{F}_C$ and $\mathbf{F}_A$ on the radius C and ulna A as shown. Determine these forces and the force $\mathbf{F}_B$ that the biceps B exerts on the radius for equilibrium. The stone has a center of mass at G. Neglect the weight of the arm.

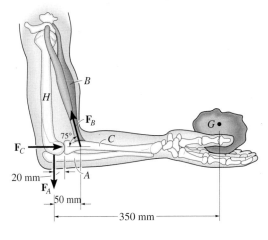

Prob. 5–21

5-22. The man is pulling a load of 40 N with one arm held as shown. Determine the force $\mathbf{F}_H$ this exerts on the humerus bone H, and the tension developed in the biceps muscle B. Neglect the weight of the man's arm.

Prob. 5–22

5-23. The ramp of a ship has a weight of 1000 N (≈ 100 kg) and a center of gravity at *G*. Determine the cable force in *CD* needed to just start lifting the ramp, (i.e., so the reaction at *B* becomes zero). Also, determine the horizontal and vertical components of force at the hinge (pin) at *A*.

5-25. Compare the force exerted on the toe and heel of a 480-N (≈ 48-kg) woman when she is wearing regular shoes and stiletto heels. Assume all her weight is placed on one foot and the reactions occur at points *A* and *B* as shown.

Prob. 5–25

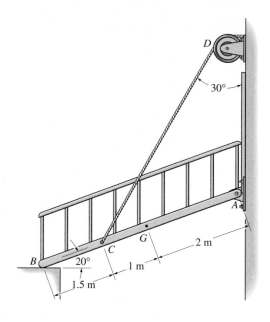

Prob. 5–23

*5-24. Determine the magnitude of force at the pin *A* and in the cable *BC* needed to support the 2000-N load. Neglect the weight of the boom *AB*.

5-26. Determine the reactions at the pins *A* and *B*. The spring has an unstretched length of 80 mm.

Prob. 5–24

Prob. 5–26

5-27. The platform assembly has a weight of 1000 N (≈ 100 kg) and center of gravity at G_1. If it is intended to support a maximum load of 1600 N placed at point G_2, determine the smallest counterweight W that should be placed at B in order to prevent the platform from tipping over.

5-29. The device is used to hold an elevator door open. If the spring has a stiffness of $k = 40$ N/m and it is compressed 0.2 m, determine the horizontal and vertical components of reaction at the pin A and the resultant force at the wheel bearing B.

Prob. 5–29

Prob. 5–27

*5-28.** Determine the tension in the cable and the horizontal and vertical components of reaction of the pin A. The pulley at D is frictionless and the cylinder weighs 80 N (≈ 8 kg).

5-30. The cutter is subjected to a horizontal force of 580 N and a normal force of 350 N. Determine the horizontal and vertical components of force acting on the pin A and the force along the hydraulic cylinder BC (a two-force member).

Prob. 5–28

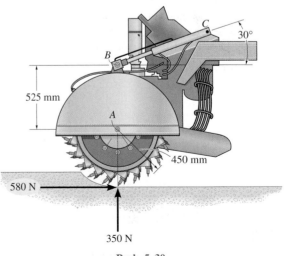

Prob. 5–30

5-31. The cantilevered jib crane is used to support the load of 780 N. If the trolley T can be placed anywhere between 0.45 m ≤ x ≤ 2.25 m, determine the maximum magnitude of reaction at the supports A and B. Note that the supports are collars that allow the crane to rotate freely about the vertical axis. The collar at B supports a force in the vertical direction, whereas the one at A does not.

Prob. 5–31

5-33. The power pole supports the three lines, each line exerting a vertical force on the pole due to its weight as shown. Determine the reactions at the fixed support D. If it is possible for wind or ice to snap the lines, determine which line(s) when removed create(s) a condition for the greatest moment reaction at D.

Prob. 5–33

*5-32.** The sports car has a mass of 1.5 Mg and mass center at G. If the front two springs each have a stiffness of $k_A = 58$ kN/m and the rear two springs each have a stiffness of $k_B = 65$ kN/m, determine their compression when the car is parked on the 30° incline. Also, what friction force $\mathbf{F}_B$ must be applied to each of the rear wheels to hold the car in equilibrium? *Hint:* First determine the normal force at A and B, then determine the compression in the springs.

Prob. 5–32

5-34. The jib crane is pin-connected at A and supported by a smooth collar at B. Determine the roller placement x of the 5000-N load so that it gives the maximum and minimum reactions at the supports. Calculate these reactions in each case. Neglect the weight of the crane. Require 1 m ≤ x ≤ 2.5 m.

Prob. 5–34

5-35. If the wheelbarrow and its contents have a mass of 60 kg and center of mass at G, determine the magnitude of the resultant force which the man must exert on *each* of the two handles in order to hold the wheelbarrow in equilibrium.

5-37. The bulk head AD is subjected to both water and soil-backfill pressures. Assuming AD is "pinned" to the ground at A, determine the horizontal and vertical reactions there and also the required tension in the ground anchor BC necessary for equilibrium. The bulk head has a mass of 800 kg.

Prob. 5–35

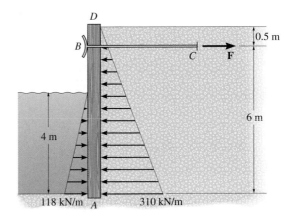

Prob. 5–37

*5-36. The pad footing is used to support the load of 12 000 N. Determine the intensities w_1 and w_2 of the distributed loading acting on the base of the footing for the equilibrium.

5-38. The telephone pole of negligible thickness is subjected to the force of 400 N directed as shown. It is supported by the cable BCD and can be assumed pinned at its base A. In order to provide clearance for a sidewalk right of way, where D is located, the strut CE is attached at C, as shown by the dashed lines (cable segment CD is removed). If the tension in CD' is to be twice the tension in BCD, determine the height h for placement of the strut CE.

Prob. 5–36

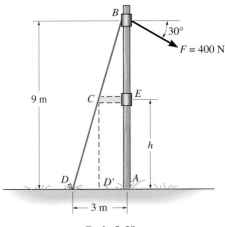

Prob. 5–38

5-39. The worker uses the hand truck to move material down the ramp. If the truck and its contents are held in the position shown and have a weight of 500 N ($\approx$ 50 kg) with center of gravity at G, determine the resultant normal force of both wheels on the ground A and the magnitude of the force required at the grip B.

5-41. The shelf supports the electric motor which has a mass of 15 kg and mass center at G_m. The platform upon which it rests has a mass of 4 kg and mass center at G_p. Assuming that a single bolt B holds the shelf up and the bracket bears against the smooth wall at A, determine this normal force at A and the horizontal and vertical components of reaction of the bolt on the bracket.

Prob. 5–39

Prob. 5–41

*5-40. The beam is subjected to the two concentrated loads as shown. Assuming that the foundation exerts a linearly varying load distribution on its bottom, determine the load intensities w_1 and w_2 for equilibrium (a) in terms of the parameters shown; (b) set $P = 500$ N, $L = 12$ m.

5-42. A cantilever beam, having an extended length of 3 m, is subjected to a vertical force of 500 N. Assuming that the wall resists this load with linearly varying distributed loads over the 0.15-m length of the beam portion inside the wall, detemine the intensities w_1 and w_2 for equilibrium.

Prob. 5–40

Prob. 5–42

5-43. The upper portion of the crane boom consists of the jib AB, which is supported by the pin at A, the guy line BC, and the backstay CD, each cable being separately attached to the mast at C. If the 5-kN load is supported by the hoist line, which passes over the pulley at B, determine the magnitude of the resultant force the pin exerts on the jib at A for equilibrium, the tension in the guy line BC, and the tension T in the hoist line. Neglect the weight of the jib. The pulley at B has a radius of 0.1 m.

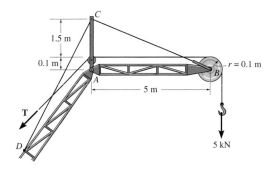

Prob. 5–43

*5-44. The mobile crane has a weight of 600 000 N ($\approx$ 60 tonne) and center of gravity at G_1; the boom has a weight of 150 000 N ($\approx$ 15 tonne) and center of gravity at G_2. Determine the smallest angle of tilt θ of the boom, without causing the crane to overturn if the suspended load is $W = 200\,000$ N. Neglect the thickness of the tracks at A and B.

5-45. The mobile crane has a weight of 600 000 N ($\approx$ 60 tonne) and center of gravity at G_1; the boom has a weight of 150 000 N ($\approx$ 15 tonne) and center of gravity at G_2. If the suspended load has a weight of $W = 80\,000$ N ($\approx$ 8 tonne), determine the normal reactions at the tracks A and B. For the calculation, neglect the thickness of the tracks and take $\theta = 30°$.

Probs. 5–44/45

5-46. The winch consists of a drum of radius 100 mm, which is pin-connected at its center C. At its outer rim is a ratchet gear having a mean radius of 150 mm. The pawl AB serves as a two-force member (short link) and holds the drum from rotating. If the suspended load is 2000 N, determine the horizontal and vertical components of reaction at the pin C.

Prob. 5–46

5-47. The crane consists of three parts, which have weights of $W_1 = 14\,000$ N ($\approx$ 1400 kg), $W_2 = 3600$ N ($\approx$ 360 kg), $W_3 = 6000$ N ($\approx$ 600 kg) and centers of gravity at G_1, G_2, and G_3, respectively. Neglecting the weight of the boom, determine (a) the reactions on each of the four tires if the load is hoisted at constant velocity and has a weight of 3200 N ($\approx$ 320 kg), and (b), with the boom held in the position shown, the maximum load the crane can lift without tipping over.

Prob. 5–47

*5-48. The boom supports the two vertical loads. Neglect the size of the collars at D and B and the thickness of the boom, and compute the horizontal and vertical components of force at the pin A and the force in cable CB. Set $F_1 = 800$ N and $F_2 = 350$ N.

5-49. The boom is intended to support two vertical loads, $\mathbf{F}_1$ and $\mathbf{F}_2$. If the cable CB can sustain a maximum load of 1500 N before it fails, determine the critical loads if $F_1 = 2F_2$. Also, what is the magnitude of the maximum reaction at pin A?

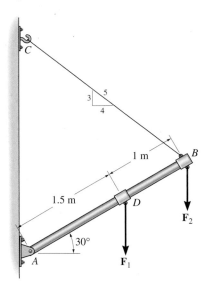

Probs. 5–48/49

5-50. Three uniform books, each having a weight W and length a, are stacked as shown. Determine the maximum distance d that the top book can extend out from the bottom one so the stack does not topple over.

Prob. 5–50

5-51. The toggle switch consists of a cocking lever that is pinned to a fixed frame at A and held in place by the spring which has an unstretched length of 200 mm. Determine the magnitude of the resultant force at A and the normal force on the peg at B when the lever is in the position shown.

Prob. 5–51

*5-52. The rigid beam of negligible weight is supported horizontally by two springs and a pin. If the springs are uncompressed when the load is removed, determine the force in each spring when the load $\mathbf{P}$ is applied. Also, compute the vertical deflection of end C. Assume the spring stiffness k is large enough so that only small deflections occur. *Hint:* The beam rotates about A so the deflections in the springs can be related.

Prob. 5–52

5-53. The uniform rod AB has a weight of 150 N ($\approx$ 15 kg) and the spring is unstretched when $\theta = 0°$. If $\theta = 30°$, determine the stiffness k of the spring so that the rod is in equilibrium.

5-55. The horizontal beam is supported by springs at its ends. Each spring has a stiffness of $k = 5$ kN/m and is originally unstretched when the beam is in the horizontal position. Determine the angle of tilt of the beam if a load of 800 N is applied at point C as shown.

*5-56. The horizontal beam is supported by springs at its ends. If the stiffness of the spring at A is $k_A = 5$ kN/m, determine the required stiffness of the spring at B so that if the beam is loaded with the 800 N it remains in the horizontal position. The springs are originally constructed so that the beam is in the horizontal position when it is unloaded.

Prob. 5–53

Probs. 5–55/56

5-54. The smooth pipe rests against the wall at the points of contact A, B, and C. Determine the reactions at these points needed to support the vertical force of 180 N. Neglect the pipe's thickness in the calculation.

5-57. Determine the distance d for placement of the load **P** for equilibrium of the smooth bar in the position θ as shown. Neglect the weight of the bar.

Prob. 5–54

Prob. 5–57

5-58. The wheelbarrow and its contents have a mass m and center of mass at G. Determine the greatest angle of tilt θ without causing the wheelbarrow to tip over.

Prob. 5–58

5-59. A boy stands out at the end of the diving board, which is supported by two springs A and B, each having a stiffness of $k = 15$ kN/m. In the position shown the board is horizontal. If the man has a mass of 40 kg, determine the angle of tilt which the board makes with the horizontal after he jumps off. Neglect the weight of the board and assume it is rigid.

Prob. 5–59

*5-60. The uniform beam has a weight W and length l and is supported by a pin at A and a cable BC. Determine the horizontal and vertical components of reaction at A and the tension in the cable necessary to hold the beam in the position shown.

Prob. 5–60

5-61. The uniform rod has a length l and weight W. It is supported at one end A by a smooth wall and the other end by a cord of length s which is attached to the wall as shown. Show that for equilibrium it is required that $h = [(s^2 - l^2)/3]^{1/2}$.

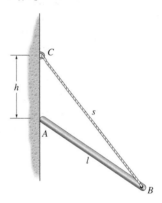

Prob. 5–61

5-62. The disk B has a mass of 20 kg and is supported on the smooth cylindrical surface by a spring having a stiffness of $k = 400$ N/m and unstretched length of $l_0 = 1$ m. The spring remains in the horizontal position since its end A is attached to the small roller guide which has negligible weight. Determine the angle θ to the nearest degree for equilibrium of the roller.

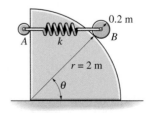

Prob. 5–62

Equilibrium in Three Dimensions

5.5 Free-Body Diagrams

The first step in solving three-dimensional equilibrium problems, as in the case of two dimensions, is to draw a free-body diagram of the body (or group of bodies considered as a system). Before we show this, however, it is necessary to discuss the types of reactions that can occur at the supports.

Support Reactions. The reactive forces and couple moments acting at various types of supports and connections, when the members are viewed in three dimensions, are listed in Table 5–2. It is important to recognize the symbols used to represent each of these supports and to understand clearly how the forces and couple moments are developed by each support. As in the two-dimensional case, *a force is developed by a support that restricts the translation of the attached member, whereas a couple moment is developed when rotation of the attached member is prevented.* For example, in Table 5–2, the ball-and-socket joint (4) prevents any translation of the connecting member; therefore, a force must act on the member at the point of connection. This force has three components having unknown magnitudes, F_x, F_y, F_z. Provided these components are known, one can obtain the magnitude of force. $F = \sqrt{F_x^2 + F_y^2 + F_z^2}$, and the force's orientation defined by the coordinate direction angles α, β, γ, Eqs. 2–7.* Since the connecting member is allowed to rotate freely about *any* axis, no couple moment is resisted by a ball-and-socket joint.

It should be noted that the *single* bearing supports (5) and (7), the *single* pin (8), and the *single* hinge (9) are shown to support both force and couple-moment components. If, however, these supports are used in conjunction with *other* bearings, pins, or hinges to hold a rigid body in equilibrium and the supports are *properly aligned* when connected to the body, then the *force reactions* at these supports *alone* may be adequate for supporting the body. In other words, the couple moments become redundant and are not shown on the free-body diagram. The reason for this should become clear after studying the examples which follow.

---

*The three unknowns may also be represented as an unknown force magnitude F and two unknown coordinate direction angles. The third direction angle is obtained using the identity $\cos^2 \alpha + \cos^2 \beta + \cos^2 \gamma = 1$, Eq. 2–10.

TABLE 5–2 • **Supports for Rigid Bodies Subjected to Three-Dimensional Force Systems**

Types of Connection	Reaction	Number of Unknowns
(1) cable	$\mathbf{F}$	One unknown. The reaction is a force which acts away from the member in the known direction of the cable.
(2) smooth surface support	$\mathbf{F}$	One unknown. The reaction is a force which acts perpendicular to the surface at the point of contact.
(3) roller	$\mathbf{F}$	One unknown. The reaction is a force which acts perpendicular to the surface at the point of contact.
(4) ball and socket	$\mathbf{F}_z$ $\mathbf{F}_y$ $\mathbf{F}_x$	Three unknowns. The reactions are three rectangular force components.
(5) single journal bearing	$\mathbf{M}_z$ $\mathbf{F}_z$ $\mathbf{M}_x$ $\mathbf{F}_x$	Four unknowns. The reactions are two force and two couple-moment components which act perpendicular to the shaft.

continued

TABLE 5–2 • *Continued*

Types of Connection	Reaction	Number of Unknown
(6) single journal bearing with square shaft	M_z F_z M_y M_x F_x	Five unknowns. The reactions are two force and three couple-moment components.
(7) single thrust bearing	M_z F_y F_z M_x F_x	Five unknowns. The reactions are three force and two couple-moment components.
(8) single smooth pin	M_z F_z F_x F_y M_y	Five unknowns. The reactions are three force and two couple-moment components.
(9) single hinge	M_z F_z F_y F_x M_x	Five unknowns. The reactions are three force and two couple-moment components.
(10) fixed support	M_z F_z F_x F_y M_y M_x	Six unknowns. The reactions are three force and three couple-moment components.

Typical examples of actual supports that are referenced to Table 5–2 are shown in the following sequence of photos.

This ball-and-socket joint provides a connection for the housing of an earth grader to its frame. (4)

This journal bearing supports the end of the shaft. (5)

This thrust bearing is used to support the drive shaft on a machine. (7)

This pin is used to support the end of the strut used on a tractor. (8)

Free-Body Diagrams.   The general procedure for establishing the free-body diagram of a rigid body has been outlined in Sec. 5.2. Essentially it requires first "isolating" the body by drawing its outlined shape. This is followed by a careful *labeling* of *all* the forces and couple moments in reference to an established *x, y, z* coordinate system. As a general rule, *components of reaction* having an *unknown magnitude* are shown acting on the free-body diagram in the *positive sense*. In this way, if any negative values are obtained, they will indicate that the components act in the negative coordinate directions.

It is a mistake to support a door using a single hinge since the hinge must develop a force $\mathbf{C}_y$ to support the weight $\mathbf{W}$ of the door and a couple moment $\mathbf{M}$ to support the moment of $\mathbf{W}$, i.e., $M = Wd$. If instead two properly aligned hinges are used, then the weight is carried by both hinges, $A_y + B_y = W$, and the moment of the door is resisted by the two hinge forces $\mathbf{F}_x$ and $-\mathbf{F}_x$. These forces form a couple, such that $F_x d' = Wd$. In other words, no couple moments are produced by the hinges on the door provided they are in *proper alignment*. Instead, the forces $\mathbf{F}_x$ and $-\mathbf{F}_x$ resist the rotation caused by $\mathbf{W}$.

E X A M P L E 5.14

Several examples of objects along with their associated free-body diagrams are shown in Fig. 5–25. In all cases, the *x, y, z* axes are established and the unknown reaction components are indicated in the positive sense. The weight of the objects is neglected.

Properly aligned journal bearings at *A, B, C*.

The force reactions developed by the bearings are sufficient for equilibrium since they prevent the shaft from rotating about each of the coordinate axes.

(a)

Fig. 5–25

200 N · m

300 N

A

C

B

Pin at *A* and cable *BC*.

$\mathbf{M}_{Az}$

$\mathbf{A}_z$

$\mathbf{M}_{Ax}$

$\mathbf{A}_x$

$\mathbf{A}_y$

z

y

x

200 N · m

300 N

T

B

(b)

Moment components are developed
by the pin on the rod to prevent
rotation about the *x* and *z* axes.

400 N

A

C

B

Properly aligned journal bearing
at *A* and hinge at *C*. Roller at *B*.

(c)

400 N

$\mathbf{A}_z$

$\mathbf{A}_x$

$\mathbf{C}_z$

$\mathbf{C}_x$

$\mathbf{C}_y$

$\mathbf{B}_z$

z

x

y

Only force reactions are developed by
the bearing and hinge on the plate to
prevent rotation about each coordinate axis.
No moments at the hinge are developed.

B

A

C

4 kN

6 kN

Thrust bearing at *A* and
cable *BC*

(d)

$\mathbf{M}_{Az}$

$\mathbf{A}_z$

$\mathbf{A}_x$

$\mathbf{A}_y$

$\mathbf{M}_{Ay}$

B

4 kN

T

C

6 kN

z

x

y

Moment components are developed
by the bearing on the rod in order to
prevent rotation about the *y* and *z* axes.

Fig. 5–25

5.6 Equations of Equilibrium

As stated in Sec. 5.1, the conditions for equilibrium of a rigid body subjected to a three-dimensional force system require that both the *resultant* force and *resultant* couple moment acting on the body be equal to *zero*.

Vector Equations of Equilibrium. The two conditions for equilibrium of a rigid body may be expressed mathematically in vector form as

$$\Sigma \mathbf{F} = \mathbf{0}$$
$$\Sigma \mathbf{M}_O = \mathbf{0} \qquad (5\text{-}5)$$

where $\Sigma \mathbf{F}$ is the vector sum of all the external forces acting on the body and $\Sigma \mathbf{M}_O$ is the sum of the couple moments and the moments of all the forces about any point O located either on or off the body.

Scalar Equations of Equilibrium. If all the applied external forces and couple moments are expressed in Cartesian vector form and substituted into Eqs. 5–5, we have

$$\Sigma \mathbf{F} = \Sigma F_x \mathbf{i} + \Sigma F_y \mathbf{j} + \Sigma F_z \mathbf{k} = \mathbf{0}$$
$$\Sigma \mathbf{M}_O = \Sigma M_x \mathbf{i} + \Sigma M_y \mathbf{j} + \Sigma M_z \mathbf{k} = \mathbf{0}$$

Since the $\mathbf{i}$, $\mathbf{j}$, and $\mathbf{k}$ components are independent from one another, the above equations are satisfied provided

$$\Sigma F_x = 0$$
$$\Sigma F_y = 0 \qquad (5\text{-}6a)$$
$$\Sigma F_z = 0$$

and

$$\Sigma M_x = 0$$
$$\Sigma M_y = 0 \qquad (5\text{-}6b)$$
$$\Sigma M_z = 0$$

These *six scalar equilibrium equations* may be used to solve for at most six unknowns shown on the free-body diagram. Equations 5–6a express the fact that the sum of the external force components acting in the x, y, and z directions must be zero, and Eqs. 5–6b require the sum of the moment components about the x, y, and z axes to be zero.

5.7 Constraints for a Rigid Body

To ensure the equilibrium of a rigid body, it is not only necessary to satisfy the equations of equilibrium, but the body must also be properly held or constrained by its supports. Some bodies may have more supports than are necessary for equilibrium, whereas others may not have enough or the supports may be arranged in a particular manner that could cause the body to collapse. Each of these cases will now be discussed.

Redundant Constraints. When a body has redundant supports, that is, more supports than are necessary to hold it in equilibrium, it becomes statically indeterminate. *Statically indeterminate* means that there will be more unknown loadings on the body than equations of equilibrium available for their solution. For example, the two-dimensional problem, Fig. 5–26a, and the three-dimensional problem, Fig. 5–26b, shown together with their free-body diagrams, are both statically indeterminate because of additional support reactions. In the two-dimensional case, there are five unknowns, that is, M_A, A_x, A_y, B_y, and C_y, for which only three equilibrium equations can be written ($\Sigma F_x = 0$, $\Sigma F_y = 0$, and $\Sigma M_O = 0$, Eqs. 5–2). The three-dimensional problem has eight unknowns, for which only six equilibrium equations can be written, Eqs. 5–6. The additional equations needed to solve indeterminate problems of the type shown in Fig. 5–26 are generally obtained from the deformation conditions at the points of support. These equations involve the physical properties of the body which are studied in subjects dealing with the mechanics of deformation, such as "mechanics of materials."*

(a)

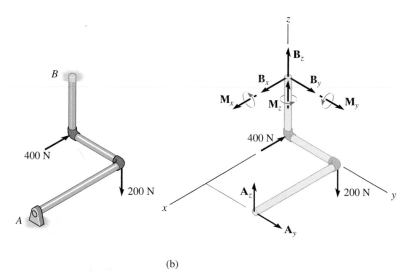

(b)

Fig. 5–26

*See R. C. Hibbeler, *Mechanics of Materials*, SI edition (Pearson Education/Prentice Hall, Inc., 2004).

Improper Constraints. In some cases, there may be as many unknown forces on the body as there are equations of equilibrium; however, *instability* of the body can develop because of *improper constraining* by the supports. In the case of three-dimensional problems, the body is improperly constrained if the support reactions *all intersect a common axis*. For two-dimensional problems, this axis is *perpendicular* to the plane of the forces and therefore appears as a point. Hence, when all the reactive forces are *concurrent* at this point, the body is improperly constrained. Examples of both cases are given in Fig. 5–27. From the free-body diagrams it is seen that the summation of moments about the *x* axis, Fig. 5–27a, or point *O*, Fig. 5–27b, will *not* be equal to zero; thus rotation about the *x* axis or point *O* will take place.* Furthermore, in both cases, it becomes *impossible* to solve *completely* for all the unknowns since one can write a moment equation that *does not* involve any of the unknown support reactions, and as a result, this reduces the number of available equilibrium equations by one.

(a)

(b)

Fig. 5–27

*For the three-dimensional problem, $\Sigma M_x = (400 \text{ N})(0.6 \text{ m}) \neq 0$, and for the two-dimensional problem, $\Sigma M_O = (100 \text{ N})(0.2 \text{ m}) \neq 0$.

(a)

Fig. 5–28

(b)

(a)

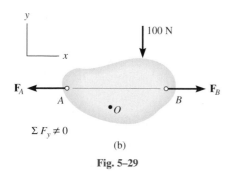

(b)

Fig. 5–29

Another way in which improper constraining leads to instability occurs when the *reactive forces* are all *parallel*. Three- and two-dimensional examples of this are shown in Fig. 5–28. In both cases, the summation of forces along the x axis will not equal zero.

In some cases, a body may have *fewer* reactive forces than equations of equilibrium that must be satisfied. The body then becomes only *partially constrained*. For example, consider the body shown in Fig. 5–29a with its corresponding free-body diagram in Fig. 5–29b. If O is a point not located on the line AB, the equation $\Sigma F_x = 0$ gives $F_A = F_B$ and $\Sigma M_O = 0$ and $\Sigma F_y = 0$, however, will not be satisfied for the loading conditions and therefore equilibrium will not be maintained.

Proper constraining therefore requires that (1) the lines of action of the reactive forces do not intersect points on a common axis, and (2) the reactive forces must not all be parallel to one another. When the minimum number of reactive forces is needed to properly constrain the body in question, the problem will be statically determinate, and therefore the equations of equilibrium can be used to determine *all* the reactive forces.

IMPORTANT POINTS

- Always draw the free-body diagram first.
- If a support *prevents translation* of a body in a specific direction, then the support exerts a *force* on the body in that direction.
- If *rotation about an axis is prevented*, then the support exerts a *couple moment* on the body about the axis.
- If a body is subjected to more unknown reactions than available equations of equilibrium, then the problem is *statically indeterminate*.
- To avoid instability of a body require that the lines of action of the reactive forces do not intersect a common axis and are not parallel to one another.

PROCEDURE FOR ANALYSIS

Three-dimensional equilibrium problems for a rigid body can be solved using the following procedure.

Free-Body Diagram

- Draw an outlined shape of the body.
- Show all the forces and couple moments acting on the body.
- Establish the origin of the x, y, z axes at a convenient point and orient the axes so that they are parallel to as many of the external forces and moments as possible.
- Label all the loadings and specify their directions relative to the x, y, z axes. In general, show all the unknown components having a positive sense along the x, y, z axes if the sense cannot be determined.
- Indicate the dimensions of the body necessary for computing the moments of forces.

Equations of Equilibrium

- If the x, y, z force and moment components seem easy to determine, then apply the six scalar equations of equilibrium; otherwise use the vector equations.
- It is not necessary that the set of axes chosen for force summation coincide with the set of axes chosen for moment summation. Also, any set of nonorthogonal axes may be chosen for this purpose.
- Choose the direction of an axis for moment summation such that it intersects the lines of action of as many unknown forces as possible. In this way, the moments of forces passing through points on this axis and forces which are parallel to the axis will then be zero.
- If the solution of the equilibrium equations yields a negative scalar for a force or couple moment magnitude, it indicates that the sense is opposite to that which was assumed on the free-body diagram.

EXAMPLE 5.15

300 N   200 N · m

1.5 m

A

2 m

3 m

C

B

(a)

z

300 N   200 N · m

981 N T_C

1 m

z'

1.5 m

1 m   1.5 m

A_z

B_x B_y y'

y

B_z

x'

x

(b)

Fig. 5–30

The homogeneous plate shown in Fig. 5–30a has a mass of 100 kg and is subjected to a force and couple moment along its edges. If it is supported in the horizontal plane by means of a roller at A, a ball-and-socket joint at B, and a cord at C, determine the components of reaction at the supports.

Solution *(Scalar Analysis)*

Free-Body Diagram. There are five unknown reactions acting on the plate, as shown in Fig. 5–30b. Each of these reactions is assumed to act in a positive coordinate direction.

Equations of Equilibrium. Since the three-dimensional geometry is rather simple, a *scalar analysis* provides a *direct solution* to this problem. A force summation along each axis yields

$$\Sigma F_x = 0; \qquad B_x = 0 \qquad\qquad\qquad Ans.$$
$$\Sigma F_y = 0; \qquad B_y = 0 \qquad\qquad\qquad Ans.$$
$$\Sigma F_z = 0; \qquad A_z + B_z + T_C - 300\ \text{N} - 981\ \text{N} = 0 \qquad (1)$$

Recall that the moment of a force about an axis is equal to the product of the force magnitude and the perpendicular distance (moment arm) from the line of action of the force to the axis. The sense of the moment is determined by the right-hand rule. Also, forces that are parallel to an axis or pass through it create no moment about the axis. Hence, summing moments of the forces on the free-body diagram, with positive moments acting along the positive x or y axis, we have

$$\Sigma M_x = 0; \qquad T_C(2\ \text{m}) - 981\ \text{N}(1\ \text{m}) + B_z(2\ \text{m}) = 0 \qquad (2)$$
$$\Sigma M_y = 0;$$
$$300\ \text{N}(1.5\ \text{m}) + 981\ \text{N}(1.5\ \text{m}) - B_z(3\ \text{m}) - A_z(3\ \text{m}) - 200\ \text{N} \cdot \text{m} = 0 \quad (3)$$

The components of force at B can be eliminated if the x', y', z' axes are used. We obtain

$$\Sigma M_{x'} = 0; \qquad 981\ \text{N}(1\ \text{m}) + 300\ \text{N}(2\ \text{m}) - A_z(2\ \text{m}) = 0 \qquad (4)$$
$$\Sigma M_{y'} = 0;$$
$$-300\ \text{N}(1.5\ \text{m}) - 981\ \text{N}(1.5\ \text{m}) - 200\ \text{N} \cdot \text{m} + T_C(3\ \text{m}) = 0 \qquad (5)$$

Solving Eqs. 1 through 3 or the more convenient Eqs. 1, 4, and 5 yields

$$A_z = 790\ \text{N} \qquad B_z = -217\ \text{N} \qquad T_C = 707\ \text{N} \qquad Ans.$$

The negative sign indicates that $\mathbf{B}_z$ acts downward.

Note that the solution of this problem does not require the use of a summation of moments about the z axis. The plate is partially constrained since the supports cannot prevent it from turning about the z axis if a force is applied to it in the x–y plane.

E X A M P L E 5.16

The windlass shown in Fig. 5–31a is supported by a thrust bearing at A and a smooth journal bearing at B, which are properly aligned on the shaft. Determine the magnitude of the vertical force **P** that must be applied to the handle to maintain equilibrium of the 100-kg bucket. Also calculate the reactions at the bearings.

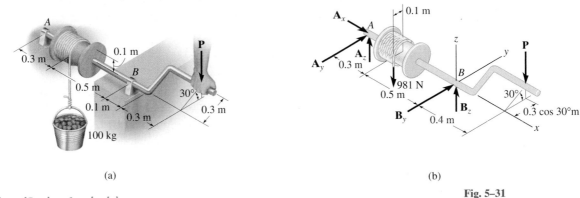

(a)   (b)

Fig. 5–31

Solution (Scalar Analysis)

Free-Body Diagram. Since the bearings at A and B are aligned correctly, *only* force reactions occur at these supports, Fig. 5–31b. Why are there no moment reactions?

Equations of Equilibrium. Summing moments about the x axis yields a direct solution for **P**. Why? For a scalar moment summation, it is necessary to determine the moment of each force as the product of the force magnitude and the *perpendicular distance* from the x axis to the line of action of the force. Using the right-hand rule and assuming positive moments act in the +**i** direction, we have

$$\Sigma M_x = 0; \quad 981 \text{ N}(0.1 \text{ m}) - P(0.3 \cos 30° \text{m}) = 0$$
$$P = 377.6 \text{ N} \qquad\qquad Ans.$$

Using this result and summing moments about the y and z axes yields

$$\Sigma M_y = 0;$$
$$-981 \text{ N}(0.5 \text{ m}) + A_z(0.8 \text{ m}) + (377.6 \text{ N})(0.4 \text{ m}) = 0$$
$$A_z = 424.3 \text{ N} \qquad\qquad Ans.$$
$$\Sigma M_z = 0; \quad -A_y(0.8 \text{ m}) = 0 \quad A_y = 0$$

The reactions at B are determined by a force summation using these results.

$$\Sigma F_x = 0; \quad A_x = 0$$
$$\Sigma F_y = 0; \quad 0 + B_y = 0 \quad B_y = 0$$
$$\Sigma F_z = 0; \quad 424.3 - 981 + B_z - 377.6 = 0 \quad B_z = 934 \text{ N} \quad Ans.$$

EXAMPLE 5.17

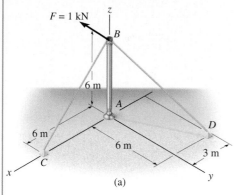

$F = 1$ kN

6 m

6 m

6 m

3 m

(a)

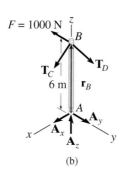

$F = 1000$ N

$\mathbf{T}_C$

$\mathbf{T}_D$

6 m $\quad \mathbf{r}_B$

$\mathbf{A}_x$ $\mathbf{A}_y$

$\mathbf{A}_z$

(b)

$\mathbf{F}_B = -\mathbf{F}_A$

B

A

$\mathbf{F}_A$

(c)

Fig. 5–32

Determine the tension in cables BC and BD and the reactions at the ball-and-socket joint A for the mast shown in Fig. 5–32a.

Solution (*Vector Analysis*)

Free-Body Diagram. There are five unknown force magnitudes shown on the free-body diagram, Fig. 5–32b.

Equations of Equilibrium. Expressing each force in Cartesian vector form, we have

$$\mathbf{F} = \{-1000\mathbf{j}\} \text{ N}$$
$$\mathbf{F}_A = A_x\mathbf{i} + A_y\mathbf{j} + A_z\mathbf{k}$$
$$\mathbf{T}_C = 0.707T_C\mathbf{i} - 0.707T_C\mathbf{k}$$
$$\mathbf{T}_D = T_D\left(\frac{\mathbf{r}_{BD}}{r_{BD}}\right) = -\frac{3}{9}T_D\mathbf{i} + \frac{6}{9}T_D\mathbf{j} - \frac{6}{9}T_D\mathbf{k}$$

Applying the force equation of equilibrium gives

$$\Sigma\mathbf{F} = \mathbf{0}; \qquad \mathbf{F} + \mathbf{F}_A + \mathbf{T}_C + \mathbf{T}_D = \mathbf{0}$$

$$\left(A_x + 0.707T_C - \frac{3}{9}T_D\right)\mathbf{i} + \left(-1000 + A_y + \frac{6}{9}T_D\right)\mathbf{j}$$
$$+ \left(A_z - 0.707T_C - \frac{6}{9}T_D\right)\mathbf{k} = \mathbf{0}$$

$$\Sigma F_x = 0; \qquad A_x + 0.707T_C - \frac{3}{9}T_D = 0 \qquad\qquad (1)$$

$$\Sigma F_y = 0; \qquad A_y + \frac{6}{9}T_D - 1000 = 0 \qquad\qquad (2)$$

$$\Sigma F_z = 0; \qquad A_z - 0.707T_C - \frac{6}{9}T_D = 0 \qquad\qquad (3)$$

Summing moments about point A, we have

$$\Sigma\mathbf{M}_A = \mathbf{0}; \qquad \mathbf{r}_B \times (\mathbf{F} + \mathbf{T}_C + \mathbf{T}_D) = \mathbf{0}$$
$$6\mathbf{k} \times (-1000\mathbf{j} + 0.707T_C\mathbf{i} - 0.707T_C\mathbf{k}$$
$$-\frac{3}{9}T_D\mathbf{i} + \frac{6}{9}T_D\mathbf{j} - \frac{6}{9}T_D\mathbf{k}) = \mathbf{0}$$

Evaluating the cross product and combining terms yields

$$(-4T_D + 6000)\mathbf{i} + (4.24T_C - 2T_D)\mathbf{j} = \mathbf{0}$$

$$\Sigma M_x = 0; \qquad -4T_D + 6000 = 0 \qquad\qquad (4)$$

$$\Sigma M_y = 0; \qquad 4.24T_C - 2T_D = 0 \qquad\qquad (5)$$

The moment equation about the z axis, $\Sigma M_z = 0$, is automatically satisfied. Why? Solving Eqs. 1 through 5 we have

$$T_C = 707 \text{ N} \qquad T_D = 1500 \text{ N} \qquad\qquad Ans.$$
$$A_x = 0 \text{ N} \qquad A_y = 0 \text{ N} \qquad A_z = 1500 \text{ N} \qquad Ans.$$

Since the mast is a two-force member, Fig. 5–32c, note that the value $A_x = A_y = 0$ could have been determined *by inspection*.

EXAMPLE 5.18

Rod AB shown in Fig. 5–33a is subjected to the 200-N force. Determine the reactions at the ball-and-socket joint A and the tension in cables BD and BE.

Solution *(Vector Analysis)*

Free-Body Diagram. Fig. 5–33b.

Equations of Equilibrium. Representing each force on the free-body diagram in Cartesian vector form, we have

$$\mathbf{F}_A = A_x\mathbf{i} + A_y\mathbf{j} + A_z\mathbf{k}$$
$$\mathbf{T}_E = T_E\mathbf{i}$$
$$\mathbf{T}_D = T_D\mathbf{j}$$
$$\mathbf{F} = \{-200\mathbf{k}\}\ \text{N}$$

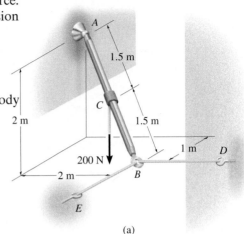

(a)

Applying the force equation of equilibrium.

$$\Sigma\mathbf{F} = \mathbf{0}; \qquad \mathbf{F}_A + \mathbf{T}_E + \mathbf{T}_D + \mathbf{F} = \mathbf{0}$$
$$(A_x + T_E)\mathbf{i} + (A_y + T_D)\mathbf{j} + (A_z - 200)\mathbf{k} = \mathbf{0}$$

$$\Sigma F_x = 0; \qquad A_x + T_E = 0 \tag{1}$$
$$\Sigma F_y = 0; \qquad A_y + T_D = 0 \tag{2}$$
$$\Sigma F_z = 0; \qquad A_z - 200 = 0 \tag{3}$$

Summing moments about point A yields

$$\Sigma\mathbf{M}_A = \mathbf{0}; \qquad \mathbf{r}_C \times \mathbf{F} + \mathbf{r}_B \times (\mathbf{T}_E + \mathbf{T}_D) = \mathbf{0}$$

Since $\mathbf{r}_C = \frac{1}{2}\mathbf{r}_B$, then

$$(0.5\mathbf{i} + 1\mathbf{j} - 1\mathbf{k}) \times (-200\mathbf{k}) + (1\mathbf{i} + 2\mathbf{j} - 2\mathbf{k}) \times (T_E\mathbf{i} + T_D\mathbf{j}) = \mathbf{0}$$

Expanding and rearranging terms gives

$$(2T_D - 200)\mathbf{i} + (-2T_E + 100)\mathbf{j} + (T_D - 2T_E)\mathbf{k} = \mathbf{0}$$

$$\Sigma M_x = 0; \qquad 2T_D - 200 = 0 \tag{4}$$
$$\Sigma M_y = 0; \qquad -2T_E + 100 = 0 \tag{5}$$
$$\Sigma M_z = 0; \qquad T_D - 2T_E = 0 \tag{6}$$

Solving Eqs. 1 through 6, we get

$$\begin{aligned}
T_D &= 100\ \text{N} &\quad &Ans.\\
T_E &= 50\ \text{N} &\quad &Ans.\\
A_x &= -50\ \text{N} &\quad &Ans.\\
A_y &= -100\ \text{N} &\quad &Ans.\\
A_z &= 200\ \text{N} &\quad &Ans.
\end{aligned}$$

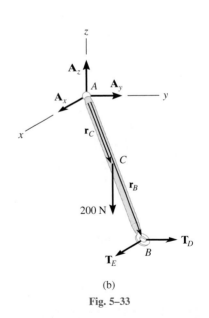

(b)

Fig. 5–33

The negative sign indicates that $\mathbf{A}_x$ and $\mathbf{A}_y$ have a sense which is opposite to that shown on the free-body diagram, Fig. 5–33b.

E X A M P L E 5.19

0.2 m

C

1 m

0.6 m

A

0.3 m

B

z

0.5 m

E

0.5 m

D

y

x

100 kg

(a)

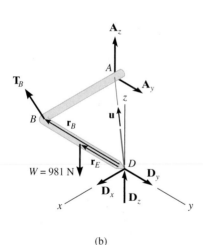

A_z

A

T_B

A_y

B

r_B

u

z

$W = 981$ N

r_E

D

D_y

D_x D_z

x

y

(b)

Fig. 5–34

The bent rod in Fig. 5–34*a* is supported at *A* by a journal bearing, at *D* by a ball-and-socket joint, and at *B* by means of cable *BC*. Using only *one equilibrium equation*, obtain a direct solution for the tension in cable *BC*. The bearing at *A* is capable of exerting force components only in the *z* and *y* directions since it is properly aligned on the shaft.

Solution (Vector Analysis)

Free-Body Diagram. As shown in Fig. 5–34*b*, there are six unknowns: three force components caused by the ball-and-socket joint, two caused by the bearing, and one caused by the cable.

Equations of Equilibrium. The cable tension $\mathbf{T}_B$ may be obtained *directly* by summing moments about an axis passing through points *D* and *A*. Why? The direction of the axis is defined by the unit vector **u**, where

$$\mathbf{u} = \frac{\mathbf{r}_{DA}}{r_{DA}} = -\frac{1}{\sqrt{2}}\mathbf{i} - \frac{1}{\sqrt{2}}\mathbf{j}$$
$$= -0.707\mathbf{i} - 0.707\mathbf{j}$$

Hence, the sum of the moments about this axis is zero provided

$$\Sigma M_{DA} = \mathbf{u} \cdot \Sigma(\mathbf{r} \times \mathbf{F}) = 0$$

Here **r** represents a position vector drawn from *any point* on the axis *DA* to any point on the line of action of force **F** (see Eq. 4–11). With reference to Fig. 5–34*b*, we can therefore write

$$\mathbf{u} \cdot (\mathbf{r}_B \times \mathbf{T}_B + \mathbf{r}_E \times \mathbf{W}) = \mathbf{0}$$
$$(-0.707\mathbf{i} - 0.707\mathbf{j}) \cdot [(-1\mathbf{j}) \times (\tfrac{0.2}{0.7}T_B\mathbf{i} - \tfrac{0.3}{0.7}T_B\mathbf{j} + \tfrac{0.6}{0.7}T_B\mathbf{k})$$
$$+ (-0.5\mathbf{j}) \times (-981\mathbf{k})] = \mathbf{0}$$
$$(-0.707\mathbf{i} - 0.707\mathbf{j}) \cdot [(-0.857T_B + 490.5)\mathbf{i} + 0.286T_B\mathbf{k}] = \mathbf{0}$$
$$- 0.707(-0.857T_B + 490.5) + 0 + 0 = 0$$
$$T_B = \frac{490.5}{0.857} = 572 \text{ N} \qquad Ans.$$

The advantage of using Cartesian vectors for this solution should be noted. It would be especially tedious to determine the perpendicular distance from the *DA* axis to the line of action of $\mathbf{T}_B$ using scalar methods.

Note: In a similar manner, we can obtain $D_z(=490.5 \text{ N})$ by summing moments about an axis passing through *AB*. Also, $A_z(=0)$ is obtained by summing moments about the *y* axis.

PROBLEMS

5-63. The uniform load has a mass of 600 kg and is lifted using a uniform 30-kg strongback beam and four wire ropes as shown. Determine the tension in each segment of rope and the force that must be applied to the sling at *A*.

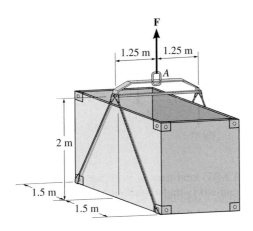

Prob. 5–63

5-65. The uniform concrete slab has a weight of 22 kN (≈ 2.2 tonne). Determine the tension in each of the three parallel supporting cables when the slab is held in the horizontal plane as shown.

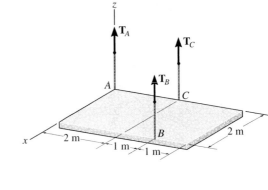

Prob. 5–65

*5-64. The wing of the jet aircraft is subjected to a thrust of *T* = 8 kN from its engine and the resultant lift force *L* = 45 kN. If the mass of the wing is 2.1 Mg and the mass center is at *G*, determine the *x*, *y*, *z* components of reaction where the wing is fixed to the fuselage at *A*.

Prob. 5–64

5-66. The air-conditioning unit is hoisted to the roof of a building using the three cables. If the tensions in the cables are T_A = 1000 N, T_B = 1200 N, and T_C = 800 N, determine the weight of the unit and the location (*x*, *y*) of its center of gravity *G*.

Prob. 5–66

5-67. The platform truck supports the three loadings shown. Determine the normal reactions on each of its three wheels.

5-69. The cart supports the uniform crate having a mass of 85 kg. Determine the vertical reactions on the three casters at A, B, and C. The caster at B is not shown. Neglect the mass of the cart and assume the casters at A and B are at the rear corner of the cart.

Prob. 5–69

Prob. 5–67

5-70. The boom AB is held in equilibrium by a ball-and-socket joint A and a pulley and cord system as shown. Determine the x, y, z components of reaction at A and the tension in cable DEC if $\mathbf{F} = \{-1500\mathbf{k}\}$ kN.

*5-68. The wrench is used to tighten the bolt at A. If the force $F = 30$ N is applied to the handle as shown, determine the magnitudes of the resultant force and moment that the bolt head exerts on the wrench. The force $\mathbf{F}$ is in a plane parallel to the x–z plane.

5-71. The cable CED can sustain a maximum tension of 800 N before it fails. Determine the greatest vertical force F that can be applied to the boom. Also, what are the x, y, z components of reaction at the ball-and-socket joint A?

Prob. 5–68

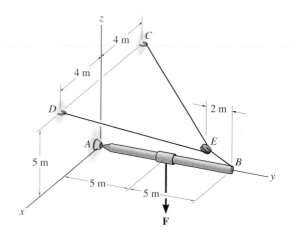

Prob. 5–70

*5-72. Determine the force components acting on the ball-and-socket at *A*, the reaction at the roller *B* and the tension on the cord *CD* needed for equilibrium of the quarter circular plate.

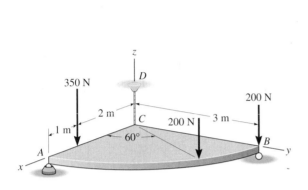

Prob. 5–72

5-74. The pole for a power line is subjected to the two cable forces of 6 kN, each force lying in a plane parallel to the *x–y* plane. If the tension in the guy wire *AB* is 8 kN, determine the *x*, *y*, *z* components of reaction at the fixed base of the pole, *O* due to these three forces.

Prob. 5–74

5-73. The windlass supports the 50-kg mass. Determine the horizontal force **P** needed to hold the handle in the position shown, and the components of reaction at the ball-and-socket joint *A* and the smooth journal bearing *B*. The bearing at *B* is in proper alignment and exerts only force reactions on the windlass.

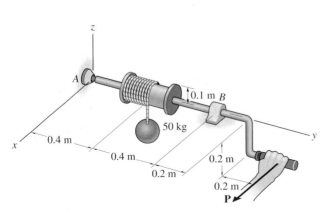

Prob. 5–73

5-75. Member *AB* is supported by a cable *BC* and at *A* by a *square* rod which fits loosely through the square hole at the end joint of the member as shown. Determine the components of reaction at *A* and the tension in the cable needed to hold the 800-N ($\approx$ 80-kg) cylinder in equilibrium.

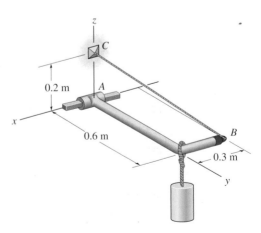

Prob. 5–75

*5-76. The pipe assembly supports the vertical loads shown. Determine the components of reaction at the ball-and-socket joint A and the tension in the supporting cables BC and BD.

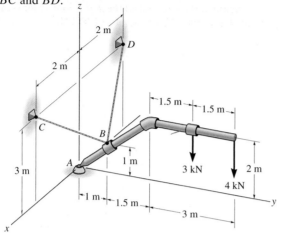

Prob. 5–76

5-77. Both pulleys are fixed to the shaft and as the shaft turns with constant angular velocity, the power of pulley A is transmitted to pulley B. Determine the horizontal tension $\mathbf{T}$ in the belt on pulley B and the x, y, z components of reaction at the journal bearing C and thrust bearing D if $\theta = 0°$. The bearings are in proper alignment and exert only force reactions on the shaft.

5-78. Both pulleys are fixed to the shaft and as the shaft turns with constant angular velocity, the power of pulley A is transmitted to pulley B. Determine the horizontal tension $\mathbf{T}$ in the belt on pulley B and the x, y, z components of reaction at the journal bearing C and thrust bearing D if $\theta = 45°$. The bearings are in proper alignment and exert only force reactions on the shaft.

Probs. 5–77/78

5-79. The bent rod is supported at A, B, and C by smooth journal bearings. Compute the x, y, z components of reaction at the bearings if the rod is subjected to forces $F_1 = 300$ N and $F_2 = 250$ N. F_1 lies in the y–z plane. The bearings are in proper alignment and exert only force reactions on the rod.

*5-80. The bent rod is supported at A, B, and C by smooth journal bearings. Determine the magnitude of $\mathbf{F}_2$ which will cause the reaction $\mathbf{C}_y$ at the bearing C to be equal to zero. The bearings are in proper alignment and exert only force reactions on the rod. Set $F_1 = 300$ N.

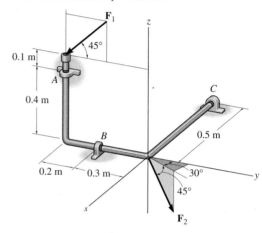

Probs. 5–79/80

5-81. The silo has a weight of 14 000 N ($\approx$ 1.4 tonne) and a center of gravity at G. Determine the vertical component of force that each of the three struts at A,B, and C exerts on the silo if it is subjected to a resultant wind loading of 1000 N which acts in the direction shown.

Prob. 5–81

5-82. Determine the tensions in the cables and the components of reaction acting on the smooth collar at A necessary to hold the 50-kN ($\approx$ 5-tonne) sign in equilibrium. The center of gravity for the sign is at G.

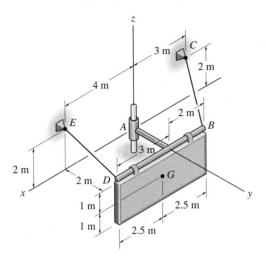

Prob. 5–82

5-83. The boom is supported by a ball-and-socket joint at A and a guy wire at B. If the 5-kN loads lie in a plane which is parallel to the x–y plane, determine the x, y, z components of reaction at A and the tension in the cable at B.

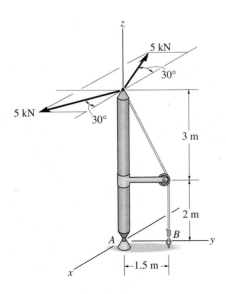

Prob. 5–83

*5-84. The boom AC is supported at A by a ball-and-socket joint and by two cables BDC and CE. Cable BDC is continuous and passes over a pulley at D. Calculate the tension in the cables and the x, y, z components of reaction at A if a crate has a weight of 80 kN.

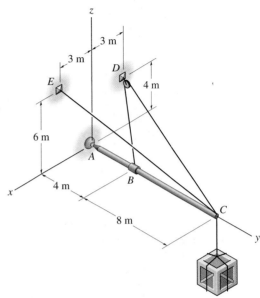

Prob. 5–84

5-85. Rod AB is supported by a ball-and-socket joint at A and a cable at B. Determine the x, y, z components of reaction at these supports if the rod is subjected to a 50-kN vertical force as shown.

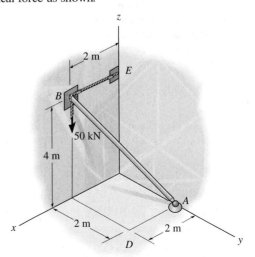

Prob. 5–85

5-86. A vertical force of 50 N acts on the crankshaft. Determine the horizontal equilibrium force **P** that must be applied to the handle and the x, y, z components of reaction at the journal bearing A and thrust bearing B. The bearings are properly aligned and exert only force reactions on the shaft.

Prob. 5–86

5-87. The platform has a mass of 3 Mg and center of mass located at G. If it is lifted with constant velocity using the three cables, determine the force in each of the cables.

*5-88. The platform has a mass of 2 Mg and center of mass located at G. If it is lifted using the three cables, determine the force in each of the cables. Solve for each force by using a single moment equation of equilibrium.

Probs. 5–87/88

5-89. The cables exert the forces shown on the pole. Assuming the pole is supported by a ball-and-socket joint at its base, detemine the components of reaction at A. The forces of 140 kN and 75 kN lie in a horizontal plane.

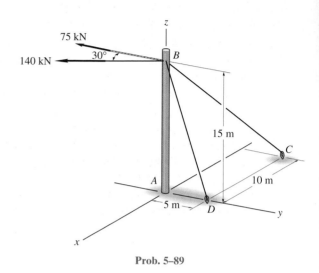

Prob. 5–89

5-90. The pole is subjected to the two forces shown. Determine the components of reaction of A assuming it to be a ball-and-socket joint. Also, compute the tension in each of the guy wires, BC and ED.

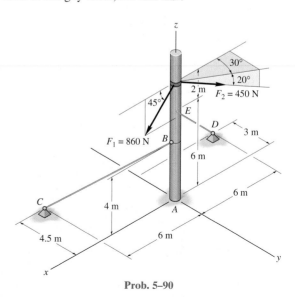

Prob. 5–90

CHAPTER REVIEW

- **Free-Body Diagram.** Before analyzing any equilibrium problem it is first necessary to draw a free-body diagram. This is an outlined shape of the body, which shows all the forces and couple moments that act on the body. Remember that a support will exert a *force* on the body in a particular direction if it prevents *translation* of the body in that direction, and it will exert a *couple moment* on the body if it prevents *rotation*. Angles used to resolve forces, and dimensions used to take moments of the forces, should also be shown on the free-body diagram.

- **Two Dimensions.** Normally the three scalar equations of equilibrium, $\Sigma F_x = 0$, $\Sigma F_y = 0$, $\Sigma M_o = 0$, can be applied when solving problems in two dimensions, since the geometry is easy to visualize. For the most direct solution, try to sum forces along an axis that will eliminate as many unknown forces as possible. Sum moments about a point O that passes through the line of action of as many unknown forces as possible.

- **Three Dimensions.** In three dimensions, it is often advantageous to use a Cartesian vector analysis when applying the equations of equilibrium. To do this, first express each known and unknown force and couple moment shown on the free-body diagram as a Cartesian vector. Then set the force summation equal to zero, $\Sigma \mathbf{F} = \mathbf{0}$. Take moments about a point O that lies on the line of action of as many unknown force components as possible. From point O direct position vectors to each force, and then use the cross product to determine the moment of each force. Require $\Sigma \mathbf{M}_o = \Sigma \mathbf{r} \times \mathbf{F} = \mathbf{0}$. The six scalar equations of equilibrium are established by setting the respective $\mathbf{i}$, $\mathbf{j}$, and $\mathbf{k}$ components of these force and moment sums equal to zero.

REVIEW PROBLEMS

5-91. The shaft assembly is supported by two smooth journal bearings A and B and a short link DC. If a couple moment is applied to the shaft as shown, determine the components of force reaction at the bearings and the force in the link. The link lies in a plane parallel to the y–z plane and the bearings are properly aligned on the shaft.

*5-92.** Determine the horizontal and vertical components of reaction at the pin A and the reaction at the roller B required to support the truss. Set $F = 600$ N.

Prob. 5–91

Prob. 5–92

5-93. If the roller at B can sustain a maximum load of 3 kN, determine the largest magnitude of each of the three forces $\mathbf{F}$ that can be supported by the truss.

5-95. The symmetrical shelf is subjected to a uniform load of 4 kPa. Support is provided by a bolt (or pin) located at each end A and A' and by the symmetrical brace arms, which bear against the smooth wall on both sides at B and B'. Determine the force resisted by each bolt at the wall and the normal force at B for equilibrium.

Prob. 5–93

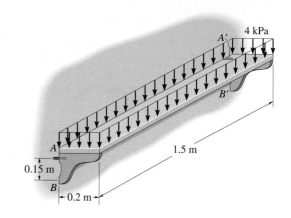

Prob. 5–95

5-94. Determine the normal reaction at the roller A and horizontal and vertical components at pin B for equilibrium of the member.

*5-96. Determine the x and z components of reaction at the journal bearing A and the tension in cords BC and BD necessary for equilibrium of the rod.

Prob. 5–94

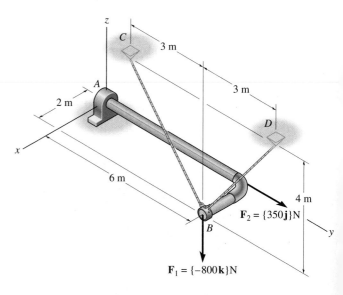

Prob. 5–96

5-97. Determine the reactions at the supports A and B for equilibrium of the beam.

Prob. 5–97

5-99. Determine the x, y, z components of reaction at the fixed wall A. The 150-N force is parallel to the z axis and the 200-N force is parallel to the y axis.

Prob. 5–99

5-98. Determine the x, y, z components of reaction at the ball supports B and C and the ball-and-socket A (not shown) for the uniformly loaded plate.

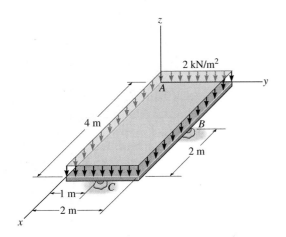

Prob. 5–98

*5-100.** The horizontal beam is supported by springs at its ends. If the stiffness of the spring at A is $k_A = 5$ kN/m, determine the required stiffness of the spring at B so that if the beam is loaded with the 800-N force, it remains in the horizontal position both before and after loading.

Prob. 5–100

The forces within the members of this truss bridge must be determined if they are to be properly designed.

Structural Analysis

- To show how to determine the forces in the members of a truss using the method of joints and the method of sections.
- To analyze the forces acting on the members of frames and machines composed of pin-connected members.

6.1 Simple Trusses

A *truss* is a structure composed of slender members joined together at their end points. The members commonly used in construction consist of wooden struts or metal bars. The joint connections are usually formed by bolting or welding the ends of the members to a common plate, called a *gusset plate,* as shown in Fig. 6–1a, or by simply passing a large bolt or pin through each of the members, Fig. 6–1b.

(a)

(b)

Fig. 6–1

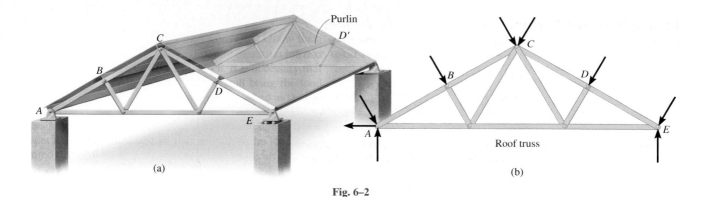

(a)

(b)

Roof truss

Fig. 6–2

Planar Trusses. *Planar* trusses lie in a single plane and are often used to support roofs and bridges. The truss *ABCDE*, shown in Fig. 6–2a, is an example of a typical roof-supporting truss. In this figure, the roof load is transmitted to the truss *at the joints* by means of a series of *purlins*, such as *DD'*. Since the imposed loading acts in the same plane as the truss, Fig. 6–2b, the analysis of the forces developed in the truss members is two-dimensional.

In the case of a bridge, such as shown in Fig. 6–3a, the load on the deck is first transmitted to *stringers,* then to *floor beams,* and finally to the *joints B, C,* and *D* of the two supporting side trusses. Like the roof truss, the bridge truss loading is also coplanar, Fig. 6–3b.

When bridge or roof trusses extend over large distances, a rocker or roller is commonly used for supporting one end, e.g., joint *E* in Figs. 6–2a and 6–3a. This type of support allows freedom for expansion or contraction of the members due to temperature or application of loads.

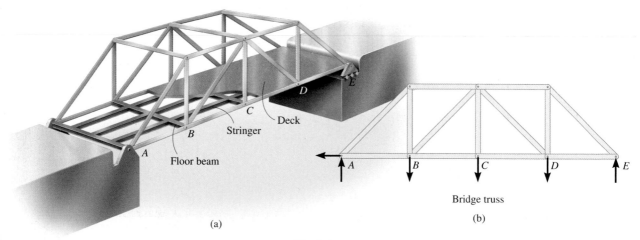

(a)

(b)

Bridge truss

Fig. 6–3

Assumptions for Design. To design both the members and the connections of a truss, it is first necessary to determine the *force* developed in each member when the truss is subjected to a given loading. In this regard, two important assumptions will be made:

1. *All loadings are applied at the joints.* In most situations, such as for bridge and roof trusses, this assumption is true. Frequently in the force analysis the weight of the members is neglected since the forces supported by the members are usually large in comparison with their weight. If the member's weight is to be included in the analysis, it is generally satisfactory to apply it as a vertical force, half of its magnitude applied at each end of the member.

2. *The members are joined together by smooth pins.* In cases where bolted or welded joint connections are used, this assumption is satisfactory provided the center lines of the joining members are *concurrent,* as in Fig. 6–1*a*.

Because of these two assumptions, *each truss member acts as a two-force member,* and therefore the forces at the ends of the member must be directed along the axis of the member. If the force tends to *elongate* the member, it is a *tensile force* (T), Fig. 6–4*a*; whereas if it tends to *shorten* the member, it is a *compressive force* (C), Fig. 6–4*b*. In the actual design of a truss it is important to state whether the nature of the force is tensile or compressive. Often, compression members must be made *thicker* than tension members because of the buckling or column effect that occurs when a member is in compression.

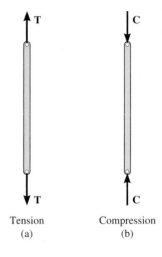

Tension
(a)

Compression
(b)

Fig. 6–4

Simple Truss. To prevent collapse, the form of a truss must be rigid. Obviously, the four-bar shape *ABCD* in Fig. 6–5 will collapse unless a diagonal member, such as *AC*, is added for support. The simplest form that is rigid or stable is a *triangle*. Consequently, a *simple truss* is constructed by *starting* with a basic triangular element, such as *ABC* in Fig. 6–6, and connecting two members (*AD* and *BD*) to form an additional element. As each additional element consisting of two members and a joint is placed on the truss, it is possible to construct a simple truss.

Fig. 6–5

Fig. 6–6

6.2 The Method of Joints

These Howe trusses are used to support the roof of the metal building. Note how the members come together at a common point on the gusset plate and how the roof purlins transmit the load to the joints.

In order to analyze or design a truss, we must obtain the force in each of its members. If we were to consider a free-body diagram of the entire truss, then the forces in the members would be *internal forces*, and they could not be obtained from an equilibrium analysis. Instead, if we consider the equilibrium of a joint of the truss then a member force becomes an *external force* on the joint's free-body diagram, and the equations of equilibrium can be applied to obtain its magnitude. This forms the basis for the *method of joints*.

Because the truss members are all straight two-force members lying in the same plane, the force system acting at each joint is *coplanar and concurrent*. Consequently, rotational or moment equilibrium is automatically satisfied at the joint (or pin), and it is only necessary to satisfy $\Sigma F_x = 0$ and $\Sigma F_y = 0$ to ensure equilibrium.

When using the method of joints, it is *first* necessary to draw the joint's free-body diagram before applying the equilibrium equations. To do this, recall that the *line of action* of each member force acting on the joint is *specified* from the geometry of the truss since the force in a member passes along the axis of the member. As an example, consider the pin at joint B of the truss in Fig. 6–7a. Three forces act on the pin, namely, the 500-N force and the forces exerted by members BA and BC. The free-body diagram is shown in Fig. 6–7b. As shown, $\mathbf{F}_{BA}$ is "pulling" on the pin, which means that member BA is in *tension*; whereas $\mathbf{F}_{BC}$ is "pushing" on the pin, and consequently member BC is in *compression*. These effects are clearly demonstrated by isolating the joint with small segments of the member connected to the pin, Fig. 6–7c. The pushing or pulling on these small segments indicates the effect of the member being either in compression or tension.

In all cases, the analysis should start at a joint having at least one known force and at most two unknown forces, as in Fig. 6–7b. In this way, application of $\Sigma F_x = 0$ and $\Sigma F_y = 0$ yields two algebraic equations which can be solved for the two unknowns. When applying these equations, the correct sense of an unknown member force can be determined using one of two possible methods:

(a)

(b)   (c)

Fig. 6–7

- *Always assume* the *unknown member forces* acting on the joint's free-body diagram to be in *tension*, i.e., "pulling" on the pin. If this is done, then numerical solution of the equilibrium equations will yield *positive scalars for members in tension and negative scalars for members in compression.* Once an unknown member force is found, use its *correct* magnitude and sense (T or C) on subsequent joint free-body diagrams.

- The *correct* sense of direction of an unknown member force can, in many cases, be determined "by inspection." For example, $\mathbf{F}_{BC}$ in Fig. 6–7b must push on the pin (compression) since its horizontal component, $F_{BC} \sin 45°$, must balance the 500-N force ($\Sigma F_x = 0$). Likewise, $\mathbf{F}_{BA}$ is a tensile force since it balances the vertical component, $F_{BC} \cos 45°$ ($\Sigma F_y = 0$). In more complicated cases, the sense of an unknown member force can be *assumed;* then, after applying the equilibrium equations, the assumed sense can be verified from the numerical results. A *positive* answer indicates that the sense is *correct,* whereas a *negative* answer indicates that the sense shown on the free-body diagram must be *reversed.* This is the method we will use in the example problems which follow.

PROCEDURE FOR ANALYSIS

The following procedure provides a typical means for analyzing a truss using the method of joints.

- Draw the free-body diagram of a joint having at least one known force and at most two unknown forces. (If this joint is at one of the supports, then it may be necessary to know the external reactions at the truss support.)

- Use one of the two methods described above for establishing the sense of an unknown force.

- Orient the x and y axes such that the forces on the free-body diagram can be easily resolved into their x and y components and then apply the two force equilibrium equations $\Sigma F_x = 0$ and $\Sigma F_y = 0$. Solve for the two unknown member forces and verify their correct sense.

- Continue to analyze each of the other joints, where again it is necessary to choose a joint having at most two unknowns and at least one known force.

- Once the force in a member is found from the analysis of a joint at one of its ends, the result can be used to analyze the forces acting on the joint at its other end. Remember that a member in *compression* "pushes" on the joint and a member in *tension* "pulls" on the joint.

E X A M P L E **6.1**

2 m

45°

2 m

(a)

(b)

(c)

$F_{BA} = 500$ N

A_x ← A → $F_{CA} = 500$ N

A_y

(d)

(e)

Fig. 6–8

Determine the force in each member of the truss shown in Fig. 6–8a and indicate whether the members are in tension or compression.

Solution

By inspection of Fig. 6–8a, there are two unknown member forces at joint B, two unknown member forces and an unknown reaction force at joint C, and two unknown member forces and two unknown reaction forces at joint A. Since we should have no more than two unknowns at the joint and at least one known force acting there, we will begin the analysis at joint B.

Joint B. The free-body diagram of the pin at B is shown in Fig. 6–8b. Applying the equations of joint equilibrium, we have

$$\xrightarrow{+} \Sigma F_x = 0; \quad 500 \text{ N} - F_{BC} \sin 45° = 0 \quad F_{BC} = 707.1 \text{ N (C)} \quad Ans.$$
$$+\uparrow \Sigma F_y = 0; \quad F_{BC} \cos 45° - F_{BA} = 0 \quad F_{BA} = 500 \text{ N} \quad (T) \quad Ans.$$

Since the force in member BC has been calculated, we can proceed to analyze joint C in order to determine the force in member CA and the support reaction at the rocker.

Joint C. From the free-body diagram of joint C, Fig. 6–8c, we have

$$\xrightarrow{+} \Sigma F_x = 0; \quad -F_{CA} + 707.1 \cos 45° \text{N} = 0 \quad F_{CA} = 500 \text{ N} \quad (T) \quad Ans.$$
$$+\uparrow \Sigma F_y = 0; \quad C_y - 707.1 \sin 45° \text{N} = 0 \quad C_y = 500 \text{ N} \quad Ans.$$

Joint A. Although it is not necessary, we can determine the support reactions at joint A using the results of $F_{CA} = 500$ N and $F_{BA} = 500$ N. From the free-body diagram, Fig. 6–8d, we have

$$\xrightarrow{+} \Sigma F_x = 0; \quad 500 \text{ N} - A_x = 0 \quad A_x = 500 \text{ N}$$
$$+\uparrow \Sigma F_y = 0; \quad 500 \text{ N} - A_y = 0 \quad A_y = 500 \text{ N}$$

The results of the analysis are summarized in Fig. 6–8e. Note that the free-body diagram of each pin shows the effects of all the connected members and external forces applied to the pin, whereas the free-body diagram of each member shows only the effects of the end pins on the member.

EXAMPLE **6.2**

Determine the forces acting in all the members of the truss shown in Fig. 6–9a.

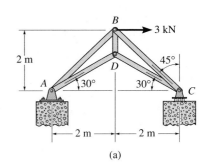

(a)

Solution

By inspection, there are more than two unknowns at each joint. Consequently, the support reactions on the truss must first be determined. Show that they have been correctly calculated on the free-body diagram in Fig. 6–9b. We can now begin the analysis at joint C. Why?

Joint C. From the free-body diagram, Fig. 6–9c,

$$\xrightarrow{+} \Sigma F_x = 0; \qquad -F_{CD} \cos 30° + F_{CB} \sin 45° = 0$$

$$+\uparrow \Sigma F_y = 0; \quad 1.5 \text{ kN} + F_{CD} \sin 30° - F_{CB} \cos 45° = 0$$

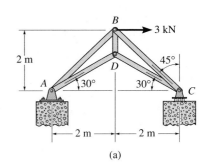

(b)

These two equations must be solved *simultaneously* for each of the two unknowns. Note, however, that a *direct solution* for one of the unknown forces may be obtained by applying a force summation along an axis that is *perpendicular* to the direction of the other unknown force. For example, summing forces along the y' axis, which is perpendicular to the direction of $\mathbf{F}_{CD}$, Fig. 6–9d, yields a direct solution for F_{CB}.

$$+\nearrow \Sigma F_{y'} = 0;$$

$$1.5 \cos 30° \text{kN} - F_{CB} \sin 15° = 0 \quad F_{CB} = 5.02 \text{ kN} \quad (\text{C}) \qquad Ans.$$

In a similar fashion, summing forces along the y'' axis, Fig. 6–9e, yields a direct solution for F_{CD}.

$$+\nearrow \Sigma F_{y''} = 0;$$

$$1.5 \cos 45° \text{kN} - F_{CD} \sin 15° = 0 \qquad F_{CD} = 4.10 \text{ kN} \quad (\text{T}) \quad Ans.$$

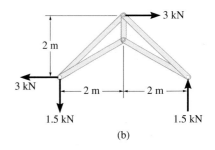

(c) (d)

Joint D. We can now proceed to analyze joint D. The free-body diagram is shown in Fig. 6–9f.

$$\xrightarrow{+} \Sigma F_x = 0; \quad -F_{DA} \cos 30° + 4.10 \cos 30° \text{kN} = 0$$

$$F_{DA} = 4.10 \text{ kN} \quad (\text{T}) \qquad Ans.$$

$$+\uparrow \Sigma F_y = 0; \qquad F_{DB} - 2(4.10 \sin 30° \text{kN}) = 0$$

$$F_{DB} = 4.10 \text{ kN} \quad (\text{T}) \qquad Ans.$$

The force in the last member, *BA*, can be obtained from joint *B* or joint *A*. As an exercise, draw the free-body diagram of joint *B*, sum the forces in the horizontal direction, and show that $F_{BA} = 0.776$ kN (C).

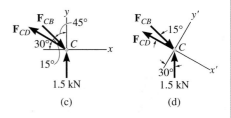

(e) (f)

Fig. 6–9

E X A M P L E 6.3

Determine the force in each member of the truss shown in Fig. 6–10*a*. Indicate whether the members are in tension or compression.

(a)   (b)

Fig. 6–10

Solution

Support Reactions. No joint can be analyzed until the support reactions are determined. Why? A free-body diagram of the entire truss is given in Fig. 6–10*b*. Applying the equations of equilibrium, we have

$$\xrightarrow{+} \Sigma F_x = 0; \qquad 600\ \text{N} - C_x = 0 \qquad C_x = 600\ \text{N}$$
$$\zeta + \Sigma M_C = 0; \quad -A_y(6\ \text{m}) + 400\ \text{N}(3\ \text{m}) + 600\ \text{N}(4\ \text{m}) = 0$$
$$A_y = 600\ \text{N}$$
$$+\uparrow \Sigma F_y = 0; \qquad 600\ \text{N} - 400\ \text{N} - C_y = 0 \qquad C_y = 200\ \text{N}$$

The analysis can now start at either joint *A* or *C*. The choice is arbitrary since there are one known and two unknown member forces acting on the pin at each of these joints.

Joint A (Fig. 6–10*c*). As shown on the free-body diagram, there are three forces that act on the pin at joint *A*. The inclination of $\mathbf{F}_{AB}$ is determined from the geometry of the truss. By inspection, can you see why this force is assumed to be compressive and $\mathbf{F}_{AD}$ tensile? Applying the equations of equilibrium, we have

$$+\uparrow \Sigma F_y = 0; \qquad 600\ \text{N} - \tfrac{4}{5}F_{AB} = 0 \qquad F_{AB} = 750\ \text{N} \quad (\text{C}) \qquad Ans.$$
$$\xrightarrow{+} \Sigma F_x = 0; \quad F_{AD} - \tfrac{3}{5}(750\ \text{N}) = 0 \qquad F_{AD} = 450\ \text{N} \quad (\text{T}) \qquad Ans.$$

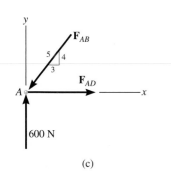

(c)

Joint D (Fig. 6–10*d*). The pin at this joint is chosen next since, by inspection of Fig. 6–10*a*, the force in *AD* is known and the unknown forces in *DB* and *DC* can be determined. Summing forces in the horizontal direction, Fig. 6–10*d*, we have

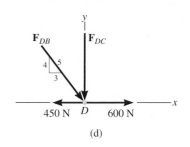

$$\xrightarrow{+} \Sigma F_x = 0; \quad -450\ \text{N} + \tfrac{3}{5}F_{DB} + 600\ \text{N} = 0 \quad F_{DB} = -250\ \text{N}$$

The negative sign indicates that $\mathbf{F}_{DB}$ acts in the *opposite sense* to that shown in Fig. 6–10*d**. Hence,

$$F_{DB} = 250\ \text{N} \qquad (\text{T}) \qquad\qquad Ans.$$

(d)

To determine $\mathbf{F}_{DC}$, we can either correct the sense of $\mathbf{F}_{DB}$ and then apply $\Sigma F_y = 0$, or apply this equation and retain the negative sign for F_{DB}, i.e.,

$$+\uparrow\Sigma F_y = 0; \quad -F_{DC} - \tfrac{4}{5}(-250\ \text{N}) = 0 \qquad F_{DC} = 200\ \text{N} \quad (\text{C}) \quad Ans.$$

Joint C (Fig. 6–10*e*).

$$\xrightarrow{+} \Sigma F_x = 0; \quad F_{CB} - 600\ \text{N} = 0 \qquad F_{CB} = 600\ \text{N} \quad (\text{C}) \qquad\qquad Ans.$$

$$+\uparrow\Sigma F_y = 0; \quad 200\ \text{N} - 200\ \text{N} \equiv 0 \quad (\text{check})$$

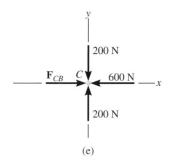

(e)

The analysis is summarized in Fig. 6–10*f*, which shows the correct free-body diagram for each pin and member.

(f)

*The proper sense could have been determined by inspection, prior to applying $\Sigma F_x = 0$.

6.3 Zero-Force Members

Truss analysis using the method of joints is greatly simplified if one is first able to determine those members which support *no loading*. These *zero-force members* are used to increase the stability of the truss during construction and to provide support if the applied loading is changed.

The zero-force members of a truss can generally be determined *by inspection* of each of its joints. For example, consider the truss shown in Fig. 6–11a. If a free-body diagram of the pin at joint A is drawn, Fig. 6–11b, it is seen that members AB and AF are zero-force members. On the other hand, notice that we could not have come to this conclusion if we had considered the free-body diagrams of joints F or B simply because there are five unknowns at each of these joints. In a similar manner, consider the free-body diagram of joint D, Fig. 6–11c. Here again it is seen that DC and DE are zero-force members. As a general rule, *if only two members form a truss joint and no external load or support reaction is applied to the joint, the members must be zero-force members*. The load on the truss in Fig. 6–11a is therefore supported by only five members as shown in Fig. 6–11d.

(a)

$$\xrightarrow{+} \Sigma F_x = 0; \quad F_{AB} = 0$$
$$+\uparrow \Sigma F_y = 0; \quad F_{AF} = 0$$

(b)

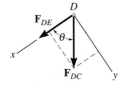

$$+\searrow \Sigma F_y = 0; F_{DC} \sin \theta = 0; \quad F_{DC} = 0 \text{ since } \sin \theta \neq 0$$
$$+\swarrow \Sigma F_x = 0; F_{DE} + 0 = 0; \quad F_{DE} = 0$$

(c)

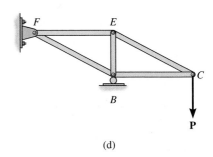

(d)

Fig. 6–11

Now consider the truss shown in Fig. 6–12*a*. The free-body diagram of the pin at joint *D* is shown in Fig. 6–12*b*. By orienting the *y* axis along members *DC* and *DE* and the *x* axis along member *DA*, it is seen that *DA* is a zero-force member. Note that this is also the case for member *CA*, Fig. 6–12*c*. In general, *if three members form a truss joint for which two of the members are collinear, the third member is a zero-force member provided no external force or support reaction is applied to the joint.* The truss shown in Fig. 6–12*d* is therefore suitable for supporting the load **P**.

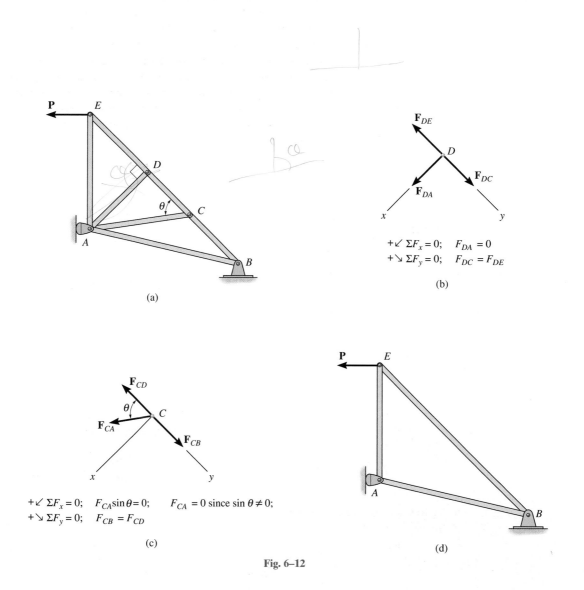

$+\swarrow \Sigma F_x = 0; \quad F_{DA} = 0$
$+\searrow \Sigma F_y = 0; \quad F_{DC} = F_{DE}$

(b)

(a)

$+\swarrow \Sigma F_x = 0; \quad F_{CA}\sin\theta = 0; \quad F_{CA} = 0 \text{ since } \sin\theta \neq 0;$
$+\searrow \Sigma F_y = 0; \quad F_{CB} = F_{CD}$

(c)

(d)

Fig. 6–12

E X A M P L E   6.4

(b)

(c)

(d)

(e)

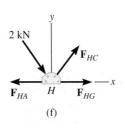

(f)

Using the method of joints, determine all the zero-force members of the *Fink roof truss* shown in Fig. 6–13a. Assume all joints are pin connected.

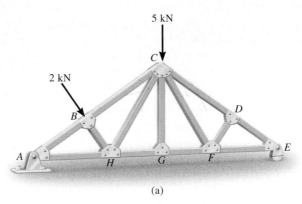

(a)

Fig. 6–13

Solution

Look for joint geometries that have three members for which two are collinear. We have

Joint G (Fig. 6–13b).

$$+\uparrow\Sigma F_y = 0; \qquad F_{GC} = 0 \qquad\qquad\qquad Ans.$$

Realize that we could not conclude that *GC* is a zero-force member by considering joint *C*, where there are five unknowns. The fact that *GC* is a zero-force member means that the 5-kN load at *C* must be supported by members *CB*, *CH*, *CF*, and *CD*.

Joint D (Fig. 6–13c).

$$+\swarrow\Sigma F_x = 0; \qquad F_{DF} = 0 \qquad\qquad\qquad Ans.$$

Joint F (Fig. 6–13d).

$$+\uparrow\Sigma F_y = 0; \quad F_{FC}\cos\theta = 0 \qquad \text{Since } \theta \neq 90°, \qquad F_{FC} = 0 \quad Ans.$$

Note that if joint *B* is analyzed, Fig. 6–13e,

$$+\searrow\Sigma F_x = 0; \quad 2\text{ kN} - F_{BH} = 0 \quad F_{BH} = 2\text{ kN} \qquad (C)$$

Note that F_{HC} must satisfy $\Sigma F_y = 0$, Fig. 6–13f, and therefore *HC* is *not* a zero-force member.

PROBLEMS

6-1. Determine the force in each member of the truss and state if the members are in tension or compression. Set $P_1 = 800$ kN and $P_2 = 400$ kN.

6-2. Determine the force on each member of the truss and state if the members are in tension or compression. Set $P_1 = 500$ kN and $P_1 = 100$ kN.

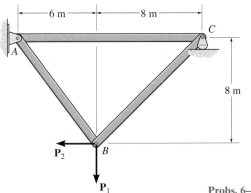

Probs. 6–1/2

6-3. The truss, used to support a balcony, is subjected to the loading shown. Approximate each joint as a pin and determine the force in each member. State whether the members are in tension or compression. Set $P_1 = 600$ kN, $P_2 = 400$ kN.

*6-4. The truss, used to support a balcony, is subjected to the loading shown. Approximate each joint as a pin and determine the force in each member. State whether the members are in tension or compression. Set $P_1 = 800$ kN, $P_2 = 0$.

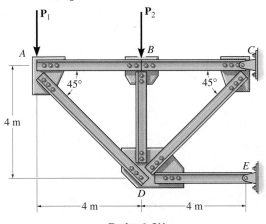

Probs. 6–3/4

6-5. Determine the force in each member of the truss and state if the members are in tension or compression. Assume each joint as a pin. Set $P = 4$ kN.

6-6. Assume that each member of the truss is made of steel having a mass per length of 4 kg/m. Set $P = 0$, determine the force in each member, and indicate if the members are in tension or compression. Neglect the weight of the gusset plates and assume each joint is a pin. Solve the problem by assuming the weight of each member can be represented as a vertical force, half of which is applied at the end of each member.

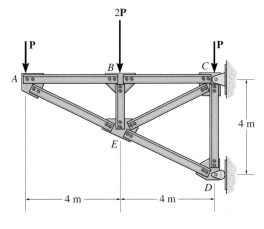

Probs. 6–5/6

6-7. Determine the force in each member of the truss and state if the members are in tension or compression.

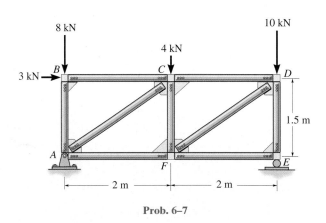

Prob. 6–7

*6-8. Determine the force in each member of the truss and state if the members are in tension or compression. Set $P_1 = 2$ kN and $P_2 = 1.5$ kN.

6-9. Determine the force in each member of the truss and state if the members are in tension or compression. Set $P_1 = P_2 = 4$ kN.

*6-12. Determine the force in each member of the truss and state if the members are in tension or compression. Set $P_1 = 10$ kN, $P_2 = 15$ kN.

6-13. Determine the force in each member of the truss and state if the members are in tension or compression. Set $P_1 = 0$, $P_2 = 20$ kN.

Probs. 6–8/9

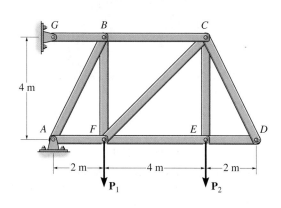

Probs. 6–12/13

6-10. Determine the force in each member of the truss and state if the members are in tension or compression. Set $P_1 = 0$, $P_2 = 100$ kN.

6-11. Determine the force in each member of the truss and state if the members are in tension or compression. Set $P_1 = 50$ kN, $P_2 = 150$ kN.

6-14. Determine the force in each member of the truss and state if the members are in tension or compression. Set $P_1 = 10$ kN, $P_2 = 20$ kN, $P_3 = 30$ kN.

6-15. Determine the force in each member of the truss and state if the members are in tension or compression. Set $P_1 = 40$ kN, $P_2 = 40$ kN, $P_3 = 0$.

Probs. 6–10/11

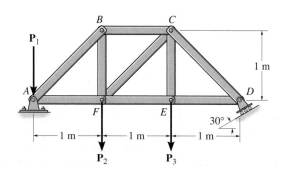

Probs. 6–14/15

*6-16. Determine the force in each member of the truss. State whether the members are in tension or compression. Set $P = 8$ kN.

6-17. If the maximum force that any member can support is 8 kN in tension and 6 kN in compression, determine the maximum force P that can be supported at joint D.

Probs. 6–16/17

6-18. Determine the force in each member of the truss and state if the members are in tension or compression. *Hint:* The horizontal force component at A must be zero. Why?

Prob. 6–18

6-19. Determine the force in each member of the truss and state if the members are in tension or compression. *Hint:* The resultant force at the pin E acts along member ED. Why?

*6-20. Each member of the truss is uniform and has a mass of 8 kg/m. Remove the external loads of 3 kN and 2 kN and determine the approximate force in each member due to the weight of the truss. State if the members are in tension or compression. Solve the problem by *assuming* the weight of each member can be represented as a vertical force, half of which is applied at each end of the member.

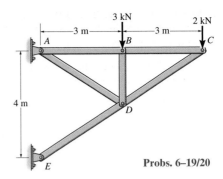

Probs. 6–19/20

6-21. Determine the force in each member of the truss in terms of the external loading and state if the members are in tension or compression.

6-22. The maximum allowable tensile force in the members of the truss is $(F_t)_{max} = 2$ kN, and the maximum allowable compressive force is $(F_c)_{max} = 1.2$ kN. Determine the maximum magnitude P of the two loads that can be applied to the truss. Take $L = 2$ m and $\theta = 30°$.

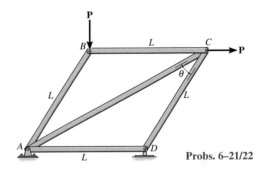

Probs. 6–21/22

6-23. Determine the force in each member of the truss and state if the members are in tension or compression.

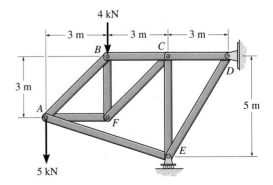

Prob. 6–23

*6-24. Determine the force in each member of the double scissors truss in terms of the load P and state if the members are in tension or compression.

Prob. 6–24

6-25. Determine the force in each member of the truss and state if the members are in tension or compression. *Hint:* The vertical component of force at C must equal zero. Why?

6-26. Each member of the truss is uniform and has a mass of 8 kg/m. Remove the external loads of 6 kN and 8 kN and determine the approximate force in each member due to the weight of the truss. State if the members are in tension or compression. Solve the problem by *assuming* the weight of each member can be represented as a vertical force, half of which is applied at each end of the member.

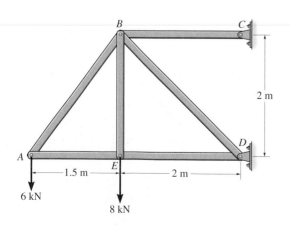

Probs. 6–25/26

6-27. Determine the force in each member of the truss in terms of the load P, and indicate whether the members are in tension or compression.

*6-28. If the maximum force that any member can support is 4 kN in tension and 3 kN in compression, determine the maximum force P that can be supported at point B. Take $d = 1$m.

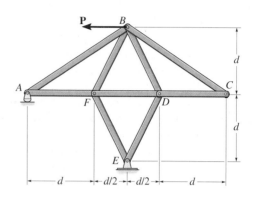

Probs. 6–27/28

■6-29. The two-member truss is subjected to the force of 300 kN. Determine the range of θ for application of the load so that the force in either member does not exceed 400 kN (T) or 200 kN (C).

Prob. 6–29

6.4 The Method of Sections

The *method of sections* is used to determine the loadings acting within a body. It is based on the principle that if a body is in equilibrium then any part of the body is also in equilibrium. For example, consider the two truss members shown on the left in Fig. 6–14. If the forces within the members are to be determined, then an imaginary section indicated by the blue line, can be used to cut each member into two parts and thereby "expose" each internal force as "external" to the free-body diagrams shown on the right. Clearly, it can be seen that equilibrium requires that the member in tension (T) be subjected to a "pull," whereas the member in compression (C) is subjected to a "push."

The method of sections can also be used to "cut" or section the members of an entire truss. If the section passes through the truss and the free-body diagram of either of its two parts is drawn, we can then apply the equations of equilibrium to that part to determine the member forces at the "cut section." Since only *three* independent equilibrium equations ($\Sigma F_x = 0$, $\Sigma F_y = 0$, $\Sigma M_O = 0$) can be applied to the isolated part of the truss, try to select a section that, in general, passes through not more than *three* members in which the forces are unknown. For example, consider the truss in Fig. 6–15*a*. If the force in member *GC* is to be determined, section *aa* would be appropriate. The free-body diagrams of the two parts are shown in Figs. 6–15*b* and 6–15*c*. In particular, note that the line of action of each member force is specified from the *geometry* of the truss, since the force in a member passes along its axis. Also, the member forces acting on one part of the truss are equal but opposite to those acting on the other part—Newton's third law. As noted above, members assumed to be in *tension* (*BC* and *GC*) are subjected to a "pull," whereas the member in *compression* (*GF*) is subjected to a "push."

Internal tensile forces

Tension

Internal compressive forces

Compression

Fig. 6–14

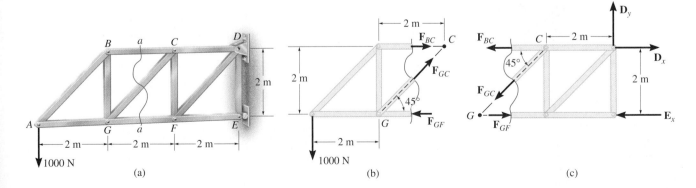

(a)

(b)

(c)

Fig. 6–15

Fig. 6–15

Two Pratt trusses are used to construct this pedestrian bridge.

The three unknown member forces $\mathbf{F}_{BC}$, $\mathbf{F}_{GC}$, and $\mathbf{F}_{GF}$ can be obtained by applying the three equilibrium equations to the free-body diagram in Fig. 6–15b. If, however, the free-body diagram in Fig. 6–15c is considered, the three support reactions $\mathbf{D}_x$, $\mathbf{D}_y$ and $\mathbf{E}_x$ will have to be determined *first*. Why? (This, of course, is done in the usual manner by considering a free-body diagram of the *entire truss.*)

When applying the equilibrium equations, one should consider ways of writing the equations so as to yield a *direct solution* for each of the unknowns, rather than having to solve simultaneous equations. For example, summing moments about C in Fig. 6–15b would yield a direct solution for $\mathbf{F}_{GF}$ since $\mathbf{F}_{BC}$ and $\mathbf{F}_{GC}$ create zero moment about C. Likewise, $\mathbf{F}_{BC}$ can be directly obtained by summing moments about G. Finally, $\mathbf{F}_{GC}$ can be found directly from a force summation in the vertical direction since $\mathbf{F}_{GF}$ and $\mathbf{F}_{BC}$ have no vertical components. This ability to *determine directly* the force in a particular truss member is one of the main advantages of using the method of sections.*

*By comparison, if the method of joints were used to determine, say, the force in member GC, it would be necessary to analyze joints A, B, and G in sequence.

As in the method of joints, there are two ways in which one can determine the correct sense of an unknown member force:

- *Always assume* that the unknown member forces at the cut section are in *tension,* i.e., "pulling" on the member. By doing this, the numerical solution of the equilibrium equations will yield *positive scalars for members in tension and negative scalars for members in compression.*

- The correct sense of an unknown member force can in many cases be determined "by inspection." For example, $\mathbf{F}_{BC}$ is a tensile force as represented in Fig. 6–15*b* since moment equilibrium about G requires that $\mathbf{F}_{BC}$ create a moment opposite to that of the 1000-N force. Also, $\mathbf{F}_{GC}$ is tensile since its vertical component must balance the 1000-N force which acts downward. In more complicated cases, the sense of an unknown member force may be *assumed.* If the solution yields a *negative* scalar, it indicates that the force's sense is *opposite* to that shown on the free-body diagram. This is the method we will use in the example problems which follow.

PROCEDURE FOR ANALYSIS

The forces in the members of a truss may be determined by the method of sections using the following procedure.

Free-Body Diagram

- Make a decision as to how to "cut" or section the truss through the members where forces are to be determined.

- Before isolating the appropriate section, it may first be necessary to determine the truss's *external* reactions. Then three equilibrium equations are available to solve for member forces at the cut section.

- Draw the free-body diagram of that part of the sectioned truss which has the least number of forces acting on it.

- Use one of the two methods described above for establishing the sense of an unknown member force.

Equations of Equilibrium

- Moments should be summed about a point that lies at the intersection of the lines of action of two unknown forces, so that the third unknown force is determined directly from the moment equation.

- If two of the unknown forces are *parallel,* forces may be summed *perpendicular* to the direction of these unknowns to determine *directly* the third unknown force.

E X A M P L E   6.5

Fig. 6–16

Determine the force in members *GE, GC,* and *BC* of the truss shown in Fig. 6–16*a*. Indicate whether the members are in tension or compression.

Solution

Section *aa* in Fig. 6–16*a* has been chosen since it cuts through the *three* members whose forces are to be determined. In order to use the method of sections, however, it is *first* necessary to determine the external reactions at *A* or *D*. Why? A free-body diagram of the entire truss is shown in Fig. 6–16*b*. Applying the equations of equilibrium, we have

$$\xrightarrow{+} \Sigma F_x = 0; \qquad 400 \text{ N} - A_x = 0 \qquad A_x = 400 \text{ N}$$
$$\curvearrowleft + \Sigma M_A = 0; \quad -1200 \text{ N}(8 \text{ m}) - 400 \text{ N}(3 \text{ m}) + D_y(12 \text{ m}) = 0$$
$$D_y = 900 \text{ N}$$
$$+\uparrow \Sigma F_y = 0; \quad A_y - 1200 \text{ N} + 900 \text{ N} = 0 \qquad A_y = 300 \text{ N}$$

Free-Body Diagram. The free-body diagram of the left portion of the sectioned truss is shown in Fig. 6–16*c*. For the analysis this diagram will be used since it involves the least number of forces.

Equations of Equilibrium. Summing moments about point *G* eliminates **F**$_{GE}$ and **F**$_{GC}$ and yields a direct solution for *F*$_{BC}$.

$$\curvearrowleft + \Sigma M_G = 0; \quad -300 \text{ N}(4 \text{ m}) - 400 \text{ N}(3 \text{ m}) + F_{BC}(3 \text{ m}) = 0$$
$$F_{BC} = 800 \text{ N} \quad \text{(T)} \qquad \qquad Ans.$$

In the same manner, by summing moments about point *C* we obtain a direct solution for *F*$_{GE}$.

$$\curvearrowleft + \Sigma M_C = 0; \quad -300 \text{ N}(8 \text{ m}) + F_{GE}(3 \text{ m}) = 0$$
$$F_{GE} = 800 \text{ N} \quad \text{(C)} \qquad \qquad Ans.$$

Since **F**$_{BC}$ and **F**$_{GE}$ have no vertical components, summing forces in the *y* direction directly yields *F*$_{GC}$, i.e.,

$$+\uparrow \Sigma F_y = 0 \qquad 300\text{N} - \tfrac{3}{5}F_{GC} = 0$$
$$F_{GC} = 500 \text{ N} \quad \text{(T)} \qquad \qquad Ans.$$

As an exercise, obtain these results by applying the equations of equilibrium to the free-body diagram of the right portion of the sectioned truss.

EXAMPLE 6.6

Determine the force in member CF of the bridge truss shown in Fig. 6–17a. Indicate whether the member is in tension or compression. Assume each member is pin-connected.

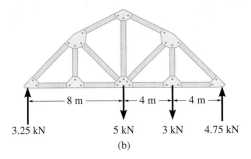

(a)   (b)

Fig. 6–17

Solution

Free-Body Diagram.   Section *aa* in Fig. 6–17a will be used since this section will "expose" the internal force in member *CF* as "external" on the free-body diagram of either the right or left portion of the truss. It is first necessary, however, to determine the external reactions on either the left or right side. Verify the results shown on the free-body diagram in Fig. 6–17b.

The free-body diagram of the right portion of the truss, which is the easiest to analyze, is shown in Fig. 6–17c. There are three unknowns, F_{FG}, F_{CF}, and F_{CD}.

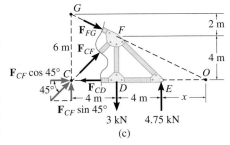

(c)

Equations of Equilibrium.   The most direct method for solving this problem requires application of the moment equation about a point that eliminates two of the unknown forces. Hence, to obtain $\mathbf{F}_{CF}$, we will eliminate $\mathbf{F}_{FG}$ and $\mathbf{F}_{CD}$ by summing moments about point O, Fig. 6–17c. Note that the location of point O measured from E is determined from proportional triangles, i.e., $4/(4 + x) = 6/(8 + x)$, $x = 4$ m. Or, stated in another manner, the slope of member GF has a drop of 2 m to a horizontal distance of 4 m. Since FD is 4 m, Fig. 6–17c, then from D to O the distance must be 8 m.

An easy way to determine the moment of $\mathbf{F}_{CF}$ about point O is to use the principle of transmissibility and move $\mathbf{F}_{CF}$ to point C, and then resolve $\mathbf{F}_{CF}$ into its two rectangular components. We have

$$\zeta + \Sigma M_O = 0;$$
$$-F_{CF} \sin 45°(12 \text{ m}) + (3 \text{ kN})(8 \text{ m}) - (4.75 \text{ kN})(4 \text{ m}) = 0$$
$$F_{CF} = 0.589 \text{ kN} \quad \text{(C)} \qquad\qquad Ans.$$

EXAMPLE 6.7

1000 N

3000 N   1000 N

1000 N

2 m — 2 m — 2 m — 2 m

4000 N   2000 N

(a)

Determine the force in member EB of the roof truss shown in Fig. 6–18a. Indicate whether the member is in tension or compression.

Solution

Free-Body Diagrams. By the method of sections, any imaginary vertical section that cuts through EB, Fig. 6–18a, will also have to cut through three other members for which the forces are unknown. For example, section aa cuts through ED, EB, FB, and AB. If the components of reaction at A are calculated first ($A_x = 0$, $A_y = 4000$ N) and a free-body diagram of the left side of this section is considered, Fig. 6–18b, it is possible to obtain $\mathbf{F}_{ED}$ by summing moments about B to eliminate the other three unknowns; however, $\mathbf{F}_{EB}$ cannot be determined from the remaining two equilibrium equations. One possible way of obtaining $\mathbf{F}_{EB}$ is first to determine $\mathbf{F}_{ED}$ from section aa, then use this result on section bb, Fig. 6–18a, which is shown in Fig. 6–18c. Here the force system is concurrent and our sectioned free-body diagram is the same as the free-body diagram for the pin at E (method of joints).

(b)

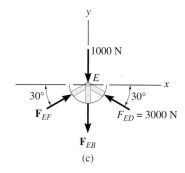

(c)

Fig. 6–18

Equations of Equilibrium. In order to determine the moment of $\mathbf{F}_{ED}$ about point B, Fig. 6–18b, we will resolve the force into its rectangular components and, by the principle of transmissibility, extend it to point C as shown. The moments of 1000 N, F_{AB}, F_{FB}, F_{EB}, and $F_{ED}\cos 30°$ are all zero about B. Therefore,

$$\zeta + \Sigma M_B = 0; \qquad 1000\ \text{N}(4\ \text{m}) + 3000\ \text{N}(2\ \text{m}) - 4000\ \text{N}(4\ \text{m})$$
$$+ F_{ED}\sin 30°(4) = 0$$
$$F_{ED} = 3000\ \text{N} \quad (C)$$

Considering now the free-body diagram of section bb, Fig. 6–18c, we have

$$\xrightarrow{+} \Sigma F_x = 0; \qquad F_{EF}\cos 30° - 3000\cos 30°\ \text{N} = 0$$
$$F_{EF} = 3000\ \text{N} \quad (C)$$
$$+\uparrow \Sigma F_y = 0;\ 2(3000\sin 30°\ \text{N}) - 1000\ \text{N} - F_{EB} = 0$$
$$F_{EB} = 2000\ \text{N} \quad (T) \qquad\qquad Ans.$$

PROBLEMS

6-30. Determine the force in members *BC*, *HC*, and *HG* of the bridge truss, and indicate whether the members are in tension or compression.

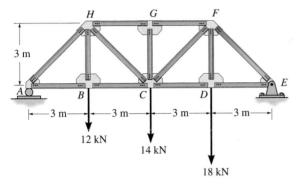

Prob. 6–30

6-31. Determine the force in members *GF*, *CF*, and *CD* of the bridge truss, and indicate whether the members are in tension or compression

Prob. 6–31

*6-32. Determine the force in members *DE*, *DF*, and *GF* of the cantilevered truss and state if the members are in tension or compression.

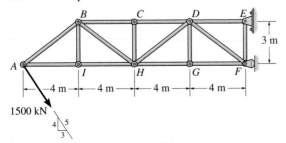

Prob. 6–32

6-33. The roof truss supports the vertical loading shown. Determine the force in members *BC*, *CK*, and *KJ* and state if these members are in tension or compression.

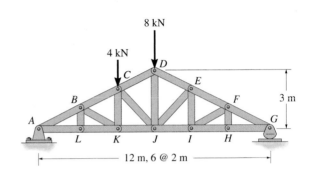

Prob. 6–33

6-34. Determine the force in members *CD*, *CJ*, *KJ*, and *DJ* of the truss which serves to support the deck of a bridge. State if these members are in tension or compression.

6-35. Determine the force in members *EI* and *JI* of the truss which serves to support the deck of a bridge. State if these members are in tension or compression.

Probs. 6–34/35

*6-36. Determine the force in members *BC, CG,* and *GF* of the *Warren truss*. Indicate if the members are in tension or compression.

6-37. Determine the force in members *CD, CF,* and *FG* of the *Warren truss*. Indicate if the members are in tension or compression.

■6-39. The truss supports the vertical load of 600 N. Determine the force in members *BC, BG,* and *HG* as the dimension *L* varies. Plot the results of *F* (ordinate with tension as positive) versus *L* (abscissa) for $0 \le L \le 3$ m.

Probs. 6–36/37

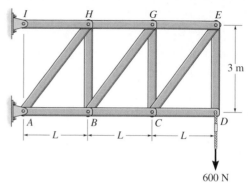

Prob. 6–39

*6-40. Determine the force in members *IC* and *CG* of the truss and state if these members are in tension or compression. Also, indicate all zero-force members.

6-41. Determine the force in members *JE* and *GF* of the truss and state if these members are in tension or compression. Also, indicate all zero-force members.

6-38. Determine the force developed in members *GB* and *GF* of the bridge truss and state if these members are in tension or compression.

Prob. 6–38

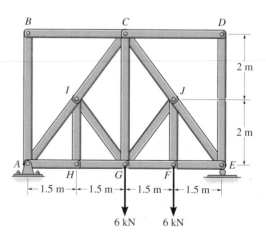

Probs. 6–40/41

6-42. Determine the force in members *BC*, *HC*, and *HG*. After the truss is sectioned use a single equation of equilibrium for the calculation of each force. State if these members are in tension or compression.

6-43. Determine the force in members *CD*, *CF*, and *CG* and state if these members are in tension or compression.

6-45. Determine the force in member *GJ* of the truss and state if this member is in tension or compression.

6-46. Determine the force in member *GC* of the truss and state if this member is in tension or compression.

Probs. 6–42/43

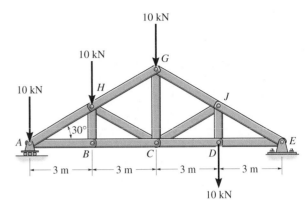

Probs. 6–45/46

*6-44. Determine the force in members *GF*, *FB*, and *BC* of the *Fink truss* and state if the members are in tension or compression.

6-47. Determine the force in members *GF*, *CF*, and *CD* of the roof truss and indicate if the members are in tension or compression.

Prob. 6–44

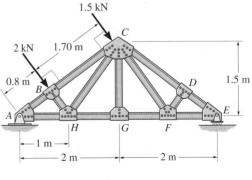

Prob. 6–47

*6-48. Determine the force in members *BG, HG,* and *BC* of the truss and state if the members are in tension or compression.

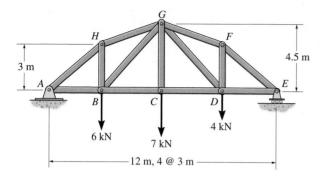

Prob. 6–48

6-49. The skewed truss carries the load shown. Determine the force in members *CB, BE,* and *EF* and state if these members are in tension or compression. Assume that all joints are pinned.

6-50. The skewed truss carries the load shown. Determine the force in members *AB, BF,* and *EF* and state if these members are in tension or compression. Assume that all joints are pinned.

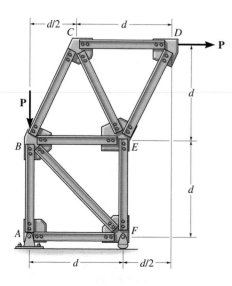

Probs. 6–49/50

6-51. Determine the force in members *CD* and *CM* of the *Baltimore bridge truss* and state if the members are in tension or compression. Also, indicate all zero-force members.

*6-52. Determine the force in members *EF, EP,* and *LK* of the *Baltimore bridge truss* and state if the members are in tension or compression. Also, indicate all zero-force members.

Probs. 6–51/52

6-53. Determine the force in members *KJ, NJ, ND,* and *CD* of the *K truss.* Indicate if the members are in tension or compression. *Hint:* Use sections *aa* and *bb.*

6-54. Determine the force in members *JI* and *DE* of the *K truss.* Indicate if the members are in tension or compression.

Probs. 6–53/54

*6.5 Space Trusses

A *space truss* consists of members joined together at their ends to form a stable three-dimensional structure. The simplest element of a space truss is a *tetrahedron*, formed by connecting six members together, as shown in Fig. 6–19. Any additional members added to this basic element would be redundant in supporting the force **P**. A *simple space truss* can be built from this basic tetrahedral element by adding three additional members and a joint, forming a system of multiconnected tetrahedrons.

Assumptions for Design.   The members of a space truss may be treated as two-force members provided the external loading is applied at the joints and the joints consist of ball-and-socket connections. These assumptions are justified if the welded or bolted connections of the joined members intersect at a common point and the weight of the members can be neglected. In cases where the weight of a member is to be included in the analysis, it is generally satisfactory to apply it as a vertical force, half of its magnitude applied at each end of the member.

Fig. 6-19

PROCEDURE FOR ANALYSIS

Either the method of joints or the method of sections can be used to determine the forces developed in the members of a simple space truss.

Method of Joints.

Generally, if the forces in *all* the members of the truss must be determined, the method of joints is most suitable for the analysis. When using the method of joints, it is necessary to solve the three scalar equilibrium equations $\Sigma F_x = 0$, $\Sigma F_y = 0$, $\Sigma F_z = 0$ at each joint. The solution of many simultaneous equations can be avoided if the force analysis begins at a joint having at least one known force and at most three unknown forces. If the three-dimensional geometry of the force system at the joint is hard to visualize, it is recommended that a Cartesian vector analysis be used for the solution.

Method of Sections.

If only a *few* member forces are to be determined, the method of sections may be used. When an imaginary section is passed through a truss and the truss is separated into two parts, the force system acting on one of the parts must satisfy the *six* scalar equilibrium equations: $\Sigma F_x = 0$, $\Sigma F_y = 0$, $\Sigma F_z = 0$, $\Sigma M_x = 0$, $\Sigma M_y = 0$, $\Sigma M_z = 0$ (Eqs. 5–6). By proper choice of the section and axes for summing forces and moments, many of the unknown member forces in a space truss can be computed *directly*, using a single equilibrium equation.

Typical roof-supporting space truss. Notice the use of ball-and-socket joints for the connections.

EXAMPLE 6.8

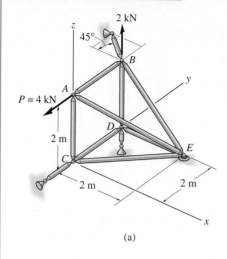

(a)

Determine the forces acting in the members of the space truss shown in Fig. 6–20a. Indicate whether the members are in tension or compression.

Solution

Since there are one known force and three unknown forces acting at joint A, the force analysis of the truss will begin at this joint.

Joint A (Fig. 6–20b).   Expressing each force that acts on the free-body diagram of joint A in vector notation, we have

$$\mathbf{P} = \{-4\mathbf{j}\} \text{ kN}, \quad \mathbf{F}_{AB} = F_{AB}\mathbf{j}, \quad \mathbf{F}_{AC} = -F_{AC}\mathbf{k},$$

$$\mathbf{F}_{AE} = F_{AE}\left(\frac{\mathbf{r}_{AE}}{r_{AE}}\right) = F_{AE}(0.577\mathbf{i} + 0.577\mathbf{j} - 0.577\mathbf{k})$$

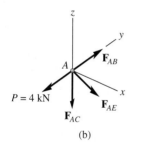

(b)

For equilibrium,

$$\Sigma \mathbf{F} = 0; \qquad \mathbf{P} + \mathbf{F}_{AB} + \mathbf{F}_{AC} + \mathbf{F}_{AE} = 0$$

$$-4\mathbf{j} + F_{AB}\mathbf{j} - F_{AC}\mathbf{k} + 0.577F_{AE}\mathbf{i} + 0.577F_{AE}\mathbf{j} - 0.577F_{AE}\mathbf{k} = 0$$

$$\Sigma F_x = 0; \qquad\qquad\qquad 0.577F_{AE} = 0$$

$$\Sigma F_y = 0; \qquad\qquad -4 + F_{AB} + 0.577F_{AE} = 0$$

$$\Sigma F_z = 0; \qquad\qquad\quad -F_{AC} - 0.577F_{AE} = 0$$

$$F_{AC} = F_{AE} = 0 \qquad\qquad\qquad Ans.$$

$$F_{AB} = 4 \text{ kN} \quad (\text{T}) \qquad Ans.$$

Since F_{AB} is known, joint B may be analyzed next.

Joint B (Fig. 6–20c).

$$\Sigma F_x = 0; \qquad -R_B \cos 45° + 0.707F_{BE} = 0$$

$$\Sigma F_y = 0; \qquad\qquad -4 + R_B \sin 45° = 0$$

$$\Sigma F_z = 0; \qquad\qquad 2 + F_{BD} - 0.707F_{BE} = 0$$

$$R_B = F_{BE} = 5.66 \text{ kN} \quad (\text{T}), \qquad F_{BD} = 2 \text{ kN} \quad (\text{C}) \qquad Ans.$$

The *scalar* equations of equilibrium may also be applied directly to the force systems on the free-body diagrams of joints D and C since the force components are easily determined. Show that

$$F_{DE} = F_{DC} = F_{CE} = 0 \qquad\qquad Ans.$$

Fig. 6–20

(c)

PROBLEMS

6-55. Determine the force in each member of the three-member space truss that supports the loading of 1000 lb and state if the members are in tension or compression.

Prob. 6–55

*6-56. Determine the force in each member of the space truss and state if the members are in tension or compression. *Hint:* The support reaction at E acts along member EB. Why?

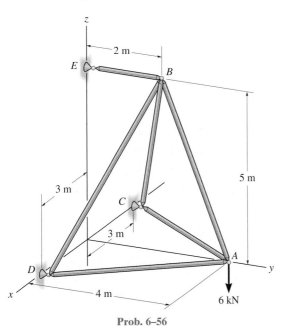

Prob. 6–56

6-57. Determine the force in each member of the space truss and state if the members are in tension or compression. The truss is supported by rollers at A, B, and C.

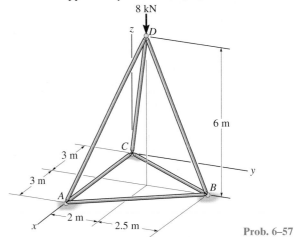

Prob. 6–57

6-58. The space truss is supported by a ball-and-socket joint at D and short links at C and E. Determine the force in each member and state if the members are in tension or compression. Take $\mathbf{F}_1 = \{-500\mathbf{k}\}$ kN and $\mathbf{F}_2 = \{400\mathbf{j}\}$ kN.

6-59. The space truss is supported by a ball-and-socket joint at D and short links at C and E. Determine the force in each member and state if the members are in tension or compression. Take $\mathbf{F}_1 = \{200\mathbf{i} + 300\mathbf{j} - 500\mathbf{k}\}$ kN and $\mathbf{F}_2 = \{400\mathbf{j}\}$ kN.

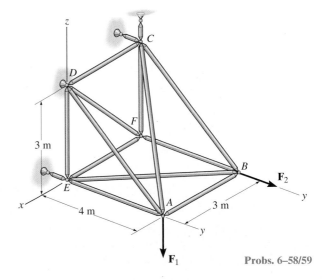

Probs. 6–58/59

*6-60. Determine the force in each member of the space truss and state if the members are in tension or compression. The truss is supported by a ball-and-socket joints at *A B*, and *E*. Set $\mathbf{F} = \{-200\mathbf{i} + 400\mathbf{j}\}$ N. *Hint:* The support reaction at *E* acts along member *E C*. Why?

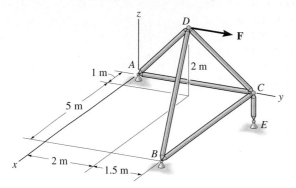

Prob. 6–60

6-61. Determine the force in each member of the space truss and state if the members are in tension or compression. The truss is supported by ball-and-socket joints at *C, D, E*, and *G*.

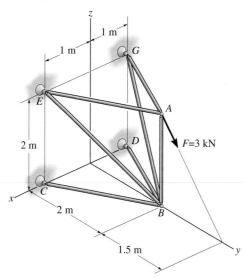

Prob. 6–61

6-62. Determine the force in members *BE, DF*, and *BC* of the space truss and state if the members are in tension or compression.

6-63. Determine the force in members *AB, CD, ED*, and *CF* of the space truss and state if the members are in tension or compression.

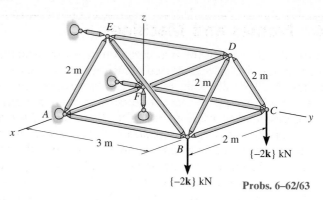

Probs. 6–62/63

*6-64. Determine the force developed in each member of the space truss and state if the members are in tension or compression. The crate has a weight of 150 kN.

Prob. 6–64

6-65. The space truss is used to support vertical forces at joints *B, C*, and *D*. Determine the force in each member and state if the members are in tension or compression.

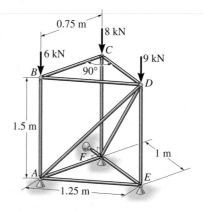

Prob. 6–65

6.6 Frames and Machines

Frames and machines are two common types of structures which are often composed of pin-connected *multiforce members*, i.e., members that are subjected to more than two forces. *Frames* are generally stationary and are used to support loads, whereas *machines* contain moving parts and are designed to transmit and alter the effect of forces. Provided a frame or machine is properly constrained and contains no more supports or members than are necessary to prevent collapse, the forces acting at the joints and supports can be determined by applying the equations of equilibrium to each member. Once the forces at the joints are obtained, it is then possible to *design* the size of the members, connections, and supports using the theory of mechanics of materials and an appropriate engineering design code.

Free-Body Diagrams. In order to determine the forces acting at the joints and supports of a frame or machine, the structure must be disassembled and the free-body diagrams of its parts must be drawn. The following important points *must* be observed:

- Isolate each part by drawing its *outlined shape*. Then show all the forces and/or couple moments that act on the part. Make sure to *label* or *identify* each known and unknown force and couple moment with reference to an established *x, y* coordinate system. Also, indicate any dimensions used for taking moments. Most often the equations of equilibrium are easier to apply if the forces are represented by their rectangular components. As usual, the sense of an unknown force or couple moment can be assumed.

- Identify all the two-force members in the structure and represent their free-body diagrams as having two equal but opposite collinear forces acting at their points of application. (See Sec. 5.4.) By recognizing the two-force members, we can avoid solving an unnecessary number of equilibrium equations.

- Forces common to any two *contacting* members act with equal magnitudes but opposite sense on the respective members. If the two members are treated as a *"system" of connected members*, then these forces are *"internal"* and are *not shown* on the *free-body diagram of the system*; however, if the free-body diagram of *each member* is drawn, the forces are *"external"* and *must* be shown on each of the free-body diagrams.

The following examples graphically illustrate application of these points in drawing the free-body diagrams of a dismembered frame or machine. In all cases, the weight of the members is neglected.

EXAMPLE 6.9

(b)

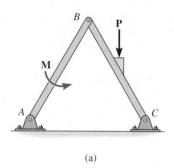

(a)

For the frame shown in Fig. 6–21a, draw the free-body diagram of (a) each member, (b) the pin at B, and (c) the two members connected together.

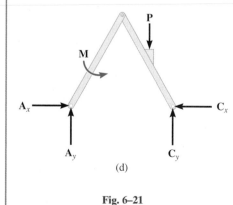

(c)

(d)

Fig. 6–21

Solution

Part (a). By inspection, members *BA* and *BC* are *not* two-force members. Instead, as shown on the free-body diagrams, Fig. 6–21b, *BC* is subjected to *not* five but *three forces*, namely, the resultant force from pins *B* and *C* and the external force **P**. Likewise, *AB* is subjected to the *resultant* forces from the pins at *A* and *B* and the external couple moment **M**.

Part (b). It can be seen in Fig. 6–21a that the pin at *B* is subjected to only *two forces*, i.e., the force of member *BC* on the pin and the force of member *AB* on the pin. For *equilibrium* these forces and therefore their respective components must be equal but opposite, Fig. 6–21c. Notice carefully how Newton's third law is applied between the pin and its contacting members, i.e., the effect of the pin on the two members, Fig. 6–21b, and the equal but opposite effect of the two members on the pin, Fig. 6–21c. Also note that $\mathbf{B}_x$ and $\mathbf{B}_y$, shown equal but opposite in Fig. 6–21b on members *AB* and *BC*, is *not* the effect of Newton's third law; instead, this results from the *equilibrium* analysis of the pin, Fig. 6–21c.

Part (c). The free-body diagram of both members connected together, yet removed from the supporting pins at *A* and *C*, is shown in Fig. 6–21d. The force components $\mathbf{B}_x$ and $\mathbf{B}_y$ are *not shown* on this diagram since they form equal but opposite collinear pairs of *internal* forces (Fig. 6–21b) and therefore cancel out. Also, to be consistent when later applying the equilibrium equations, the unknown force components at *A* and *C* must act in the *same sense* as those shown in Fig. 6–21b. Here the couple moment **M** can be applied at any point on the frame in order to determine the reactions at *A* and *C*. Note, however, that it must act on member *AB* in Fig. 6–21b and *not* on member *BC*.

EXAMPLE 6.10

A constant tension in the conveyor belt is maintained by using the device shown in Fig. 6–22a. Draw the free-body diagrams of the frame and the cylinder which supports the belt. The suspended block has a weight of W.

(a)

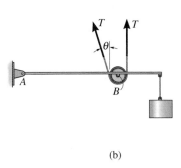

(b)

Fig. 6–22

Solution

The idealized model of the device is shown in Fig. 6–22b. Here the angle θ is assumed to be known. Notice that the tension in the belt is the same on each side of the cylinder, since the cylinder is free to turn. From this model, the free-body diagrams of the frame and cylinder are shown in Figs. 6–22c and 6–22d, respectively. Note that the force that the pin at B exerts on the cylinder can be represented by either its horizontal and vertical components $\mathbf{B}_x$ and $\mathbf{B}_y$, which can be determined by using the force equations of equilibrium applied to the cylinder, or by the two components T, which provide equal but opposite couple moments on the cylinder and thus keep it from turning. Also, realize that once the pin reactions at A have been determined, half of their values act on each side of the frame since pin connections occur on each side, Fig. 6–22a.

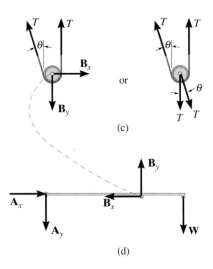

(c)

(d)

E X A M P L E   **6.11**

Draw the free-body diagram of each part of the smooth piston and link mechanism used to crush recycled cans, which is shown in Fig. 6–23a.

(a)

Solution

By inspection, member *AB* is a two-force member. The free-body diagrams of the parts are shown in Fig. 6–23b. Since the pins at *B* and *D connect only two parts together,* the forces there are shown as equal but opposite on the separate free-body diagrams of their connected members. In particular, four components of force act on the piston: $\mathbf{D}_x$ and $\mathbf{D}_y$ represent the effect of the pin (or lever *EBD*), $\mathbf{N}_w$ is the *resultant force* of the floor, and **P** is the resultant compressive force caused by the can *C*.

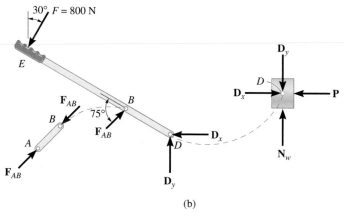

(b)

Fig. 6–23

EXAMPLE 6.12

For the frame shown in Fig. 6–24a, draw the free-body diagrams of (a) the entire frame including the pulleys and cords, (b) the frame without the pulleys and cords, and (c) each of the pulleys.

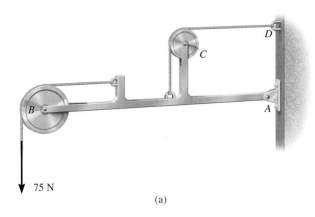

(a)

Solution

Part (a). When the entire frame including the pulleys and cords is considered, the interactions at the points where the pulleys and cords are connected to the frame become pairs of *internal forces* which cancel each other and therefore are not shown on the free-body diagram, Fig. 6–24b.

Part (b). When the cords and pulleys are removed, their effect *on the frame* must be shown, Fig. 6–24c.

Part (c). The force components $\mathbf{B}_x$, $\mathbf{B}_y$, $\mathbf{C}_x$, $\mathbf{C}_y$ of the pins on the pulleys, Fig. 6–24d, are equal but opposite to the force components exerted by the pins on the frame, Fig. 6–24c. Why?

Fig. 6–24

EXAMPLE 6.13

(a)

Fig. 6–25

Draw the free-body diagrams of the bucket and the vertical boom of the back hoe shown in the photo, Fig. 6–25a. The bucket and its contents have a weight W. Neglect the weight of the members.

Solution

The idealized model of the assembly is shown in Fig. 6–25b. Not shown are the required dimensions and angles that must be obtained, along with the location of the center of gravity G of the load. By inspection, members AB, BC, BE, and HI are all two-force members since they are pin connected at their end points and no other forces act on them. The free-body diagrams of the bucket and the boom are shown in Fig. 6–25c. Note that pin C is subjected to only two forces, the force of link BC and the force of the boom. For equilibrium, these forces must be equal in magnitude but opposite in direction, Fig. 6–25d. The pin at B is subjected to three forces, Fig. 6–25e. The force $\mathbf{F}_{BE}$ is caused by the hydraulic cylinder, and the forces $\mathbf{F}_{BA}$ and $\mathbf{F}_{BC}$ are caused by the links. These three forces are related by the two equations of force equilibrium applied to the pin.

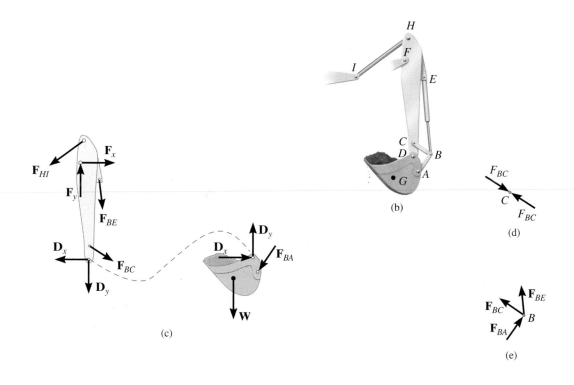

Before proceeding, it is recommended to cover the solutions to the previous examples and attempt to draw the requested free-body diagrams. When doing so, make sure the work is neat and that all the forces and couple moments are properly labeled.

Equations of Equilibrium.   Provided the structure (frame or machine) is properly supported and contains no more supports or members than are necessary to prevent its collapse, then the unknown forces at the supports and connections can be determined from the equations of equilibrium. If the structure lies in the x–y plane, then for *each* free-body diagram drawn the loading must satisfy $\Sigma F_x = 0$, $\Sigma F_y = 0$, and $\Sigma M_O = 0$. The selection of the free-body diagrams used for the analysis is *completely arbitrary*. They may represent each of the members of the structure, a portion of the structure, or its entirety. For example, consider finding the six components of the pin reactions at A, B, and C for the frame shown in Fig. 6–26a. If the frame is dismembered, as it is in Fig. 6–26b, these unknowns can be determined by applying the three equations of equilibrium to each of the two members (total of six equations). The free-body diagram of the *entire frame* can also be used for part of the analysis, Fig. 6–26c. Hence, if so desired, all six unknowns can be determined by applying the three equilibrium equations to the entire frame, Fig. 6–26c, and also to either one of its members. Furthermore, the answers can be checked in part by applying the three equations of equilibrium to the remaining "second" member. In general, then, this problem can be solved by writing *at most* six equilibrium equations using free-body diagrams of the members and/or the combination of connected members. Any more than six equations written would *not* be unique from the original six and would only serve to check the results.

(a)

(b)

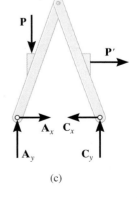

(c)

Fig. 6–26

PROCEDURE FOR ANALYSIS

The joint reactions on frames or machines (structures) composed of multiforce members can be determined using the following procedure.

Free-Body Diagram.

- Draw the free-body diagram of the entire structure, a portion of the structure, or each of its members. The choice should be made so that it leads to the most direct solution of the problem.

- When the free-body diagram of a group of members of a structure is drawn, the forces at the connected parts of this group are internal forces and are not shown on the free-body diagram of the group.

- Forces common to two members which are in contact act with equal magnitude but opposite sense on the respective free-body diagrams of the members.

- Two-force members, regardless of their shape, have equal but opposite collinear forces acting at the ends of the member.

- In many cases it is possible to tell by inspection the proper sense of the unknown forces acting on a member; however, if this seems difficult, the sense can be assumed.

- A couple moment is a free vector and can act at any point on the free-body diagram. Also, a force is a sliding vector and can act at any point along its line of action.

Equations of Equilibrium.

- Count the number of unknowns and compare it to the total number of equilibrium equations that are available. In two dimensions, there are three equilibrium equations that can be written for each member.

- Sum moments about a point that lies at the intersection of the lines of action of as many unknown forces as possible.

- If the solution of a force or couple moment magnitude is found to be negative, it means the sense of the force is the reverse of that shown on the free-body diagrams.

EXAMPLE 6.14

Determine the horizontal and vertical components of force which the pin at C exerts on member CB of the frame in Fig. 6–27a.

Solution I

Free-Body Diagrams. By inspection it can be seen that AB is a two-force member. The free-body diagrams are shown in Fig. 6–27b.

Equations of Equilibrium. The *three unknowns*, C_x, C_y, and F_{AB}, can be determined by applying the three equations of equilibrium to member CB.

$\zeta + \Sigma M_C = 0$; $2000\text{ N}(2\text{ m}) - (F_{AB}\sin 60°)(4\text{ m}) = 0$ $F_{AB} = 1154.7\text{ N}$

$\xrightarrow{+} \Sigma F_x = 0$; $1154.7\cos 60°\text{N} - C_x = 0$ $C_x = 577\text{ N}$ *Ans.*

$+\uparrow\Sigma F_y = 0$; $1154.7\sin 60°\text{N} - 2000\text{ N} + C_y = 0$ $C_y = 1000\text{ N}$ *Ans.*

Solution II

Free-Body Diagrams. If one does not recognize that AB is a two-force member, then more work is involved in solving this problem. The free-body diagrams are shown in Fig. 6–27c.

Equations of Equilibrium. The *six unknowns*, A_x, A_y, B_x, B_y, C_x, C_y, are determined by applying the three equations of equilibrium to each member.

   Member AB

$\zeta + \Sigma M_A = 0$; $B_x(3\sin 60°\text{ m}) - B_y(3\cos 60°\text{ m}) = 0$ (1)

$\xrightarrow{+} \Sigma F_x = 0$; $A_x - B_x = 0$ (2)

$+\uparrow\Sigma F_y = 0$; $A_y - B_y = 0$ (3)

   Member BC

$\zeta + \Sigma M_C = 0$; $2000\text{ N}(2\text{ m}) - B_y(4\text{ m}) = 0$ (4)

$\xrightarrow{+} \Sigma F_x = 0$; $B_x - C_x = 0$ (5)

$+\uparrow\Sigma F_y = 0$; $B_y - 2000\text{ N} + C_y = 0$ (6)

The results for C_x and C_y can be determined by solving these equations in the following sequence: 4, 1, 5, then 6. The results are

$$B_y = 1000\text{ N}$$
$$B_x = 577\text{ N}$$
$$C_x = 577\text{ N} \qquad Ans.$$
$$C_y = 1000\text{ N} \qquad Ans.$$

By comparison, Solution I is simpler since the requirement that F_{AB} in Fig. 6–27b be equal, opposite, and collinear at the ends of member AB automatically satisfies Eqs. 1, 2, and 3 above and therefore eliminates the need to write these equations. *As a result, always identify the two-force members before starting the analysis!*

(a)

(b)

(c)

Fig. 6–27

E X A M P L E 6.15

The compound beam shown in Fig. 6–28a is pin connected at B. Determine the reactions at its supports. Neglect its weight and thickness.

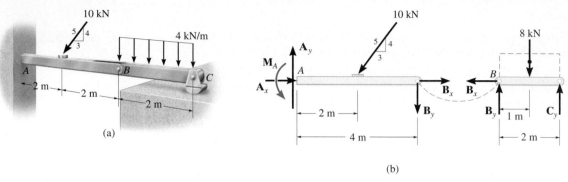

(a)

(b)

Fig. 6–28

Solution

Free-Body Diagrams. By inspection, if we consider a free-body diagram of the entire beam *ABC*, there will be three unknown reactions at *A* and one at *C*. These four unknowns cannot all be obtained from the three equations of equilibrium, and so it will become necessary to dismember the beam into its two segments as shown in Fig. 6–28b.

Equations of Equilibrium. The six unknowns are determined as follows:

Segment BC

$$\xrightarrow{+} \Sigma F_x = 0; \qquad\qquad\qquad B_x = 0$$

$$\downarrow + \Sigma M_B = 0; \qquad -8 \text{ kN}(1 \text{ m}) + C_y(2 \text{ m}) = 0$$

$$+\uparrow \Sigma F_y = 0; \qquad\qquad B_y - 8 \text{ kN} + C_y = 0$$

Segment AB

$$\xrightarrow{+} \Sigma F_x = 0; \qquad\qquad A_x - (10 \text{ kN})(\tfrac{3}{5}) + B_x = 0$$

$$\downarrow + \Sigma M_A = 0; \qquad M_A - (10 \text{ kN})(\tfrac{4}{5})(2 \text{ m}) - B_y(4 \text{ m}) = 0$$

$$+\uparrow \Sigma F_y = 0; \qquad\qquad A_y - (10 \text{ kN})(\tfrac{4}{5}) - B_y = 0$$

Solving each of these equations successively, using previously calculated results, we obtain

$$A_x = 6 \text{ kN} \qquad A_y = 12 \text{ kN} \qquad M_A = 32 \text{ kN} \cdot \text{m} \qquad\qquad Ans.$$

$$B_x = 0 \qquad\qquad B_y = 4 \text{ kN}$$

$$C_y = 4 \text{ kN} \qquad\qquad\qquad\qquad\qquad\qquad\qquad\qquad Ans.$$

EXAMPLE 6.16

Determine the horizontal and vertical components of force which the pin at C exerts on member $ABCD$ of the frame shown in Fig. 6–29a.

(a)

Solution

Free-Body Diagrams. By inspection, the three components of reaction that the supports exert on $ABCD$ can be determined from a free-body diagram of the entire frame, Fig. 6–29b. Also, the free-body diagram of each frame member is shown in Fig. 6–29c. Notice that member BE is a two-force member. As shown by the colored dashed lines, the forces at B, C, and E have equal magnitudes but opposite directions on the separate free-body diagrams.

Equations of Equilibrium. The six unknowns A_x, A_y, F_B, C_x, C_y, and D_x will be determined from the equations of equilibrium applied to the entire frame and then to member CEF. We have

Entire Frame

$$\zeta + \Sigma M_A = 0; \quad -981 \text{ N}(2 \text{ m}) + D_x(2.8 \text{ m}) = 0 \qquad D_x = 700.7 \text{ N}$$

$$\xrightarrow{+} \Sigma F_x = 0; \qquad A_x - 700.7 \text{ N} = 0 \qquad A_x = 700.7 \text{ N}$$

$$+\uparrow \Sigma F_y = 0; \qquad A_y - 981 \text{ N} = 0 \qquad A_y = 981 \text{ N}$$

Member CEF

$$\zeta + \Sigma M_C = 0; \quad -981 \text{ N}(2 \text{ m}) - (F_B \sin 45°)(1.6 \text{ m}) = 0$$
$$F_B = -1734.2 \text{ N}$$

$$\xrightarrow{+} \Sigma F_x = 0; \qquad -Cx - (-1734.2 \cos 45° \text{ N}) = 0$$
$$C_x = 1226 \text{ N} \qquad\qquad\qquad \textit{Ans.}$$

$$+\uparrow \Sigma F_y = 0; \quad C_y - (-1734.2 \sin 45° \text{ N}) - 981 \text{ N} = 0$$
$$C_y = -245 \text{ N} \qquad\qquad\qquad \textit{Ans.}$$

(b)

Since the magnitudes of $\mathbf{F}_B$ and $\mathbf{C}_y$ were calculated as negative quantities, they were assumed to be acting in the wrong sense on the free-body diagrams, Fig. 6–29c. The correct sense of these forces might have been determined "by inspection" *before* applying the equations of equilibrium to member CEF. As shown in Fig. 6–29c, moment equilibrium about point E on member CEF indicates that $\mathbf{C}_y$ must actually act *downward* to counteract the moment created by the 981-N force about E. Similarly, summing moments about C, it is seen that the vertical component of $\mathbf{F}_B$ must actually act *upward*, and so $\mathbf{F}_B$ must act upward and to the right.

The above calculations can be checked by applying the three equilibrium equations to member $ABCD$, Fig. 6–29c.

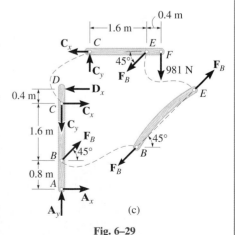

(c)

Fig. 6–29

EXAMPLE **6.17**

The smooth disk shown in Fig. 6–30a is pinned at D and has a weight of 20 N. Neglecting the weights of the other members, determine the horizontal and vertical components of reaction at pins B and D.

(a)

Solution

Free-Body Diagrams. By inspection, the three components of reaction at the supports can be determined from a free-body diagram of the entire frame, Fig. 6–30b. Also, free-body diagrams of the members are shown in Fig. 6–30c.

Equations of Equilibrium. The eight unknowns can of course be obtained by applying the eight equilibrium equations to each member—three to member AB, three to member BCD, and two to the disk. (Moment equilibrium is automatically satisfied for the disk.) If this is done, however, all the results can be obtained only from a simultaneous solution of some of the equations. (Try it and find out.) To avoid this situation, it is best to first determine the three support reactions on the *entire* frame; then, using these results, the remaining five equilibrium equations can be applied to two other parts in order to solve successively for the other unknowns.

(b)

(c)

Fig. 6–30

Entire Frame

$\zeta+\Sigma M_A = 0;$ $-20 \text{ N}(3 \text{ cm}) + C_x(3.5 \text{ cm}) = 0$ $C_x = 17.1 \text{ N}$

$\xrightarrow{+} \Sigma F_x = 0;$ $A_x - 17.1 \text{ N} = 0$ $A_x = 17.1 \text{ N}$

$+\uparrow\Sigma F_y = 0;$ $A_y - 20 \text{ N} = 0$ $A_y = 20 \text{ N}$

Member AB

$\xrightarrow{+} \Sigma F_x = 0;$ $17.1 \text{ N} - B_x = 0$ $B_x = 17.1 \text{ N}$ *Ans.*

$\zeta+\Sigma M_B = 0;$ $-20 \text{ N}(6 \text{ cm}) + N_D(3 \text{ cm}) = 0$ $N_D = 40 \text{ N}$

$+\uparrow\Sigma F_y = 0;$ $20 \text{ N} - 40 \text{ N} + B_y = 0$ $B_y = 20 \text{ N}$ *Ans.*

Disk

$\xrightarrow{+} \Sigma F_x = 0;$ $D_x = 0$ *Ans.*

$+\uparrow\Sigma F_y = 0;$ $40 \text{ N} - 20 \text{ N} - D_y = 0$ $D_y = 20 \text{ N}$ *Ans.*

E X A M P L E 6.18

Determine the tension in the cables and also the force **P** required to support the 600-N force using the frictionless pulley system shown in Fig. 6–31*a*.

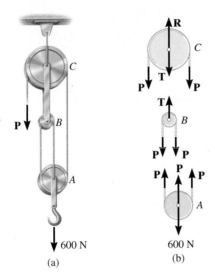

Fig. 6–31

Solution

Free-Body Diagram. A free-body diagram of each pulley *including* its pin and a portion of the contacting cable is shown in Fig. 6–31*b*. Since the cable is *continuous* and the pulleys are frictionless, the cable has a *constant tension P* acting throughout its length (see Example 5.7). The link connection between pulleys *B* and *C* is a two-force member, and therefore it has an unknown tension *T* acting on it. Notice that the *principle of action, equal but opposite reaction* must be carefully observed for forces **P** and **T** when the *separate* free-body diagrams are drawn.

Equations of Equilibrium. The three unknowns are obtained as follows:

Pulley A

$+\uparrow \Sigma F_y = 0;$ $3P - 600 \text{ N} = 0$ $P = 200 \text{ N}$ *Ans.*

Pulley B

$+\uparrow \Sigma F_y = 0;$ $T - 2P = 0$ $T = 400 \text{ N}$ *Ans.*

Pulley C

$+\uparrow \Sigma F_y = 0;$ $R - 2P - T = 0$ $R = 800 \text{ N}$ *Ans.*

EXAMPLE 6.19

(a)

(b)

(c)

Fig. 6–32

A man having a weight of 750 N ($\approx$ 75 kg) supports himself by means of the cable and pulley system shown in Fig. 6–32a. If the seat has a weight of 75 N ($\approx$ 7.5 kg), determine the force that he must exert on the cable at A and the force he exerts on the seat. Neglect the weight of the cables and pulleys.

Solution I

Free-Body Diagrams. The free-body diagrams of the man, seat, and pulley C are shown in Fig. 6–32b. The *two* cables are subjected to tensions T_A and T_E, respectively. The man is subjected to three forces: his weight, the tension T_A of cable AC, and the reaction N_s of the seat.

Equations of Equilibrium. The three unknowns are obtained as follows:

Man

$$+\uparrow \Sigma F_y = 0; \qquad T_A + N_s - 750 \text{ N} = 0 \qquad (1)$$

Seat

$$+\uparrow \Sigma F_y = 0; \qquad T_E + N_s - 75 \text{ N} = 0 \qquad (2)$$

Pulley C

$$+\uparrow \Sigma F_y = 0; \qquad 2T_E - T_A = 0 \qquad (3)$$

Here T_E can be determined by adding Eqs. 1 and 2 to eliminate N_s and then using Eq. 3. The other unknowns are then obtained by resubstitution of T_E.

$$T_A = 550 \text{ N}$$
$$T_E = 275 \text{ N} \qquad \qquad Ans.$$
$$N_E = 200 \text{ N}$$
$$\qquad \qquad \qquad Ans.$$

Solution II

Free-Body Diagrams. By using the blue section shown in Fig. 6–32a, the man, pulley, and seat can be considered as a *single system*, Fig. 6–32c. Here N_s and T_A are *internal* forces and hence are not included on this "combined" free-body diagram.

Equations of Equilibrium. Applying $\Sigma F_y = 0$ yields a *direct* solution for T_E.

$$+\uparrow \Sigma F_y = 0; \qquad 3T_E - 75 \text{ N} - 750 \text{ N} = 0 \qquad T_E = 275 \text{ N}$$

The other unknowns can be obtained from Eqs. 2 and 3.

E X A M P L E   **6.20**

The hand exerts a force of 35 N on the grip of the spring compressor shown in Fig. 6–33a. Determine the force in the spring needed to maintain equilibrium of the mechanism.

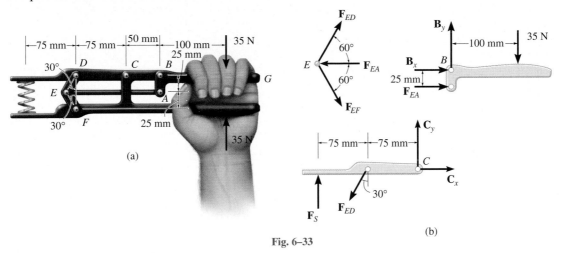

Fig. 6–33

Solution

Free-Body Diagrams.   By inspection, members *EA, ED,* and *EF* are all two-force members. The free-body diagrams for parts *DC* and *ABG* are shown in Fig. 6–33b. The pin at *E* has also been included here since *three* force interactions occur on this pin. They represent the effects of members *ED, EA,* and *EF*. Note carefully how equal and opposite force reactions occur between each of the parts.

Equations of Equilibrium.   By studying the free-body diagrams, the most direct way to obtain the spring force is to apply the equations of equilibrium in the following sequence:

Lever ABG

$$\zeta + \Sigma M_B = 0; \quad F_{EA}(25 \text{ mm}) - 35 \text{ N}(100 \text{ mm}) = 0 \qquad F_{EA} = 140 \text{ N}$$

Pin E

$$+\uparrow \Sigma F_y = 0; \quad F_{ED} \sin 60° - F_{EF} \sin 60° = 0 \quad F_{ED} = F_{EF} = F$$
$$\xrightarrow{+} \Sigma F_x = 0; \quad 2F \cos 60° - 140 \text{ N} = 0 \qquad F = 140 \text{ N}$$

Arm DC

$$\zeta + \Sigma M_C = 0; \quad -F_s(150 \text{ mm}) + 140 \cos 30° \text{ N}(75 \text{ mm}) = 0$$
$$F_s = 60.62 \text{ N} \qquad\qquad Ans.$$

E X A M P L E 6.21

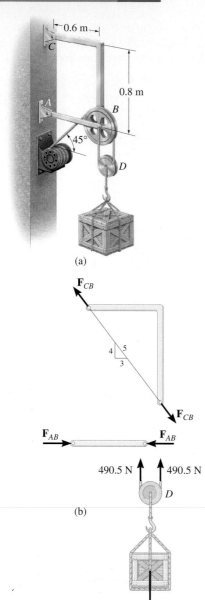

(a)

(b)

F_{CB}

4 5
  3

F_{CB}

F_{AB} F_{AB}

490.5 N   490.5 N

D

981 N

The 100-kg block is held in equilibrium by means of the pulley and continuous cable system shown in Fig. 6–34a. If the cable is attached to the pin at B, compute the forces which this pin exerts on each of its connecting members.

Solution

Free-Body Diagrams. A free-body diagram of each member of the frame is shown in Fig. 6–34b. By inspection, members AB and CB are two-force members. Furthermore, the cable must be subjected to a force of 490.5 N in order to hold pulley D and the block in equilibrium. A free-body diagram of the pin at B is needed since *four interactions* occur at this pin. These are caused by the attached cable (490.5 N), member AB ($\mathbf{F}_{AB}$), member CB ($\mathbf{F}_{CB}$), and pulley B ($\mathbf{B}_x$ and $\mathbf{B}_y$).

Equations of Equilibrium. Applying the equations of force equilibrium to pulley B, we have

$$\xrightarrow{+} \Sigma F_x = 0; \quad B_x - 490.5 \cos 45° \text{ N} = 0 \quad B_x = 346.8 \text{ N} \qquad Ans.$$

$$+\uparrow \Sigma F_y = 0; \quad B_y - 490.5 \sin 45° \text{ N} - 490.5 \text{ N} = 0$$

$$B_y = 837.3 \text{ N} \qquad Ans.$$

Using these results, equilibrium of the pin requires that

$$+\uparrow \Sigma F_y = 0; \quad \tfrac{4}{5}F_{CB} - 837.3 \text{ N} - 490.5 \text{ N} \quad F_{CB} = 1660 \text{ N} \quad Ans.$$

$$\xrightarrow{+} \Sigma F_x = 0; \quad F_{AB} - \tfrac{3}{5}(1660 \text{ N}) - 346.8 \text{ N} = 0 \quad F_{AB} = 1343 \text{ N} \quad Ans.$$

It may be noted that the two-force member CB is subjected to bending as caused by the force $\mathbf{F}_{CB}$. From the standpoint of design, it would be better to make this member *straight* (from C to B) so that the force $\mathbf{F}_{CB}$ would create only tension in the member.

(c)

Fig. 6–34

Before solving the following problems, it is suggested that a brief review be made of all the previous examples. This may be done by covering each solution, trying to locate the two-force members, drawing the free-body diagrams, and conceiving ways of applying the equations of equilibrium to obtain the solution.

PROBLEMS

6-66. In each case, determine the force **P** required to maintain equilibrium. The block weighs 100 N ($\approx$ 10 kg).

*6-68. Determine the force **P** needed to support the 100-N ($\approx$ 10-kg) weight. Each pulley has a weight of 10 N ($\approx$ 1 kg). Also, what are the cord reactions at *A* and *B*?

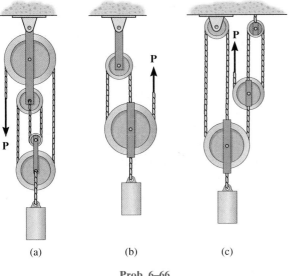

(a)   (b)   (c)

Prob. 6–66

Prob. 6–68

6-67. The eye hook has a positive locking latch when it supports the load because its two parts are pin-connected at *A* and they bear against one another along the smooth surface at *B*. Determine the resultant force at the pin and the normal force at *B* when the eye hook supports a load of 800 N.

6-69. The link is used to hold the rod in place. Determine the required axial force on the screw at *E* if the largest force to be exerted on the rod at *B*, *C* or *D* is to be 100 N. Also, find the magnitude of the force reaction at pin *A*. Assume all surfaces of contact are smooth.

Prob. 6–67

Prob. 6–69

6-70. The principles of a *differential chain block* are indicated schematically in the figure. Determine the magnitude of force **P** needed to support the 800-N force. Also, find the distance *x* where the cable must be attached to bar *AB* so the bar remains horizontal. All pulleys have a radius of 60 mm.

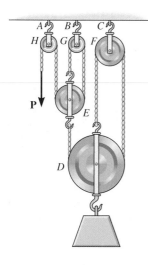

Prob. 6–70

6-71. Determine the force *P* needed to support the 20-kg mass using the *Spanish Burton rig*. Also, what are the reactions at the supporting hooks *A*, *B*, and *C*?

Prob. 6–71

***6-72.** The compound beam is fixed at *A* and supported by a rocker at *B* and *C*. There are hinges (pins) at *D* and *E*. Determine the reactions at the supports.

Prob. 6–72

6-73. The compound beam is pin-supported at *C* and supported by a roller at *A* and *B*. There is a hinge (pin) at *D*. Determine the reactions at the supports. Neglect the thickness of the beam.

Prob 6–73

6-74. Determine the greatest force *P* that can be applied to the frame if the largest force resultant acting at *A* can have a magnitude of 2 kN.

Prob. 6–74

6-75. Determine the horizontal and vertical components force at pins *A* and *C* of the two-member frame.

Prob. 6–75

6-77. Determine the horizontal and vertical components of force at pins *A*, *B*, and *C*, and the reactions to the fixed support *D* of the three-member frame.

Prob. 6–77

*6-76. The three-hinged arch supports the loads $F_1 = 8$ kN and $F_2 = 5$ kN. Determine the horizontal and vertical components of reaction at the pin supports *A* and *B*. Take $h = 2$ m.

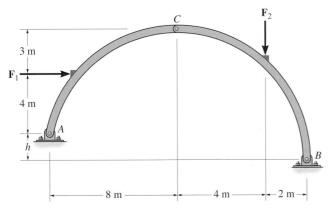

Prob. 6–76

6-78. Determine the horizontal and vertical components of force at *C* which member *ABC* exerts on member *CEF*.

Prob. 6–78

6-79. Determine the horizontal and vertical components of force that the pins at *A*, *B*, and *C* exert on their connecting members.

6-81. Determine the force *P* on the cord, and the angle θ that the pulley-supporting link *AB* makes with the vertical. Neglect the mass of the pulleys and the link. The block has a weight of 200 N ($\approx$ 20 kg) and the cord is attached to the pin at *B*. The pulleys have radii of $r_1 = 2$ cm, and $r_2 = 1$ cm.

Prob. 6–79

Prob. 6–81

***6-80.** The hoist supports the 125-kg engine. Determine the force the load creates in member *DB* and in member *FB*, which contains the hydraulic cylinder *H*.

6-82. The front of the car is to be lifted using a smooth, rigid 3.5 m long board. The car has a weight of 17.5 kN and a center of gravity at *G*. Determine the position *x* of the fulcrum so that an applied force of 500 N at *E* will lift the front wheels of the car.

Prob. 6–80

Prob. 6–82

6-83. The wall crane supports a load of 700 N. Determine the horizontal and vertical components of reaction at the pins A and D. Also, what is the force in the cable at the winch W?

6-85. Determine the horizontal and vertical components of force which the pins exert on member ABC.

Prob. 6–85

Prob. 6–83

6-86. The engine hoist is used to support the 200-kg engine. Determine the force acting in the hydraulic cylinder AB, the horizontal and vertical components of force at the pin C, and the reactions at the fixed support D.

*6-84. Determine the force that the smooth roller C exerts on beam AB. Also, what are the horizontal and vertical components of reaction at pin A? Neglect the weight of the frame and roller.

Prob. 6–84

Prob. 6–86

6-87. Determine the horizontal and vertical components of force at pins B and C.

Prob. 6–87

6-89. Determine the horizontal and vertical components of force at each pin. The suspended cylinder has a weight of 800 N ($\approx$ 80 kg).

Prob. 6–89

*6-88. The pipe cutter is clamped around the pipe P. If the wheel at A exerts a normal force of $F_A = 80$ N on the pipe, determine the normal forces of wheels B and C on the pipe. Also compute the pin reaction on the wheel at C. The three wheels each have a radius of 7 mm and the pipe has an outer radius of 10 mm.

Prob. 6–88

6-90. The toggle clamp is subjected to a force $\mathbf{F}$ at the handle. Determine the vertical clamping force acting at E.

Prob. 6–90

6-91. Determine the horizontal and vertical components of force which the pins at *A, B,* and *C* exert on member *ABC* of the frame.

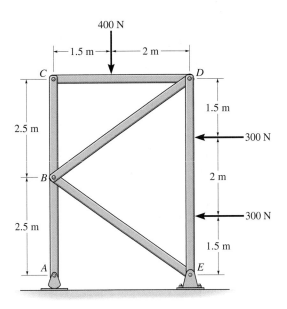

Prob. 6–91

*6-92. The derrick is pin-connected to the pivot at *A.* Determine the largest mass that can be supported by the derrick if the maximum force that can be sustained by the pin at *A* is 18 kN.

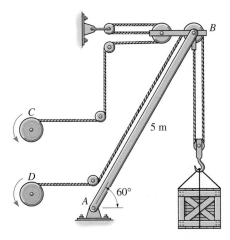

Prob. 6–92

6-93. Determine the required mass of the suspended cylinder if the tension in the chain wrapped around the freely turning gear is to be 2 kN. Also, what is the magnitude of the resultant force on pin *A*?

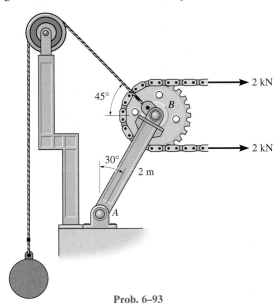

Prob. 6–93

6-94. The pumping unit is used to recover oil. When the walking beam *ABC* is horizontal, the force acting in the wireline at the well head is 1000 N. Determine the torque **M** which must be exerted by the motor in order to overcome this load. The horse-head *C* weighs 240 N and has a center of gravity at G_C. The walking beam *ABC* has a weight of 520 N and a center of gravity at G_B, and the counterweight has a weight of 800 N and a center of gravity at G_W. The pitman, *AD,* is pin-connected at its ends and has negligible weight.

Prob. 6–94

6-95. Determine the force P on the cable if the spring is compressed 10 mm when the mechanism is in the position shown. The spring has a stiffness of $k = 12$ kN/m.

Prob. 6–95

6-97. The compound arrangement of the pan scale is shown. If the mass on the pan is 4 kg, determine the horizontal and vertical components at pins A, B, and C and the distance x of the 25-g mass to keep the scale in balance.

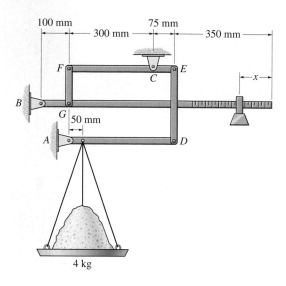

Prob. 6–97

*6-96. Determine the force that the jaws J of the metal cutters exert on the smooth cable C if 100-N forces are applied to the handles. The jaws are pinned at E and A, and D and B. There is also a pin at F.

Prob. 6–96

6-98. The scissors lift consists of *two* sets of cross members and *two* hydraulic cylinders, DE, symmetrically located on *each side* of the platform. The platform has a uniform mass of 60 kg, with a center of gravity at G_1. The load of 85 kg, with center of gravity at G_2, is centrally located between each side of the platform. Determine the force in each of the hydraulic cylinders for equilibrium. Rollers are located at B and D.

Prob. 6–98

6-99. Determine the horizontal and vertical components of force that the pins at *A, B,* and *C* exert on the frame. The cylinder has a mass of 80 kg. The pulley has a radius of 0.1 m.

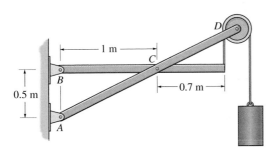

Prob. 6–99

*6-100. By squeezing on the hand brake of the bicycle, the rider subjects the brake cable to a tension of 200 N. If the caliper mechanism is pin-connected to the bicycle frame at *B,* determine the normal force each brake pad exerts on the rim of the wheel. Is this the force that stops the wheel from turning? Explain.

Prob. 6–100

6-101. If a force of *P* = 30 N is applied perpendicular to the handle of the mechanism, determine the magnitude of force **F** for equilibrium. The members are pin-connected at *A, B, C,* and *D.*

Prob. 6–101

6-102. The bucket of the backhoe and its contents have a weight of 3000 N (≈ 300 kg) and a center of gravity at *G.* Determine the forces of the hydraulic cylinder *AB* and in links *AC* and *AD* in order to hold the load in the position shown. The bucket is pinned at *E.*

Prob. 6–102

6-103. Two smooth tubes A and B, each having the same weight, W, are suspended from a common point O by means of equal-length cords. A third tube, C, is placed between A and B. Determine the greatest weight of C without upsetting equilibrium.

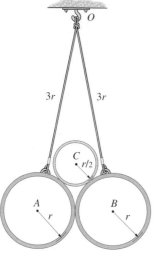

Prob. 6–103

6-105. The compound beam is fixed supported at C and supported by rockers at A and B. If there are hinges (pins) at D and E, determine the components of reaction at the supports. Neglect the thickness of the beam.

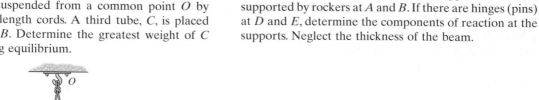

Prob. 6–105

*6-104. The double link grip is used to lift the beam. If the beam weighs 4 kN, determine the horizontal and vertical components of force acting on the pin at A and the horizontal and vertical components of force that the flange of the beam exerts on the jaw at B.

Prob. 6–104

6-106. Determine the horizontal and vertical components of force at pin B and the normal force the pin at C exerts on the smooth slot. Also, determine the moment and horizontal and vertical reactions of force at A. There is a pulley at E.

Prob. 6–106

6-107. The symmetric coil tong supports the coil which has a mass of 800 kg and center of mass at G. Determine the horizontal and vertical components of force the linkage exerts on plate $DEIJH$ at points D and E. The coil exerts only vertical reactions at K and L.

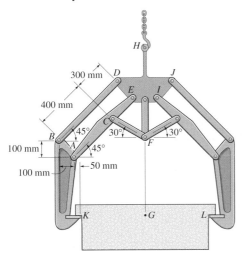

Prob. 6–107

*6-108. If a force of 40 N is applied to the grip of the clamp, determine the compressive force F that the wood block exerts on the clamp.

Prob. 6–108

6-109. If each of the three uniform links of the mechanism has a length L and weight W, determine the angle θ for equilibrium. The spring, which always remains vertical, is unstretched when $\theta = 0°$.

Prob. 6–109

6-110. The flat-bed trailer has a weight of 35 kN (≈ 3500 kg) and center of gravity at G_T. It is pin-connected to the cab at D. The cab has a weight of 30 kN (≈ 3000 kg) and center of gravity at G_C. Determine the range of values x for the position of the 10-kN (≈ 1000-kg) load L so that when it is placed over the rear axle, no axle is subjected to more than 27.5 kN. The load has a center of gravity at G_L.

Prob. 6–110

6-111. The three pin-connected members shown in the *top view* support a downward force of 60 N at *G*. If only vertical forces are supported at the connections *B*, *C*, *E* and pad supports *A*, *D*, *F*, determine the reactions at each pad.

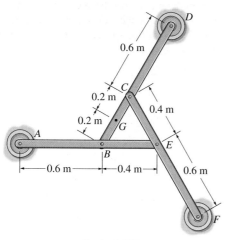

Prob. 6–111

*6-112. The aircraft-hangar door opens and closes slowly by means of a motor which draws in the cable *AB*. If the door is made in two sections (bifold) and each section has a uniform weight *W* and length *L*, determine the force in the cable as a function of the door's position *θ*. The sections are pin-connected at *C* and *D* and the bottom is attached to a roller that travels along the vertical track.

Prob. 6–112

6-113. A man having a weight of 750 N ($\approx$ 75 kg) attempts to lift himself using one of the two methods shown. Determine the total force he must exert on bar *AB* in each case and the normal reaction he exerts on the platform at *C*. Neglect the weight of the platform.

6-114. A man having a weight of 750 N ($\approx$ 75 kg) attempts to lift himself using one of the two methods shown. Determine the total force he must exert on bar *AB* in each case and the normal reaction he exerts on the platform at *C*. The platform has a weight of 150 N ($\approx$ 15 kg).

(a)   (b)

Probs. 6–113/114

6-115. The piston *C* moves vertically between the two smooth walls. If the spring has a stiffness of $k = 15$ N/cm, and is unstretched when $\theta = 0°$, determine the couple **M** that must be applied to *AB* to hold the mechanism in equilibrium when $\theta = 30°$.

Prob. 6–115

*6-116. The two-member frame supports the 200-N (≈ 20-kg) cylinder and 250 N·m couple moment. Determine the force of the roller at B on member AC and the horizontal and vertical components of force which the pin at C exerts on member CB and the pin at A exerts on member AC. The roller C does not contact member CB.

Prob. 6–116

6-117. The tractor boom supports the uniform mass of 500 kg in the bucket which has a center of mass at G. Determine the force in each hydraulic cylinder AB and CD and the resultant force at pins E and F. The load is supported equally on each side of the tractor by a similar mechanism.

Prob. 6–117

6-118. The mechanism is used to hide kitchen appliances under a cabinet by allowing the shelf to rotate downward. If the mixer weighs 50 N (≈ 5 kg), is centered on the shelf, and has a mass center at G, determine the stretch in the spring necessary to hold the shelf in the equilibrium position shown. There is a similar mechanism on each side of the shelf, so that each mechanism supports 25 N (≈ 2.5 kg) of the load. The springs each have a stiffness of $k = 1$ N/mm spring.

Prob. 6–118

6-119. The linkage for a hydraulic jack is shown. If the load on the jack is 10 kN, determine the pressure acting on the fluid when the jack is in the position shown. All lettered points are pins. The piston at H has a cross-sectional area of $A = 1250$ mm^2. *Hint:* First find the force F acting along link EH. The pressure in the fluid is $p = F/A$.

Prob. 6–119

*6-120. Determine the required force P that must be applied at the blade of the pruning shears so that the blade exerts a normal force of 100 N on the twig at E.

Prob. 6–120

6-121. The three power lines exert the forces shown on the truss joints, which in turn are pin-connected to the poles AH and EG. Determine the force in the guy cable AI and the pin reaction at the support H.

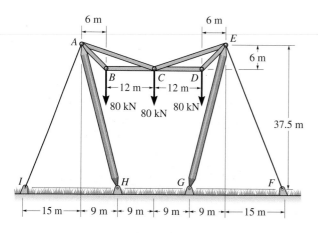

Prob. 6–121

6-122. The hydraulic crane is used to lift the 1400-N ($\approx$ 140-kg) load. Determine the force in the hydraulic cylinder AB and the force in links AC and AD when the load is held in the position shown.

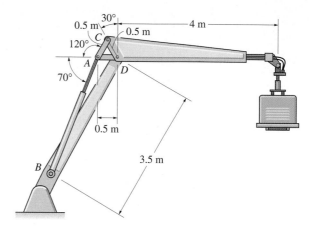

Prob. 6–122

6-123. The kinetic sculpture requires that each of the three pinned beams be in perfect balance at all times during its slow motion. If each member has a uniform weight of 2 kN/m and length of 3 m, determine the necessary counterweights W_1, W_2, and W_3 which must be added to the ends of each member to keep the system in balance for any position. Neglect the size of the counterweights.

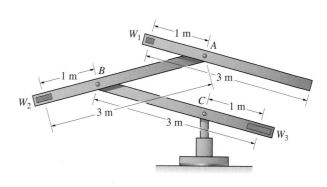

Prob. 6–123

*6-124. The three-member frame is connected at its ends using ball-and-socket joints. Determine the *x, y, z* components of reaction at *B* and the tension in member *ED*. The force acting at *D* is **F** = {135**i** + 200**j** − 180**k**} kN.

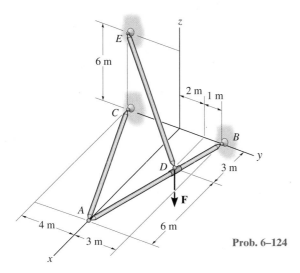

Prob. 6–124

6-125. The four-member "A" frame is supported at *A* and *E* by smooth collars and at *G* by a pin. All the other joints are ball-and-sockets. If the pin at *G* will fail when the resultant force there is 800 N, determine the largest vertical force *P* that can be supported by the frame. Also, what are the *x, y, z* force components which member *BD* exerts on members *EDC* and *ABC*? The collars at *A* and *E* and the pin at *G* only exert force components on the frame.

Prob. 6–125

6-126. The structure is subjected to the loading shown. Member *AD* is supported by a cable *AB* and roller at *C* and fits through a smooth circular hole at *D*. Member *ED* is supported by a roller at *D* and a pole that fits in a smooth snug circular hole at *E*. Determine the *x, y, z* components of reaction at *E* and the tension in cable *AB*.

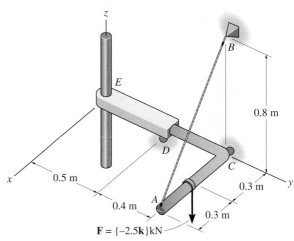

Prob. 6–126

6-127. The structure is subjected to the force of 45 kN which lies in a plane parallel to the *y*-*z* plane. Member *AB* is supported by a ball-and-socket joint at *A* and fits through a snug hole at *B*. Member *CD* is supported by a pin at *C*. Determine the *x, y, z* components of reaction at *A* and *C*.

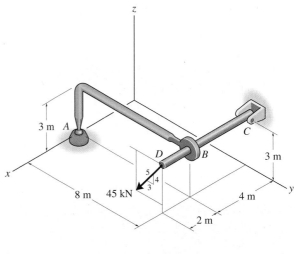

Prob. 6–127

CHAPTER REVIEW

- **Truss Analysis.** A simple truss consists of triangular elements connected together by pin joints. The forces within it members can be determined by assuming the members are all two-force members, connected concurrently at each joint.

- **Method of Joints.** If a truss is in equilibrium, then each of its joints is also in equilibrium. For a coplanar truss, the concurrent force system at each joint must satisfy force equilibrium, $\Sigma F_x = 0$, $\Sigma F_y = 0$. To obtain a numerical solution for the forces in the members, select a joint that has a free-body diagram with at most two unknown forces and one known force. (This may require first finding the reactions at the supports.) Once a member force is determined, use its value and apply it to an adjacent joint. Remember that forces that are found to *pull* on the joint are in *tension*, and those that *push* on the joint are in *compression*. To avoid a simultaneous solution of two equations, try to sum forces in a direction that is perpendicular to one of the unknowns. This will allow a direct solution for the other unknown. To further simplify the analysis, first identify all the zero-force members.

- **Method of Sections.** If a truss is in equilibrium, then each section of the truss is also in equilibrium. Pass a section through the member whose force is to be determined. Then draw the free-body diagram of the sectioned part having the least number of forces on it. Sectioned members subjected to *pulling* are in *tension*, and those that are subjected to *pushing* are in *compression*. If the force system is coplanar, then three equations of equilibrium are available to determine the unknowns. If possible, sum forces in a direction that is perpendicular to two of the three unknown forces. This will yield a direct solution for the third force. Likewise, sum moments about a point that passes through the line of action of two of the three unknown forces, so that the third unknown force can be determined directly.

- **Frames and Machines.** The forces acting at the joints of a frame or machine can be determined by drawing the free-body diagrams of each of its members or parts. The principle of action-reaction should be carefully observed when drawing these forces on each adjacent member or pin. For a coplanar force system, there are three equilibrium equations available for each member.

REVIEW PROBLEMS

*6-128. Determine the resultant forces at pins B and C on member ABC of the four-member frame.

Prob. 6–128

6-129. The mechanism consists of identical meshed gears A and B and arms which are fixed to the gears. The spring attached to the ends of the arms has an unstretched length of 100 mm and a stiffness of $k = 250$ N/m. If a torque of $M = 6$ N·m is applied to gear A, determine the angle θ through which each arm rotates. The gears are each pinned to fixed supports at their centers.

Prob. 6–129

6-130. Determine the horizontal and vertical components of force at pins A and C of the two-member frame.

Prob. 6–130

6-131. The spring has an unstretched length of 0.3 m. Determine the angle θ for equilibrium if the uniform links each have a mass of 5 kg.

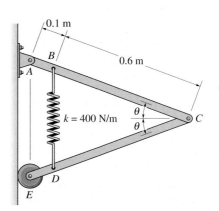

Prob. 6–131

*6-132. The spring has an unstretched length of 0.3 m. Determine the mass m of each uniform link if the angle $\theta = 20°$ for equilibrium.

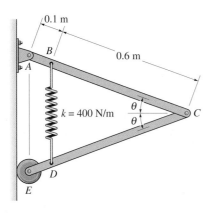

Prob. 6–132

6-133. Determine the horizontal and vertical components of force that the pins A and B exert on the two-member frame. Set $F = 0$.

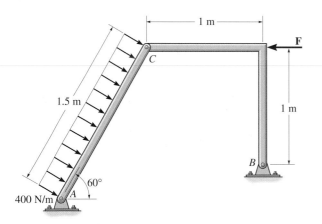

Prob. 6–133

6-134. Determine the horizontal and vertical components of force that pins A and B exert on the two-member frame. Set $F = 500$ N.

Prob. 6–134

6-135. The two-bar mechanism consists of a lever arm AB and smooth link CD, which has a fixed collar at its end C and a roller at the other end D. Determine the force **P** needed to hold the lever in the position θ. The spring has a stiffness k and unstretched length $2L$. The roller contacts either the top or bottom portion of the horizontal guide.

Prob. 6–135

*6-136. Determine the force in each member of the truss and state if the members are in tension or compression.

6-137. Determine the force in members *AB, AD,* and *AC* of the space truss and state if the members are in tension or compression.

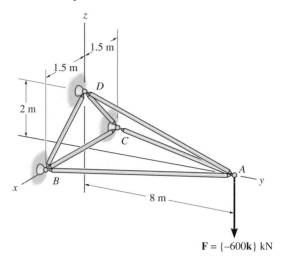

Prob. 6–136

Prob. 6–137

DESIGN PROJECTS

6–1D   DESIGN OF A BRIDGE TRUSS

A bridge having a horizontal top cord is to span between the two piers *A* and *B* having an arbitrary height. It is required that a pin-connected truss be used, consisting of steel members bolted together to steel gusset plates, such as the one shown in the figure. The end supports are assumed to be a pin at *A* and a roller at *B*. A vertical loading of 5 kN is to be supported within the middle 3m of the span. This load can be applied in part to several joints on the top cord within this region, or to a single joint at the middle of the top cord. The force of the wind and the weight of the members are to be neglected.

Assume the maximum tensile force in each member cannot exceed 4.25 kN; and regardless of the length of the member, the maximum compressive force cannot exceed 3.5 kN. Design the most economical truss that will support the loading. The members cost $3.50/m, and the gusset plates cost $8.00 each. Submit your cost analysis for the materials, along with a scaled drawing of the truss, identifying on this drawing the tensile and compressive force in each member. Also, include your calculations of the complete force analysis.

Prob. 6–1D

6–2D   DESIGN OF A CART LIFT

A hand cart is used to move a load from one loading dock to another. Any dock will have a different elevation relative to the bed of a truck that backs up to it. It is necessary that the loading platform on the hand cart will bring the load resting on it up to the elevation of each truck bed as shown. The maximum elevation difference between the frame of the hand cart and a truck bed is 0.3 m. Design a hand-operated mechanical system that will allow the load to be lifted this distance from the frame of the hand cart. Assume the operator can exert a (comfortable) force of 100 N to make the lift, and that the maximum load, centered on the loading platform, is 2000 N. Submit a scaled drawing of your design, and explain how it works based on a force analysis.

Prob. 6–2D

6–3D   DESIGN OF A PULLEY SYSTEM

The steel beam AB, having a length of 5 m and a mass of 700 kg is to be hoisted in its horizontal position to a height of 4 m. Design a pulley-and-rope system, which can be suspended from the overhead beam CD, that will allow a single worker to hoist the beam. Assume that the maximum (comfortable) force that he can apply to the rope is 180 N. Submit a drawing of your design, specify its approximate material cost, and discuss the safety aspects of its operation. Rope costs $1.25/m and each pulley costs $3.00.

Prob. 6–3D

6–4D DESIGN OF A TOOL USED TO POSITION A SUSPENDED LOAD

Heavy loads are suspended from an overhead pulley and each load must be positioned over a depository. Design a tool that can be used to shorten or lengthen the pulley cord AB a small amount in order to make the location adjustment. Assume the worker can apply a maximum (comfortable) force of 100 N to the tool, and the maximum force allowed in cord AB is 2000 N. Submit a scaled drawing of the tool, and a brief paragraph to explain how it works using a force analysis. Include a discussion on the safety aspects of its use.

Prob. 6–4D

6–5D DESIGN OF A FENCE-POST REMOVER

A farmer wishes to remove several fence posts. Each post is buried 0.5 m in the ground and will require a maximum vertical pulling force of 700 N to remove it. He can use his truck to develop the force, but he needs to devise a method for their removal without breaking the posts. Design a method that can be used, considering that the only materials available are a strong rope and several pieces of wood having various sizes and lengths. Submit a sketch of your design and discuss the safety and reliability of its use. Also, provide a force analysis to show how it works and why it will cause minimal damage to a post when it is removed.

700 N maximum

Prob. 6–5D

The design and analysis of any structural member requires knowledge of the internal loadings acting within it, not only when it is in place and subjected to service loads, but also when it is being hoisted as shown here. In this chapter, we will discuss how engineers determine these loadings.

Internal Forces

- To show how to use the method of sections for determining the internal loadings in a member.

- To generalize this procedure by formulating equations that can be plotted so that they describe the internal shear and moment throughout a member.

- To analyze the forces and study the geometry of cables supporting a load.

7.1 Internal Forces Developed in Structural Members

The design of any structural or mechanical member requires an investigation of the loading acting within the member in order to be sure the material can resist this loading. These internal loadings can be determined by using the *method of sections*. To illustrate the procedure, consider the "simply supported" beam shown in Fig. 7–1a, which is subjected to the forces $\mathbf{F}_1$ and $\mathbf{F}_2$ and the *support reactions* $\mathbf{A}_x$, $\mathbf{A}_y$, and $\mathbf{B}_y$, Fig. 7–1b. If the *internal loadings* acting on the cross section at C are to be determined, then an imaginary section is passed through the beam, cutting it into two segments. By doing this the internal loadings at the section become *external* on the free-body diagram of each segment,

(a)

(b)

Fig. 7–1

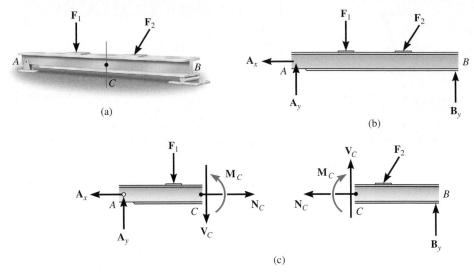

(a)

(b)

(c)

Fig. 7–1

Fig. 7–1c. Since both segments (*AC* and *CB*) were in equilibrium *before* the beam was sectioned, equilibrium of each segment is maintained provided rectangular force components $\mathbf{N}_C$ and $\mathbf{V}_C$ and a resultant couple moment $\mathbf{M}_C$ are developed at the section. Note that these loadings must be equal in magnitude and opposite in direction on each of the segments (Newton's third law). The magnitude of each of these loadings can now be determined by applying the three equations of equilibrium to either segment *AC* or *CB*. A *direct solution* for $\mathbf{N}_C$ is obtained by applying $\Sigma F_x = 0$; $\mathbf{V}_C$ is obtained directly from $\Sigma F_y = 0$; and $\mathbf{M}_C$ is determined by summing moments about point *C*, $\Sigma M_C = 0$, in order to eliminate the moments of the unknowns $\mathbf{N}_C$ and $\mathbf{V}_C$.

To save on material the beams used to support the roof of this shelter were tapered since the roof loading will produce a larger internal moment at the beams' centers than at their ends.

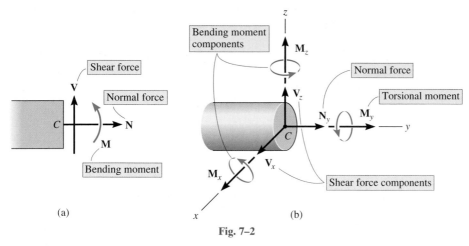

(a)

(b)

Fig. 7–2

In mechanics, the force components **N**, acting normal to the beam at the cut section, and **V**, acting tangent to the section, are termed the *normal or axial force* and the *shear force*, respectively. The couple moment **M** is referred to as the *bending moment*, Fig. 7–2a. In three dimensions, a general internal force and couple moment resultant will act at the section. The *x, y, z* components of these loadings are shown in Fig. 7–2b. Here $\mathbf{N}_y$ is the *normal force*, and $\mathbf{V}_x$ and $\mathbf{V}_z$ are *shear force components*. $\mathbf{M}_y$ is a *torsional or twisting moment*, and $\mathbf{M}_x$ and $\mathbf{M}_z$ are *bending moment components*. For most applications, these *resultant loadings* will act at the geometric center or centroid (*C*) of the section's cross-sectional area. Although the magnitude for each loading generally will be different at various points along the axis of the member, the method of sections can always be used to determine their values.

Free-Body Diagrams. Since frames and machines are composed of *multiforce members*, each of these members will generally be subjected to internal normal, shear, and bending loadings. For example, consider the frame shown in Fig. 7–3a. If the blue section is passed through the frame to determine the internal loadings at points *H, G*, and *F*, the resulting free-body diagram of the top portion of this section is shown in Fig. 7–3b. At each point where a member is sectioned there is an unknown normal force, shear force, and bending moment. As a result, we cannot apply the *three* equations of equilibrium to this section in order to obtain these *nine unknowns*.*Instead, to solve this problem we must *first dismember* the frame and determine the reactions at the connections of the members using the techniques of Sec. 6.6. Once this is done, *each member* may then be sectioned at its appropriate point, and the three equations of equilibrium can be applied to determine **N**, **V**, and **M**. For example, the free-body diagram of segment *DG*, Fig. 7–3c, can be used to determine the internal loadings at *G* provided the reactions of the pin, $\mathbf{D}_x$ and $\mathbf{D}_y$, are known.

*Recall that this method of analysis worked well for trusses since truss members are *straight two-force members* which support only an axial or normal load.

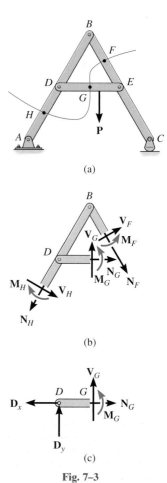

(a)

(b)

(c)

Fig. 7–3

In each case, the link on the backhoe is a two-force member. In the top photo it is subjected to both bending and axial load at its center. By making the member straight, as in the bottom photo, then only an axial force acts within the member.

PROCEDURE FOR ANALYSIS

The method of sections can be used to determine the internal loadings at a specific location in a member using the following procedure.

Support Reactions.

- Before the member is "cut" or sectioned, it may first be necessary to determine the member's support reactions, so that the equilibrium equations are used only to solve for the internal loadings when the member is sectioned.

- If the member is part of a frame or machine, the reactions at its connections are determined using the methods of Sec. 6.6.

Free-Body Diagram.

- Keep all distributed loadings, couple moments, and forces acting on the member in their *exact locations*, then pass an imaginary section through the member, perpendicular to its axis at the point where the internal loading is to be determined.

- After the section is made, draw a free-body diagram of the segment that has the least number of loads on it, and indicate the *x, y, z* components of the force and couple moment resultants at the section.

- If the member is subjected to a *coplanar* system of forces, only **N**, **V**, and **M** act at the section.

- In many cases it may be possible to tell by inspection the proper sense of the unknown loadings; however, if this seems difficult, the sense can be assumed.

Equations of Equilibrium.

- Moments should be summed at the section about axes passing through the *centroid* or geometric center of the member's cross-sectional area in order to eliminate the unknown normal and shear forces and thereby obtain direct solutions for the moment components.

- If the solution of the equilibrium equations yields a negative scalar, the assumed sense of the quantity is opposite to that shown on the free-body diagram.

EXAMPLE 7.1

The bar is fixed at its end and is loaded as shown in Fig. 7–4a. Determine the internal normal force at points B and C.

Solution

Fig. 7–4

(a)   (b)   (c)

Support Reactions. A free-body diagram of the entire bar is shown in Fig. 7–4b. By inspection, only a normal force $\mathbf{A}_y$ acts at the fixed support since the loads are applied symmetrically along the bar's axis. ($A_x = 0$, $M_A = 0$.)

$$+\uparrow \Sigma F_y = 0; \quad A_y - 16\text{ kN} + 12\text{ kN} - 4\text{ kN} = 0 \quad A_y = 8\text{ kN}$$

Free-Body Diagrams. The internal forces at B and C will be found using the free-body diagrams of the sectioned bar shown in Fig. 7–4c. No shear or moment act on the sections since they are not required for equilibrium. In particular, segments AB and DC will be chosen here, since they contain the *least* number of forces.

Equations of Equilibrium.
   Segment AB

$$+\uparrow \Sigma F_y = 0; \quad 8\text{ kN} - N_B = 0 \quad N_B = 8\text{ kN} \qquad \textit{Ans.}$$

   Segment DC

$$+\uparrow \Sigma F_y = 0; \quad N_C - 4\text{ kN} = 0 \quad N_C = 4\text{ kN} \qquad \textit{Ans.}$$

Try working this problem in the following manner: Determine N_B from segment BD. (Note that this approach *does not require* solution for the support reaction at A.) Using the result for N_B, isolate segment BC to determine N_C.

EXAMPLE 7.2

The circular shaft is subjected to three concentrated torques as shown in Fig. 7–5a. Determine the internal torques at points B and C.

Solution

(a)

Support Reactions. Since the shaft is subjected only to collinear torques, a torque reaction occurs at the support, Fig. 7–5b. Using the right-hand rule to define the positive directions of the torques, we require

$$\Sigma M_x = 0; \quad -10 \text{ N} \cdot \text{m} + 15 \text{ N} \cdot \text{m} + 20 \text{ N} \cdot \text{m} - T_D = 0$$

$$T_D = 25 \text{ N} \cdot \text{m}$$

(b)

Fig. 7–5

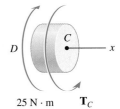

(c)

Free-Body Diagrams. The internal torques at B and C will be found using the free-body diagrams of the shaft segments AB and CD shown in Fig. 7–5c.

Equations of Equilibrium. Applying the equation of moment equilibrium along the shaft's axis, we have

   Segment AB

$$\Sigma M_x = 0; \quad -10 \text{ N} \cdot \text{m} + 15 \text{ N} \cdot \text{m} - T_B = 0 \quad T_B = 5 \text{ N} \cdot \text{m} \quad Ans.$$

   Segment CD

$$\Sigma M_x = 0; \quad T_C - 25 \text{ N} \cdot \text{m} = 0 \quad T_C = 25 \text{ N} \cdot \text{m} \quad Ans.$$

Try to solve for T_C by using segment CA. Note that this approach *does not require* a solution for the support reaction at D.

E X A M P L E 7.3

The beam supports the loading shown in Fig. 7–6a. Determine the internal normal force, shear force, and bending moment acting just to the left, point B, and just to the right, point C, of the 6-kN force.

(a)

Solution

Support Reactions. The free-body diagram of the beam is shown in Fig. 7–6b. When determining the *external reactions*, realize that the 9-kN · m couple moment is a free vector and therefore it can be placed *anywhere* on the free-body diagram of the entire beam. Here we will only determine $\mathbf{A}_y$, since segments AB and AC will be used for the analysis.

$$\zeta+\Sigma M_D = 0; \quad 9\,\text{kN} \cdot \text{m} + (6\,\text{kN})(6\,\text{m}) - A_y(9\,\text{m}) = 0$$

$$A_y = 5\,\text{kN}$$

(b)

Free-Body Diagrams. The free-body diagrams of the left segments AB and AC of the beam are shown in Figs. 7–6c and 7–6d. In this case the 9-kN · m couple moment is *not included* on these diagrams since it must be kept in its *original position* until *after* the section is made and the appropriate body is isolated. In other words, the free-body diagrams of the left segments of the beam do not show the couple moment since this moment does not actually act on these segments.

(c)

(d)

Fig. 7–6

Equations of Equilibrium.

Segment AB

$$\xrightarrow{+} \Sigma F_x = 0; \qquad\qquad N_B = 0 \qquad\qquad Ans.$$
$$+\uparrow \Sigma F_y = 0; \qquad 5\,\text{kN} - V_B = 0 \qquad V_B = 5\,\text{kN} \qquad Ans.$$
$$\zeta+\Sigma M_B = 0; \quad -(5\,\text{kN})(3\,\text{m}) + M_B = 0 \qquad M_B = 15\,\text{kN} \cdot \text{m} \; Ans.$$

Segment AC

$$\xrightarrow{+} \Sigma F_x = 0; \qquad\qquad N_C = 0 \qquad\qquad Ans.$$
$$+\uparrow \Sigma F_y = 0; \qquad 5\,\text{kN} - 6\,\text{kN} + V_C = 0 \qquad V_C = 1\,\text{kN} \; Ans.$$
$$\zeta+\Sigma M_C = 0; \quad -(5\,\text{kN})(3\,\text{m}) + M_C = 0 \qquad M_C = 15\,\text{kN} \cdot \text{m} \quad Ans.$$

Here the moment arm for the 5-kN force in both cases is approximately 3 m since B and C are "almost" coincident.

EXAMPLE 7.4

(a)

Determine the internal normal force, shear force, and bending moment acting at point B of the two-member frame shown in Fig. 7–7a.

Solution

Support Reactions. A free-body diagram of each member is shown in Fig. 7–7b. Since CD is a two-force member, the equations of equilibrium need to be applied only to member AC.

$$\zeta + \Sigma M_A = 0; \quad -400 \text{ kN } (4 \text{ m}) + (\tfrac{3}{5})F_{DC}(8 \text{ m}) = 0 \quad F_{DC} = 333.3 \text{ kN}$$

$$\xrightarrow{+} \Sigma F_x = 0; \quad -A_x + (\tfrac{4}{5})(333.3 \text{ kN}) = 0 \quad A_x = 266.7 \text{ kN}$$

$$+\uparrow \Sigma F_y = 0; \quad A_y - 400 \text{ kN} + \tfrac{3}{5}(333.3 \text{ kN}) = 0 \quad A_y = 200 \text{ kN}$$

Free-Body Diagrams. Passing an imaginary section perpendicular to the axis of member AC through point B yields the free-body diagrams of segments AB and BC shown in Fig. 7–7c. When constructing these diagrams it is important to keep the distributed loading exactly as it is until *after* the section is made. Only then can it be replaced by a single resultant force. Why? Also, notice that $\mathbf{N}_B$, $\mathbf{V}_B$, and $\mathbf{M}_B$ act with equal magnitude but opposite direction on each segment—Newton's third law.

(b)

Equations of Equilibrium. Applying the equations of equilibrium to segment AB, we have

$$\xrightarrow{+} \Sigma F_x = 0; \quad N_B - 266.7 \text{ kN} = 0 \quad N_B = 267 \text{ kN} \quad Ans.$$

$$+\uparrow \Sigma F_y = 0; \quad 200 \text{ kN} - 200 \text{ kN} - V_B = 0 \quad V_B = 0 \quad Ans.$$

$$\zeta + \Sigma M_B = 0; \quad M_B - 200 \text{ kN } (4 \text{ m}) - 200 \text{ kN } (2 \text{ m}) = 0$$

$$M_B = 400 \text{ kN} \cdot \text{m} \quad Ans.$$

As an exercise, try to obtain these same results using segment BC.

Fig. 7–7

(c)

E X A M P L E 7.5

Determine the normal force, shear force, and bending moment acting at point E of the frame loaded as shown in Fig. 7–8a.

(a)

(b)

(c)

Fig. 7–8

Solution

Support Reactions.   By inspection, members AC and CD are two-force members, Fig. 7–8b. In order to determine the internal loadings at E, we must first determine the force $\mathbf{R}$ at the end of member AC. To do this we must analyze the equilibrium of the pin at C. Why?

Summing forces in the vertical direction on the pin, Fig. 7–8b, we have

$$+\uparrow \Sigma F_y = 0; \qquad R \sin 45° - 600 \text{ N} = 0 \qquad R = 848.5 \text{ N}$$

Free-Body Diagram.   The free-body diagram of segment CE is shown in Fig. 7–8c.

Equations of Equilibrium.

$$\xrightarrow{+} \Sigma F_x = 0; \quad 848.5 \cos 45° \text{ N} - V_E = 0 \qquad V_E = 600 \text{ N } Ans.$$

$$+\uparrow \Sigma F_y = 0; \quad -848.5 \sin 45° \text{ N} + N_E = 0 \qquad N_E = 600 \text{ N } Ans.$$

$$\mathord{\llcorner}+\Sigma M_E = 0; \ 848.5 \cos 45° \text{N}(0.5 \text{ m}) - M_E = 0 \quad M_E = 300 \text{ N} \cdot \text{m } Ans.$$

These results indicate a poor design. Member AC should be *straight* (from A to C) so that bending within the member is *eliminated*. If AC is straight then the internal force would only create tension in the member. See Example 6.21.

E X A M P L E 7.6

The uniform sign shown in Fig. 7–9a has a mass of 650 kg and is supported on the fixed column. Design codes indicate that the expected maximum uniform wind loading that will occur in the area where it is located is 900 Pa. Determine the internal loadings at A.

(a)

(b)

Solution

The idealized model for the sign is shown in Fig. 7–9b. Here the necessary dimensions are indicated. We can consider the free-body diagram of a section above point A since it does not involve the support reactions.

Free-Body Diagram. The sign has a weight of $W = 650(9.81) = 6.376$ kN, and the wind creates a resultant force of $F_w = 900$ N/m^2(6m)(2.5m) $= 13.5$ kN perpendicular to the face of the sign. These loadings are shown on the free-body diagram, Fig. 7–9c.

Equations of Equilibrium. Since the problem is three dimensional, a vector analysis will be used.

$$\Sigma \mathbf{F} = \mathbf{0};\qquad \mathbf{F}_A - 13.5\mathbf{i} - 6.376\mathbf{k} = \mathbf{0}$$

$$\mathbf{F}_A = \{13.5\mathbf{i} + 6.38\mathbf{k}\}\text{ kN} \qquad\qquad Ans.$$

$$\Sigma \mathbf{M}_A = \mathbf{0};\qquad \mathbf{M}_A + \mathbf{r} \times (\mathbf{F}_w + \mathbf{W}) = \mathbf{0}$$

$$\mathbf{M}_A + \begin{vmatrix} \mathbf{i} & \mathbf{j} & \mathbf{k} \\ 0 & 3 & 5.25 \\ -13.5 & 0 & 6.376 \end{vmatrix} = \mathbf{0}$$

$$M_A = \{-19.1\mathbf{i} + 70.9\mathbf{j} + 40.5\mathbf{k}\}\text{ kN}\cdot\text{m} \qquad Ans.$$

Here $\mathbf{F}_{A_z} = \{6.38\mathbf{k}\}$ kN represents the normal force N, whereas $\mathbf{F}_{A_x} = \{13.5\mathbf{i}\}$ kN is the shear force. Also, the torsional moment is $\mathbf{M}_{A_z} = \{40.5\mathbf{k}\}$ kN $\cdot$ m, and the bending moment is determined from its components $\mathbf{M}_{A_x} = \{-19.1\mathbf{i}\}$ kN $\cdot$ m and $\mathbf{M}_{A_y} = \{-70.9\mathbf{j}\}$ kN $\cdot$ m; i.e., $M_b = \sqrt{M_x^2 + M_y^2}$.

(c)

Fig. 7–9

PROBLEMS

7-1. The column is fixed to the floor and is subjected to the loads shown. Determine the internal normal force, shear force, and moment at points A and B.

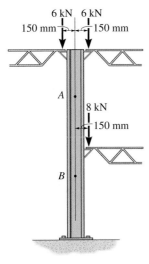

6 kN 6 kN

150 mm — 150 mm

A

8 kN

150 mm

B

Prob. 7–1

7-2. The rod is subjected to the forces shown. Determine the internal normal force at points A, B, and C.

550 N

150 N A 150 N

350 N B 350 N

C

Prob. 7–2

7-3. The forces act on the shaft shown. Determine the internal normal force at points A, B, and C.

5 kN   4 kN   7 kN   4 kN

A

4 kN B C

Prob. 7–3

*7-4. The shaft is supported by the two smooth bearings A and B. The four pulleys attached to the shaft are used to transmit power to adjacent machinery. If the torques applied to the pulleys are as shown, determine the internal torques at points C, D, and E.

A

C

40 N·m

15 N·m

D

65 N·m

E

10 N·m

B

Prob. 7–4

7-5. The shaft is supported by a journal bearing at A and a thrust bearing at B. Determine the normal force, shear force, and moment at a section passing through (a) point C, which is just to the right of the bearing at A, and (b) point D, which is just to the left of the 3000-N force.

2500 N   75 N/m   3000 N

C D B

A

2 m   4 m

0.5 m

Prob. 7–5

7-6. Determine the internal normal force and shear force, and the bending moment in the beam at points C and D. Assume the support at B is a roller. Point C is located just to the right of the 8-kN load.

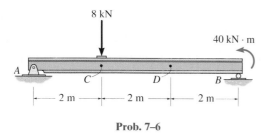

Prob. 7–6

7-7. Determine the shear force and moment at points C and D.

Prob. 7–7

*7-8. Determine the normal force, shear force, and moment at a section passing through point C. Assume the support at A can be approximated by a pin and B as a roller.

Prob. 7–8

7-9. Determine the normal force, shear force, and moment at a section passing through point D. Take $w = 150$ N/m.

7-10. The beam AB will fail if the maximum internal moment at D reaches 800 N · m or the normal force in member BC becomes 1500 N. Determine the largest load w it can support.

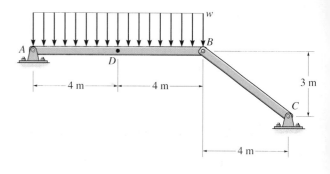

Probs. 7–9/10

7-11. Determine the shear force and moment acting at a section passing through point C in the beam.

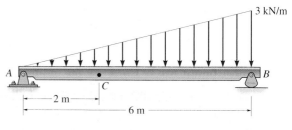

Prob. 7–11

*7-12. The boom *DF* of the jib crane and the column *DE* have a uniform weight of 750 N/m. If the hoist and load weigh 1500 N, determine the normal force, shear force, and moment in the crane at sections passing through points *A*, *B*, and *C*. *Hint*: (Treat the boom tip, beyond the hoist, as weightless.)

Prob. 7–12

7-13. Determine the internal normal force, shear force, and moment acting at point *C* and at point *D*, which is located just to the right of the roller support at *B*.

Prob. 7–13

7-14. Determine the normal force, shear force, and moment at a section passing through point *D* of the two-member frame.

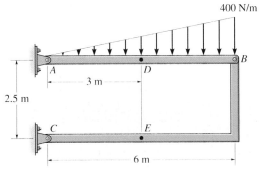

Prob. 7–14

7-15. Determine the normal force, shear force, and moment at a section passing through point *E* of the two-member frame.

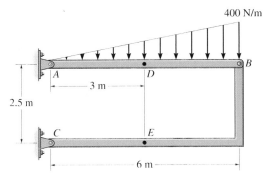

Prob. 7–15

*7-16. The strongback or lifting beam is used for materials handling. If the suspended load has a weight of 2 kN and a center of gravity of *G*, determine the placement *d* of the padeyes on the top of the beam so that there is no moment developed within the length *AB* of the beam. The lifting bridle has two legs that are positioned at 45°, as shown.

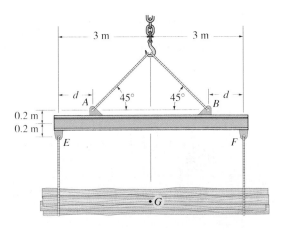

Prob. 7–16

7-17. Determine the normal force, shear force, and moment acting at a section passing through point *C*.

7-18. Determine the normal force, shear force, and moment acting at a section passing through point *D*.

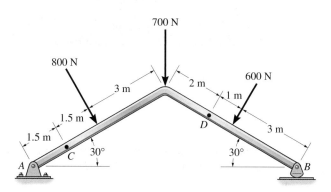

Probs. 7–17/18

7-19. Determine the normal force, shear force, and moment at a section passing through point *C*. Take *P* = 8 kN.

*7-20. The cable will fail when subjected to a tension of 2 kN. Determine the largest vertical load *P* the frame will support and calculate the internal normal force, shear force, and moment at a section passing through point *C* for this loading.

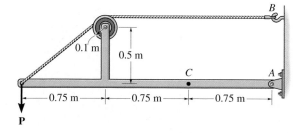

Probs. 7–19/20

7-21. Determine the internal normal force, shear force, and bending moment in the beam at point *B*.

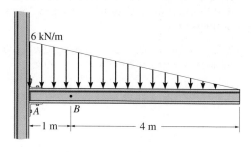

Prob. 7–21

7-22. Determine the ratio of *a/b* for which the shear force will be zero at the midpoint *C* of the beam.

Prob. 7–22

7-23. Determine the internal normal force, shear force, and bending moment at point *C*.

Prob. 7–23

*7-24. The jack *AB* is used to straighten the bent beam *DE* using the arrangement shown. If the axial compressive force in the jack is 20 kN, determine the internal moment developed at point *C* of the top beam. Neglect the weight of the beams.

7-25. Solve Prob. 7–24 assuming that each beam has a uniform weight of 2400 N/m (≈ 240 kg/m).

Probs. 7–24/25

7-26. Determine the normal force, shear force, and moment in the beam at sections passing through points *D* and *E*. Point *E* is just to the right of the 3-kN load.

Prob. 7–26

7-27. Determine the normal force, shear force, and moment at a section passing through point *D* of the two-member frame.

Prob. 7–27

*7-28. Determine the normal force, shear force, and moment at sections passing through points *E* and *F*. Member *BC* is pinned at *B* and there is a smooth slot in it at *C*. The pin at *C* is fixed to member *CD*.

Prob. 7–28

7-29. Determine the internal normal force, shear force, and the moment at points *C* and *D*.

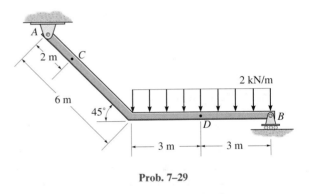

Prob. 7–29

7-30. Determine the normal force, shear force, and moment acting at sections passing through points *B* and *C* on the curved rod.

Prob. 7–30

7-31. The cantilevered rack is used to support each end of a smooth pipe that has a total weight of 300 N ($\approx$ 30 kg). Determine the normal force, shear force, and moment that act in the arm at its fixed support A along a vertical section.

7-33. Determine the internal normal force, shear force, and bending moment in the beam at points D and E. Point E is just to the right of the 20-kN load. Assume A is a roller support, the splice at B is a pin, and C is a fixed support.

Prob. 7–33

Prob. 7–31

7-34. Determine the internal normal force, shear force, and bending moment at points E and F of the frame.

*7-32. Determine the normal force, shear force, and moment at a section passing through point D of the two-member frame.

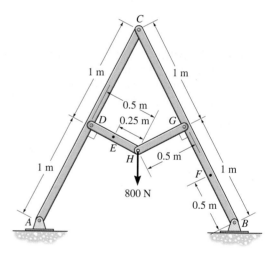

Prob. 7–34

7-35. Determine the distance a as a fraction of the beam's length L for locating the roller support so that the moment in the beam at B is zero.

Prob. 7–32

Prob. 7–35

*7-36. The semicircular arch is subjected to a uniform distributed load along its axis of w_0 per unit length. Determine the internal normal force, shear force, and moment in the arch at $\theta = 45°$.

7-37. Solve Prob. 7–36 for $\theta = 120°$.

Probs. 7–36/37

7-38. Determine the x, y, z components of internal loading at a section passing through point C in the pipe assembly. Neglect the weight of the pipe. Take $\mathbf{F}_1 = \{350\mathbf{j} - 400\mathbf{k}\}$ N and $\mathbf{F}_2 = \{150\mathbf{j} - 300\mathbf{k}\}$ N.

7-39. Determine the x, y, z components of internal loading at a section passing through point C in the pipe assembly. Neglect the weight of the pipe. Take $\mathbf{F}_1 = \{80\mathbf{i} + 200\mathbf{j} - 300\mathbf{k}\}$ N and $\mathbf{F}_2 = \{250\mathbf{i} - 150\mathbf{j} - 200\mathbf{k}\}$ N.

*7-40. Determine the x, y, z components of force and moment at point C in the pipe assembly. Neglect the weight of the pipe. The load acting at $(0, 3.5\text{ m}, 3\text{ m})$ is $\mathbf{F}_1 = \{-24\mathbf{i} - 10\mathbf{k}\}$ N and $\mathbf{M} = \{-30\mathbf{k}\}$ N · m and at point $(0, 3.5\text{ m}, 0)$ $\mathbf{F}_2 = \{-80\mathbf{i}\}$ N.

Prob. 7–40

7-41. Determine the x, y, z components of force and moment at point C in the pipe assembly. Neglect the weight of the pipe. Take $\mathbf{F}_1 = \{350\mathbf{i} - 400\mathbf{j}\}$ N and $\mathbf{F}_2 = \{-200\mathbf{j} + 150\mathbf{k}\}$ N.

Probs. 7–38/39

Prob. 7–41

*7.2 Shear and Moment Equations and Diagrams

Beams are structural members which are designed to support loadings applied perpendicular to their axes. In general, beams are long, straight bars having a constant cross-sectional area. Often they are classified as to how they are supported. For example, a *simply supported beam* is pinned at one end and roller-supported at the other, Fig. 7–10, whereas a *cantilevered beam* is fixed at one end and free at the other. The actual design of a beam requires a detailed knowledge of the *variation* of the internal shear force V and bending moment M acting at *each point* along the axis of the beam. After this force and bending-moment analysis is complete, one can then use the theory of mechanics of materials and an appropriate engineering design code to determine the beam's required cross-sectional area.

The *variations* of V and M as functions of the position x along the beam's axis can be obtained by using the method of sections discussed in Sec. 7.1. Here, however, it is necessary to section the beam at an arbitrary distance x from one end rather than at a specified point. If the results are plotted, the graphical variations of V and M as functions of x are termed the *shear diagram* and *bending-moment diagram*, respectively.

In general, the internal shear and bending-moment functions generally will be discontinuous, or their slopes will be discontinuous at points where a distributed load changes or where concentrated forces or couple moments are applied. Because of this, these functions must be determined for *each segment* of the beam located between any two discontinuities of loading. For example, sections located at x_1, x_2, and x_3 will have to be used to describe the variation of V and M throughout the length of the beam in Fig. 7–10. These functions will be valid *only* within regions from O to a for x_1, from a to b for x_2, and from b to L for x_3.

The internal normal force will not be considered in the following discussion for two reasons. In most cases, the loads applied to a beam act perpendicular to the beam's axis and hence produce only an internal shear force and bending moment. For design purposes, the beam's resistance to shear, and particularly to bending, is more important than its ability to resist a normal force.

The designer of this shop crane realized the need for additional reinforcement around the joint in order to prevent severe internal bending of the joint when a large load is suspended from the chain hoist.

Fig. 7–10

Sign Convention. Before presenting a method for determining the shear and bending moment as functions of x and later plotting these functions (shear and bending-moment diagrams), it is first necessary to establish a *sign convention* so as to define a "positive" and "negative" shear force and bending moment acting in the beam. [This is analogous to assigning coordinate directions x positive to the right and y positive upward when plotting a function $y = f(x)$.] Although the choice of a sign convention is arbitrary, here we will choose the one used for the majority of engineering applications. It is illustrated in Fig. 7–11. Here the positive directions are denoted by an internal *shear force* that causes *clockwise rotation* of the member on which it acts, and by an internal *moment* that causes *compression or pushing on the upper part* of the member. Also, positive moment would tend to bend the member if it were elastic, concave upward. Loadings that are opposite to these are considered negative.

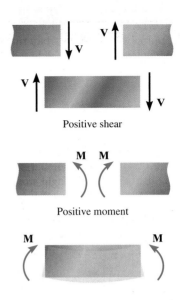

Positive shear

Positive moment

Beam sign convention

Fig. 7–11

PROCEDURE FOR ANALYSIS

The shear and bending-moment diagrams for a beam can be constructed using the following procedure.

Support Reactions.

- Determine all the reactive forces and couple moments acting on the beam and resolve all the forces into components acting perpendicular and parallel to the beam's axis.

Shear and Moment Functions.

- Specify separate coordinates x having an origin at the beam's *left end* and extending to regions of the beam *between* concentrated forces and/or couple moments, or where there is no discontinuity of distributed loading.

- Section the beam perpendicular to its axis at each distance x and draw the free-body diagram of one of the segments. Be sure **V** and **M** are shown acting in their *positive sense*, in accordance with the sign convention given in Fig. 7–11.

- The shear V is obtained by summing forces perpendicular to the beam's axis.

- The moment M is obtained by summing moments about the sectioned end of the segment.

Shear and Moment Diagrams.

- Plot the shear diagram (V versus x) and the moment diagram (M versus x). If computed values of the functions describing V and M are *positive*, the values are plotted above the x axis, whereas *negative* values are plotted below the x axis.

- Generally, it is convenient to plot the shear and bending-moment diagrams directly below the free-body diagram of the beam.

EXAMPLE 7.7

(a)

2.5 kN

$0 \le x < 2$ m

(b)

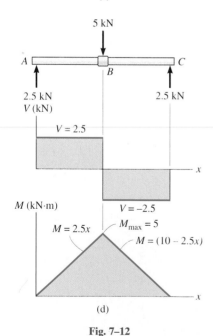

$V = 2.5$

x

$V = -2.5$

M (kN·m)

$M = 2.5x$

$M_{max} = 5$

$M = (10 - 2.5x)$

x

(d)

Fig. 7–12

Draw the shear and bending-moment diagrams for the shaft shown in Fig. 7–12a. The support at A is a thrust bearing and the support at C is a journal bearing.

Solution

Support Reactions. The support reactions have been computed, as shown on the shaft's free-body diagram, Fig. 7–12d.

Shear and Moment Functions. The shaft is sectioned at an arbitrary distance x from point A, extending within the region AB, and the free-body diagram of the left segment is shown in Fig. 7–12b. The unknowns **V** and **M** are assumed to act in the *positive sense* on the right-hand face of the segment according to the established sign convention. Why? Applying the equilibrium equations yields

$$+\uparrow \Sigma F_y = 0; \qquad\qquad V = 2.5 \text{ kN} \qquad\qquad (1)$$

$$\zeta + \Sigma M = 0; \qquad\qquad M = 2.5x \text{ kN} \cdot \text{m} \qquad\qquad (2)$$

A free-body diagram for a left segment of the shaft extending a distance x within the region BC is shown in Fig. 7–12c. As always, **V** and **M** are shown acting in the positive sense. Hence,

$$+\uparrow \Sigma F_y = 0; \qquad 2.5 \text{ kN} - 5 \text{ kN} - V = 0$$
$$V = -2.5 \text{ kN} \qquad\qquad (3)$$

$$\zeta + \Sigma M = 0; \qquad M + 5 \text{ kN}(x - 2\text{m}) - 2.5 \text{ kN}(x) = 0$$
$$M = (10 - 2.5x) \text{ kN} \cdot \text{m} \qquad\qquad (4)$$

Shear and Moment Diagrams. When Eqs. 1 through 4 are plotted within the regions in which they are valid, the shear and bending-moment diagrams shown in Fig. 7–12d are obtained. The shear diagram indicates that the internal shear force is always 2.5 kN (positive) within shaft segment AB. Just to the right of point B, the shear force changes sign and remains at a constant value of -2.5 kN for segment BC. The moment diagram starts at zero, increases linearly to point B at $x = 2$ m, where $M_{max} = 2.5 \text{ kN}(2 \text{ m}) = 5 \text{ kN} \cdot \text{m}$, and thereafter decreases back to zero.

It is seen in Fig. 7–12d that the graph of the shear and moment diagrams is discontinuous at points of concentrated force, i.e., points A, B, and C. For this reason, as stated earlier, it is necessary to express both the shear and bending-moment functions separately for regions between concentrated loads. It should be realized, however, that all loading discontinuities are mathematical, arising from the *idealization of a concentrated force and couple moment.* Physically, loads are always applied over a finite area, and if the load variation could actually be accounted for, the shear and bending-moment diagrams would then be continuous over the shaft's entire length.

EXAMPLE 7.8

Draw the shear and bending-moment diagrams for the beam shown in Fig. 7–13a.

Solution

Support Reactions.   The support reactions have been computed as shown on the beam's free-body diagram, Fig. 7–13c.

Shear and Moment Functions.   A free-body diagram for a left segment of the beam having a length x is shown in Fig. 7–13b. The distributed loading acting on this segment has an intensity of $\frac{2}{3}x$ at its end and is replaced by a resultant force *after* the segment is isolated as a free-body diagram. The *magnitude* of the resultant force is equal to $\frac{1}{2}(x)(\frac{2}{3}x) = \frac{1}{3}x^2$. This force *acts through the centroid* of the distributed loading area, a distance $\frac{1}{3}x$ from the right end. Applying the two equations of equilibrium yields

(a)

$$+\uparrow\Sigma F_y = 0; \qquad 9 - \frac{1}{3}x^2 - V = 0$$

$$V = \left(9 - \frac{x^2}{3}\right) \text{ kN} \qquad (1)$$

(b)

$$\zeta + \Sigma M = 0; \qquad M + \frac{1}{3}x^2\left(\frac{x}{3}\right) - 9x = 0$$

$$M = \left(9x - \frac{x^3}{9}\right) \text{ kN} \cdot \text{m} \qquad (2)$$

Shear and Moment Diagrams.   The shear and bending-moment diagrams shown in Fig. 7–13c are obtained by plotting Eqs. 1 and 2. The point of *zero shear* can be found using Eq. 1:

$$V = 9 - \frac{x^2}{3} = 0$$

$$x = 5.20 \text{ m}$$

It will be shown in Sec. 7.3 that this value of x happens to represent the point on the beam where the *maximum moment* occurs. Using Eq. (2), we have

$$M_{\text{max}} = \left(9(5.20) - \frac{(5.20)^3}{9}\right) \text{ kN} \cdot \text{m}$$

$$= 31.2 \text{ kN} \cdot \text{m}$$

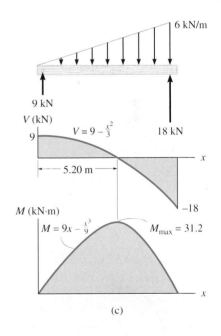

(c)

Fig. 7–13

PROBLEMS

For each of the following problems, establish the x axis with the origin at the left side of the beam, and obtain the internal shear and moment as a function of x. Use these results to plot the shear and moment diagrams.

7-42. Draw the shear and moment diagrams for the beam (a) in terms of the parameters shown; (b) set $P = 600$ kN, $a = 5$ m, $b = 7$ m.

Prob. 7–42

7-43. Draw the shear and moment diagrams for the cantilevered beam.

Prob. 7–43

*****7-44.** The suspender bar supports the 3000-N (≈ 300-kg) engine. Draw the shear and moment diagrams for the bar.

Prob. 7–44

7-45. Draw the shear and moment diagrams for the beam (a) in terms of the parameters shown; (b) set $M_0 = 500$ N $\cdot$ m, $L = 8$ m.

7-46. If $L = 9$ m, the beam will fail when the maximum shear force is $V_{max} = 5$ kN or the maximum bending moment is $M_{max} = 2$ kN $\cdot$ m. Determine the magnitude M_0 of the largest couple moments it will support.

Probs. 7–45/46

7-47. The shaft is supported by a thrust bearing at A and a journal bearing at B. If $L = 10$ m, the shaft will fail when the maximum moment is $M_{\max} = 5$ kN · m. Determine the largest uniform distributed load w the shaft will support.

Prob. 7–47

***7-48.** Draw the shear and moment diagrams for the beam.

Prob. 7–48

7-49. Draw the shear and bending-moment diagrams for the beam.

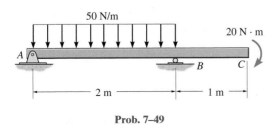

Prob. 7–49

7-50. Draw the shear and moment diagrams for the beam.

Prob. 7–50

7-51. Draw the shear and moment diagrams for the beam.

Prob. 7–51

***7-52.** Draw the shear and moment diagrams for the beam.

Prob. 7–52

7-53. Draw the shear and bending-moment diagrams for each of the two segments of the compound beam.

Prob. 7–53

7-54. Draw the shear and bending-moment diagrams for beam ABC. Note that there is a pin at B.

Prob. 7–54

7-55. Draw the shear and moment diagrams for the compound beam. The beam is pin-connected at E and F.

Prob. 7–55

*7-56. Draw the shear and moment diagrams for the beam.

Prob. 7–56

7-57. If $L = 1.8$ m, the beam will fail when the maximum shear force is $V_{max} = 80$ N, or the maximum moment is $M_{max} = 12$ N · m. Determine the largest intensity w of the distributed loading it will support.

Prob. 7–57

7-58. The beam will fail when the maximum internal moment is M_{max}. Determine the position x of the concentrated force $\mathbf{P}$ and its smallest magnitude that will cause failure.

Prob. 7–58

7-59. Draw the shear and moment diagrams for the beam.

Prob. 7–59

*7-60. Draw the shear and bending-moment diagrams for the beam.

300 N/m

A   B   C

3 m   4 m

Prob. 7–60

7-61. Draw the shear and moment diagrams for the beam.

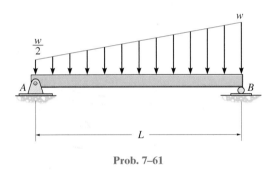

w

$\frac{w}{2}$

A   B

L

Prob. 7–61

7-62. Draw the shear and moment diagrams for the beam (a) in terms of the parameters shown; (b) set $P = 800$ N, $a = 5$ m, $L = 12$ m.

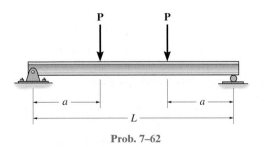

P   P

a   a

L

Prob. 7–62

7-63. Express the x, y, z components of internal loading in the rod as a function of y, where $0 \le y \le 4$ m.

z

800 N/m

x   y

4 m

y

2 m

1500 N

Prob. 7–63

*7-64. Determine the normal force, shear force, and moment in the curved rod as a function of θ.

r

θ

5 3
  4

P

Prob. 7–64

7-65. Express the internal shear and moment components acting in the rod as a function of y, where $0 \le y \le 4$ m.

z

4 kN/m

x

y

4 m

2 m

y

Prob. 7–65

*7.3 Relations between Distributed Load, Shear, and Moment

In cases where a beam is subjected to several concentrated forces, couple moments, and distributed loads, the method of constructing the shear and bending-moment diagrams discussed in Sec. 7.2 may become quite tedious. In this section a simpler method for constructing these diagrams is discussed—a method based on differential relations that exist between the load, shear, and bending moment.

Distributed Load. Consider the beam AD shown in Fig. 7–14a, which is subjected to an arbitrary load $w = w(x)$ and a series of concentrated forces and couple moments. In the following discussion, the *distributed load* will be considered *positive* when the *loading acts downward* as shown. A free-body diagram for a small segment of the beam having a length Δx is chosen at a point x along the beam which is *not* subjected to a concentrated force or couple moment, Fig. 7–14b. Hence any results obtained will not apply at points of concentrated loading. The internal shear force and bending moment shown on the free-body diagram are assumed to act in the *positive sense* according to the established sign convention. Note that both the shear force and moment acting on the right-hand face must be increased by a small, finite amount in order to keep the segment in equilibrium. The distributed loading has been replaced by a resultant force $\Delta F = w(x) \Delta x$ that acts at a fractional distance $k (\Delta x)$ from the right end, where $0 < k < 1$ [for example, if $w(x)$ is *uniform*, $k = \frac{1}{2}$]. Applying the equations of equilibrium, we have

$$+\uparrow \Sigma F_y = 0; \qquad V - w(x) \Delta x - (V + \Delta V) = 0$$
$$\Delta V = -w(x) \Delta x$$
$$\zeta+\Sigma M_O = 0; \quad -V\Delta x - M + w(x) \Delta x[k(\Delta x)] + (M + \Delta M) = 0$$
$$\Delta M = V\Delta x - w(x)k(\Delta x)^2$$

Dividing by Δx and taking the limit as $\Delta x \rightarrow 0$, these two equations become

w F_1 F_2 $w = w(x)$

A $\bullet B$ $\bullet C$ D x

M_1 M_2

x Δx

(a)

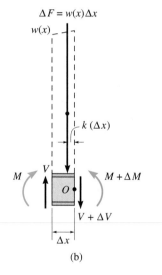

$\Delta F = w(x)\Delta x$

$w(x)$

$k (\Delta x)$

M V $M + \Delta M$

O

$V + \Delta V$

Δx

(b)

Fig. 7–14

$$\boxed{\dfrac{dV}{dx} = -w(x)}$$

$$\begin{array}{c} \text{Slope of} \\ \text{shear diagram} \end{array} = \begin{array}{c} \text{Negative of distributed} \\ \text{load intensity} \end{array}$$

(7–1)

$$\boxed{\dfrac{dM}{dx} = V}$$

$$\begin{array}{c} \text{Slope of} \\ \text{moment diagram} \end{array} = \text{Shear}$$

(7–2)

These two equations provide a convenient means for plotting the shear and moment diagrams for a beam. At a specific point in a beam, Eq. 7–1 states that the *slope of the shear diagram is equal to the negative of the intensity of the distributed load*, while Eq. 7–2 states that the *slope of the moment diagram is equal to the shear*. In particular, if the shear is equal to zero, $dM/dx = 0$, and therefore *a point of zero shear corresponds to a point of maximum (or possibly minimum) moment*.

Equations 7–1 and 7–2 may also be rewritten in the form $dV = -w(x)\,dx$ and $dM = V\,dx$. Noting that $w(x)\,dx$ and $V\,dx$ represent differential areas under the distributed-loading and shear diagrams, respectively, we can integrate these areas between two points B and C along the beam, Fig. 7–14a, and write

This concrete beam is used to support the roof. Its size and the placement of steel reinforcement within it can be determined once the shear and moment diagrams have been established.

$$\Delta V_{BC} = -\int w(x)\,dx$$

$$\frac{\text{Change}}{\text{in shear}} = \frac{\text{Negative of area under}}{\text{loading curve}}$$

(7–3)

and

$$\Delta M_{BC} = \int V\,dx$$

$$\frac{\text{Change}}{\text{in moment}} = \frac{\text{Area under}}{\text{shear diagram}}$$

(7–4)

Equation 7–3 states that the *change in shear between points B and C is equal to the negative of the area under the distributed-loading curve between these points*. Similarly, from Eq. 7–4, the *change in moment between B and C is equal to the area under the shear diagram within region BC*. Because two integrations are involved, first to determine the change in shear, Eq. 7–3, then to determine the change in moment, Eq. 7–4, we can state that if the loading curve $w = w(x)$ is a polynomial of degree n, then $V = V(x)$ will be a curve of degree $n + 1$, and $M = M(x)$ will be a curve of degree $n + 2$.

As stated previously, the above equations do not apply at points where a *concentrated* force or couple moment acts. These two special cases create *discontinuities* in the shear and moment diagrams, and as a result, each deserves separate treatment.

(a)

(b)

Fig. 7–15

Force. A free-body diagram of a small segment of the beam in Fig. 7–14a, taken from under one of the forces, is shown in Fig. 7–15a. Here it can be seen that force equilibrium requires

$$+\uparrow \Sigma F_y = 0; \qquad\qquad \Delta V = -F \qquad\qquad (7\text{–}5)$$

Thus, the *change in shear is negative*, so that on the shear diagram the shear will "jump" *downward when* **F** *acts downward* on the beam. Likewise, the jump in shear (ΔV) is upward when **F** acts upward.

Couple Moment. If we remove a segment of the beam in Fig. 7–14a that is located at the couple moment, the free-body diagram shown in Fig. 7–15b results. In this case letting $\Delta x \rightarrow 0$, moment equilibrium requires

$$\zeta + \Sigma M = 0; \qquad\qquad \Delta M = M_0 \qquad\qquad (7\text{–}6)$$

Thus, the *change in moment is positive*, or the moment diagram will "jump" *upward if* **M**$_0$ *is clockwise*. Likewise, the jump ΔM is downward when **M**$_0$ is counterclockwise.

The examples which follow illustrate application of the above equations for the construction of the shear and moment diagrams. After working through these examples, it is recommended that Examples 7–7 and 7–8 be solved using this method.

Each outrigger such as AB supporting this crane acts as a beam which is fixed to the frame of the crane at one end and subjected to a force **F** on the footing at its other end. A proper design requires that the outrigger is able to resist its maximum internal shear and moment. The shear and moment diagrams indicate that the shear will be constant throughout its length and the maximum moment occurs at the support A.

IMPORTANT POINTS

- The slope of the shear diagram is equal to the negative of the intensity of the distributed loading, where positive distributed loading is downward, i.e., $dV/dx = -w(x)$.

- If a concentrated force acts downward on the beam, the shear will jump downward by the amount of the force.

- The change in the shear ΔV between two points is equal to *the negative of the area* under the distributed-loading curve between the points.

- The slope of the moment diagram is equal to the shear, i.e., $dM/dx = V$.

- The change in the moment ΔM between two points is equal to the *area* under the shear diagram between the two points.

- If a *clockwise* couple moment acts on the beam, the shear will not be affected, however, the moment diagram will jump *upward* by the amount of the moment.

- Points of *zero shear* represent points of *maximum or minimum moment* since $dM/dx = 0$.

EXAMPLE 7.9

Draw the shear and moment diagrams for the beam shown in Fig. 7–16a.

(b)

(a)

(c)

Fig. 7–16

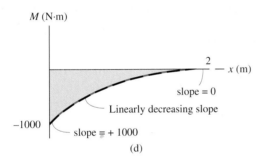

(d)

Solution

Support Reactions. The reactions at the fixed support have been calculated and are shown on the free-body diagram of the beam, Fig. 7–16b.

Shear Diagram. The shear at the end points is plotted first, Fig. 7–16c. From the sign convention, Fig. 7–11, $V = +1000$ at $x = 0$ and $V = 0$ at $x = 2$. Since $dV/dx = -w = -500$ a straight, *negative* sloping line connects the end points.

Moment Diagram. From our sign convention, Fig. 7–11, the moments at the beam's end points, $M = -1000$ at $x = 0$ and $M = 0$ at $x = 2$, are plotted first, Fig. 7–16d. Successive values of shear taken from the shear diagram, Fig. 7–16c, indicate that the *slope $dM/dx = V$* of the moment diagram, Fig. 7–16d, is always positive yet *linearly decreasing* from $dM/dx = 1000$ at $x = 0$ to $dM/dx = 0$ at $x = 2$. Thus, due to the integrations, w a constant yields V a sloping line (first-degree curve) and M a parabola (second-degree curve).

E X A M P L E **7.10**

(a)

(b)

(c)

(d)

(e)

Fig. 7–17

Draw the shear and moment diagrams for the cantilevered beam shown in Fig. 7–17a.

Solution

Support Reactions. The reactions at the fixed support have been calculated and are shown on the free-body diagram of the beam, Fig. 7–17b.

Shear Diagram. Using the established sign convention, Fig. 7–11, the shear at the ends of the beam is plotted first; i.e., $x = 0$, $V = +1080$; $x = 2$, $V = +600$, Fig. 7–17c.

Since the uniform distributed load is downward and *constant*, the slope of the shear diagram is $dV/dx = -w = -400$ for $0 \le x < 1.2$ as indicated.

The magnitude of shear at $x = 1.2$ is $V = +600$. This can be determined by first finding the area under the load diagram between $x = 0$ and $x = 1.2$. This represents the change in shear. That is, $\Delta V = -\int w(x)\,dx = -400(1.2) = -480$. Thus $V\mid_{x=1.2} = V\mid_{x=0} + (-480) = 1080 - 480 = 600$. Also, we can obtain this value by using the method of sections, Fig. 7–17e, where for equilibrium $V = +600$.

Since the load between $1.2 < x \le 2$ is $w = 0$, the slope $dV/dx = 0$ as indicated. This brings the shear to the required value of $V = +600$ at $x = 2$.

Moment Diagram. Again, using the established sign convention, the moments at the ends of the beam are plotted first; i.e., $x = 0$, $M = -1588$; $x = 2$, $M = -100$, Fig. 7–17d.

Each value of shear gives the slope of the moment diagram since $dM/dx = V$. As indicated, at $x = 0$, $dM/dx = +1080$; and at $x = 1.2$, $dM/dx = +600$. For $0 \le x < 1.2$, specific values of the shear diagram are positive but linearly decreasing. Hence, the moment diagram is parabolic with a linearly decreasing positive slope.

The magnitude of moment at $x = 1.2$ is -580. This can be found by first determining the trapezoidal area under the shear diagram, which represents the change in moment, $\Delta M = \int V\,dx = 600(1.2) + \frac{1}{2}(1080 - 600)(1.2) = +1008$. Thus, $M\mid_{x=1.2} = M\mid_{x=0} + 1008 = -1588 + 1008 = -580$. The more "basic" method of sections can also be used, where equilibrium at $x = 1.2$ requires $M = -580$, Fig. 7–17e.

The moment diagram has a constant slope for $1.2 < x \le 2$ since, from the shear diagram, $dM/dx = V = +600$. This brings the value of $M = -100$ at $x = 2$, as required.

EXAMPLE 7.11

Draw the shear and moment diagrams for the shaft in Fig. 7–18a. The support at A is a thrust bearing and the support at B is a journal bearing.

(a)

(b)

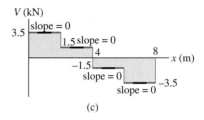

(c)

Solution

Support Reactions. The reactions at the supports are shown on the free-body diagram in Fig. 7–18b.

Shear Diagram. The end points $x = 0$, $V = +3.5$ and $x = 8$, $V = -3.5$ are plotted first, as shown in Fig. 7–18c.

Since there is no distributed load on the shaft, the slope of the shear diagram throughout the shaft's length is zero; i.e., $dV/dx = -w = 0$. There is a discontinuity or "jump" of the shear diagram, however, at each concentrated force. From Eq. 7–5, $\Delta V = -F$, the change in shear is negative when the force acts downward and positive when the force acts upward. Stated another way, the "jump" follows the force, i.e., a downward force causes a downward jump, and vice versa. Thus, the 2-kN force at $x = 2$ m changes the shear from 3.5 kN to 1.5 kN; the 3-kN force at $x = 4$ m changes the shear from 1.5 kN to -1.5 kN, etc. We can *also* obtain numerical values for the shear at a specified point in the shaft by using the method of sections, as for example, $x = 2^+$ m, $V = 1.5$ kN in Fig. 7–18e.

Moment Diagram. The end points $x = 0$, $M = 0$ and $x = 8$, $M = 0$ are plotted first, as shown in Fig. 7–18d.

Since the shear is constant in each region of the shaft, the moment diagram has a corresponding constant positive or negative slope as indicated on the diagram. Numerical values for the change in moment at any point can be computed from the *area* under the shear diagram. For example, at $x = 2$ m, $\Delta M = \int V\,dx = 3.5(2) = 7$. Thus, $M|_{x=2} = M|_{x=0} + 7 = 0 + 7 = 7$. Also, by the method of sections, we can determine the moment at a specified point, as for example, $x = 2^+$ m, $M = 7$ kN · m, Fig. 7–18e.

(d)

(e)

Fig. 7–18

E X A M P L E **7.12**

(b)

(c)

(d)

(e)

(f)

Fig. 7–19

Sketch the shear and moment diagrams for the beam shown in Fig. 7–19a.

(a)

Solution

Support Reactions. The reactions are calculated and indicated on the free-body diagram, Fig. 7–19b.

Shear Diagram. As in Example 7.11, the shear diagram can be constructed by "following the load" on the free-body diagram. In this regard, beginning at A, the reaction is up so $V_A = +100$ kN. Fig. 7–19c. No load acts between A and C, so the shear remains constant; i.e., $dV/dx = -w(x) = 0$. At C the 600-kN force acts downward, so the shear jumps down 600 kN, from 100 kN to -500 kN. Again the shear is constant (no load) and ends at -500 kN, point B. Notice that no jump or discontinuity in shear occurs at D, the point where the 4000-kN · m couple moment is applied, Fig. 7–19b. This is because, for force equilibrium, $\Delta V = 0$ in Fig. 7–15b.

Moment Diagram. The moment at each end of the beam is zero. These two points are plotted first, Fig. 7–19d. The slope of the moment diagram from A to C is constant since $dM/dx = V = +100$. The value of the moment at C can be determined by the method of sections, Fig. 7–19e where $M_C = +1000$ kN · m, or by first computing the rectangular area under the shear diagram between A and C to obtain the change in moment $\Delta M_{AC} = (100 \text{ kN})(10 \text{ m}) = 1000$ kN · m. Since $M_A = 0$, then $M_C = 0 + 1000$ kN · m $= 1000$ kN · m. From C to D the slope of the moment diagram is $dM/dx = V = -500$, Fig. 7–19c. The area under the shear diagram between points C and D is $\Delta M_{CD} = (-500 \text{ kN})(5 \text{ m}) = -2500$ kN · m, so that $M_D = M_C + \Delta M_{CD} = 1000 - 2500 = -1500$ kN · m. A jump in the moment diagram occurs at point D, which is caused by the concentrated couple moment of 4000 kN · m. From Eq. 7–6, the jump is *positive* since the couple moment is *clockwise*. Thus, at $x = 15^+$ m, the moment is $M_D = -1500 + 4000 = 2500$ kN · m. This value can *also* be determined by the method of sections, Fig. 7–19f. From point D the slope of $dM/dx = -500$ is maintained until the diagram closes to zero at B, Fig. 7–19d.

PROBLEMS

7-66. Draw the shear and moment diagrams for the beam.

Prob. 7–66

7-67. Draw the shear and moment diagrams for the beam *ABCDE*. All pulleys have a radius of 0.1 m. Neglect the weight of the beam and pulley arrangement. The load weighs 500 N.

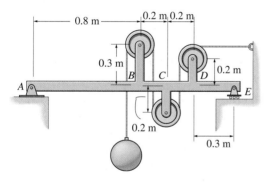

Prob. 7–67

*7-68. Draw the shear and moment diagrams for the beam.

Prob. 7–68

7-69. Draw the shear and moment diagrams for the beam.

Prob. 7–69

7-70. Draw the shear and moment diagrams for the beam.

Prob. 7–70

7-71. Draw the shear and moment diagrams for the beam.

Prob. 7–71

*7-72. Draw the shear and moment diagrams for the shaft. The support at A is a journal bearing and at B it is a thrust bearing.

Prob. 7–72

7-73. Draw the shear and moment diagrams for the beam.

Prob. 7–73

7-74. Draw the shear and moment diagrams for the shaft. The support at A is a journal bearing and at B it is a thrust bearing.

Prob. 7–74

7-75. Draw the shear and moment diagrams for the beam.

Prob. 7–75

*7-76. Draw the shear and moment diagrams for the shaft. The support at A is a thrust bearing and at B it is a journal bearing.

Prob. 7–76

7-77. Draw the shear and moment diagrams for the beam.

Prob. 7–77

7-78. The beam will fail when the maximum moment is $M_{max} = 30$ kN · m or the maximum shear is $V_{max} = 8$ kN. Determine the largest distributed load w the beam will support.

Prob. 7–78

7-79. The beam consists of two segments pin connected at B. Draw the shear and moment diagrams for the beam.

Prob. 7–79

*7-80. Draw the shear and moment diagrams for the beam.

Prob. 7–80

7-81. The beam consists of two segments pin-connected at B. Draw the shear and moment diagrams for the beam.

Prob. 7–81

7-82. Draw the shear and moment diagrams for the beam.

Prob. 7–82

7-83. Draw the shear and moment diagrams for the beam.

Prob. 7–83

*7-84. Draw the shear and moment diagrams for the beam.

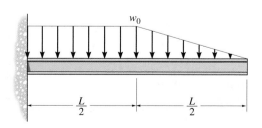

Prob. 7–84

7-85. Draw the shear and moment diagrams for the beam.

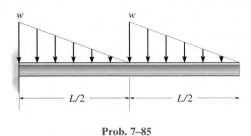

Prob. 7–85

7-86. Draw the shear and moment diagrams for the beam.

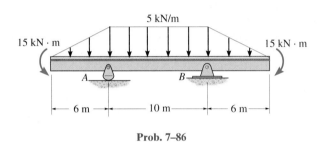

Prob. 7–86

7-87. Draw the shear and moment diagrams for the beam.

Prob. 7–87

*7-88. Draw the shear and moment diagrams for the beam.

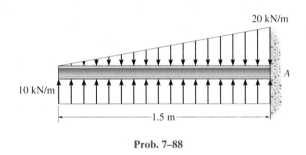

Prob. 7–88

*7.4   Cables

Flexible cables and chains are often used in engineering structures for support and to transmit loads from one member to another. When used to support suspension bridges and trolley wheels, cables form the main load-carrying element of the structure. In the force analysis of such systems, the weight of the cable itself may be neglected because it is often small compared to the load it carries. On the other hand, when cables are used as transmission lines and guys for radio antennas and derricks, the cable weight may become important and must be included in the structural analysis. Three cases will be considered in the analysis that follows: (1) a cable subjected to concentrated loads; (2) a cable subjected to a distributed load; and (3) a cable subjected to its own weight. Regardless of which loading conditions are present, provided the loading is coplanar with the cable, the requirements for equilibrium are formulated in an identical manner.

When deriving the necessary relations between the force in the cable and its slope, we will make the assumption that the cable is *perfectly flexible* and *inextensible*. Due to its flexibility, the cable offers no resistance to bending, and therefore, the tensile force acting in the cable is always tangent to the cable at points along its length. Being inextensible, the cable has a constant length both before and after the load is applied. As a result, once the load is applied, the geometry of the cable remains fixed, and the cable or a segment of it can be treated as a rigid body.

Cable Subjected to Concentrated Loads. When a cable of negligible weight supports several concentrated loads, the cable takes the form of several straight-line segments, each of which is subjected to a constant tensile force. Consider, for example, the cable shown in Fig. 7–20, where the distances h, L_1, L_2, and L_3 and the loads P_1 and P_2 are known. The problem here is to determine the *nine unknowns* consisting of the tension in each of the *three* segments, the *four* components of reaction at A and B, and the sags y_C and y_D at the *two* points C and D. For the solution we can write *two* equations of force equilibrium at each of points A, B, C, and D. This results in a total of *eight equations*.* To complete the solution, it will be necessary to know something about the geometry of the cable in order to obtain the necessary ninth equation. For example, if the cable's total *length L* is specified, then the Pythagorean theorem can be used to relate each of the three segmental lengths, written in terms of h, y_C, y_D, L_1, L_2, and L_3, to the total length L. Unfortunately, this type of problem cannot be solved easily by hand. Another possibility, however, is to specify one of the sags, either y_C or y_D, instead of the cable length. By doing this, the equilibrium equations are then sufficient for obtaining the unknown forces and the remaining sag. Once the sag at each point of loading is obtained, the length of the cable can be determined by trigonometry. The following example illustrates a procedure for performing the equilibrium analysis for a problem of this type.

Each of the cable segments remains approximately straight as they support the weight of these traffic lights.

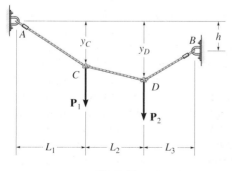

Fig. 7–20

*As will be shown in the following example, the eight equilibrium equations can *also* be written for the entire cable, or any part thereof. But *no more* than *eight* equations are available.

E X A M P L E 7.13

Determine the tension in each segment of the cable shown in Fig. 7–21a.

(a)

(b)

(c)

Fig. 7–21

Solution

By inspection, there are four unknown external reactions (A_x, A_y, E_x, and E_y) and four unknown cable tensions, one in each cable segment. These eight unknowns along with the two unknown sags y_B and y_D can be determined from *ten* available equilibrium equations. One method is to apply these equations as force equilibrium ($\Sigma F_x = 0$, $\Sigma F_y = 0$) to each of the five points A through E. Here, however, we will take a more direct approach.

Consider the free-body diagram for the entire cable, Fig. 7–21b. Thus,

$$\xrightarrow{+} \Sigma F_x = 0; \qquad -A_x + E_x = 0$$
$$\curvearrowleft + \Sigma M_E = 0; \quad -A_y(18\,\text{m}) + 4\,\text{kN}(15\,\text{m}) + 15\,\text{kN}(10\,\text{m}) + 3\,\text{kN}(2\,\text{m}) = 0$$
$$A_y = 12\,\text{kN}$$
$$+\uparrow \Sigma F_y = 0; \quad 12\,\text{kN} - 4\,\text{kN} - 15\,\text{kN} - 3\,\text{kN} + E_y = 0$$
$$E_y = 10\,\text{kN}$$

Since the sag $y_C = 12$ m is known, we will now consider the leftmost section, which cuts cable BC, Fig. 7–21c.

$$\curvearrowleft + \Sigma M_C = 0; \; A_x(12 \text{ m}) - 12 \text{ kN}(8 \text{ m}) + 4 \text{ kN}(5 \text{ m}) = 0$$
$$A_x = E_x = 6.33 \text{ kN}$$
$$\xrightarrow{+} \Sigma F_x = 0; \qquad T_{BC} \cos \theta_{BC} - 6.33 \text{ kN} = 0$$
$$+\uparrow \Sigma F_y = 0; \qquad 12 \text{ kN} - 4 \text{ kN} - T_{BC} \sin \theta_{BC} = 0$$

Thus,

$$\theta_{BC} = 51.6°$$
$$T_{BC} = 10.2 \text{ kN} \qquad\qquad Ans.$$

(d)   (e)   (f)

Proceeding now to analyze the equilibrium of points A, C, and E in sequence, we have

Point A.   (Fig. 7–21d)

$\xrightarrow{+} \Sigma F_x = 0;$ $T_{AB} \cos \theta_{AB} - 6.33 \text{ kN} = 0$

$+\uparrow \Sigma F_y = 0;$ $-T_{AB} \sin \theta_{AB} + 12 \text{ kN} = 0$

$$\theta_{AB} = 62.2°$$
$$T_{AB} = 13.6 \text{ kN} \qquad\qquad Ans.$$

Point C.   (Fig. 7–21e)

$\xrightarrow{+} \Sigma F_x = 0;$ $T_{CD} \cos \theta_{CD} - 10.2 \cos 51.6° \text{ kN} = 0$

$+\uparrow \Sigma F_y = 0;$ $T_{CD} \sin \theta_{CD} + 10.2 \sin 51.6° \text{ kN} - 15 \text{ kN} = 0$

$$\theta_{CD} = 47.9°$$
$$T_{CD} = 9.44 \text{ kN} \qquad\qquad Ans.$$

Point E.   (Fig. 7–21f)

$\xrightarrow{+} \Sigma F_x = 0;$ $6.33 \text{ kN} - T_{ED} \cos \theta_{ED} = 0$

$+\uparrow \Sigma F_y = 0;$ $10 \text{ kN} - T_{ED} \sin \theta_{ED} = 0$

$$\theta_{ED} = 57.7°$$
$$T_{ED} = 11.8 \text{ kN} \qquad\qquad Ans.$$

By comparison, the maximum cable tension is in segment AB since this segment has the greatest slope (θ) and it is required that for any left-hand cable segment the horizontal component $T \cos \theta = A_x$ (a constant). Also, since the slope angles that the cable segments make with the horizontal have now been determined, it is possible to determine the sags y_B and y_D, Fig. 7–21a, using trigonometry.

Cable Subjected to a Distributed Load. Consider the weightless cable shown in Fig. 7–22a, which is subjected to a loading function $w = w(x)$ *as measured in the x direction.* The free-body diagram of a small segment of the cable having a length Δs is shown in Fig. 7–22b. Since the tensile force in the cable changes continuously in both magnitude and direction along the cable's length, this change is denoted on the free-body diagram by ΔT. The distributed load is represented by its resultant

(a)

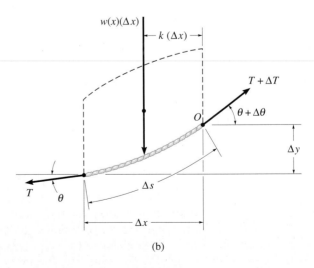

(b)

Fig. 7–22

force $w(x)(\Delta x)$, which acts at a fractional distance $k(\Delta x)$ from point O, where $0 < k < 1$. Applying the equations of equilibrium yields

$$\xrightarrow{+} \Sigma F_x = 0; \qquad -T\cos\theta + (T + \Delta T)\cos(\theta + \Delta\theta) = 0$$

$$+\uparrow\Sigma F_y = 0; \qquad -T\sin\theta - w(x)(\Delta x) + (T + \Delta T)\sin(\theta + \Delta\theta) = 0$$

$$\downarrow+\Sigma M_O = 0; \qquad w(x)(\Delta x)k(\Delta x) - T\cos\theta\,\Delta y + T\sin\theta\,\Delta x = 0$$

Dividing each of these equations by Δx and taking the limit as $\Delta x \to 0$, and hence $\Delta y \to 0$, $\Delta\theta \to 0$, and $\Delta T \to 0$, we obtain

$$\frac{d(T\cos\theta)}{dx} = 0 \tag{7–7}$$

$$\frac{d(T\sin\theta)}{dx} - w(x) = 0 \tag{7–8}$$

$$\frac{dy}{dx} = \tan\theta \tag{7–9}$$

Integrating Eq. 7–7, we have

$$T\cos\theta = \text{constant} = F_H \tag{7–10}$$

Here F_H represents the horizontal component of tensile force at *any point* along the cable.

Integrating Eq. 7–8 gives

$$T\sin\theta = \int w(x)\,dx \tag{7–11}$$

Dividing Eq. 7–11 by Eq. 7–10 eliminates T. Then, using Eq. 7–9, we can obtain the slope

$$\tan\theta = \frac{dy}{dx} = \frac{1}{F_H}\int w(x)\,dx$$

Performing a second integration yields

$$\boxed{y = \frac{1}{F_H}\int\left(\int w(x)\,dx\right)dx} \tag{7–12}$$

This equation is used to determine the curve for the cable, $y = f(x)$. The horizontal force component F_H and the two constants, say C_1 and C_2, resulting from the integration are determined by applying the boundary conditions for the cable.

The cable and suspenders are used to support the uniform load of a gas pipe which crosses the river.

EXAMPLE 7.14

The cable of a suspension bridge supports half of the uniform road surface between the two columns at A and B, as shown in Fig. 7–23a. If this distributed loading is w_0, determine the maximum force developed in the cable and the cable's required length. The span length L and, sag h are known.

(a)

Fig. 7–23

Solution

We can determine the unknowns in the problem by first finding the curve that defines the shape of the cable by using Eq. 7–12. For reasons of symmetry, the origin of coordinates has been placed at the cable's center. Noting that $w(x) = w_0$, we have

$$y = \frac{1}{F_H} \int \left(\int w_0 \, dx \right) dx$$

Performing the two integrations gives

$$y = \frac{1}{F_H} \left(\frac{w_0 x^2}{2} + C_1 x + C_2 \right) \tag{1}$$

The constants of integration may be determined by using the boundary conditions $y = 0$ at $x = 0$ and $dy/dx = 0$ at $x = 0$. Substituting into Eq. 1 yields $C_1 = C_2 = 0$. The curve then becomes

$$y = \frac{w_0}{2F_H} x^2 \tag{2}$$

This is the equation of a *parabola*. The constant F_H may be obtained by using the boundary condition $y = h$ at $x = L/2$. Thus,

$$F_H = \frac{w_0 L^2}{8h} \qquad (3)$$

Therefore, Eq. 2 becomes

$$y = \frac{4h}{L^2} x^2 \qquad (4)$$

Since F_H is known, the tension in the cable may be determined using Eq. 7–10, written as $T = F_H/\cos\theta$. For $0 \le \theta < \pi/2$, the maximum tension will occur when θ is *maximum*, i.e., at point B, Fig. 7–23a. From Eq. 2, the slope at this point is

$$\left.\frac{dy}{dx}\right|_{x=L/2} = \tan\theta_{\max} = \left.\frac{w_0}{F_H} x\right|_{x=L/2}$$

or

$$\theta_{\max} = \tan^{-1}\left(\frac{w_0 L}{2F_H}\right) \qquad (5)$$

Therefore,

$$T_{\max} = \frac{F_H}{\cos(\theta_{\max})} \qquad (6)$$

Using the triangular relationship shown in Fig. 7–23b, which is based on Eq. 5, Eq. 6 may be written as

$$T_{\max} = \frac{\sqrt{4F_H^2 + w_0^2 L^2}}{2}$$

Substituting Eq. 3 into the above equation yields

$$T_{\max} = \frac{w_0 L}{2}\sqrt{1 + \left(\frac{L}{4h}\right)^2} \qquad Ans.$$

(b)

Fig. 7–23

For a differential segment of cable length ds, we can write

$$ds = \sqrt{(dx)^2 + (dy)^2} = \sqrt{1 + \left(\frac{dy}{dx}\right)^2}\, dx$$

Hence, the total length of the cable, $\mathscr{L}$, can be determined by integration. Using Eq. 4, we have

$$\mathscr{L} = \int ds = 2\int_0^{L/2} \sqrt{1 + \left(\frac{8h}{L^2} x\right)^2}\, dx \qquad (7)$$

Integrating yields

$$\mathscr{L} = \frac{L}{2}\left[\sqrt{1 + \left(\frac{4h}{L}\right)^2} + \frac{L}{4h}\sinh^{-1}\left(\frac{4h}{L}\right)\right] \qquad Ans.$$

Over time the forces the cables exert on this telephone pole have caused it to tilt. Proper bracing should be required.

Cable Subjected to Its Own Weight. When the weight of the cable becomes important in the force analysis, the loading function along the cable becomes a function of the arc length s rather than the projected length x. A generalized loading function $w = w(s)$ acting along the cable is shown in Fig. 7–24a. The free-body diagram for a segment of the cable is shown in Fig. 7–24b. Applying the equilibrium equations to the force system on this diagram, one obtains relationships identical to those given by Eqs. 7–7 through 7–9, but with ds replacing dx. Therefore, it may be shown that

$$T \cos \theta = F_H$$

$$T \sin \theta = \int w(s)\, ds \qquad (7\text{–}13)$$

$$\frac{dy}{dx} = \frac{1}{F_H} \int w(s)\, ds \qquad (7\text{–}14)$$

To perform a direct integration of Eq. 7–14, it is necessary to replace dy/dx by ds/dx. Since

$$ds = \sqrt{dx^2 + dy^2}$$

then

$$\frac{dy}{dx} = \sqrt{\left(\frac{ds}{dx}\right)^2 - 1}$$

Therefore,

$$\frac{ds}{dx} = \left\{ 1 + \frac{1}{F_H^2} \left(\int w(s)\, ds \right)^2 \right\}^{1/2}$$

Separating the variables and integrating yields

$$x = \int \frac{ds}{\left\{ 1 + \dfrac{1}{F_H^2} \left(\displaystyle\int w(s)\, ds \right)^2 \right\}^{1/2}} \qquad (7\text{–}15)$$

The two constants of integration, say C_1 and C_2, are found using the boundary conditions for the cable.

(a)

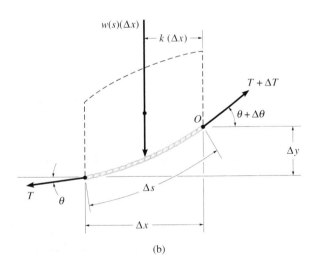

(b)

Fig. 7–24

E X A M P L E 7.15

Fig. 7-25

Determine the deflection curve, the length, and the maximum tension in the uniform cable shown in Fig. 7–25. The cable weighs $w_0 = 5$ N/m.

Solution

For reasons of symmetry, the origin of coordinates is located at the center of the cable. The deflection curve is expressed as $y = f(x)$. We can determine it by first applying Eq. 7–15, where $w(s) = w_0$.

$$x = \int \frac{ds}{[1 + (1/F_H^2)(\int w_0\, ds)^2]^{1/2}}$$

Integrating the term under the integral sign in the denominator, we have

$$x = \int \frac{ds}{[1 + (1/F_H^2)(w_0 s + C_1)^2]^{1/2}}$$

Substituting $u = (1/F_H)(w_0 s + C_1)$ so that $du = (w_0/F_H)\, ds$, a second integration yields

$$x = \frac{F_H}{w_0}(\sinh^{-1} u + C_2)$$

or

$$x = \frac{F_H}{w_0}\left\{ \sinh^{-1}\left[\frac{1}{F_H}(w_0 s + C_1) \right] + C_2 \right\} \qquad (1)$$

To evaluate the constants note that, from Eq. 7–14,

$$\frac{dy}{dx} = \frac{1}{F_H}\int w_0\, ds \quad \text{or} \quad \frac{dy}{dx} = \frac{1}{F_H}(w_0 s + C_1)$$

Since $dy/dx = 0$ at $s = 0$, then $C_1 = 0$. Thus,

$$\frac{dy}{dx} = \frac{w_0 s}{F_H} \qquad (2)$$

The constant C_2 may be evaluated by using the condition $s = 0$ at $x = 0$ in Eq. 1, in which case $C_2 = 0$. To obtain the deflection curve, solve for s in Eq. 1, which yields

$$s = \frac{F_H}{w_0}\sinh\left(\frac{w_0}{F_H}x \right) \qquad (3)$$

Now substitute into Eq. 2, in which case

$$\frac{dy}{dx} = \sinh\left(\frac{w_0}{F_H}x \right)$$

Hence

$$y = \frac{F_H}{w_0} \cosh\left(\frac{w_0}{F_H} x\right) + C_3 \qquad (4)$$

If the boundary condition $y = 0$ at $x = 0$ is applied, the constant $C_3 = -F_H/w_0$, and therefore the deflection curve becomes

$$y = \frac{F_H}{w_0}\left[\cosh\left(\frac{w_0}{F_H} x\right) - 1\right]$$

This equation defines the shape of a *catenary curve*. The constant F_H is obtained by using the boundary condition that $y = h$ at $x = L/2$, in which case

$$h = \frac{F_H}{w_0}\left[\cosh\left(\frac{w_0 L}{2 F_H}\right) - 1\right] \qquad (5)$$

Since $w_0 = 5$ N/m, $h = 6$ m, and $L = 20$ m, Eqs. 4 and 5 become

$$y = \frac{F_H}{5 \text{ N/m}}\left[\cosh\left(\frac{5 \text{ N/m}}{F_H} x\right) - 1\right] \qquad (6)$$

$$6 \text{ m} = \frac{F_H}{5 \text{ N/m}}\left[\cosh\left(\frac{50 \text{ N}}{F_H}\right) - 1\right] \qquad (7)$$

Equation 7 can be solved for F_H by using a trial-and-error procedure. The result is

$$F_H = 45.9 \text{ N}$$

and therefore the deflection curve, Eq. 6, becomes

$$y = 9.19[\cosh(0.109x) - 1] \text{ m} \qquad \textit{Ans.}$$

Using Eq. 3, with $x = 10$ m, the half-length of the cable is

$$\frac{\mathcal{L}}{2} = \frac{45.9 \text{ N}}{5 \text{ N/m}} \sinh\left[\frac{5 \text{ N/m}}{45.9 \text{ N}}(10 \text{ m})\right] = 12.1 \text{ m}$$

Hence,

$$\mathcal{L} = 24.2 \text{ m} \qquad \textit{Ans.}$$

Since $T = F_H/\cos\theta$, Eq. 7–13, the maximum tension occurs when θ is maximum, i.e., at $s = \mathcal{L}/2 = 12.1$ m. Using Eq. 2 yields

$$\left.\frac{dy}{dx}\right|_{s=12.1 \text{ m}} = \tan\theta_{max} = \frac{5 \text{ N/m}(12.1 \text{ m})}{45.9 \text{ N}} = 1.32$$

$$\theta_{max} = 52.8°$$

Thus,

$$T_{max} = \frac{F_H}{\cos\theta_{max}} = \frac{45.9 \text{ N}}{\cos 52.8°} = 75.9 \text{ N} \qquad \textit{Ans.}$$

PROBLEMS

Neglect the weight of the cable in the following problems, unless specified.

7-89. Determine the tension in each segment of the cable and the cable's total length.

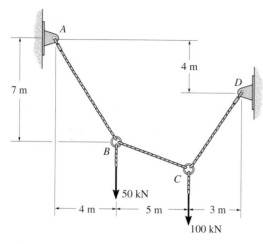

Prob. 7–89

7-90. Determine the tension in each segment of the cable and the cable's total length.

Prob. 7–90

7-91. The cable supports the three loads shown. Determine the sags y_B and y_D of points B and D. Take $P_1 = 400$ N, $P_2 = 250$ N.

*7-92. The cable supports the three loads shown. Determine the magnitude of $\mathbf{P}_1$ if $P_2 = 300$ N and $y_B = 2$ m, Also find the sag y_D.

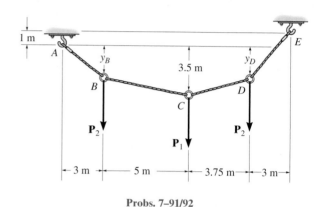

Probs. 7–91/92

7-93. The cable supports the loading shown. Determine the distance x_B the force at point B acts from A. Set $P = 40$ kN.

7-94. The cable supports the loading shown. Determine the magnitude of the horizontal force $\mathbf{P}$ so that $x_B = 6$ m.

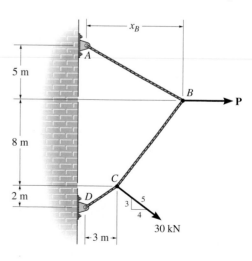

Probs. 7–93/94

7-95. Determine the forces P_1 and P_2 needed to hold the cable in the position shown, i.e., so segment CD remains horizontal. Also, compute the sag y_D and the maximum tension in the cable.

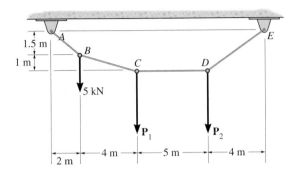

Prob. 7–95

7-97. Determine the maximum uniform loading w, measured in kN/m, that the cable can support if it is capable of sustaining a maximum tension of 3000 kN before it will break.

7-98. The cable is subjected to a uniform loading of $w = 250$ kN/m. Determine the maximum and minimum tension in the cable.

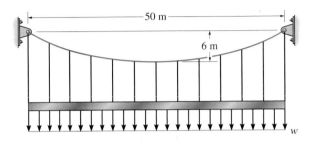

Probs. 7–97/98

***7-96.** The cable supports the three loads shown. Determine the sags y_B and y_D of points B and D and the tension in each segment of the cable.

7-99. The cable AB is subjected to a uniform loading of 200 N/m. If the weight of the cable is neglected and the slope angles at points A and B are 30° and 60°, respectively, determine the curve that defines the cable shape and the maximum tension developed in the cable.

Prob. 7–96

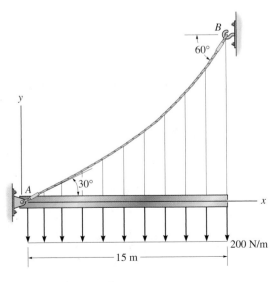

Prob. 7–99

*7-100. The cable supports a girder which weighs 850 kN/m. Determine the tension in the cable at points A, B, and C.

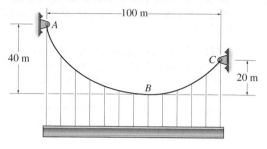

Prob. 7–100

7-101. The cable is subjected to the triangular loading. If the slope of the cable at point O is zero, determine the equation of the curve $y = f(x)$ which defines the cable shape OB, and the maximum tension developed in the cable.

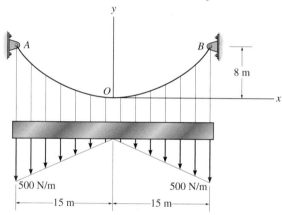

Prob. 7–101

7-102. The cable is subjected to the parabolic loading $w = 150(1 - (x/50)^2)$ kN/m, where x is in m. Determine the equation $y = f(x)$ which defines the cable shape AB and the maximum tension in the cable.

Prob. 7–102

7-103. The cable will break when the maximum tension reachs $T_{max} = 10$ kN. Determine the minimum sag h if it supports the uniform distributed load of $w = 600$ N/m.

Prob. 7–103

*7-104. Determine the maximum tension developed in the cable if it is subjected to a uniform load of 600 N/m.

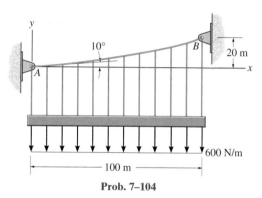

Prob. 7–104

■ **7-105.** A cable has a weight of 5 N/m. If it can span 300 m and has a sag of 15 m, determine the length of the cable. The ends of the cable are supported at the same elevation.

7-106. Show that the deflection curve of the cable discussed in Example 7.15 reduces to Eq. (4) in Example 7.14 when the *hyperbolic cosine function* is expanded in terms of a series and only the first two terms are retained. (The answer indicates that the *catenary* may be replaced by a *parabola* in the analysis of problems in which the sag is small. In this case, the cable weight is assumed to be uniformly distributed along the horizontal.)

7-107. A uniform cord is suspended between two points having the same elevation. Determine the sag-to-span ratio so that the maximum tension in the cord equals the cord's total weight.

■ ***7-108.** A cable has a weight of 2 N/m ($\approx$ 0.2 kg/m). If it can span 100 m and has a sag of 12 m, determine the length of the cable. The ends of the cable are supported from the same elevation.

7-109. The transmission cable having a weight of 20 N/m is strung across the river as shown. Determine the required force that must be applied to the cable at its points of attachment to the towers at *B* and *C*.

Prob. 7–109

7-110. The cable weighs 90 N/m and is 45 m in length. Determine the sag *h* so that the cable spans 30 m. Find the minimum tension in the cable.

Prob. 7–110

7-111. A cable stretches between two points which are 150 m apart and at the same elevation. The line sags 5 m and the cable has a weight of 0.3 kN/m ($\approx$ 30 kg/m). Determine the length of the cable and the maximum tension in the cable.

■*7-112. The cable has a mass of 0.5 kg/m and is 25 m long. Determine the vertical and horizontal components of force it exerts on the top of the tower.

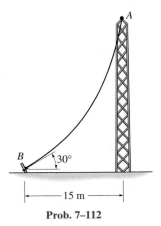

Prob. 7–112

■7-113. A 50-m cable is suspended between two points a distance of 15 m apart and at the same elevation. If the minimum tension in the cable is 200 N, determine the total weight of the cable and the maximum tension developed in the cable.

7-114. The man picks up the 15.6-m chain and holds it just high enough so it is completely off the ground. The chain has points of attachment *A* and *B* that are 15 m apart. If the chain has a weight of 45 N/m, and the man weighs 675 N ($\approx$ 67.5 kg), determine the force he exerts on the ground. Also, how high *h* must he lift the chain? *Hint:* The slopes at *A* and *B* are zero.

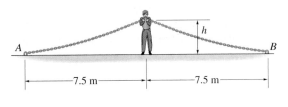

Prob. 7–114

7-115. The balloon is held in place using a 120-m cord that weighs 12 N/m and makes a 60° angle with the horizontal. If the tension in the cord at point *A* is 675 N, determine the length of the cord, *l*, that is lying on the ground and the height h. *Hint:* Establish the coordinate system at *B* as shown.

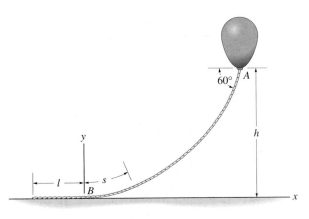

Prob. 7–115

CHAPTER REVIEW

- **Internal Loadings.** If a coplanar force system acts on a member, then in general a resultant internal *normal force N, shear force V*, and *bending moment M* will act at any cross section along the member. These resultants are determined using the method of sections. To find them, the member is sectioned at the point where the internal loadings are to be determined. A free-body diagram of one of the sectioned parts is then drawn. The normal force is determined by summing forces normal to the cross section. The shear force is found by summing forces tangent to the cross section, and the bending moment is found by summing moments about the centroid of the cross-sectional area. If the member is subjected to a three-dimensional loading, then, in general, a *torsional loading* will also act on the cross section. It can be determined by summing moments about an axis that is perpendicular to the cross section and passes through its centroid.

- **Shear and Moment Diagrams as Functions of x.** To construct the shear and moment diagrams for a member, it is necessary to section the member at an arbitrary point, located a distance x from one end. The unknown shear and moment are indicated on the cross section in the positive direction according to the established sign convention. Application of the equilibrium equations will give these loadings as a function of x, which can then be plotted. If the external loading consists of changes in the distributed load, or a series of concentrated forces and couple moments act on the member, then different expressions for V and M must be determined within regions between these different loadings.

- **Graphical Methods for Establishing Shear and Moment Diagrams.** It is possible to plot the shear and moment diagrams quickly by using differential relationships that exist between the distributed loading w and V and M. The slope of the shear diagram is equal to the distributed loading at any point, $dV/dx = -w$; and the slope of the moment diagram is equal to the shear at any point, $V = dM/dx$. Also, the change in shear between any two points is equal to the area under the distributed loading between the points, $\Delta V = \int w\,dx$, and the change in the moment is equal to the area under the shear diagram between the points, $\Delta M = \int V\,dx$.

- **Cables.** When a flexible and inextensible cable is subjected to a series of concentrated forces, then the analysis of the cable can be performed by using the equations of equilibrium applied to free-body diagrams of either segments or points of application of the loading. If external distributed loads or the weight of the cable are to be considered, then the forces and shape of the cable must be determined by first analyzing the forces on a differential segment of the cable and then integrating this result.

REVIEW PROBLEMS

*■7-116. A 450-N (≈ 45-kg) cable is attached between two points at a distance 15 m apart having equal elevations. If the maximum tension developed in the cable is 337.5 N, determine the length of the cable and the sag.

7-117. Determine the distance a between the supports in terms of the beam's length L so that the moment in the *symmetric* beam is zero at the beam's center.

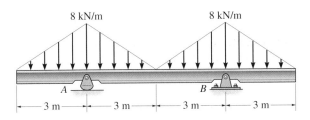

Prob. 7–117

7-118. Draw the shear and moment diagrams for the beam.

Prob. 7–118

7-119. Draw the shear and moment diagrams for the beam ABC.

Prob. 7–119

*7-120. Draw the shear and moment diagrams for the beam.

Prob. 7–120

7-121. Determine the normal force, shear force, and moment at points B and C of the beam.

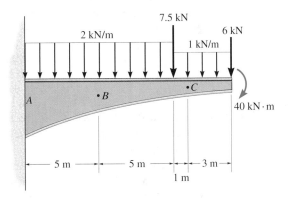

Prob. 7–121

7-122. A chain is suspended between points at the same elevation and spaced a distance of 60 m apart. If it has a weight of 0.5 kN/m (≈ 50 kg/m) and the sag is 3 m, determine the maximum tension in the chain.

7-123. Draw the shear and moment diagrams for the beam.

Prob. 7–123

The effective design of a brake system, such as the one for this bicycle, requires efficient capacity for the mechanism to resist frictional forces. In this chapter we will study the nature of friction and show how these forces are considered in engineering analysis.

Friction

- To introduce the concept of dry friction and show how to analyze the equilibrium of rigid bodies subjected to this force.
- To present specific applications of frictional force analysis on wedges, screws, belts, and bearings.
- To investigate the concept of rolling resistance.

8.1 Characteristics of Dry Friction

The heat generated by the abrasive action of friction can be noticed when using this grinder to sharpen a metal blade.

Friction may be defined as a force of resistance acting on a body which prevents or retards slipping of the body relative to a second body or surface with which it is in contact. This force always acts *tangent* to the surface at points of contact with other bodies and is directed so as to oppose the possible or existing motion of the body relative to these points.

In general, two types of friction can occur between surfaces. *Fluid friction* exists when the contacting surfaces are separated by a film of fluid (gas or liquid). The nature of fluid friction is studied in fluid mechanics since it depends upon knowledge of the velocity of the fluid and the fluid's ability to resist shear force. In this book only the effects of *dry friction* will be presented. This type of friction is often called *Coulomb friction* since its characteristics were studied extensively by C. A. Coulomb in 1781. Specifically, dry friction occurs between the contacting surfaces of bodies in the absence of a lubricating fluid.

Rough surface

(a)

(b)

(c)

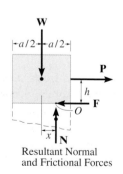

Resultant Normal
and Frictional Forces

(d)

Fig. 8–1

Theory of Dry Friction. The theory of dry friction can best be explained by considering what effects are caused by pulling horizontally on a block of uniform weight **W** which is resting on a rough horizontal surface, Fig. 8–1a. To properly develop a full understanding of the nature of friction, it is necessary to consider the surfaces of contact to be *nonrigid or deformable.* The other portion of the block, however, will be considered rigid. As shown on the free-body diagram of the block, Fig. 8–1b, the floor exerts a *distribution* of both *normal force* $\Delta\mathbf{N}_n$ and *frictional force* $\Delta\mathbf{F}_n$ along the contacting surface. For equilibrium, the normal forces must act *upward* to balance the block's weight **W**, and the frictional forces act to the left to prevent the applied force **P** from moving the block to the right. Close examination of the contacting surfaces between the floor and block reveals how these frictional and normal forces develop, Fig. 8–1c. It can be seen that many microscopic irregularities exist between the two surfaces and, as a result, reactive forces $\Delta\mathbf{R}_n$ are developed at each of the protuberances.* These forces act at all points of contact, and, as shown, each reactive force contributes both a frictional component $\Delta\mathbf{F}_n$ and a normal component $\Delta\mathbf{N}_n$.

Equilibrium. For simplicity in the following analysis, the effect of the *distributed* normal and frictional loadings will be indicated by their *resultants* **N** and **F**, which are represented on the free-body diagram as shown in Fig. 8-1d. Clearly, the distribution of $\Delta\mathbf{F}_n$ in Fig. 8–1b indicates that **F** always acts *tangent to the contacting surface, opposite* to the direction of **P**. On the other hand, the normal force **N** is determined from the distribution of $\Delta\mathbf{N}_n$ in Fig. 8–1b and is directed upward to balance the block's weight **W**. Notice that **N** acts a distance x to the right of the line of action of **W**, Fig. 8–1d. This location, which coincides with the centroid or geometric center of the loading diagram in Fig. 8–1b, is necessary in order to balance the "tipping effect" caused by **P**. For example, if **P** is applied at a height h from the surface, Fig. 8–1d, then moment equilibrium about point O is satisfied if $Wx = Ph$ or $x = Ph/W$. In particular, the block will be on the verge of *tipping* if N acts at the right corner of the block, $x = a/2$.

---

*Besides mechanical interactions as explained here, which is referred to as a classical approach, a detailed treatment of the nature of frictional forces must also include the effects of temperature, density, cleanliness, and atomic or molecular attraction between the contacting surfaces. See J. Krim, *Scientific American*, October, 1996.

Equilibrium

(e)

Impending Motion. In cases where h is small or the surfaces of contact are rather "slippery," the frictional force **F** may *not* be great enough to balance **P**, and consequently the block will tend to slip *before* it can tip. In other words, as P is slowly increased, F correspondingly increases until it attains a certain *maximum value* F_s, called the *limiting static frictional force*, Fig. 8–1e. When this value is reached, the block is in *unstable equilibrium* since any further increase in P will cause deformations and fractures at the points of surface contact, and consequently the block will begin to move. Experimentally, it has been determined that the limiting static frictional force F_s is *directly proportional* to the resultant normal force N. This may be expressed mathematically as

$$F_s = \mu_s N \qquad (8\text{–}1)$$

where the constant of proportionality, μ_s (mu "sub" s), is called the *coefficient of static friction*.

Thus, when the block is on the *verge of sliding*, the normal force **N** and frictional force $\mathbf{F}_s$ combine to create a resultant $\mathbf{R}_s$, Fig. 8–1e. The angle ϕ_s that $\mathbf{R}_s$ makes with **N** is called the *angle of static friction*. From the figure,

$$\phi_s = \tan^{-1}\left(\frac{F_s}{N}\right) = \tan^{-1}\left(\frac{\mu_s N}{N}\right) = \tan^{-1} \mu_s$$

Tabular Values of μ_s. Typical values for μ_s, found in many engineering handbooks, are given in Table 8–1. Although this coefficient is generally less than 1, be aware that in some cases it is possible, as in the case of aluminum on aluminum, for μ_s to be greater than 1. Physically this means, of course, that in this case the frictional force is greater than the corresponding normal force. Furthermore, it should be noted that μ_s is dimensionless and depends only on the characteristics of the two surfaces in contact. A wide range of values is given for each value of μ_s since experimental testing was done under variable conditions of roughness and cleanliness of the contacting surfaces. For applications, therefore, it is important that both caution and judgment be exercised when selecting a coefficient of friction for a given set of conditions. When a more accurate calculation of F_s is required, the coefficient of friction should be determined directly by an experiment that involves the two materials to be used.

TABLE 8–1
Typical Values for μ_s

Contact Materials	Coefficient of Static Friction (μ_s)
Metal on ice	0.03–0.05
Wood on wood	0.30–0.70
Leather on wood	0.20–0.50
Leather on metal	0.30–0.60
Aluminum on aluminum	1.10–1.70

(a)

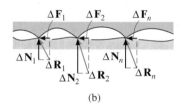

(b)

Fig. 8–2

Motion. If the magnitude of **P** acting on the block is increased so that it becomes greater than F_s, the frictional force at the contacting surfaces drops slightly to a smaller value F_k, called the *kinetic frictional force*. The block will *not* be held in equilibrium ($P > F_k$); instead, it will begin to slide with increasing speed, Fig. 8–2a. The drop made in the frictional force magnitude, from F_s (static) to F_k (kinetic), can be explained by again examining the surfaces of contact, Fig. 8–2b. Here it is seen that when $P > F_s$, then P has the capacity to shear off the peaks at the contact surfaces and cause the block to "lift" somewhat out of its settled position and "ride" on top of these peaks. Once the block begins to slide, high local temperatures at the points of contact cause momentary adhesion (welding) of these points. The continued shearing of these welds is the dominant mechanism creating friction. Since the resultant contact forces $\Delta \mathbf{R}_n$ are aligned slightly more in the vertical direction than before, they thereby contribute *smaller* frictional components, $\Delta \mathbf{F}_n$, than when the irregularities are meshed.

Experiments with sliding blocks indicate that the magnitude of the resultant frictional force $\mathbf{F}_k$ is directly proportional to the magnitude of the resultant normal force $\mathbf{N}$. This may be expressed mathematically as

$$\boxed{F_k = \mu_k N} \tag{8–2}$$

Here the constant of proportionality, μ_k, is called the *coefficient of kinetic friction*. Typical values for μ_k are approximately 25 percent *smaller* than those listed in Table 8–1 for μ_s.

As shown in Fig. 8–2a, in this case, the resultant $\mathbf{R}_k$ has a line of action defined by ϕ_k. This angle is referred to as the *angle of kinetic friction*, where

$$\phi_k = \tan^{-1}\left(\frac{F_k}{N}\right) = \tan^{-1}\left(\frac{\mu_k N}{N}\right) = \tan^{-1} \mu_k$$

By comparison, $\phi_s \geq \phi_k$.

The above effects regarding friction can be summarized by reference to the graph in Fig. 8–3, which shows the variation of the frictional force F versus the applied load P. Here the frictional force is categorized in three different ways: namely, F is a *static-frictional force* if equilibrium is maintained; F is a *limiting static-frictional force* F_s when it reaches a maximum value needed to maintain equilibrium; and finally, F is termed a *kinetic-frictional force* F_k when sliding occurs at the contacting surface. Notice also from the graph that for very large values of P or for high speeds, because of aerodynamic effects, F_k and likewise μ_k begin to decrease.

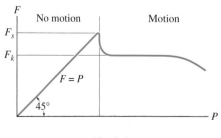

Fig. 8–3

Characteristics of Dry Friction. As a result of *experiments* that pertain to the foregoing discussion, the following rules which apply to bodies subjected to dry friction may be stated.

- The frictional force acts *tangent* to the contacting surfaces in a direction *opposed* to the *relative motion* or tendency for motion of one surface against another.

- The maximum static frictional force F_s that can be developed is independent of the area of contact, provided the normal pressure is not very low nor great enough to severely deform or crush the contacting surfaces of the bodies.

- The maximum static frictional force is generally greater than the kinetic frictional force for any two surfaces of contact. However, if one of the bodies is moving with a *very low velocity* over the surface of another, F_k becomes approximately equal to F_s, i.e., $\mu_s \approx \mu_k$.

- When *slipping* at the surface of contact is *about to occur*, the maximum static frictional force is proportional to the normal force, such that $F_s = \mu_s N$.

- When *slipping* at the surface of contact is *occurring*, the kinetic frictional force is proportional to the normal force, such that $F_k = \mu_k N$.

8.2 Problems Involving Dry Friction

If a rigid body is in equilibrium when it is subjected to a system of forces that includes the effect of friction, the force system must satisfy not only the equations of equilibrium but *also* the laws that govern the frictional forces.

Types of Friction Problems. In general, there are three types of mechanics problems involving dry friction. They can easily be classified once the free-body diagrams are drawn and the total number of unknowns are identified and compared with the total number of available equilibrium equations. Each type of problem will now be explained and illustrated graphically by examples. In all these cases the geometry and dimensions for the problem are assumed to be known.

(a)

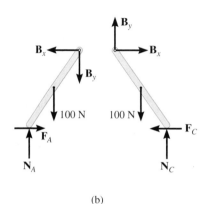

(b)

Fig. 8–4

Equilibrium. Problems in this category are strictly equilibrium problems which require *the total number of unknowns to be equal to the total number of available equilibrium equations.* Once the frictional forces are determined from the solution, however, their numerical values must be checked to be sure they satisfy the inequality $F \leq \mu_s N$; otherwise, slipping will occur and the body will not remain in equilibrium. A problem of this type is shown in Fig. 8–4a. Here we must determine the frictional forces at A and C to check if the equilibrium position of the two-member frame can be maintained. If the bars are uniform and have known weights of 100 N each, then the free-body diagrams are as shown in Fig. 8–4b. There are six unknown force components which can be determined *strictly* from the six equilibrium equations (three for each member). Once F_A, N_A, F_C, and N_C are determined, then the bars will remain in equilibrium provided $F_A \leq 0.3N_A$ and $F_C \leq 0.5N_C$ are satisfied.

Impending Motion at All Points. In this case *the total number of unknowns will equal the total number of available equilibrium equations plus the total number of available frictional equations, $F = \mu N$.* In particular, if *motion is impending* at the points of contact, then $F_s = \mu_s N$; whereas if the body is *slipping*, then $F_k = \mu_k N$. For example, consider the problem of finding the smallest angle θ at which the 100-N bar in Fig. 8–5a can be placed against the wall without slipping. The free-body diagram is shown in Fig. 8–5b. Here there are *five* unknowns: F_A, N_A, F_B, N_B, θ. For the solution there are *three* equilibrium equations and *two* static frictional equations which apply at *both* points of contact, so that $F_A = 0.3N_A$ and $F_B = 0.4N_B$.

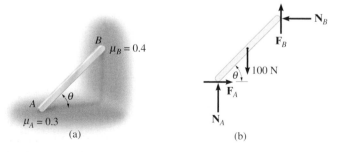

Fig. 8–5

Impending Motion at Some Points. Here *the total number of unknowns will be less than the number of available equilibrium equations plus the total number of frictional equations or conditional equations for tipping.* As a result, several possibilities for motion or impending motion will exist and the problem will involve a determination of the kind of motion which actually occurs. For example, consider the two-member frame shown in Fig. 8–6a. In this problem we wish to determine the horizontal force P needed to cause movement. If each member has a weight of 100 N, then the free-body diagrams are as shown in Fig. 8–6b. There are *seven* unknowns: N_A, F_A, N_C, F_C, B_x, B_y, P. For a unique solution we must satisfy the *six* equilibrium equations (three for each member) and only *one* of two possible static frictional equations. This means that as P increases it will either cause slipping at A and no slipping at C, so that $F_A = 0.3N_A$ and $F_C \leq 0.5N_C$; or slipping occurs at C and no slipping at A, in which case $F_C = 0.5N_C$ and $F_A \leq 0.3N_A$. The actual situation can be determined by calculating P for each case and then choosing the case for which P is *smaller.* If in both cases the *same value* for P is calculated, which in practice would be highly improbable, then slipping at both points occurs simultaneously; i.e., the *seven unknowns* will satisfy *eight equations*.

(a)

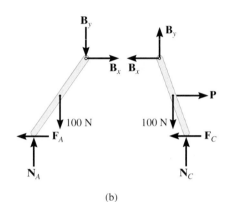

(b)

Fig. 8–6

Consider pushing on the uniform crate that has a weight W and sits on the rough surface. As shown on the first free-body diagram, if the magnitude of **P** is small, the crate will remain in equilibrium. As P increases the crate will either be on the verge of slipping on the surface ($F = \mu_s W$), or if the surface is very rough (large μ_s) then the resultant normal force will shift to the corner, $x = b/2$, as shown on the second free-body diagram, and the crate will tip over. The crate has a greater chance of tipping if **P** is applied at a greater height h above the surface, or if the crate's width b is smaller.

Equilibrium Versus Frictional Equations. It was stated earlier that the frictional force *always* acts so as to either oppose the relative motion or impede the motion of a body over its contacting surface. Realize, however, that we can *assume* the sense of the frictional force in problems which require F to be an "equilibrium force" and satisfy the inequality $F < \mu_s N$. The correct sense is made known *after* solving the equations of equilibrium for F. For example, if F is a negative scalar the sense of **F** is the reverse of that which was assumed. This convenience of *assuming* the sense of **F** is possible because the equilibrium equations equate to zero the *components of vectors* acting in the *same direction*. In cases where the frictional equation $F = \mu N$ is used in the solution of a problem, however, the convenience of *assuming* the sense of **F** is *lost*, since the frictional equation relates only the *magnitudes* of two perpendicular vectors. Consequently, **F** *must always* be shown acting with its *correct sense* on the free-body diagram whenever the frictional equation is used for the solution of a problem.

PROCEDURE FOR ANALYSIS

Equilibrium problems involving dry friction can be solved using the following procedure.

Free-Body Diagrams.

- Draw the necessary free-body diagrams, and unless it is stated in the problem that impending motion or slipping occurs, *always* show the frictional forces as unknowns; i.e., *do not assume $F = \mu N$.*

- Determine the number of unknowns and compare this with the number of available equilibrium equations.

- If there are more unknowns than equations of equilibrium, it will be necessary to apply the frictional equation at some, if not all, points of contact to obtain the extra equations needed for a complete solution.

- If the equation $F = \mu N$ is to be used, it will be necessary to show **F** acting in the proper direction on the free-body diagram.

Equations of Equilibrium and Friction.

- Apply the equations of equilibrium and the necessary frictional equations (or conditional equations if tipping is possible) and solve for the unknowns.

- If the problem involves a three-dimensional force system such that it becomes difficult to obtain the force components or the necessary moment arms, apply the equations of equilibrium using Cartesian vectors.

E X A M P L E   8.1

The uniform crate shown in Fig. 8–7a has a mass of 20 kg. If a force $P = 80$ N is applied to the crate, determine if it remains in equilibrium. The coefficient of static friction is $\mu = 0.3$.

(a)

Fig. 8–7

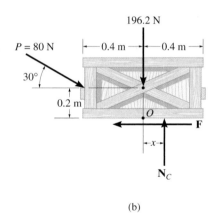

(b)

Solution

Free-Body Diagram. As shown in Fig. 8–7b, the *resultant* normal force $\mathbf{N}_C$ must act a distance x from the crate's center line in order to counteract the tipping effect caused by $\mathbf{P}$. There are *three unknowns*, F, N_C, and x, which can be determined strictly from the *three* equations of equilibrium.

Equations of Equilibrium.

$$\xrightarrow{+} \Sigma F_x = 0; \qquad 80 \cos 30° \text{ N} - F = 0$$
$$+\uparrow \Sigma F_y = 0; \qquad -80 \sin 30° \text{ N} + N_C - 196.2 \text{ N} = 0$$
$$\downarrow + \Sigma M_O = 0; \quad 80 \sin 30° \text{ N}(0.4 \text{ m}) - 80 \cos 30° \text{ N}(0.2 \text{ m}) + N_C(x) = 0$$

Solving,

$$F = 69.3 \text{ N}$$
$$N_C = 236 \text{ N}$$
$$x = -0.00908 \text{ m} = -9.08 \text{ mm}$$

Since x is negative it indicates the *resultant* normal force acts (slightly) to the *left* of the crate's center line. No tipping will occur since $x \le 0.4$ m. Also, the *maximum* frictional force which can be developed at the surface of contact is $F_{max} = \mu_s N_C = 0.3(236 \text{ N}) = 70.8$ N. Since $F = 69.3$ N < 70.8 N, the crate will *not slip*, although it is very close to doing so.

E X A M P L E 8.2

It is observed that when the bed of the dump truck is raised to an angle of $\theta = 25°$ the vending machines begin to slide off the bed, Fig. 8–8a. Determine the static coefficient of friction between them and the surface of the truck.

(a)

Solution

An idealized model of a vending machine resting on the bed of the truck is shown in Fig. 8–8b. The dimensions have been measured and the center of gravity has been located. We will assume that the machine weighs W.

Free-Body Diagram. As shown in Fig. 8–8c, the dimension x is used to locate the position of the resultant normal force $\mathbf{N}$. There are four unknowns, N, F, μ_s, and x.

Equations of Equilibrium.

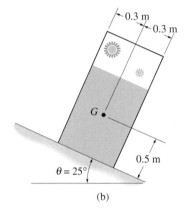

(b)

$$+\searrow \Sigma F_x = 0; \qquad\qquad W \sin 25° - F = 0 \qquad\qquad (1)$$

$$+\nearrow \Sigma F_y = 0; \qquad\qquad N - W \cos 25° = 0 \qquad\qquad (2)$$

$$\curvearrowright + \Sigma M_O = 0; \quad -W \sin \theta\, (0.5 \text{ m}) + W \cos \theta\, (x) = 0 \qquad (3)$$

Since slipping impends at $\theta = 25°$, using the first two equations, we have

$$F_s = \mu_s N; \qquad\qquad W \sin 25° = \mu_s (W \cos 25°)$$

$$\mu_s = \tan 25° = 0.466 \qquad\qquad\qquad Ans.$$

The angle of $\theta = 25°$ is referred to as the *angle of repose*, and by comparison, it is equal to the angle of static friction $\theta = \phi_s$. Notice from the calculation that θ is independent of the weight of the vending machine, and so knowing θ provides a convenient method for determining the coefficient of static friction.

From Eq. 3, with $\theta = 25°$, we find $x = 0.233$ m. Since 0.233 m $<$ 0.5 m, indeed the vending machine will slip before it can tip as observed in Fig. 8–8a.

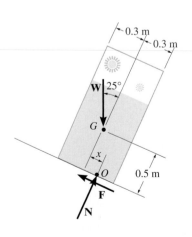

Fig. 8–8

EXAMPLE 8.3

The uniform rod having a weight W and length l is supported at its ends against the surface at A and B in Fig. 8–9a. If the rod is on the verge of slipping when $\theta = 30°$, determine the coefficient of static friction μ_s at A and B. Neglect the thickness of the rod for the calculation.

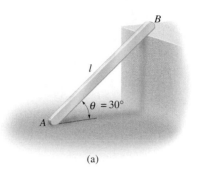

(a)

Solution

Free-Body Diagram. As shown in Fig. 8–9b, there are *five* unknowns: F_A, N_A, F_B, N_B, and μ_s. These can be determined from the *three* equilibrium equations and *two* frictional equations applied at points A and B. The frictional forces must be drawn with their correct sense so that they oppose the tendency for motion of the rod. Why? (Refer to p. 386.)

Equations of Friction and Equilibrium. Writing the frictional equations,

$$F = \mu_s N; \qquad\qquad F_A = \mu_s N_A$$
$$F_B = \mu_s N_B$$

Using these results and applying the equations of equilibrium yields

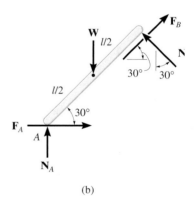

(b)

Fig. 8–9

$$\xrightarrow{+} \Sigma F_x = 0; \quad \mu_s N_A + \mu_s N_B \cos 30° - N_B \sin 30° = 0 \qquad (1)$$
$$+\uparrow \Sigma F_y = 0; \quad N_A - W + N_B \cos 30° + \mu_s N_B \sin 30° = 0 \qquad (2)$$
$$\left\lfloor + \Sigma M_A = 0; \quad N_B l - W\left(\frac{l}{2}\right)\cos 30° = 0 \qquad (3)\right.$$

$$N_B = 0.4330\,W$$

From Eqs. 1 and 2,

$$\mu_s N_A = 0.2165\,W - (0.3750\,W)\mu_s$$
$$N_A = 0.6250\,W - (0.2165\,W)\mu_s$$

By division,

$$0.6250\mu_s - 0.2165\mu_s^2 = 0.2165 - 0.375\mu_s$$

or,

$$\mu_s^2 - 4.619\,\mu_s + 1 = 0$$

Solving for the smallest root,

$$\mu_s = 0.228 \qquad\qquad Ans.$$

EXAMPLE 8.4

(a)

(b)

Fig. 8–10

The concrete pipes are stacked in the yard as shown in Fig. 8–10a. Determine the minimum coefficient of static friction at each point of contact so that the pile does not collapse.

Solution

Free-body Diagrams. Recognize that the coefficient of static friction between two pipes, at A and B, and between a pipe and the ground, at C, will be different since the contacting surfaces are different. We will assume each pipe has an outer radius r and weight W. The free-body diagrams for two of the pipes are shown in Fig. 8–10b. There are six unknowns, N_A, F_A, N_B, F_B, N_C, F_C. (Note that when collapse is about to occur the normal force at D is zero.) Since only the six equations of equilibrium are necessary to obtain the unknowns, the sense of direction of the frictional forces can be verified from the solution.

Equations of Equilibrium. For the top pipe we have

$$\zeta + \Sigma M_O = 0; \quad -F_A(r) + F_B(r) = 0; \quad F_A = F_B = F$$

$$\xrightarrow{+} \Sigma F_x = 0; \quad N_A \sin 30° - F \cos 30° - N_B \sin 30° + F \cos 30° = 0$$

$$N_A = N_B = N$$

$$+\uparrow \Sigma F_y = 0; \quad 2N \cos 30° + 2F \sin 30° - W = 0 \qquad (1)$$

For the bottom pipe, using $F_A = F$ and $N_A = N$, we have,

$$\zeta + \Sigma M_{O'} = 0; \quad F_C(r) - F(r) = 0; \quad F_C = F$$

$$\xrightarrow{+} \Sigma F_x = 0; \quad -N \sin 30° + F \cos 30° + F = 0 \qquad (2)$$

$$+\uparrow \Sigma F_y = 0; \quad N_C - W - N \cos 30° - F \sin 30° = 0 \qquad (3)$$

From Eq. 2, $F = 0.2679 N$, so that between the pipes

$$(\mu_s)_{min} = \frac{F}{N} = 0.268 \qquad \qquad Ans.$$

Using this result in Eq. 1.

$$N = 0.5 W$$

From Eq. 3,

$$N_C - W - (0.5 W) \cos 30° - 0.2679 (0.5 W) \sin 30° = 0$$

$$N_C = 1.5 W$$

At the ground, the smallest required coefficient of static friction would be

$$(\mu'_s)_{min} = \frac{F}{N_C} = \frac{0.2679(0.5 W)}{1.5 W} = 0.0893 \qquad Ans.$$

Hence a greater coefficient of static friction is required between the pipes than that required at the ground; and so it is likely that if slipping would occur between the pipes the bottom two pipes would roll away from one another without slipping as the top pipe falls downward.

E X A M P L E   8.5

Beam AB is subjected to a uniform load of 200 N/m and is supported at B by post BC, Fig. 8–11a. If the coefficients of static friction at B and C are $\mu_B = 0.2$ and $\mu_C = 0.5$, determine the force $\mathbf{P}$ needed to pull the post out from under the beam. Neglect the weight of the members and the thickness of the post.

200 N/m

4 m

B

0.75 m

P

C

0.25 m

(a)

Solution

Free-Body Diagrams. The free-body diagram of beam AB is shown in Fig. 8–11b. Applying $\Sigma M_A = 0$, we obtain $N_B = 400$ N. This result is shown on the free-body diagram of the post, Fig. 8–11c. Referring to this member, the *four* unknowns F_B, P, F_C, and N_C are determined from the *three* equations of equilibrium and *one* frictional equation applied either at B or C.

Equations of Equilibrium and Friction.

$$\overset{+}{\rightarrow} \Sigma F_x = 0; \qquad P - F_B - F_C = 0 \qquad (1)$$
$$+\uparrow \Sigma F_y = 0; \qquad N_C - 400 \text{ N} = 0 \qquad (2)$$
$$\underset{+}{\curvearrowright} \Sigma M_C = 0; \qquad -P(0.25 \text{ m}) + F_B(1 \text{ m}) = 0 \qquad (3)$$

(Post Slips Only at B) This requires $F_C \leq \mu_C N_C$ and

$$F_B = \mu_B N_B; \qquad F_B = 0.2(400 \text{ N}) = 80 \text{ N}$$

Using this result and solving Eqs. 1 through 3, we obtain

$$P = 320 \text{ N}$$
$$F_C = 240 \text{ N}$$
$$N_C = 400 \text{ N}$$

Since $F_C = 240 \text{ N} > \mu_C N_C = 0.5(400 \text{ N}) = 200 \text{ N}$, the other case of movement must be investigated.

(Post Slips Only at C.) Here $F_B \leq \mu_B N_B$ and

$$F_C = \mu_C N_C; \qquad F_C = 0.5 N_C \qquad (4)$$

Solving Eqs. 1 through 4 yields

$$P = 267 \text{ N} \qquad \qquad \textit{Ans.}$$
$$N_C = 400 \text{ N}$$
$$F_C = 200 \text{ N}$$
$$F_B = 66.7 \text{ N}$$

Obviously, this case occurs first since it requires a *smaller* value for P.

800 N

A_x

A

F_B

2 m | 2 m

A_y

$N_B = 400$ N

(b)

400 N

B

F_B

0.75 m

P

0.25 m

F_C

C

N_C

(c)

Fig. 8–11

E X A M P L E | **8.6**

400 mm

B

A

20°

(a)

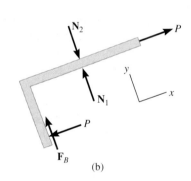

N_2

P

y

x

N_1

P

F_B

(b)

981 N

20°

O

400 mm

P

F_B

F_A

N_A

(c)

Fig. 8–12

Determine the normal force P that must be exerted on the rack to begin pushing the 100-kg pipe shown in Fig. 8–12a up the 20° incline. The coefficients of static friction at the points of contact are $(\mu_s)_A = 0.15$, and $(\mu_s)_B = 0.4$.

Solution

Free-Body Diagram. As shown in Fig. 8–12b, the rack must exert a force P on the pipe due to force equilibrium in the x direction. There are four unknowns P, F_A, N_A, and F_B acting on the pipe Fig. 8–12c. These can be determined from the *three* equations of equilibrium and *one* frictional equation, which apply either at A or B. If slipping begins to occur only at B, the pipe will begin to roll up the incline; whereas if slipping occurs only at A, the pipe will begin to *slide* up the incline. Here we must find N_B.

Equations of Equilibrium and Friction (for Fig. 8–12c)

$$+\nearrow\Sigma F_x = 0; \qquad -F_A + P - 981 \sin 20° \text{ N} = 0 \qquad (1)$$

$$+\nwarrow\Sigma F_y = 0; \qquad N_A - F_B - 981 \cos 20° \text{ N} = 0 \qquad (2)$$

$$\downarrow+\Sigma M_O = 0; \qquad F_B(400 \text{ mm}) - F_A(400 \text{ mm}) = 0 \qquad (3)$$

(Pipe Rolls up Incline.) In this case $F_A \leq 0.15 N_A$ and

$$(F_s)_B = (\mu_s)_B N_B; \qquad F_B = 0.4P \qquad (4)$$

The direction of the frictional force at B must be specified correctly. Why? Since the spool is being forced up the incline, $\mathbf{F}_B$ acts downward to prevent any clockwise rolling motion of the pipe, Fig. 8–12c. Solving Eqs. 1 through 4, we have

$$N_A = 1146 \text{ N} \quad F_A = 224 \text{ N} \quad F_B = 224 \text{ N} \quad P = 559 \text{ N}$$

The assumption regarding no slipping at A should be checked.

$$F_A \overset{?}{\leq} (\mu_s)_A N_A; \qquad 224 \text{ N} \overset{?}{\leq} 0.15(1146 \text{ N}) = 172 \text{ N}$$

The inequality does *not apply*, and therefore slipping occurs at A and not at B. Hence, the other case of motion will occur.

(Pipe Slides up Incline.) In this case, $P \leq 0.4 N_B$ and

$$(F_s)_A = (\mu_s)_A N_A; \qquad F_A = 0.15 N_A \qquad (5)$$

Solving Eqs. 1 through 3 and 5 yields

$$N_A = 1085 \text{ N} \quad F_A = 163 \text{ N} \quad F_B = 163 \text{ N} \quad P = 498 \text{ N} \qquad \textit{Ans.}$$

The validity of the solution ($P = 498$ N) can be checked by testing the assumption that indeed no slipping occurs at B.

$$F_B \leq (\mu_s)_B P; \qquad 163 \text{ N} < 0.4(498 \text{ N}) = 199 \text{ N} \qquad \text{(check)}$$

PROBLEMS

8-1. The mine car and its contents have a total mass of 6 Mg and a center of gravity at G. If the coefficient of static friction between the wheels and the tracks is $\mu_s = 0.4$ when the wheels are locked, find the normal force acting on the front wheels at B and the rear wheels at A when the brakes at both A and B are locked. Does the car move?

10 kN

0.9 m

$\bullet G$

B A 0.15 m

← 0.6 m →

1.5 m

Prob. 8–1

8-2. If the horizontal force $P = 80$ N, determine the normal and frictional forces acting on the 300-N ($\approx$ 30-kg) crate. Take $\mu_s = 0.3$, $\mu_k = 0.2$.

P

20°

Prob. 8–2

8-3. The uniform pole has a weight of 150 N ($\approx$ 15 kg) and a length of 7.8 m. If it is placed against the smooth wall and on the rough floor in the position $d = 3$ m, will it remain in this position when it is released? The coefficient of static friction is $\mu_s = 0.3$.

*8-4. The uniform pole has a weight of 150 N ($\approx$ 15 kg) and a length of 7.8 m. Determine the maximum distance d it can be placed from the smooth wall and not slip. The coefficient of static friction between the floor and the pole is $\mu_s = 0.3$.

B

7.8 m

A

d

Probs. 8–3/4

8-5. The uniform 100-N ($\approx$ 10-kg) ladder rests on the rough floor for which the coefficient of static friction is $\mu_s = 0.8$ and against the smooth wall at B. Determine the horizontal force P the man must exert on the ladder in order to cause it to move.

8-6. The uniform 100-N ($\approx$ 10-kg) ladder rests on the rough floor for which the coefficient of static friction is $\mu_s = 0.4$ and against the smooth wall at B. Determine the horizontal force P the man must exert on the ladder in order to cause it to move.

B

2.5 m

4 m

P

2.5 m

A

3 m

Probs. 8–5/6

8-7. An axial force of $T = 4000$ N is applied to the bar. If the coefficient of static friction at the jaws C and D is $\mu_s = 0.5$, determine the smallest normal force that the screw at A must exert on the smooth surface of the links at B and C in order to hold the bar stationary. The links are pin-connected at F and G.

Prob. 8–7

*8-8. The winch on the truck is used to hoist the garbage bin onto the bed of the truck. If the loaded bin has a weight of 40 000 N ($\approx$ 4 tonne) and center of gravity at G, determine the force in the cable needed to begin the lift. The coefficients of static friction at A and B are $\mu_A = 0.3$ and $\mu_B = 0.2$, respectively. Neglect the height of the support at A.

Prob. 8–8

8-9. The 5-m ladder has a uniform weight of 400 N ($\approx$ 40 kg) and rests against the smooth wall at B. If the coefficient of static friction at A is $\mu_s = 0.4$, determine if the ladder will slip. Take $\theta = 60°$.

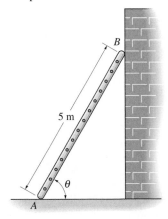

Prob. 8–9

8-10. The block brake is used to stop the wheel from rotating when the wheel is subjected to a couple moment $\mathbf{M}_0$. If the coefficient of static friction between the wheel and the block is μ_s, determine the smallest force P that should be applied.

8-11. Show that the brake in Prob. 8–10 is self locking, i.e., $P \leq 0$, provided $b/c \leq \mu_s$.

*8-12. Solve Prob. 8–10 if the couple moment $\mathbf{M}_0$ is applied counterclockwise.

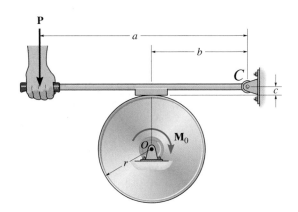

Probs. 8–10/11/12

8-13. The block brake consists of a pin-connected lever and friction block at B. The coefficient of static friction between the wheel and the lever is $\mu_s = 0.3$, and a torque of 5 N·m is applied to the wheel. Determine if the brake can hold the wheel stationary when the force applied to the lever is (a) $P = 30$ N, (b) $P = 70$ N.

8-14. Solve Prob. 8–13 if the 5-N·m torque is applied counter-clockwise.

Probs. 8–13/14

8-15. The tractor has a weight of 22 500 N (≈ 2250 kg) with center of gravity at *G*. The driving traction is developed at the rear wheels *B*, while the front wheels at *A* are free to roll. If the coefficient of static friction between the wheels at *B* and the ground is $\mu_s = 0.5$, determine if it is possible to pull at *P* = 6000 N without causing the wheels at *B* to slip or the front wheels at *A* to lift off the ground.

8-17. The drum has a weight of 500 N (≈ 50 kg) and rests on the floor for which the coefficient of static friction is $\mu_s = 0.6$. If $a = 0.6$ m and $b = 0.9$ m, determine the smallest magnitude of the force **P** that will cause impending motion of the drum.

8-18. The drum has a weight of 500 N (≈ 50 kg) and rests on the floor for which the coefficient of static friction is $\mu_s = 0.5$. If $a = 0.9$ m and $a = 1.2$ m, determine the smallest magnitude of the force **P** that will cause impending motion of the drum.

Prob. 8–15

Probs. 8–17/18

***8-16.** The car has a mass of 1.6 Mg and center of mass at *G*. If the coefficient of static friction between the shoulder of the road and the tires is $\mu_s = 0.4$, determine the greatest slope θ the shoulder can have without causing the car to slip or tip over if the car travels along the shoulder at constant velocity.

8-19. The coefficient of static friction between the shoes at *A* and *B* of the tongs and the pallet is $\mu_s' = 0.5$, and between the pallet and the floor $\mu_s = 0.4$. If a horizontal towing force of *P* = 300 N is applied to the tongs, determine the largest mass that can be towed.

Prob. 8–16

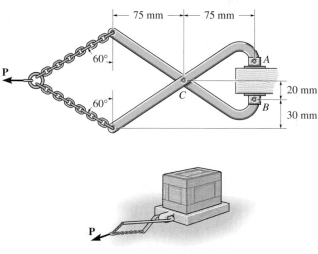

Prob. 8–19

*8-20. The pipe is hoisted using the tongs. If the coefficient of static friction at A and B is μ_s. determine the smallest dimension b so that any pipe of inner diameter d can be lifted.

Prob. 8–20

8-21. Determine the maximum weight W the man can lift with constant velocity using the pulley system, without and then with the "leading block" or pulley at A. The man has a weight of 800 N ($\approx$ 80 kg) and the coefficient of static friction between his feet and the ground is $\mu_s = 0.6$.

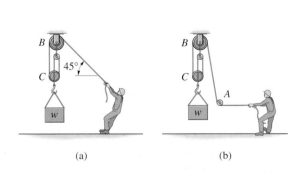

(a)   (b)

Prob. 8–21

8-22. The uniform dresser has a weight of 360 N ($\approx$ 36 kg) and rests on a tile floor for which $\mu_s = 0.25$. If the man pushes on it in the horizontal direction $\theta = 0°$, determine the smallest magnitude of force $\mathbf{F}$ needed to move the dresser. Also, if the man has a weight of 600 N ($\approx$ 60 kg), determine the smallest coefficient of static friction between his shoes and the floor so that he does not slip.

8-23. The uniform dresser has a weight of 360 N ($\approx$ 36 kg) and rests on a tile floor for which $\mu_s = 0.25$. If the man pushes on it in the direction $\theta = 30°$, determine the smallest magnitude of force $\mathbf{F}$ needed to move the dresser. Also, if the man has a weight of 600 N ($\approx$ 60 kg), determine the smallest coefficient of static friction between his shoes and the floor so that he does not slip.

Probs. 8–22/23

*8-24. The 5-kg cylinder is suspended from two equal-length cords. The end of each cord is attached to a ring of negligible mass, which passes along a horizontal shaft. If the coefficient of static friction between each ring and the shaft is $\mu_s = 0.5$, determine the greatest distance d by which the rings can be separated and still support the cylinder.

600 mm   600 mm

Prob. 8–24

8-25. The board can be adjusted vertically by tilting it up and sliding the smooth pin A along the vertical guide G. When placed horizontally, the bottom C then bears along the edge of the guide, where $\mu_s = 0.4$. Determine the largest dimension d which will support any applied force $\mathbf{F}$ without causing the board to slip downward.

Prob. 8–25

8-26. The homogeneous semicylinder has a mass m and mass center at G. Determine the largest angle θ of the inclined plane upon which it rests so that it does not slip down the plane. The coefficient of static friction between the plane and the cylinder is $\mu_s = 0.3$. Also, what is the angle ϕ for this case?

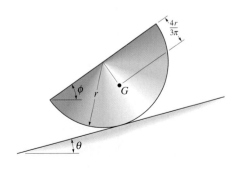

Prob. 8–26

8-27. Car A has a mass of 1.4 Mg and mass center at G. If car B exerts a horizontal force on A of 2 kN, determine if this force is great enough to move car A. The coefficients of static and kinetic friction between the tires and the road are $\mu_s = 0.5$ and $\mu_k = 0.35$. Assume B's bumper is smooth.

Prob. 8–27

*8-28. A 35-kg disk rests on an inclined surface for which $\mu_s = 0.2$. Determine the maximum vertical force $\mathbf{P}$ that may be applied to link AB without causing the disk to slip at C.

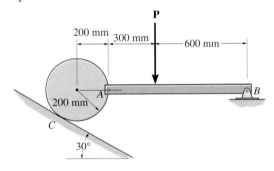

Prob. 8–28

8-29. The crate has a weight W and the coefficient of static friction at the surface is $\mu_s = 0.3$. Determine the orientation of the cord and the smallest possible force $\mathbf{P}$ that has to be applied to the cord so that the crate is on the verge of moving.

Prob. 8–29

8-30. The 4000-N ($\approx$ 400-kg) concrete pipe is being lowered from the truck bed when it is in the position shown. If the coefficient of static friction at the points of support A and B is $\mu_s = 0.4$, determine where it begins to slip first: at A or B, or both at A and B.

Prob. 8–30

8-31. The friction pawl is pinned at A and rests against the wheel at B. It allows freedom of movement when the wheel is rotating counterclockwise about C. Clockwise rotation is prevented due to friction of the pawl which tends to bind the wheel. If $(\mu_s)_B = 0.6$, determine the design angle θ which will prevent clockwise motion for any value of applied moment M. *Hint:* Neglect the weight of the pawl so that it becomes a two-force member.

Prob. 8–31

*8-32. The semicylinder of mass m and radius r lies on the rough inclined plane for which $\phi = 10°$ and the coefficient of static friction is $\mu_s = 0.3$. Determine if the semicylinder slides down the plane, and if not, find the angle of tip θ of its base AB.

8-33. The semicylinder of mass m and radius r lies on the rough inclined plane. If the inclination $\phi = 15°$, determine the smallest coefficient of static friction which will prevent the semicylinder from slipping.

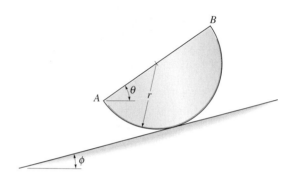

Probs. 8–32/33

8-34. The door brace AB is to be designed to prevent opening the door. If the brace forms a pin connection under the doorknob and the coefficient of static friction with the floor is $\mu_s = 0.5$, determine the largest length L the brace can have to prevent the door from being opened. Neglect the weight of the brace.

Prob. 8–34

8-35. The man has a weight of 900 N ($\approx$ 90 kg), and the coefficient of static friction between his shoes and the floor is $\mu_s = 0.5$. Determine where he should position his center of gravity G at d in order to exert the maximum horizontal force on the door. What is this force?

1 m

d

Prob. 8–35

*8-36. The 400-N ($\approx$ 40-kg) boy stands on the beam and pulls on the cord with a force large enough to just cause him to slip. If $(\mu_s)_D = 0.4$ between his shoes and the beam, determine the reactions at A and B. The beam is uniform and has a weight of 500 N ($\approx$ 50 kg). Neglect the size of the pulleys and the thickness of the beam.

8-37. The 400-N ($\approx$ 40-kg) boy stands on the beam and pulls with a force of 200 N. If $(\mu_s)_D = 0.4$, determine the frictional force between his shoes and the beam and the reactions at A and B. The beam is uniform and has a weight of 500 N ($\approx$ 50 kg). Neglect the size of the pulleys and the thickness of the beam.

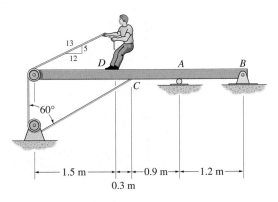

13 / 5 / 12

D A B

C

60°

1.5 m 0.9 m 1.2 m

0.3 m

Probs. 8–36/37

8-38. Two blocks A and B have a weight of 10 N ($\approx$ 1 kg) and 6 N ($\approx$ 0.6 kg), respectively. They are resting on the incline for which the coefficients of static friction are $\mu_A = 0.15$ and $\mu_B = 0.25$. Determine the incline angle θ for which both blocks begin to slide. Also find the required stretch or compression in the connecting spring for this to occur. The spring has a stiffness of $k = 2$ N/m.

8-39. Two blocks A and B have a weight of 10 N ($\approx$ 1 kg) and 6 N ($\approx$ 0.6 kg), respectively. They are resting on the incline for which the coefficients of static friction are $\mu_A = 0.15$ and $\mu_B = 0.25$. Determine the angle θ which will cause motion of one of the blocks. What is the friction force under each of the blocks when this occurs? The spring has a stiffness of $k = 2$ N/m and is originally unstretched.

$k = 2$ N/m B

A

θ

Probs. 8–38/39

*8-40. Determine the smallest force the man must exert on the rope in order to move the 80-kg crate. Also, what is the angle θ at this moment? The coefficient of static friction between the crate and the floor is $\mu_s = 0.3$.

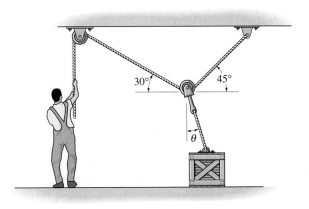

30° 45°

θ

Prob. 8–40

8-41. The three bars have a weight of $W_A = 20$ N (≈ 2 kg), $W_B = 40$ N (≈ 4 kg) and $W_C = 60$ N (≈ 6 kg), respectively. If the coefficients of static friction at the surfaces of contact are as shown, determine the smallest horizontal force P needed to move block A.

Prob. 8–41

8-42. The friction hook is made from a fixed frame which is shown colored and a cylinder of negligible weight. A piece of paper is placed between the smooth wall and the cylinder. If $\theta = 20°$, determine the smallest coefficient of static friction μ at all points of contact so that any weight W of paper p can be held.

Prob. 8–42

8-43. The refrigerator has a weight of 900 N (≈ 90 kg) and rests on a tile floor for which $\mu_s = 0.25$. If the man pushes horizontally on the refrigerator in the direction shown, determine the smallest magnitude of force needed to move it. Also, if the man has a weight of 750 N (≈ 75 kg), determine the smallest coefficient of friction between his shoes and the floor so that he does not slip.

*8-44. The refrigerator has a weight of 900 N (≈ 90 kg) and rests on a tile floor for which $\mu_s = 0.25$. Also, the man has a weight of 750 N (≈ 75 kg) and the coefficient of static friction between the floor and his shoes is $\mu_s = 0.6$. If he pushes horizontally on the refrigerator, determine if he can move it. If so does the refrigerator slip or tip?

Probs. 8–43/44

8-45. The wheel weighs 100 N (≈ 10 kg) and rests on a surface for which $\mu_B = 0.2$. A cord wrapped around it is attached to the top of the 150-N (≈ 15-kg) homogeneous block. If the coefficient of static friction at D is $\mu_D = 0.3$, determine the smallest vertical force that can be applied tangentially to the wheel which will cause motion to impend.

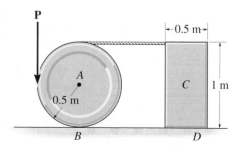

Prob. 8–45

8-46. Each of the cylinders has a mass of 50 kg. If the coefficients of static friction at the points of contact are $\mu_A = 0.5$, $\mu_B = 0.5$, $\mu_C = 0.5$, and $\mu_D = 0.6$, determine the couple moment M needed to rotate cylinder E.

Prob. 8–46

8-47. The beam AB has a negligible mass and thickness and is subjected to a triangular distributed loading. It is supported at one end by a pin and at the other end by a post having a mass of 50 kg and negligible thickness. Determine the minimum force P needed to move the post. The coefficients of static friction at B and C are $\mu_B = 0.4$ and $\mu_C = 0.2$, respectively.

*8-48.** The beam AB has a negligible mass and thickness and is subjected to a triangular distributed loading. It is supported at one end by a pin and at the other end by a post having a mass of 50 kg and negligible thickness. Determine the two coefficients of static friction at B and at C so that when the magnitude of the applied force is increased to $P = 150$ N, the post slips at both B and C simultaneously.

8-49. The block of weight W is being pulled up the inclined plane of slope α using a force $\mathbf{P}$. If $\mathbf{P}$ acts at the angle ϕ as shown, show that for slipping to occur, $P = W \sin(\alpha + \theta)/\cos(\phi - \theta)$, where θ is the angle of friction; $\theta = \tan^{-1} \mu$.

8-50. Determine the angle ϕ at which $\mathbf{P}$ should act on the block so that the magnitude of $\mathbf{P}$ is as small as possible to begin pushing the block up the incline. What is the corresponding value of P? The block weighs W and the slope α is known.

Probs. 8–49/50

8-51. The beam AB has a negligible mass and thickness and is subjected to a force of 200 N. It is supported at one end by a pin and at the other end by a spool having a mass of 40 kg. If a cable is wrapped around the inner core of the spool, determine the minimum cable force P needed to move the spool. The coefficients of static friction at B and D are $\mu_B = 0.4$ and $\mu_D = 0.2$, respectively.

Probs. 8–47/48

Prob. 8–51

*8-52. Block *C* has a mass of 50 kg and is confined between two walls by smooth rollers. If the block rests on top of the 40-kg spool, determine the minimum cable force *P* needed to move the spool. The cable is wrapped around the spool's inner core. The coefficients of static friction at *A* and *B* are $\mu_A = 0.3$ and $\mu_B = 0.6$.

Prob. 8–52

8-53. The uniform 60-kg crate *C* rests uniformly on a 10-kg dolly *D*. If the front casters of the dolly at *A* are locked to prevent rolling while the casters at *B* are free to roll, determine the maximum force **P** that may be applied without causing motion of the crate. The coefficient of static friction between the casters and the floor is $\mu_f = 0.35$ and between the dolly and the crate, $\mu_d = 0.5$.

Prob. 8–53

8-54. Two blocks *A* and *B*, each having a mass of 6 kg, are connected by the linkage shown. If the coefficients of static friction at the contacting surfaces are $\mu_A = 0.2$ and $\mu_B = 0.8$, determine the largest vertical force *P* that may be applied to pin *C* without causing the blocks to slip. Neglect the weight of the links.

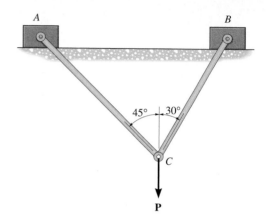

Prob. 8–54

8-55. The uniform beam has a weight *W* and length 4*a*. It rests on the fixed rails at *A* and *B*. If the coefficient of static friction at the rails is μ_s, determine the horizontal force *P*, applied perpendicular to the face of the beam, which will cause the beam to move.

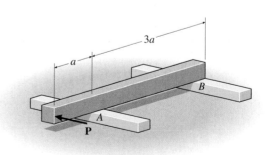

Prob. 8–55

*8-56. The uniform 6-kg slender rod rests on the top center of the 3-kg block. If the coefficients of static friction at the points of contact are $\mu_A = 0.4$, $\mu_B = 0.6$, and $\mu_C = 0.3$, determine the largest couple moment M which can be applied to the rod without causing motion of the rod.

Prob. 8–56

8-57. The carpenter slowly pushes the uniform board horizontally over the top of the saw horse. The board has a uniform weight of 50 N/m, and the saw horse has a weight of 75 N (≈ 7.5 kg) and a center of gravity at G. Determine if the saw horse will stay in position, slip, or tip if the board is pushed forward when $d = 3$ m. The coefficients of static friction are shown in the figure.

8-58. The carpenter slowly pushes the uniform board horizontally over the top of the saw horse. The board has a uniform weight of 50 N/m, and the saw horse has a weight of 75 N (≈ 7.5 kg) and a center of gravity at G. Determine if the saw horse will stay in position, slip, or tip if the board is pushed forward when $d = 4.2$ m. The coefficients of static friction are shown in the figure.

Probs. 8–57/58

8-59. The 45-kg disk rests on the surface for which the coefficient of static friction is $\mu_A = 0.2$. Determine the largest couple moment M that can be applied to the bar without causing motion.

*8-60. The 45-kg disk rests on the surface for which the coefficient of static friction is $\mu_A = 0.15$. If $M = 50$ N · m, determine the friction force at A.

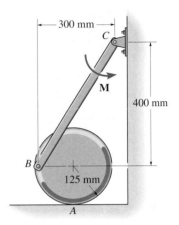

Probs. 8–59/60

8-61. The end C of the two-bar linkage rests on the top center of the 50-kg cylinder. If the coefficients of static friction at C and E are $\mu_C = 0.6$ and $\mu_E = 0.3$, determine the largest vertical force P which can be applied at B without causing motion. Neglect the mass of the bars.

Prob. 8–61

8.3 Wedges

Wedges are often used to adjust the elevation of structural or mechanical parts. Also, they provide stability for objects such as this tank.

A *wedge* is a simple machine which is often used to transform an applied force into much larger forces, directed at approximately right angles to the applied force. Also, wedges can be used to give small displacements or adjustments to heavy loads.

Consider, for example, the wedge shown in Fig. 8–13a, which is used to *lift* a block of weight **W** by applying a force **P** to the wedge. Free-body diagrams of the block and wedge are shown in Fig. 8–13b. Here we have excluded the weight of the wedge since it is usually *small* compared to the weight of the block. Also, note that the frictional forces F_1 and F_2 must oppose the motion of the wedge. Likewise, the frictional force F_3 of the wall on the block must act downward so as to oppose the block's upward motion. The locations of the resultant normal forces are not important in the force analysis since neither the block nor wedge will "tip." Hence the moment equilibrium equations will not be considered. There are seven unknowns consisting of the applied force **P**, needed to cause motion of the wedge, and six normal and frictional forces. The seven available equations consist of two force equilibrium equations ($\Sigma F_x = 0$, $\Sigma F_y = 0$) applied to the wedge and block (four equations total) and the frictional equation $F = \mu N$ applied at each surface of contact (three equations total).

If the block is to be *lowered*, the frictional forces will all act in a sense opposite to that shown in Fig. 8–13b. The applied force **P** will act to the right as shown if the coefficient of friction is very *small* or the wedge angle θ is *large*. Otherwise, **P** may have the reverse sense of direction in order to *pull* on the wedge to remove it. If **P** is *not* applied, or **P** = **0**, and friction forces hold the block in place, then the wedge is referred to as *self-locking*.

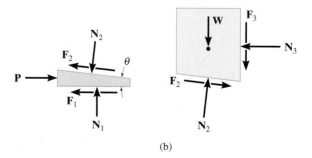

(b)

Fig. 8–13

EXAMPLE 8.7

The uniform stone in Fig. 8–14a has a mass of 500 kg and is held in the horizontal position using a wedge at B. If the coefficient of static friction is $\mu_s = 0.3$ at, the surfaces of contact, determine the minimum force $\mathbf{P}$ needed to remove the wedge. Is the wedge self-locking? Assume that the stone does not slip at A.

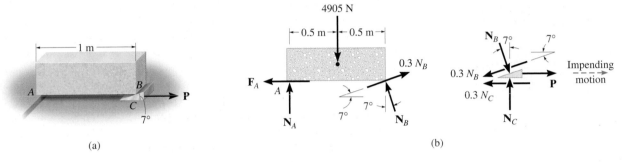

Fig. 8–14

Solution

The minimum force P requires $F = \mu_s N$ at the surfaces of contact with the wedge. The free-body diagrams of the stone and wedge are shown in Fig. 8–14b. On the wedge the friction force opposes the motion, and on the stone at A, $F_A \leq \mu_s N_A$, since slipping does not occur there. There are five unknowns F_A, N_A, N_B, N_C, and P. Three equilibrium equations for the stone and two for the wedge are available for solution. From the free-body diagram of the stone,

$$\zeta + \Sigma M_A = 0; \quad -4905\ \text{N}(0.5\ \text{m}) + (N_B \cos 7°\ \text{N})(1\ \text{m})$$
$$+(0.3 N_B \sin 7°\ \text{N})(1\ \text{m}) = 0$$
$$N_B = 2383.1\ \text{N}$$

Using this result for the wedge, we have

$$\xrightarrow{+} \Sigma F_x = 0; \quad 2383.1 \sin 7°\ \text{N} - 0.3(2383.1 \cos 7°\ \text{N}) + P - 0.3 N_C = 0$$
$$+\uparrow \Sigma F_y = 0;$$
$$N_C - 2383.1 \cos 7°\ \text{N} - 0.3(2383.1 \sin 7°\ \text{N}) = 0$$
$$N_C = 2452.5\ \text{N}$$
$$P = 1154.9\ \text{N} = 1.15\ \text{kN} \qquad\qquad Ans.$$

Since P is positive, indeed the wedge must be pulled out. If P was zero, the wedge would remain in place (self-locking) and the frictional forces developed at B and C would satisfy $F_B < \mu_s N_B$ and $F_C < \mu_s N_C$.

8.4 Frictional Forces on Screws

Square-threaded screws find applications on valves, jacks, and vises, where particularly large forces must be developed along the axis of the screw.

In most cases screws are used as fasteners; however, in many types of machines they are incorporated to transmit power or motion from one part of the machine to another. A *square-threaded screw* is commonly used for the latter purpose, especially when large forces are applied along its axis. In this section we will analyze the forces acting on square-threaded screws. The analysis of other types of screws, such as the V-thread, is based on these same principles.

A *screw* may be thought of simply as an inclined plane or wedge wrapped around a cylinder. A nut initially at position A on the screw shown in Fig. 8–15a will move up to B when rotated 360° around the screw. This rotation is equivalent to translating the nut up an inclined plane of height l and length $2\pi r$, where r is the mean radius of the thread, Fig. 8–15b. The rise l for a single revolution is referred to as the *lead* of the screw, where the *lead angle* is given by $\theta = \tan^{-1}(l/2\pi r)$.

Frictional Analysis. When a screw is subjected to large axial loads, the frictional forces developed on the thread become important if we are to determine the moment **M*** needed to turn the screw. Consider, for example, the square-threaded jack screw shown in Fig. 8–16, which supports the vertical load **W**. The reactive forces of the jack to this load are actually distributed over the circumference of the screw thread in contact with the screw hole in the jack, that is, within region h shown in Fig. 8–16. For simplicity, this portion of thread can be imagined as being unwound from the screw and represented as a simple block resting on an inclined plane having the screw's lead angle θ, Fig. 8–17a. Here the inclined plane represents the inside *supporting thread* of the jack base. Three forces act on the block or screw. The force **W** is the total axial load applied to the screw. The horizontal force **S** is caused by the applied moment **M**, such that by summing moments about the axis of the screw, $M = Sr$, where r is the screw's mean radius. As a result of **W** and **S**, the inclined plane exerts a resultant force **R** on the block, which is shown to have components acting normal, **N**, and tangent, **F**, to the contacting surfaces.

(a)

(b)

Fig. 8–15

*For applications, **M** is developed by applying a horizontal force **P** at a right angle to the end of a lever that would be fixed to the screw.

Fig. 8–16

Upward Screw Motion. Provided M is great enough, the screw (and hence the block) can either be brought to the verge of upward impending motion or motion can be occurring. Under these conditions, R acts at an angle $(\theta + \phi)$ from the vertical as shown in Fig. 8–17a, where $\phi = \tan^{-1}(F/N) = \tan^{-1}(\mu N/N) = \tan^{-1}\mu$. Applying the two force equations of equilibrium to the block, we obtain

$$\xrightarrow{+} \Sigma F_x = 0; \qquad\qquad S - R\sin(\theta + \phi) = 0$$
$$+\uparrow\Sigma F_y = 0; \qquad\qquad R\cos(\theta + \phi) - W = 0$$

Eliminating R and solving for S, then substituting this value into the equation $M = Sr$, yields

$$\boxed{M = Wr\tan(\theta + \phi)} \qquad\qquad (8\text{–}3)$$

As indicated, M is the moment necessary to cause upward impending motion of the screw, provided $\phi = \phi_s = \tan^{-1}\mu_s$ (the angle of static friction). If ϕ is replaced by $\phi_k = \tan^{-1}\mu_k$ (the angle of kinetic friction), Eq. 8–3 will give a smaller value M necessary to maintain uniform upward motion of the screw.

Downward Screw Motion $(\theta > \phi)$. If the surface of the screw is very *slippery*, it may be possible for the screw to rotate downward if the magnitude of the moment is reduced to, say, $M' < M$. As shown in Fig. 8–17b, this causes the effect of $\mathbf{M'}$ to become $\mathbf{S'}$, and it requires the angle ϕ (ϕ_s or ϕ_k) to lie on the opposite side of the normal n to the plane supporting the block, such that $\theta > \phi$. For this case, Eq. 8–3 becomes

$$\boxed{M' = Wr\tan(\theta - \phi)} \qquad\qquad (8\text{–}4)$$

Upward screw motion

(a)

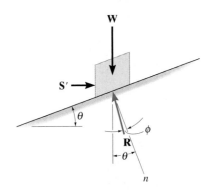

Downward screw motion $(\theta > \phi)$

(b)

Fig. 8–17

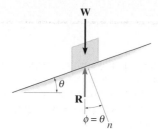

Self-locking screw ($\theta = \phi$)
(on the verge of rotating downward)

(c)

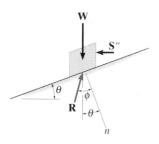

Downward screw motion ($\theta < \phi$)

(d)

Fig. 8–17

Self-Locking Screw. If the moment **M** (or its effect **S**) is *removed*, the screw will remain *self-locking*; i.e., it will support the load **W** by *friction forces alone* provided $\phi \geq \theta$. To show this, consider the necessary limiting case when $\phi = \theta$, Fig. 8–17c. Here vertical equilibrium is maintained since **R** is vertical and thus balances **W**.

Downward Screw Motion ($\theta < \phi$). When the surface of the screw is *very rough*, the screw will not rotate downward as stated above. Instead, the direction of the applied moment must be *reversed* in order to cause the motion. The free-body diagram shown in Fig. 8–17d is representative of this case. Here **S″** is caused by the applied (reverse) moment **M″**. Hence Eq. 8–3 becomes

$$M'' = Wr \tan(\phi - \theta) \qquad (8\text{–}5)$$

Each of the above cases should be thoroughly understood before proceeding to solve problems.

E X A M P L E   8.8

2 kN

M

2 kN

Fig. 8–18

The turnbuckle shown in Fig. 8–18 has a square thread with a mean radius of 5 mm and a lead of 2 mm. If the coefficient of static friction between the screw and the turnbuckle is $\mu_s = 0.25$, determine the moment **M** that must be applied to draw the end screws closer together. Is the turnbuckle self-locking?

Solution

The moment may be obtained by using Eq. 8–3. Why? Since friction at *two screws* must be overcome, this requires

$$M = 2[Wr \tan(\theta + \phi)] \qquad (1)$$

Here $W = 2000$ N, $r = 5$ mm, $\phi_s = \tan^{-1} \mu_s = \tan^{-1}(0.25) = 14.04°$, and $\theta = \tan^{-1}(l/2\pi r) = \tan^{-1}(2 \text{ mm}/[2\pi(5 \text{ mm})]) = 3.64°$. Substituting these values into Eq. 1 and solving gives

$$M = 2[(2000 \text{ N})(5 \text{ mm}) \tan(14.04° + 3.64°)]$$
$$= 6374.7 \text{ N} \cdot \text{mm} = 6.37 \text{ N} \cdot \text{m} \qquad \qquad Ans.$$

When the moment is *removed*, the turnbuckle will be self-locking; i.e., it will not unscrew since $\phi_s > \theta$.

PROBLEMS

8-62. Determine the minimum applied force **P** required to move wedge *A* to the right. The spring is compressed a distance of 175 mm. Neglect the weight of *A* and *B*. The coefficient of static friction for all contacting surfaces is $\mu_s = 0.35$. Neglect friction at the rollers.

*8-64. The wedge has a negligible weight and a coefficient of static friction $\mu_s = 0.35$ with all contacting surfaces. Determine the largest angle θ so that it is "self-locking." This requires no slipping for any magnitude of the force **P** applied to the joint.

Prob. 8–64

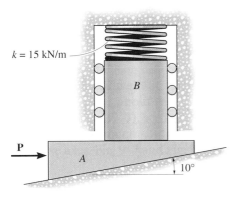

Prob. 8–62

8-65. If the spring is compressed 60 mm and the coefficient of static friction between the tapered stub *S* and the slider *A* is $\mu_{SA} = 0.5$, determine the horizontal force **P** needed to move the slider forward. The stub is free to move without friction within the fixed collar *C*. The coefficient of static friction between *A* and surface *B* is $\mu_{AB} = 0.4$. Neglect the weights of the slider and stub.

8-63. Determine the largest weight of the wedge that can be placed between the 8-kN ($\approx$ 800-kg) cylinder and the wall without upsetting equilibrium. The coefficient of static friction at *A* and *C* is $\mu_s = 0.5$ and at *B*, $\mu_s' = 0.6$.

Prob. 8–63

Prob. 8–65

8-66. The coefficient of static friction between wedges B and C is $\mu_s = 0.6$ and between the surfaces of contact B and A and C and D, $\mu_s' = 0.4$. If the spring is compressed 200 mm when in the position shown, determine the smallest force P needed to move wedge C to the left. Neglect the weight of the wedges.

8-67. The coefficient of static friction between the wedges B and C is $\mu_s = 0.6$ and between the surfaces of contact B and A and C and D, $\mu_s' = 0.4$. If $P = 50$ N, determine the largest allowable compression of the spring without causing wedge C to move to the left. Neglect the weight of the wedges.

Probs. 8–66/67

*8-68. The wedge blocks are used to hold the specimen in a tension testing machine. Determine the design angle θ of the wedges so that the specimen will not slip regardless of the applied load. The coefficients of static friction are $\mu_A = 0.1$ at A and $\mu_B = 0.6$ at B. Neglect the weight of the blocks.

Prob. 8–68

8-69. The beam is adjusted to the horizontal position by means of a wedge located at its right support. If the coefficient of static friction between the wedge and the two surfaces of contact is $\mu_s = 0.25$, determine the horizontal force **P** required to push the wedge forward. Neglect the weight and size of the wedge and the thickness of the beam.

Prob. 8–69

8-70. If the beam AD is loaded as shown, determine the horizontal force P which must be applied to the wedge in order to remove it from under the beam. The coefficients of static friction at the wedge's top and bottom surfaces are $\mu_{CA} = 0.25$ and $\mu_{CB} = 0.35$, respectively. If $P = 0$, is the wedge self-locking? Neglect the weight and size of the wedge and the thickness of the beam.

Prob. 8–70

8-71. The column is used to support the upper floor. If a force $F = 80$ N is applied perpendicular to the handle to tighten the screw, determine the compressive force in the column. The square-threaded screw on the jack has a coefficient of static friction of $\mu_s = 0.4$, mean diameter of 25 mm, and a lead of 3 mm.

***8-72.** If the force **F** is removed from the handle of the jack in Prob. 8-71, determine if the screw is self-locking.

Probs. 8–71/72

8-73. The square-threaded screw has a mean diameter of 20 mm and a lead of 4 mm. If the weight of the plate A is 50 N (≈ 5 kg), determine the smallest coefficient of static friction between the screw and the plate so that the plate does not travel down the screw when the plate is suspended as shown.

Prob. 8–73

8-74. The square threaded screw of the clamp has a mean diameter of 14 mm and a lead of 6 mm. If $\mu_s = 0.2$ for the threads, and the torque applied to the handle is 1.5 N·m, determine the compressive force **F** on the block.

Prob. 8–74

8-75. The device is used to pull the battery cable terminal C from the post of a battery. If the required pulling force is 340 N, determine the torque **M** that must be applied to the handle on the screw to tighten it. The screw has square threads, a mean diameter of 5 mm, a lead of 2 mm, and the coefficient of static friction is $\mu_s = 0.5$.

Prob. 8–75

*8-76. The automobile jack is subjected to a vertical load of $F = 8$ kN. If a square-threaded screw, having a lead of 5 mm and a mean diameter of 10 mm, is used in the jack, determine the force that must be applied perpendicular to the handle to (a) raise the load, and (b) lower the load; $\mu_s = 0.2$. The supporting plate exerts only vertical forces at A and B.

Prob. 8–76

8-77. Determine the clamping force on the board A if the screw of the "C" clamp is tightened with a twist of $M = 8$ N·m. The single square-threaded screw has a mean radius of 10 mm, a lead of 3 mm, and the coefficient of static friction is $\mu_s = 0.35$.

8-78. If the required clamping force at the board A is to be 50 N, determine the torque M that must be applied to the handle of the "C" clamp to tighten it down. The single square-threaded screw has a mean radius of 10 mm, a lead of 3 mm, and the coefficient of static friction is $\mu_s = 0.35$.

Probs. 8–77/78

8-79. The shaft has a square-threaded screw with a lead of 8 mm and a mean radius of 15 mm. If it is in contact with a plate gear having a mean radius of 30 mm, determine the resisting torque M on the plate gear which can be overcome if a torque of 7 N·m is applied to the shaft. The coefficient of static friction at the screw is $\mu_B = 0.2$. Neglect friction of the bearings located at A and B.

Prob. 8–79

*8-80. The braking mechanism consists of two pinned arms and a square-threaded screw with left and righthand threads. Thus when turned, the screw draws the two arms together. If the lead of the screw is 4 mm, the mean diameter 12 mm, and the coefficient of static friction is $\mu_s = 0.35$, determine the tension in the screw when a torque of 5 N·m is applied to tighten the screw. If the coefficient of static friction between the brake pads A and B and the circular shaft is $\mu'_s = 0.5$, determine the maximum torque M the brake can resist.

Prob. 8–80

8-81. The fixture clamp consist of a square-threaded screw having a coefficient of static friction of $\mu_s = 0.3$, mean diameter of 3 mm, and a lead of 1 mm. The five points indicated are pin connections. Determine the clamping force at the smooth blocks D and E when a torque of $M = 0.08$ N·m is applied to the handle of the screw.

8-83. A turnbuckle, similar to that shown in Fig. 8–18, is used to tension member AB of the truss. The coefficient of the static friction between the square threaded screws and the turnbuckle is $\mu_s = 0.5$. The screws have a mean radius of 6 mm and a lead of 3 mm. If a torque of $M = 10$ N·m is applied to the turnbuckle, to draw the screws closer together, determine the force in each member of the truss. No external forces act on the truss.

Prob. 8–81

Prob. 8–83

8-82. The clamp provides pressure from several directions on the edges of the board. If the square-threaded screw has a lead of 3 mm, radius of 10 mm, and the coefficient of static friction is $\mu_s = 0.4$, determine the horizontal force developed on the board at A and the vertical forces developed at B and C if a torque of $M = 1.5$ N·m is applied to the handle to tighten it further. The blocks at B and C are pin-connected to the board.

*8-84. A turnbuckle, similar to that shown in Fig. 8–18, is used to tension member AB of the truss. The coefficient of the static friction between the square-threaded screws and the turnbuckle is $\mu_s = 0.5$. The screws have a mean radius of 6 mm and a lead of 3 mm. Determine the torque M which must be applied to the turnbuckle to draw the screws closer together, so that the compressive force of 500 N is developed in member BC.

Prob. 8–82

Prob. 8–84

8.5 Frictional Forces on Flat Belts

Flat or V-belts are often used to transmit the torque developed by a motor to a fan or blower.

Whenever belt drives or band brakes are designed, it is necessary to determine the frictional forces developed between the belt and its contacting surface. In this section we will analyze the frictional forces acting on a flat belt, although the analysis of other types of belts, such as the V-belt, is based on similar principles.

Here we will consider the flat belt shown in Fig. 8–19a, which passes over a fixed curved surface, such that the total angle of belt to surface contact in radians is β and the coefficient of friction between the two surfaces is μ. We will determine the tension T_2 in the belt which is needed to pull the belt counterclockwise over the surface and thereby overcome both the frictional forces at the surface of contact and the known tension T_1. Obviously, $T_2 > T_1$.

Frictional Analysis. A free-body diagram of the belt segment in contact with the surface is shown in Fig. 8–19b. Here the normal force **N** and the frictional force **F**, acting at different points along the belt, will vary both in magnitude and direction. Due to this *unknown* force distribution, the analysis of the problem will proceed on the basis of initially studying the forces acting on a differential element of the belt.

A free-body diagram of an element having a length ds is shown in Fig. 8–19c. Assuming either impending motion or motion of the belt, the magnitude of the frictional force $dF = \mu\, dN$. This force opposes the sliding motion of the belt and thereby increases the magnitude of the tensile force acting in the belt by dT. Applying the two force equations of equilibrium, we have

$$\searrow + \Sigma F_x = 0; \qquad T \cos\left(\frac{d\theta}{2}\right) + \mu\, dN - (T + dT) \cos\left(\frac{d\theta}{2}\right) = 0$$

$$+ \nearrow \Sigma F_y = 0; \qquad dN - (T + dT) \sin\left(\frac{d\theta}{2}\right) - T \sin\left(\frac{d\theta}{2}\right) = 0$$

(a)

(b)

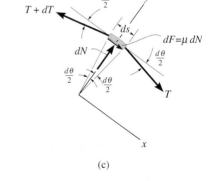

(c)

Fig. 8–19

Since $d\theta$ is of *infinitesimal size*, $\sin(d\theta/2)$ and $\cos(d\theta/2)$ can be replaced by $d\theta/2$ and 1, respectively. Also, the *product* of the two infinitesimals dT and $d\theta/2$ may be neglected when compared to infinitesimals of the first order. The above two equations therefore reduce to

$$\mu \, dN = dT$$

and

$$dN = T \, d\theta$$

Eliminating dN yields

$$\frac{dT}{T} = \mu \, d\theta$$

Integrating this equation between all the points of contact that the belt makes with the drum, and noting that $T = T_1$ at $\theta = 0$ and $T = T_2$ at $\theta = \beta$, yields

$$\int_{T_1}^{T_2} \frac{dT}{T} = \mu \int_{0}^{\beta} d\theta$$

$$\ln \frac{T_2}{T_1} = \mu\beta$$

Solving for T_2, we obtain

$$\boxed{T_2 = T_1 e^{\mu\beta}} \tag{8–6}$$

where T_2, T_1 = belt tensions; T_1 opposes the direction of motion (or impending motion) of the belt measured relative to the surface, while T_2 acts in the direction of the relative belt motion (or impending motion); because of friction, $T_2 > T_1$.

μ = coefficient of static or kinetic friction between the belt and the surface of contact

β = angle of belt to surface contact, measured in radians

e = 2.718..., base of the natural logarithm

Note that T_2 is *independent* of the *radius* of the drum and instead it is a function of the angle of belt to surface contact, β. Furthermore, as indicated by the integration, this equation is valid for flat belts placed on *any shape* of contacting surface. For application, Eq. 8–6 is valid only when *impending motion* or *motion* occurs.

E X A M P L E 8.9

The maximum tension that can be developed in the cord shown in Fig. 8–20a is 500 N. If the pulley at A is free to rotate and the coefficient of static friction at the fixed drums B and C is $\mu_s = 0.25$, determine the largest mass of the cylinder that can be lifted by the cord. Assume that the force **F** applied at the end of the cord is directed vertically downward, as shown.

(a)

Solution

Lifting the cylinder, which has a weight $W = mg$, causes the cord to move counterclockwise over the drums at B and C; hence, the maximum tension T_2 in the cord occurs at D. Thus, $T_2 = 500$ N. A section of the cord passing over the drum at B is shown in Fig. 8–20b. Since $180° = \pi$ rad, the angle of contact between the drum and the cord is $\beta = (135°/180°)\pi = 3\pi/4$ rad. Using Eq. 8–6, we have

$$T_2 = T_1 e^{\mu_s \beta}; \qquad 500 \text{ N} = T_1 e^{0.25[(3/4)\pi]}$$

Hence,

$$T_1 = \frac{500 \text{ N}}{e^{0.25[(3/4)\pi]}} = \frac{500 \text{ N}}{1.80} = 277.4 \text{ N}$$

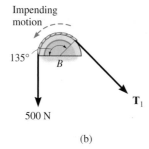

(b)

Since the pulley at A is free to rotate, equilibrium requires that the tension in the cord remains the *same* on both sides of the pulley.

The section of the cord passing over the drum at C is shown in Fig. 8–20c. The weight $W < 277.4$ N. Why? Applying Eq. 8–6, we obtain

$$T_2 = T_1 e^{\mu_s \beta}; \qquad 277.4 \text{ N} = W e^{0.25[(3/4)\pi]}$$
$$W = 153.9 \text{ N}$$

so that

$$m = \frac{W}{g} = \frac{153.9 \text{ N}}{9.81 \text{ m/s}^2}$$
$$= 15.7 \text{ kg} \qquad\qquad Ans.$$

(c)

Fig. 8–20

PROBLEMS

8-85. A "hawser" is wrapped around a fixed "capstan" to secure a ship for docking. If the tension in the rope, caused by the ship, is 7500 N, determine the least number of complete turns the rope must be rapped around the capstan in order to prevent slipping of the rope. The greatest horizontal force that a longshoreman can exert on the rope is 250 N. The coefficient of static friction is $\mu_s = 0.3$.

250 N

7500 N

Prob. 8–85

8-86. The truck, which has a mass of 3.4 Mg, is to be lowered down the slope by a rope that is wrapped around a tree. If the wheels are free to roll and the man at A can resist a pull of 300 N, determine the minimum number of turns the rope should be wrapped around the tree to lower the truck at a constant speed. The coefficient of kinetic friction between the tree and rope is $\mu_k = 0.3$.

A

20°

Prob. 8–86

8-87. Determine the maximum and the minimum values of weight W which may be applied without causing the 50-N ($\approx$ 5-kg) block to slip. The coefficient of static friction between the block and the plane is $\mu_s = 0.2$, and between the rope and the drum $D \ \mu_s' = 0.3$.

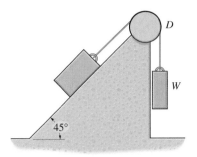

D

W

45°

Prob. 8–87

*8-88. A cylinder having a mass of 250 kg is to be supported by the cord which wraps over the pipe. Determine the smallest vertical force **F** needed to support the load if the cord passes (a) once over the pipe, $\beta = 180°$, and (b) two times over the pipe, $\beta = 540°$. Take $\mu_s = 0.2$.

8-89. A cylinder having a mass of 250 kg is to be supported by the cord which wraps over the pipe. Determine the largest vertical force **F** that can be applied to the cord without moving the cylinder. The cord passes (a) once over the pipe, $\beta = 180°$, and (b) two times over the pipe, $\beta = 540°$. Take $\mu_s = 0.2$.

F

Probs. 8–88/89

8-90. The boat has a weight of 2500 N ($\approx$ 250 kg) and is held in position off the side of a ship by the spars at A and B. A man having a weight of 650 N ($\approx$ 65 kg) gets in the boat, wraps a rope around an overhead boom at C, and ties it to the end of the boat as shown. If the boat is disconnected from the spars, determine the *minimum number of half turns* the rope must make around the boom so that the boat can be safely lowered into the water at constant velocity. Also, what is the normal force between the boat and the man? The coefficient of kinetic friction between the rope and the boom is $\mu_s = 0.15$. *Hint:* The problem requires that the normal force between the man's feet and the boat be as small as possible.

Prob. 8–90

8-91. Determine the smallest lever force P needed to prevent the wheel from rotating if it is subjected to a torque of $M = 250$ N $\cdot$ m. The coefficient of static friction between the belt and the wheel is $\mu_s = 0.3$. The wheel is pin-connected at its center, B.

*8-92. Determine the torque M that can be resisted by the band brake if a force of $P = 30$ N is applied to the handle of the lever. The coefficient of static friction between the belt and the wheel is $\mu_s = 0.3$. The wheel is pin-connected at its center, B.

Probs. 8–91/92

8-93. Blocks A and B weigh 500 N ($\approx$ 50 kg) and 300 N ($\approx$ 30 kg), respectively. Using the coefficients of static friction indicated, determine the greatest weight of block D without causing motion.

8-94. Blocks A and B weigh 750 N ($\approx$ 75 kg) each, and D weighs 300 N ($\approx$ 30 kg). Using the coefficients of static friction indicated, determine the frictional force between blocks A and B and between block A and the floor C.

Probs. 8–93/94

8-95. Show that the frictional relationship between the belt tensions, the coefficient of friction μ, and the angular contacts α and β for the V-belt is $T_2 = T_1 e^{\mu\beta/\sin(\alpha/2)}$.

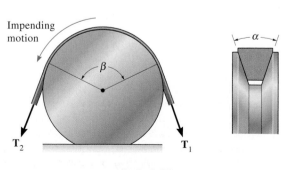

Prob. 8–95

*8-96. The smooth beam is being hoisted using a rope which is wrapped around the beam and passes through a ring at A as shown. If the end of the rope is subjected to a tension **T** and the coefficient of static friction between the rope and ring is $\mu_s = 0.3$, determine the angle of θ for equilibrium.

Prob. 8–96

8-97. The 20-kg motor has a center of gravity at G and is pin-connected at C to maintain a tension in the drive belt. Determine the smallest counterclockwise twist or torque **M** that must be supplied by the motor to turn the disk B if wheel A locks and causes the belt to slip over the disk. No slipping occurs at A. The coefficient of static friction between the belt and the disk is $\mu_s = 0.3$.

Prob. 8–97

8-98. The simple band brake is constructed so that the ends of the friction strap are connected to the pin at A and the lever arm at B. If the wheel is subjected to a torque of $M = 200$ N $\cdot$ m, determine the smallest force P applied to the lever that is required to hold the wheel stationary. The coefficient of static friction between the strap and wheel is $\mu_s = 0.5$.

Prob. 8–98

8-99. The cylinder weighs 10 N ($\approx$ 1 kg) and is held in equilibrium by the belt and wall. If slipping does not occur at the wall, determine the minimum vertical force P which must be applied to the belt for equilibrium. The coefficient of static friction between the belt and the cylinder is $\mu_s = 0.25$.

Prob. 8–99

*8-100. The uniform concrete pipe has a weight of 4 kN ($\approx$ 400 kg) and is unloaded slowly from the truck bed using the rope and skids shown. If the coefficient of kinetic friction between the rope and pipe is $\mu_k = 0.3$, determine the force the worker must exert on the rope to lower the pipe at constant speed. There is a pulley at B, and the pipe does not slip on the skids. The lower portion of the rope is parallel to the skids.

Prob. 8–100

8-101. A cord having a weight of 6 N/m ($\approx$ 0.6 kg/m) and a total length of 3 m is suspended over a peg P as shown. If the coefficient of static friction between the peg and cord is $\mu_s = 0.5$, determine the longest length h which one side of the suspended cord can have without causing motion. Neglect the size of the peg and the length of cord draped over it.

8-103. Blocks A and B have a mass of 7 kg and 10 kg, respectively. Using the coefficients of static friction indicated, determine the largest vertical force P which can be applied to the cord without causing motion.

Prob. 8–103

Prob. 8–101

8-102. A conveyer belt is used to transfer granular material and the frictional resistance on the top of the belt is $F = 500$ N. Determine the smallest stretch of the spring attached to the moveable axle of the idle pulley B so that the belt does not slip at the drive pulley A when the torque **M** is applied. What minimum torque **M** is required to keep the belt moving? The coefficient of static friction between the belt and the wheel at A is $\mu_s = 0.2$.

***8-104.** The belt on the portable dryer wraps around the drum D, idler pulley A, and motor pulley B. If the motor can develop a maximum torque of $M = 0.80$ N $\cdot$ m, determine the smallest spring tension required to hold the belt from slipping. The coefficient of static friction between the belt and the drum and motor pulley is $\mu_s = 0.3$.

Prob. 8–102

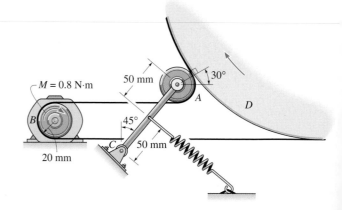

Prob. 8–104

8-105. Block A has a mass of 50 kg and rests on surface for which $\mu_s = 0.25$. If the coefficient of static friction between the cord and the fixed peg at C is $\mu'_s = 0.3$, determine the greatest mass of the suspended cylinder D without causing motion.

8-106. Block A rests on the surface for which $\mu_s = 0.25$. If the mass of the suspended cylinder D is 4 kg, determine the smallest mass of block A so that it does not slip or tip. The coefficient of static friction between the cord and the fixed peg at C is $\mu'_s = 0.3$.

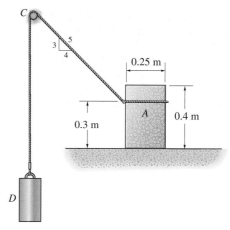

Probs. 8–105/106

*8.6 Frictional Forces on Collar Bearings, Pivot Bearings, and Disks

Pivot and *collar bearings* are commonly used in machines to support an *axial load* on a rotating shaft. These two types of support are shown in Fig. 8–21. Provided the bearings are not lubricated, or are only partially lubricated, the laws of dry friction may be applied to determine the moment **M** needed to turn the shaft when it supports an axial force **P**.

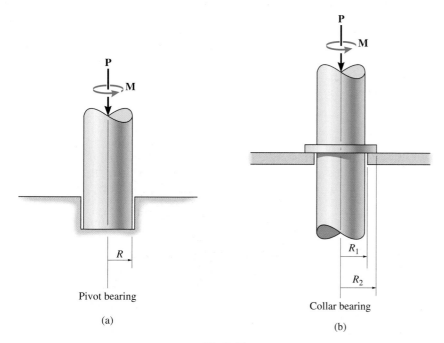

Pivot bearing

(a)

Collar bearing

(b)

Fig. 8–21

Fig. 8–22

Frictional Analysis. The collar bearing on the shaft shown in Fig. 8–22 is subjected to an axial force **P** and has a total bearing or contact area $\pi(R_2^2 - R_1^2)$. In the following analysis, the normal pressure p is considered to be *uniformly distributed* over this area—a reasonable assumption provided the bearing is new and evenly supported. Since $\Sigma F_z = 0$, then p, measured as a force per unit area, is $p = P/\pi(R_2^2 - R_1^2)$.

The moment needed to cause impending rotation of the shaft can be determined from moment equilibrium about the z axis. A small area element $dA = (r\,d\theta)(dr)$, shown in Fig. 8–22, is subjected to both a normal force $dN = p\,dA$ and an associated frictional force,

$$dF = \mu_s\,dN = \mu_s p\,dA = \frac{\mu_s P}{\pi(R_2^2 - R_1^2)}\,dA$$

The normal force does not create a moment about the z axis of the shaft; however, the frictional force does; namely, $dM = r\,dF$. Integration is needed to compute the total moment created by all the frictional forces acting on differential areas dA. Therefore, for impending rotational motion,

$$\Sigma M_z = 0; \qquad M - \int_A r\,dF = 0$$

Substituting for dF and dA and integrating over the entire bearing area yields

$$M = \int_{R_1}^{R_2}\int_0^{2\pi} r\left[\frac{\mu_s P}{\pi(R_2^2 - R_1^2)}\right](r\,d\theta\,dr) = \frac{\mu_s P}{\pi(R_2^2 - R_1^2)}\int_{R_1}^{R_2} r^2\,dr\int_0^{2\pi} d\theta$$

or

$$M = \tfrac{2}{3}\mu_s P\left(\frac{R_2^3 - R_1^3}{R_2^2 - R_1^2}\right) \qquad (8\text{–}7)$$

This equation gives the magnitude of moment required for impending rotation of the shaft. The frictional moment developed at the end of the shaft, when it is *rotating* at constant speed, can be found by substituting μ_k for μ_s in Eq. 8–7.

When $R_2 = R$ and $R_1 = 0$, as in the case of a pivot bearing, Fig. 8–21a, Eq. 8–7 reduces to

$$M = \tfrac{2}{3}\mu_s PR \qquad (8\text{–}8)$$

Recall from the initial assumption that both Eqs. 8–7 and 8–8 apply only for bearing surfaces subjected to *constant pressure*. If the pressure is not uniform, a variation of the pressure as a function of the bearing area must be determined before integrating to obtain the moment. The following example illustrates this concept.

Frictional forces acting on the disk of this sanding machine must be overcome by the torque developed by the motor which turns it.

EXAMPLE 8.10

The uniform bar shown in Fig. 8–23a has a total mass m. If it is assumed that the normal pressure acting at the contacting surface varies linearly along the length of the bar as shown, determine the couple moment **M** required to rotate the bar. Assume that the bar's width a is negligible in comparison to its length l. The coefficient of static friction is equal to μ_s.

Solution

A free-body diagram of the bar is shown in Fig. 8–23b. Since the bar has a total weight of $W = mg$, the intensity w_0 of the distributed lead at the center $(x = 0)$ is determined from vertical force equilibrium, Fig. 8–23a.

$$+\uparrow \Sigma F_z = 0; \quad -mg + 2\left[\frac{1}{2}\left(\frac{l}{2}\right)w_0\right] = 0 \qquad w_0 = \frac{2\,mg}{l}$$

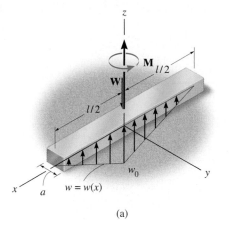

(a)

Since $w = 0$ at $x = l/2$, the distributed load expressed as a function of x is

$$w = w_0\left(1 - \frac{2\,x}{l}\right) = \frac{2\,mg}{l}\left(1 - \frac{2\,x}{l}\right)$$

The magnitude of the normal force acting on a segment of area having a length dx is therefore

$$dN = w\,dx = \frac{2\,mg}{l}\left(1 - \frac{2\,x}{l}\right)dx$$

The magnitude of the frictional force acting on the same element of area is

$$dF = \mu_s\,dN = \frac{2\,\mu_s mg}{l}\left(1 - \frac{2\,x}{l}\right)dx$$

Hence, the moment created by this force about the z axis is

$$dM = x\,dF = \frac{2\,\mu_s mg}{l}\,x\left(1 - \frac{2\,x}{l}\right)dx$$

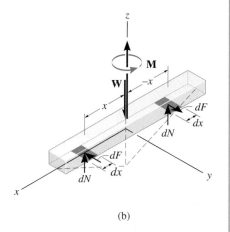

(b)

Fig. 8–23

The summation of moments about the z axis of the bar is determined by integration, which yields

$$\Sigma M_z = 0; \qquad M - 2\int_0^{l/2}\frac{2\,\mu_s mg}{l}\,x\left(1 - \frac{2\,x}{l}\right)dx = 0$$

$$M = \frac{4\,\mu_s mg}{l}\left(\frac{x^2}{2} - \frac{2\,x^3}{3\,l}\right)\Big|_0^{l/2}$$

$$M = \frac{\mu_s mgl}{6} \qquad\qquad Ans.$$

8.7 Frictional Forces on Journal Bearings

Unwinding the cable from this spool requires overcoming friction from the supporting shaft.

When a shaft or axle is subjected to lateral loads, a *journal bearing* is commonly used for support. Well-lubricated journal bearings are subjected to the laws of fluid mechanics, in which the viscosity of the lubricant, the speed of rotation, and the amount of clearance between the shaft and bearing are needed to determine the frictional resistance of the bearing. When the bearing is not lubricated or is only partially lubricated, however, a reasonable analysis of the frictional resistance can be based on the laws of dry friction.

Frictional Analysis. A typical journal-bearing support is shown in Fig. 8–24a. As the shaft rotates in the direction shown in the figure, it rolls up against the wall of the bearing to some point A where slipping occurs. If the lateral load acting at the end of the shaft is **P**, it is necessary that the bearing reactive force **R** acting at A be equal and opposite to **P**, Fig. 8–24b. The moment needed to maintain constant rotation of the shaft can be found by summing moments about the z axis of the shaft; i.e.,

$$\Sigma M_z = 0; \qquad M - (R \sin \phi_k)r = 0$$

or

$$M = Rr \sin \phi_k \qquad (8\text{–}9)$$

where ϕ_k is the angle of kinetic friction defined by $\tan \phi_k = F/N = \mu_k N/N = \mu_k$. In Fig. 8–24c, it is seen that $r \sin \phi_k = r_f$. The dashed circle with radius r_f is called the *friction circle*, and as the shaft rotates, the reaction **R** will always be tangent to it. If the bearing is partially lubricated, μ_k is small, and therefore $\mu_k = \tan \phi_k \approx \sin \phi_k \approx \phi_k$. Under these conditions, a reasonable *approximation* to the moment needed to overcome the frictional resistance becomes

$$M \approx Rr\mu_k \qquad (8\text{–}10)$$

The following example illustrates a common application of this analysis.

Rotation

z

A

(a)

Fig. 8–24

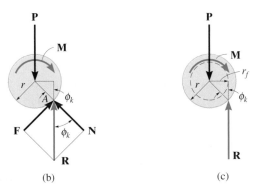

(b)

(c)

EXAMPLE 8.11

The 100-mm-diameter pulley shown in Fig. 8–25a fits loosely on a 10-mm-diameter shaft for which the coefficient of static friction is $\mu_s = 0.4$. Determine the minimum tension T in the belt needed to (a) raise the 100-kg block and (b) lower the block. Assume that no slipping occurs between the belt and pulley and neglect the weight of the pulley.

50 mm

$r = 5$ mm

100 kg   T   (a)

Solution

Part (a)   A free-body diagram of the pulley is shown in Fig. 8–25b. When the pulley is subjected to belt tensions of 981 N each, it makes contact with the shaft at point P_1. As the tension T is *increased*, the pulley will roll around the shaft to point P_2 before motion impends. From the figure, the friction circle has a radius $r_f = r \sin \phi_s$. Using the simplification that $\sin \phi_s \approx$ (tan $\phi_s \approx \phi_s$), then $r_f \approx r\mu_s = (5 \text{ mm})(0.4) = 2$ mm, so that summing moments about P_2 gives

$\zeta + \Sigma M_{p_2} = 0;$ 981 N(52 mm) $- T(48$ mm$) = 0$
$$T = 1063 \text{ N} = 1.06 \text{ kN} \qquad Ans.$$

If a more exact analysis is used, then $\phi_s = \tan^{-1} 0.4 = 21.8°$. Thus, the radius of the friction circle would be $r_f = r \sin \phi_s = 5 \sin 21.8° = 1.86$ mm. Therefore,

$\zeta + \Sigma M_{P_2} = 0;$
     981 N(50 mm $+ 1.86$ mm) $- T(50$ mm $- 1.86$ mm$) = 0$
$$T = 1057 \text{ N} = 1.06 \text{ kN} \qquad Ans.$$

Part (b)   When the block is lowered, the resultant force **R** acting on the shaft passes through point P_3, as shown in Fig. 8–25c. Summing moments about this point yields

$\zeta + \Sigma M_{P_3} = 0;$ 981 N(48 mm) $- T(52$ mm$) = 0$
$$T = 906 \text{ N} \qquad Ans.$$

Impending motion

r_f ϕ_s

P_1 P_2

981 N   **R**   T

52 mm 48 mm

(b)

ϕ_s

r_f Impending motion

P_3

981 N   **R**   T

48 mm 52 mm

(c)

Fig. 8–25

*8.8 Rolling Resistance

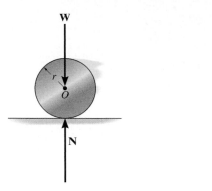

W

r
O

N

Rigid surface of contact

(a)

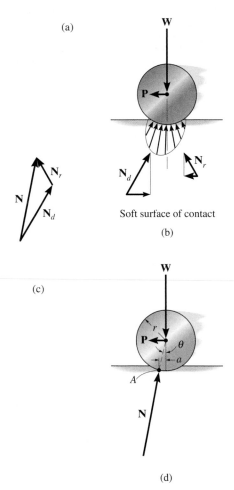

W

P

N_d

N_r

Soft surface of contact

(b)

N_r

N

N_d

(c)

W

P θ
a

A

N

(d)

Fig. 8–26

If a *rigid* cylinder of weight **W** rolls at constant velocity along a *rigid* surface, the normal force exerted by the surface on the cylinder acts at the tangent point of contact, as shown in Fig. 8–26*a*. Under these conditions, provided the cylinder does not encounter frictional resistance from the air, motion will continue indefinitely. Actually, however, no materials are perfectly rigid, and therefore the reaction of the surface on the cylinder consists of a distribution of normal pressure. For example, consider the cylinder to be made of a very hard material, and the surface on which it rolls to be relatively soft. Due to its weight, the cylinder compresses the surface underneath it, Fig. 8–26*b*. As the cylinder rolls, the surface material in front of the cylinder *retards* the motion since it is being *deformed*, whereas the material in the rear is *restored* from the deformed state and therefore tends to *push* the cylinder forward. The normal pressures acting on the cylinder in this manner are represented in Fig. 8–26*b* by their resultant forces N_d and N_r. Unfortunately, the magnitude of the force of *deformation*, N_d, and its horizontal component is *always greater* than that of *restoration*, N_r, and consequently a horizontal driving force **P** must be applied to the cylinder to maintain the motion. Fig. 8–26*b*.*

Rolling resistance is caused primarily by this effect, although it is also, to a smaller degree, the result of surface adhesion and relative micro-sliding between the surfaces of contact. Because the actual force **P** needed to overcome these effects is difficult to determine, a simplified method will be developed here to explain one way engineers have analyzed this phenomenon. To do this, we will consider the resultant of the *entire* normal pressure, $N = N_d + N_r$, acting on the cylinder, Fig. 8–26*c*. As shown in Fig. 8–26*d*, this force acts at an angle θ with the vertical. To keep the cylinder in equilibrium, i.e., rolling at a constant rate, it is necessary that **N** be *concurrent* with the driving force **P** and the weight **W**. Summing moments about point A gives $Wa = P(r \cos \theta)$. Since the deformations are generally very small in relation to the cylinder's radius, $\cos \theta \approx 1$; hence,

$$Wa \approx Pr$$

or

$$\boxed{P \approx \frac{Wa}{r}} \qquad (8\text{–}11)$$

*Actually, the deformation force N_d causes *energy* to be stored in the material as its magnitude is increased, whereas the restoration force N_r, as its magnitude is decreased, allows some of this energy to be released. The remaining energy is *lost* since it is used to heat up the surface, and if the cylinder's weight is very large, it accounts for permanent deformation of the surface. Work must be done by the horizontal force **P** to make up for this loss.

The distance a is termed the *coefficient of rolling resistance,* which has the dimension of length. For instance, $a \approx 0.5$ mm for a wheel rolling on a rail, both of which are made of mild steel. For hardened steel ball bearings on steel, $a \approx 0.1$ mm. Experimentally, though, this factor is difficult to measure, since it depends on such parameters as the rate of rotation of the cylinder, the elastic properties of the contacting surfaces, and the surface finish. For this reason, little reliance is placed on the data for determining a. The analysis presented here does, however, indicate why a heavy load (W) offers greater resistance to motion (P) than a light load under the same conditions. Furthermore, since the ratio Wa/r is generally very small compared to $\mu_k W$, the force needed to *roll* the cylinder over the surface will be much less than that needed to *slide* the cylinder across the surface. Hence, the analysis indicates why roller or ball bearings are often used to minimize the frictional resistance between moving parts.

Rolling resistance of railroad wheels on the rails is small since steel is very stiff. By comparison, the rolling resistance of the wheels of a tractor in a wet field is very large.

E X A M P L E 8.12

A 10-kg steel wheel shown in Fig. 8–27a has a radius of 100 mm and rests on an inclined plans made of wood. If θ is increased so that the wheel begins to roll-down the incline with constant velocity when $\theta = 1.2°$, determine the coefficient of rolling resistance.

(a)

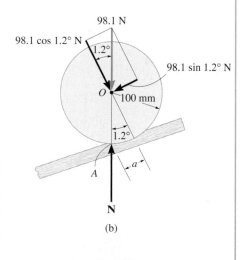

(b)

Fig. 8–27

Solution

As shown on the free-body diagram, Fig. 8–27b, when the wheel has impending motion, the normal reaction **N** acts at point A defined by the dimension a. Resolving the weight into components parallel and perpendicular to the incline, and summing moments about point A, yields (approximately)

$$\zeta + \Sigma M_A = 0; \quad 98.1 \cos 1.2° \, N(a) - 98.1 \sin 1.2° \, N(100 \text{ mm}) = 0$$

Solving, we obtain

$$a = 2.09 \text{ mm} \qquad \qquad Ans.$$

PROBLEMS

8-107. The collar bearing uniformly supports an axial force of $P = 2000$ N. If the coefficient of static friction is $\mu_s = 0.3$, determine the torque M required to overcome friction.

***8-108.** The collar bearing uniformly supports an axial force of $P = 2000$ N. If a torque of $M = 4$ N · m is applied to the shaft and causes it to rotate at constant velocity, determine the coefficient of kinetic friction at the surface of contact.

Probs. 8–107/108

8-109. The *disk clutch* is used in standard transmissions of automobiles. If four springs are used to force the two plates A and B together, determine the force in each spring required to transmit a moment of $M = 900$ N · m across the plates. The coefficient of static friction between A and B is $\mu_s = 0.3$.

Prob. 8–109

8-110. The annular ring bearing is subjected to a thrust of 4000 N. If $\mu_s = 0.35$, determine the torque M that must be applied to overcome friction.

Prob. 8–110

8-111. The floor-polishing machine rotates at a constant angular velocity. If it has a weight of 400 N ($\approx$ 40 kg), determine the couple forces F the operator must apply to the handles to hold the machine stationary. The coefficient of kinetic friction between the floor and brush is $\mu_k = 0.3$. Assume the brush exerts a uniform pressure on the floor.

Prob. 8–111

*8-112. The plate clutch consists of a flat plate A that slides over the rotating shaft S. The shaft is fixed to the driving plate gear B. If the gear C, which is in mesh with B, is subjected to a torque of $M = 0.8 \text{ N} \cdot \text{m}$, determine the smallest force P, that must be applied via the control arm, to stop the rotation. The coefficient of static friction between the plates A and D is $\mu_s = 0.4$. Assume the bearing pressure between A and D to be uniform.

Prob. 8–112

8-113. A tube has a total weight of 1000 N ($\approx$ 100 kg), length $l = 3.2$ m, and radius $= 0.3$ m. If it rests in sand for which the coefficient of static friction it is $\mu_s = 0.23$, determine the torque **M** needed to turn it. Assume that the pressure distribution along the length of the tube is defined by $p = p_0 \sin \theta$. For the solution it is necessary to determine p_0, the peak pressure, in terms of the weight and tube dimensions.

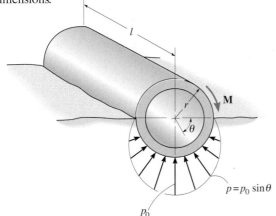

Prob. 8–113

8-114. Because of wearing at the edges, the pivot bearing is subjected to a conical pressure distribution at its surface of contact. Determine the torque M required to overcome friction and turn the shaft, which supports an axial force **P**. The coefficient of static friction is μ_s. For the solution, it is necessary to determine the peak pressure p_0 in terms of P and the bearing radius R.

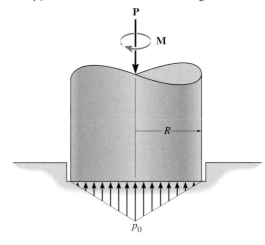

Prob. 8–114

8-115. The pivot bearing is subjected to a pressure distribution at its surface of contact which varies as shown. If the coefficient of static friction is μ, determine the torque M required to overcome friction if the shaft supports an axial force **P**.

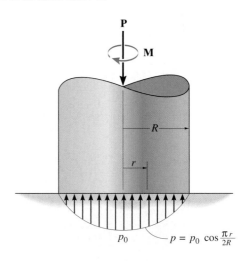

Prob. 8–115

*8-116. The tractor is used to push the 7 500-N (≈ 750-kg) pipe. To do this it must overcome the frictional forces at the ground, caused by sand. Assuming that the sand exerts a pressure on the bottom of the pipe as shown, and the coefficient of static friction between the pipe and the sand is $\mu_s = 0.3$, determine the force required to push the pipe forward. Also, determine the peak pressure p_0.

Prob. 8–116

8-117. A 200-mm diameter post is driven 3 m into sand for which $\mu_s = 0.3$. If the normal pressure acting *completely around the post* varies linearly with depth as shown, determine the frictional torque **M** that must be overcome to rotate the post.

Prob. 8–117

8-118. A pulley having a diameter of 80 mm and mass of 1.25 kg is supported loosely on a shaft having a diameter of 20 mm. Determine the torque M that must be applied to the pulley to cause it to rotate with constant motion. The coefficient of kinetic friction between the shaft and pulley is $\mu_k = 0.4$. Also calculate the angle θ which the normal force at the point of contact makes with the horizontal. The shaft itself cannot rotate.

Prob. 8–118

8-119. The pulley has a radius of 3 cm and fits loosely on the 0.5-cm-diameter shaft. If the loadings acting on the belt cause the pulley to rotate with constant angular velocity, determine the frictional force between the shaft and the pulley and compute the coefficient of kinetic friction. The pulley weighs 18 N.

*8-120. The pulley has a radius of 3 cm and fits loosely on the 0.5-cm-diameter shaft. If the loadings acting on the belt cause the pulley to rotate with constant angular velocity, determine the frictional force between the shaft and the pulley and compute the coefficient of kinetic friction. Neglect the weight of the pulley.

Probs. 8–119/120

8-121. Determine the tension **T** in the belt needed to overcome the tension of 200 N created on the other side. Also, what are the normal and frictional components of force developed on the collar bushing? The coefficient of static friction is $\mu_s = 0.21$.

8-122. If a tension force $T = 215$ N is required to pull the 200-N force around the collar bushing, determine the coefficient of static friction at the contacting surface. The belt does not slip on the collar.

Probs. 8–121/122

8-123. A disk having an outer diameter of 120 mm fits loosely over a fixed shaft having a diameter of 30 mm. If the coefficient of static friction between the disk and the shaft is $\mu_s = 0.15$ and the disk has a mass of 50 kg, determine the smallest vertical force **F** acting on the rim which must be applied to the disk to cause it to slip over the shaft.

Prob. 8–123

*8-124.** The weight of the body on the tibiotalar joint J is 500 N (≈ 50 kg). If the radius of curvature of the talus surface of the ankle is 35 mm, and the coefficient of static friction between the bones is $\mu_s = 0.1$, determine the force T developed in the Achilles tendon necessary to rotate the joint.

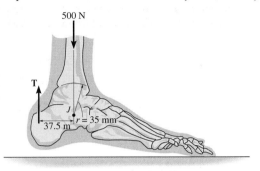

Prob. 8–124

8-125. The collar fits *loosely* around a fixed shaft that has a radius of 50 mm. If the coefficient of kinetic friction between the shaft and the collar is $\mu_k = 0.3$, determine the force P on the horizontal segment of the belt so that the collar rotates counterclockwise with a constant angular velocity. Assume that the belt does not slip on the collar; rather, the collar slips on the shaft. Neglect the weight and thickness of the belt and collar. The radius, measured from the center of the collar to the mean thickness of the belt, is 56.25 mm.

8-126. The collar fits *loosely* around a fixed shaft that has a radius of 50 mm. If the coefficient of kinetic friction between the shaft and the collar is $\mu_k = 0.3$, determine the force P on the horizontal segment of the belt so that the collar rotates clockwise with a constant angular velocity. Assume that the belt does not slip on the collar; rather, the collar slips on the shaft. Neglect the weight and thickness of the belt and collar. The radius, measured from the center of the collar to the mean thickness of the belt, is 56.25 mm.

Probs. 8–125/126

8-127. The connecting rod is attached to the piston by a 20-mm-diameter pin at B and to the crank shaft by a 50-mm-diameter bearing A. If the piston is moving downwards, and the coefficient of static friction at these points is $\mu_s = 0.2$, determine the radius of the friction circle at each connection.

*****8-128.** The connecting rod is attached to the piston by a 20-mm-diameter pin at B and to the crank shaft by a 50-mm-diameter bearing A. If the piston is moving upwards, and the coefficient of static friction at these points is $\mu_s = 0.3$, determine the radius of the friction circle at each connection.

Probs. 8–127/128

8-129. The vehicle has a weight of 13 kN ($\approx$ 1300 kg) and center of gravity at G. Determine the horizontal force **P** that must be applied to overcome the rolling resistance of the wheels. The coefficient of rolling resistance is 12 mm. The tires have a diameter of 0.825 m.

Prob. 8–129

8-130. The hand cart has wheels with a diameter of 80 mm. If a crate having a mass of 500 kg is placed on the cart so that each wheel carries an equal load, determine the horizontal force P that must be applied to the handle to overcome the rolling resistance. The coefficient of rolling resistance is 2 mm. Neglect the mass of the cart.

Prob. 8–130

8-131. The cylinder is subjected to a load that has a weight W. If the coefficients of rolling resistance for the cylinder's top and bottom surfaces are a_A and a_B, respectively, show that a force having a magnitude of $P = [W(a_A + a_B)]/2\,r$ is required to move the load and thereby roll the cylinder forward. Neglect the weight of the cylinder.

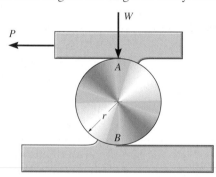

Prob. 8–131

*****8-132.** A large crate having a mass of 200 kg is moved along the floor using a series of 150-mm-diameter rollers for which the coefficient of rolling resistance is 3 mm at the ground and 7 mm at the bottom surface of the crate. Determine the horizontal force **P** needed to push the crate forward at a constant speed. *Hint:* Use the result of Prob. 8-131.

Prob. 8–132

8-133. The lawn roller weighs 1500 N ($\approx$ 150 kg). If the rod *BA* is held at an angle of 30° from the horizontal and the coefficient of rolling resistance for the roller is 50 mm, determine the force **F** needed to push the roller at constant speed. Neglect friction developed at the axle and assume that the resultant force acting on the handle is applied along *BA*.

225 mm

30°

A

F

B

Prob. 8–133

CHAPTER REVIEW

- ***Dry Friction.*** Frictional forces exist at rough surfaces of contact. They act on a body so as to oppose the motion or tendency of motion of the body. A static friction force approaches a maximum value of $F_s = \mu_s N$, where μ_s is the *coefficient of static friction*. In this case motion between the contacting surfaces is about to impend. If slipping occurs, then the friction force remains essentially constant and equal to a value of $F_k = \mu_k N$. Here μ_k is the *coefficient of kinetic friction*. The solution of a problem involving friction requires first drawing the free-body diagram of the body. If the unknowns cannot be determined strictly from the equations of equilibrium, and the possibility of slipping can occur, then the friction equation should be applied at the appropriate points of contact in order to complete the solution. It may also be possible for slender objects to tip over, and this situation should also be investigated.

- ***Wedges, Screws, Belts, and Bearings.*** A frictional analysis of these objects can be performed by applying the friction equation at the points of contact and then using the equations of equilibrium to relate the frictional force to the other external forces acting on the object. By combining the resulting equations, the force of friction can then be eliminated from the analysis, so that the force needed to overcome the effects of friction can be determined.

- ***Rolling Resistance.*** The resistance of a wheel to roll over a surface is caused by *deformation* between the two materials of contact. This effect causes the resultant normal force acting on the rolling body to be inclined so that it provides a component that acts in the opposite direction of the force causing the motion. The effect is characterized using the *coefficient of rolling resistance*, which is determined from experiment.

REVIEW PROBLEMS

8-134. A single force **P** is applied to the handle of the drawer. If friction is neglected at the bottom side and the coefficient of static friction along the sides is $\mu_s = 0.4$, determine the largest spacing s between the symmetrically placed handles so that the drawer does not bind at the corners A and B when the force **P** is applied to one of the handles.

Prob. 8–134

8-135. The truck has a mass of 1.25 Mg and a center of mass at G. Determine the greatest load it can pull if (a) the truck has rear-wheel drive while the front wheels are free to roll, and (b) the truck has four-wheel drive. The coefficient of static friction between the wheels and the ground is $\mu_s = 0.5$, and between the crate and the ground, it is $\mu_s' = 0.4$.

*8-136. Solve Prob. 8-135 if the truck and crate are traveling up a 10° incline.

Probs. 8–135/136

8-137. The cam or short link is pinned at A and is used to hold mops or brooms against a wall. If the coefficient of static friction between the broomstick and the cam is $\mu_s = 0.2$, determine if it is possible to support the broom having a weight W. The surface at B is smooth. Neglect the weight of the cam.

Prob. 8–137

8-138. The carton clamp on the forklift has a coefficient of static friction of $\mu_s = 0.5$ with any cardboard carton, whereas a cardboard carton has a coefficient of static friction of $\mu_s' = 0.4$ with any other cardboard carton. Compute the smallest horizontal force P the clamp must exert on the sides of a carton so that two cartons A and B each weighing 150 N ($\approx$ 15 kg) can be lifted. What smallest clamping force P' is required to lift three 150 N ($\approx$ 15 kg) cartons? The third carton C is placed between A and B.

Prob. 8–138

8-139. The tractor pulls on the fixed tree stump. Determine the torque that must be applied by the engine to the rear wheels to cause them to slip. The front wheels are free to roll. The tractor weighs 17 500 N ($\approx$ 1750 kg) and has a center of gravity at G. The coefficient of static friction between the rear wheels and the ground is $\mu_s = 0.5$.

*8-140. The tractor pulls on the fixed tree stump. If the coefficient of static friction between the rear wheels and the ground is $\mu_s = 0.6$, determine if the rear wheels slip or the front wheels lift off the ground as the engine provides torque to the rear wheels. What is the torque needed to cause the motion? The front wheels are free to roll. The tractor weighs 12 500 N ($\approx$ 1250 kg) and has a center of gravity at G.

Probs. 8–139/140

DESIGN PROJECTS

8–1D DESIGN OF A ROPE-AND-PULLEY SYSTEM FOR PULLING A CRATE UP AN INCLINE.

A large 300-kg packing crate is to be hoisted up the 25° incline. The coefficient of static friction between the incline and the crate is $\mu_s = 0.5$, and the coefficient of kinetic friction is $\mu_k = 0.4$. Using a system of ropes and pulleys, design a method that will allow a single worker to pull the crate up the ramp. Pulleys can be attached to any point on the wall AB. Assume the worker can exert a maximum (comfortable) pull of 200 N on a rope. Submit a drawing of your design and a force analysis to show how it operates. Estimate the material cost required for its construction. Assume rope costs $0.75/m and a pulley costs $1.80.

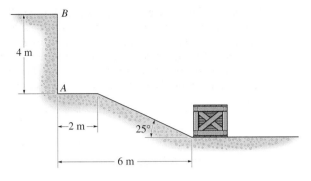

Fig. 8–1D

8–2D DESIGN OF A DEVICE FOR LIFTING STAINLESS-STEEL PIPES.

Stainless-steel pipes are stacked vertically in a manufacturing plant and are to be moved by an overhead crane from one point to another. The pipes have inner diameters ranging from 100 mm $\leq d \leq$ 250 mm and the maximum mass of any pipe is 500 kg. Design a device that can be connected to the hook and used to lift each pipe. The device should be made of structural steel and should be able to grip the pipe only from its inside surface, since the outside surface is required not to be scratched or damaged. Assume the smallest coefficient of static friction between the two steels is $\mu_s = 0.25$. Submit a scaled drawing of your device, along with a brief explanation of how it works based on a force analysis.

Fig. 8–2D

8–3D DESIGN OF A TOOL USED TO TURN PLASTIC PIPE.

PVC plastic is often used for sewer pipe. If the outer diameter of any pipe ranges from 100 mm $\leq$ d $\leq$ 200 mm, design a tool that can be used by a worker in order to turn the pipe when it is subjected to a maximum anticipated ground resistance of 120 N · m. The device is to be made of steel and should be designed so that it does not cut into the pipe and leave any significant marks on its surface. Assume a worker can apply a maximum (comfortable) force of 200 N, and take the minimum coefficient of static friction between the PVC and the steel to be $\mu_s = 0.35$. Submit a scaled drawing of the device, and a brief paragraph to explain how it works based on a force analysis.

Fig. 8–3D

When a pressure vessel is designed, it is important to be able to determine the center of gravity of its component parts, calculate its volume and surface area, and reduce three-dimensional distributed loadings to their resultants. These topics are discussed in this chapter.

Center of Gravity and Centroid

- To discuss the concept of the center of gravity, center of mass, and the centroid.
- To show how to determine the location of the center of gravity and centroid for a system of discrete particles and a body of arbitrary shape.
- To use the theorems of Pappus and Guldinus for finding the area and volume for a surface of revolution.
- To present a method for finding the resultant of a general distributed loading and show how it applies to finding the resultant of a fluid.

9.1 Center of Gravity and Center of Mass for a System of Particles

Center of Gravity. The *center of gravity* G is a point which locates the resultant weight of a system of particles. To show how to determine this point consider the system of n particles fixed within a region of space as shown in Fig. 9–1a. The weights of the particles comprise a system of parallel forces* which can be replaced by a single (equivalent) resultant weight having the defined point G of application. To find the $\bar{x}, \bar{y}, \bar{z}$ coordinates of G, we must use the principles outlined in Sec. 4.9.

---

*This is not true in the exact sense, since the weights are not parallel to each other; rather they are all *concurrent* at the earth's center. Furthermore, the acceleration of gravity g is actually different for each particle since it depends on the distance from the earth's center to the particle. For all practical purposes, however, both of these effects can generally be neglected.

(a)

(b)

Fig. 9–1

This requires that the resultant weight be equal to the total weight of all n particles; that is,

$$W_R = \Sigma W$$

The sum of the moments of the weights of all the particles about the x, y, and z axes is then equal to the moment of the resultant weight about these axes. Thus, to determine the $\bar{x}$ coordinate of G, we can sum moments about the y axis. This yields

$$\bar{x} W_R = \tilde{x}_1 W_1 + \tilde{x}_2 W_2 + \cdots + \tilde{x}_n W_n$$

Likewise, summing moments about the x axis, we can obtain the $\bar{y}$ coordinate; i.e.,

$$\bar{y} W_R = \tilde{y}_1 W_1 + \tilde{y}_2 W_2 + \cdots + \tilde{y}_n W_n$$

Although the weights do not produce a moment about the z axis, we can obtain the $\bar{z}$ coordinate of G by imagining the coordinate system, with the particles fixed in it, as being rotated $90°$ about the x (or y) axis, Fig. 9–1b. Summing moments about the x axis, we have

$$\bar{z} W_R = \tilde{z}_1 W_1 + \tilde{z}_2 W_2 + \cdots + \tilde{z}_n W_n$$

We can generalize these formulas, and write them symbolically in the form

$$\bar{x} = \frac{\Sigma \tilde{x} W}{\Sigma W} \qquad \bar{y} = \frac{\Sigma \tilde{y} W}{\Sigma W} \qquad \bar{z} = \frac{\Sigma \tilde{z} W}{\Sigma W} \tag{9–1}$$

Here

$\bar{x}, \bar{y}, \bar{z}$ represent the coordinates of the center of gravity G of the system of particles.

$\tilde{x}, \tilde{y}, \tilde{z}$ represent the coordinates of each particle in the system.

ΣW is the resultant sum of the weights of all the particles in the system.

These equations are easily remembered if it is kept in mind that they simply represent a balance between the sum of the moments of the weights of each particle of the system and the moment of the *resultant* weight for the system.

Center of Mass.   To study problems concerning the motion of *matter* under the influence of force, i.e., dynamics, it is necessary to locate a point called the *center of mass*. Provided the acceleration due to gravity g for every particle is constant, then $W = mg$. Substituting into Eqs. 9–1 and canceling g from both the numerator and denominator yields

$$\bar{x} = \frac{\Sigma \tilde{x} m}{\Sigma m} \qquad \bar{y} = \frac{\Sigma \tilde{y} m}{\Sigma m} \qquad \bar{z} = \frac{\Sigma \tilde{z} m}{\Sigma m} \tag{9–2}$$

By comparison, then, the location of the center of gravity *coincides* with that of the center of mass.* Recall, however, that particles have "weight" only when under the influence of a gravitational attraction, whereas the center of mass is independent of gravity. For example, it would be meaningless to define the center of gravity of a system of particles representing the planets of our solar system, while the center of mass of this system is important.

9.2 Center of Gravity, Center of Mass, and Centroid for a Body

Center of Gravity. A rigid body is composed of an infinite number of particles, and so if the principles used to determine Eqs. 9–1 are applied to the system of particles composing a rigid body, it becomes necessary to use integration rather than a discrete summation of the terms. Considering the arbitrary particle located at $(\tilde{x}, \tilde{y}, \tilde{z})$ and having a weight dW, Fig. 9–2, the resulting equations are

$$\overline{x} = \frac{\int \tilde{x} \, dW}{\int dW} \qquad \overline{y} = \frac{\int \tilde{y} \, dW}{\int dW} \qquad \overline{z} = \frac{\int \tilde{z} \, dW}{\int dW} \qquad (9\text{–}3)$$

In order to apply these equations properly, the differential weight dW must be expressed in terms of its associated volume dV. If γ represents the *specific weight* of the body, measured as a weight per unit volume, then $dW = \gamma \, dV$ and therefore

$$\overline{x} = \frac{\int_V \tilde{x}\gamma \, dV}{\int_V \gamma \, dV} \qquad \overline{y} = \frac{\int_V \tilde{y}\gamma \, dV}{\int_V \gamma \, dV} \qquad \overline{z} = \frac{\int_V \tilde{z}\gamma \, dV}{\int_V \gamma \, dV} \qquad (9\text{–}4)$$

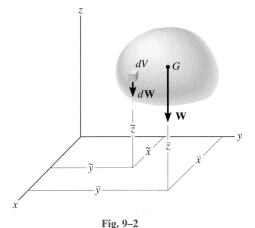

Fig. 9–2

Here integration must be performed throughout the entire volume of the body.

Center of Mass. The *density* ρ, or mass per unit volume, is related to γ by the equation $\gamma = \rho g$, where g is the acceleration due to gravity. Substituting this relationship into Eqs. 9–4 and canceling g from both the numerators and denominators yields similar equations (with ρ replacing γ) that can be used to determine the body's *center of mass*.

---

*This is true as long as the gravity field is assumed to have the same magnitude and direction everywhere. That assumption is appropriate for most engineering applications, since gravity does not vary appreciably between, for instance, the bottom and the top of a building.

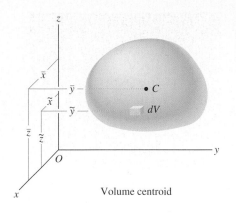

Volume centroid

Fig. 9–3

Centroid. The *centroid C* is a point which defines the *geometric center* of an object. Its location can be determined from formulas similar to those used to determine the body's center of gravity or center of mass. In particular, if the material composing a body is uniform or *homogeneous*, the *density or specific weight* will be *constant* throughout the body, and therefore this term will factor out of the integrals and *cancel* from both the numerators and denominators of Eqs. 9–4. The resulting formulas define the centroid of the body since they are independent of the body's weight and instead depend only on the body's geometry. Three specific cases will be considered.

Volume. If an object is subdivided into volume elements dV, Fig. 9–3, the location of the centroid $C(\overline{x}, \overline{y}, \overline{z})$ for the volume of the object can be determined by computing the "moments" of the elements about each of the coordinate axes. The resulting formulas are

$$\overline{x} = \frac{\int_V \tilde{x}\, dV}{\int_V dV} \qquad \overline{y} = \frac{\int_V \tilde{y}\, dV}{\int_V dV} \qquad \overline{z} = \frac{\int_V \tilde{z}\, dV}{\int_V dV} \qquad (9\text{–}5)$$

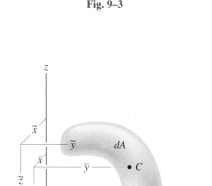

Area centroid

Fig. 9–4

Area. In a similar manner, the centroid for the surface area of an object, such as a plate or shell, Fig. 9–4, can be found by subdividing the area into differential elements dA and computing the "moments" of these area elements about each of the coordinate axes, namely,

$$\overline{x} = \frac{\int_A \tilde{x}\, dA}{\int_A dA} \qquad \overline{y} = \frac{\int_A \tilde{y}\, dA}{\int_A dA} \qquad \overline{z} = \frac{\int_A \tilde{z}\, dA}{\int_A dA} \qquad (9\text{–}6)$$

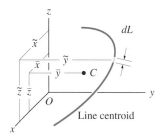

Line centroid

Fig. 9–5

Line. If the geometry of the object, such as a thin rod or wire, takes the form of a line, Fig. 9–5, the balance of moments of the differential elements dL about each of the coordinate axes yields

$$\overline{x} = \frac{\int_L \tilde{x}\, dL}{\int_L dL} \qquad \overline{y} = \frac{\int_L \tilde{y}\, dL}{\int_L dL} \qquad \overline{z} = \frac{\int_L \tilde{z}\, dL}{\int_L dL} \qquad (9\text{–}7)$$

Remember that when applying Eqs. 9–4 through 9–7 it is best to choose a coordinate system that simplifies as much as possible the equation used to describe the object's boundary. For example, polar coordinates are generally appropriate for areas having circular boundaries. Also, the terms $\tilde{x}, \tilde{y}, \tilde{z}$ in the equations refer to the "moment arms" or coordinates of the *center of gravity or centroid for the differential element* used. If possible, this differential element should be chosen such that it has a differential size or thickness in only *one direction*. When this is done, only a single integration is required to cover the entire region.

Symmetry. The *centroids* of some shapes may be partially or completely specified by using conditions of *symmetry*. In cases where the shape has an axis of symmetry, the centroid of the shape will lie along that axis. For example, the centroid C for the line shown in Fig. 9–6 must lie along the y axis since for every elemental length dL at a distance $+\tilde{x}$ to the right of the y axis there is an identical element at a distance $-\tilde{x}$ to the left. The total moment for all the elements about the axis of symmetry will therefore cancel; i.e., $\int \tilde{x}\, dL = 0$ (Eq. 9–7), so that $\bar{x} = 0$. In cases where a shape has two or three axes of symmetry, it follows that the centroid lies at the intersection of these axes, Fig. 9–7 and Fig. 9–8.

Integration must be used to determine the location of the center of gravity of this goal post.

Fig. 9–6

Fig. 9–7

Fig. 9–8

IMPORTANT POINTS

- The centroid represents the geometric center of a body. This point coincides with the center of mass or the center of gravity only if the material composing the body is uniform or homogeneous.

- Formulas used to locate the center of gravity or the centroid simply represent a balance between the sum of moments of all the parts of the system and the moment of the "resultant" for the system.

- In some cases the centroid is located at a point that is not on the object, as in the case of a ring, where the centroid is at its center. Also, this point will lie on any axis of symmetry for the body.

PROCEDURE FOR ANALYSIS

The center of gravity or centroid of an object or shape can be determined by single integrations using the following procedure.

Differential Element.

- Select an appropriate coordinate system, specify the coordinate axes, and then choose a differential element for integration.
- For lines the element dL is represented as a differential line segment.
- For areas the element dA is generally a rectangle having a finite length and differential width.
- For volumes the element dV is either a circular disk having a finite radius and differential thickness, or a shell having a finite length and radius and a differential thickness.
- Locate the element at an arbitrary point (x, y, z) on the curve that defines the shape.

Size and Moment Arms.

- Express the length dL, area dA, or volume dV of the element in terms of the coordinates of the curve used to define the geometric shape.
- Determine the coordinates or moment arms $\tilde{x}, \tilde{y}, \tilde{z}$ for the centroid or center of gravity of the element.

Integrations.

- Substitute the formulations for $\tilde{x}, \tilde{y}, \tilde{z}$ and dL, dA, or dV into the appropriate equations (Eqs. 9–4 through 9–7) and perform the integrations.[*]
- Express the function in the integrand in terms of the *same variable as the differential thickness of the element* in order to perform the integration.
- The limits of the integral are defined from the two extreme locations of the element's differential thickness, so that when the elements are "summed" or the integration performed, the entire region is covered.

[*]Formulas for integration are given in Appendix A.

EXAMPLE 9.1

Locate the centroid of the rod bent into the shape of a parabolic arc, shown in Fig. 9–9.

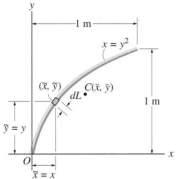

Fig. 9–9

Solution

Differential Element. The differential element is shown in Fig. 9–9. It is located on the curve at the *arbitrary point* (x, y).

Area and Moment Arms. The differential length of the element dL can be expressed in terms of the differentials dx and dy by using the Pythagorean theorem.

$$dL = \sqrt{(dx)^2 + (dy)^2} = \sqrt{\left(\frac{dx}{dy}\right)^2 + 1}\, dy$$

Since $x = y^2$, then $dx/dy = 2y$. Therefore, expressing dL in terms of y and dy, we have

$$dL = \sqrt{(2y)^2 + 1}\, dy$$

The centroid is located at $\tilde{x} = x$, $\tilde{y} = y$.

Integrations. Applying Eqs. 9–7 and integrating with respect to y using the formulas in Appendix A, we have

$$\bar{x} = \frac{\displaystyle\int_L \tilde{x}\, dL}{\displaystyle\int_L dL} = \frac{\displaystyle\int_0^1 x\sqrt{4y^2 + 1}\, dy}{\displaystyle\int_0^1 \sqrt{4y^2 + 1}\, dy} = \frac{\displaystyle\int_0^1 y^2\sqrt{4y^2 + 1}\, dy}{\displaystyle\int_0^1 \sqrt{4y^2 + 1}\, dy}$$

$$= \frac{0.6063}{1.479} = 0.410 \text{ m} \qquad\qquad Ans.$$

$$\bar{y} = \frac{\displaystyle\int_L \tilde{y}\, dL}{\displaystyle\int_L dL} = \frac{\displaystyle\int_0^1 y\sqrt{4y^2 + 1}\, dy}{\displaystyle\int_0^1 \sqrt{4y^2 + 1}\, dy} = \frac{0.8484}{1.479} = 0.574 \text{ m} \quad Ans.$$

E X A M P L E **9.2**

Locate the centroid of the circular wire segment shown in Fig. 9–10.

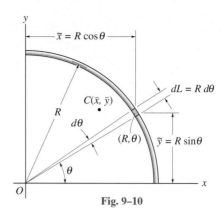

Fig. 9–10

Solution

Polar coordinates will be used to solve this problem since the arc is circular.

Differential Element. A differential circular arc is selected as shown in the figure. This element intersects the curve at (R, θ).

Length and Moment Arm. The differential length of the element is $dL = R\, d\theta$, and its centroid is located at $\tilde{x} = R \cos \theta$ and $\tilde{y} = R \sin \theta$.

Integrations. Applying Eqs. 9–7 and integrating with respect to θ, we obtain

$$\bar{x} = \frac{\displaystyle\int_L \tilde{x}\, dL}{\displaystyle\int_L dL} = \frac{\displaystyle\int_0^{\pi/2} (R \cos \theta) R\, d\theta}{\displaystyle\int_0^{\pi/2} R\, d\theta} = \frac{R^2 \displaystyle\int_0^{\pi/2} \cos \theta\, d\theta}{R \displaystyle\int_0^{\pi/2} d\theta} = \frac{2R}{\pi} \quad Ans.$$

$$\bar{y} = \frac{\displaystyle\int_L \tilde{y}\, dL}{\displaystyle\int_L dL} = \frac{\displaystyle\int_0^{\pi/2} (R \sin \theta) R\, d\theta}{\displaystyle\int_0^{\pi/2} R\, d\theta} = \frac{R^2 \displaystyle\int_0^{\pi/2} \sin \theta\, d\theta}{R \displaystyle\int_0^{\pi/2} d\theta} = \frac{2R}{\pi} \quad Ans.$$

E X A M P L E **9.3**

Determine the distance $\overline{y}$ from the x axis to the centroid of the area of the triangle shown in Fig. 9–11.

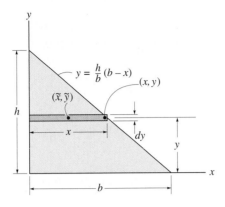

Fig. 9–11

Solution

Differential Element. Consider a rectangular element having thickness dy which intersects the boundary at (x, y), Fig. 9–11.

Area and Moment Arms. The area of the element is $dA = x\,dy = \frac{b}{h}(h - y)\,dy$, and its centroid is located a distance $\tilde{y} = y$ from the x axis.

Integrations. Applying the second of Eqs. 9–6 and integrating with respect to y yields

$$\overline{y} = \frac{\displaystyle\int_A \tilde{y}\,dA}{\displaystyle\int_A dA} = \frac{\displaystyle\int_0^h y\frac{b}{h}(h - y)\,dy}{\displaystyle\int_0^h \frac{b}{h}(h - y)\,dy} = \frac{\frac{1}{6}bh^2}{\frac{1}{2}bh}$$

$$= \frac{h}{3} \qquad\qquad\qquad \textit{Ans.}$$

E X A M P L E 9.4

Locate the centroid for the area of a quarter circle shown in Fig. 9–12a.

Solution I

Differential Element. Polar coordinates will be used since the boundary is circular. We choose the element in the shape of a *triangle*, Fig. 9–12a. (Actually the shape is a circular sector; however, neglecting higher-order differentials, the element becomes triangular.) The element intersects the curve at point (R, θ).

Area and Moment Arms. The area of the element is

$$dA = \tfrac{1}{2}(R)(R\,d\theta) = \frac{R^2}{2}d\theta$$

and using the results of Example 9.3, the centroid of the (triangular) element is located at $\tilde{x} = \tfrac{2}{3}R\cos\theta$, $\tilde{y} = \tfrac{2}{3}R\sin\theta$.

Integrations. Applying Eqs. 9–6 and integrating with respect to θ, we obtain

$$\overline{x} = \frac{\displaystyle\int_A \tilde{x}\,dA}{\displaystyle\int_A dA} = \frac{\displaystyle\int_0^{\pi/2}\left(\frac{2}{3}R\cos\theta\right)\frac{R^2}{2}d\theta}{\displaystyle\int_0^{\pi/2}\frac{R^2}{2}d\theta}$$

$$= \frac{\left(\dfrac{2}{3}R\right)\displaystyle\int_0^{\pi/2}\cos\theta\,d\theta}{\displaystyle\int_0^{\pi/2}d\theta} = \frac{4R}{3\pi} \qquad Ans.$$

$$\overline{y} = \frac{\displaystyle\int_A \tilde{y}\,dA}{\displaystyle\int_A dA} = \frac{\displaystyle\int_0^{\pi/2}\left(\frac{2}{3}R\sin\theta\right)\frac{R^2}{2}d\theta}{\displaystyle\int_0^{\pi/2}\frac{R^2}{2}d\theta}$$

$$= \frac{\left(\dfrac{2}{3}R\right)\displaystyle\int_0^{\pi/2}\sin\theta\,d\theta}{\displaystyle\int_0^{\pi/2}d\theta} = \frac{4R}{3\pi} \qquad Ans.$$

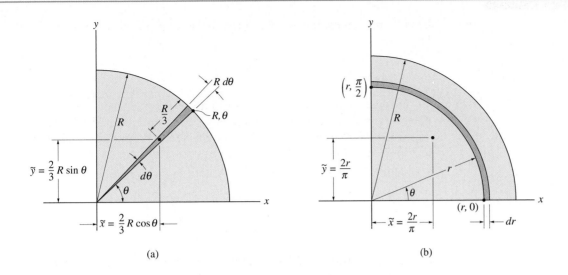

Fig. 9–12

Solution II

Differential Element.   The differential element may be chosen in the form of a *circular arc* having a thickness dr as shown in Fig. 9–12b. The element intersects the axes at points $(r, 0)$ and $(r, \pi/2)$.

Area and Moment Arms.   The area of the element is $dA = (2\pi r/4)\, dr$. Since the centroid of a 90° circular arc was determined in Example 9.2, then for the element $\widetilde{x} = 2r/\pi$, $\widetilde{y} = 2r/\pi$.

Integrations.   Using Eqs. 9–6 and integrating with respect to r, we obtain

$$\overline{x} = \frac{\displaystyle\int_A \widetilde{x}\, dA}{\displaystyle\int_A dA} = \frac{\displaystyle\int_0^R \frac{2r}{\pi}\left(\frac{2\pi r}{4}\right) dr}{\displaystyle\int_0^R \frac{2\pi r}{4}\, dr} = \frac{\displaystyle\int_0^R r^2\, dr}{\displaystyle\frac{\pi}{2}\int_0^R r\, dr} = \frac{4}{3}\frac{R}{\pi} \qquad Ans.$$

$$\overline{y} = \frac{\displaystyle\int_A \widetilde{y}\, dA}{\displaystyle\int_A dA} = \frac{\displaystyle\int_0^R \frac{2r}{\pi}\left(\frac{2\pi r}{4}\right) dr}{\displaystyle\int_0^R \frac{2\pi r}{4}\, dr} = \frac{\displaystyle\int_0^R r^2\, dr}{\displaystyle\frac{\pi}{2}\int_0^R r\, dr} = \frac{4}{3}\frac{R}{\pi} \qquad Ans.$$

E X A M P L E 9.5

Locate the centroid of the area shown in Fig. 9–13a.

Solution I

Differential Element. A differential element of thickness dx is shown in Fig. 9–13a. The element intersects the curve at the *arbitrary point* (x, y), and so it has a height y.

Area and Moment Arms. The area of the element is $dA = y\,dx$, and its centroid is located at $\widetilde{x} = x$, $\widetilde{y} = y/2$.

Integrations. Applying Eqs. 9–6 and integrating with respect to x yields

$$\overline{x} = \frac{\displaystyle\int_A \widetilde{x}\,dA}{\displaystyle\int_A dA} = \frac{\displaystyle\int_0^1 xy\,dx}{\displaystyle\int_0^1 y\,dx} = \frac{\displaystyle\int_0^1 x^3\,dx}{\displaystyle\int_0^1 x^2\,dx} = \frac{0.250}{0.333} = 0.75 \text{ m} \qquad Ans.$$

$$\overline{y} = \frac{\displaystyle\int_A \widetilde{y}\,dA}{\displaystyle\int_A dA} = \frac{\displaystyle\int_0^1 (y/2)y\,dx}{\displaystyle\int_0^1 y\,dx} = \frac{\displaystyle\int_0^1 (x^2/2)x^2\,dx}{\displaystyle\int_0^1 x^2\,dx} = \frac{0.100}{0.333} = 0.3 \text{ m } Ans.$$

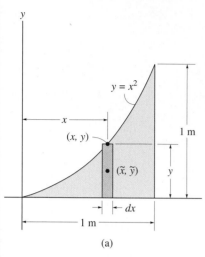

(a)

Solution II

Differential Element. The differential element of thickness dy is shown in Fig. 9–13b. The element intersects the curve at the *arbitrary point* (x, y), and so it has a length $(1 - x)$.

Area and Moment Arms. The area of the element is $dA = (1 - x)\,dy$, and its centroid is located at

$$\widetilde{x} = x + \left(\frac{1 - x}{2}\right) = \frac{1 + x}{2}, \quad \widetilde{y} = y$$

(b)

Fig. 9–13

Integrations. Applying Eqs. 9–6 and integrating with respect to y, we obtain

$$\widetilde{x} = \frac{\displaystyle\int_A \widetilde{x}\,dA}{\displaystyle\int_A dA} = \frac{\displaystyle\int_0^1 [(1 + x)/2](1 - x)\,dy}{\displaystyle\int_0^1 (1 - x)\,dy} = \frac{\dfrac{1}{2}\displaystyle\int_0^1 (1 - y)\,dy}{\displaystyle\int_0^1 (1 - \sqrt{y})\,dy} = \frac{0.250}{0.333} = 0.75 \text{ m} \qquad Ans.$$

$$\overline{y} = \frac{\displaystyle\int_A \widetilde{y}\,dA}{\displaystyle\int_A dA} = \frac{\displaystyle\int_0^1 y(1 - x)\,dy}{\displaystyle\int_0^1 (1 - x)\,dy} = \frac{\displaystyle\int_0^1 (y - y^{3/2})\,dy}{\displaystyle\int_0^1 (1 - \sqrt{y})\,dy} = \frac{0.100}{0.333} = 0.3 \text{ m} \qquad Ans.$$

E X A M P L E   **9.6**

Locate the $\bar{x}$ centroid of the shaded area bounded by the two curves $y = x$ and $y = x^2$, Fig. 9–14.

Solution I

Differential Element. A differential element of thickness dx is shown in Fig. 9–14a. The element intersects the curves at *arbitrary points* (x, y_1) and (x, y_2), and so it has a height $(y_2 - y_1)$.

Area and Moment Arm. The area of the element is $dA = (y_2 - y_1)\,dx$, and its centroid is located at $\tilde{x} = x$.

Integration. Applying Eq. 9–6, we have

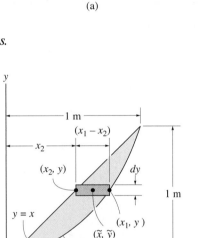

(a)

$$\bar{x} = \frac{\displaystyle\int_A \tilde{x}\,dA}{\displaystyle\int_A dA} = \frac{\displaystyle\int_0^1 x(y_2 - y_1)\,dx}{\displaystyle\int_0^1 (y_2 - y_1)\,dx} = \frac{\displaystyle\int_0^1 x(x - x^2)\,dx}{\displaystyle\int_0^1 (x - x^2)\,dx} = \frac{\frac{1}{12}}{\frac{1}{6}} = 0.5 \text{ m} \quad Ans.$$

Solution II

Differential Element. A differential element having a thickness dy is shown in Fig. 9–14b. The element intersects the curves at the *arbitrary points* (x_2, y) and (x_1, y), and so it has a length $(x_1 - x_2)$.

Area and Moment Arm. The area of the element is $dA = (x_1 - x_2)\,dy$, and its centroid is located at

$$\tilde{x} = x_2 + \frac{x_1 - x_2}{2} = \frac{x_1 + x_2}{2}$$

Integration. Applying Eq. 9–6, we have

Fig. 9–14

$$\bar{x} = \frac{\displaystyle\int_A \tilde{x}\,dA}{\displaystyle\int_A dA} = \frac{\displaystyle\int_0^1 [(x_1 + x_2)/2](x_1 - x_2)\,dy}{\displaystyle\int_0^1 (x_1 - x_2)\,dy} = \frac{\displaystyle\int_0^1 [(\sqrt{y} + y)/2](\sqrt{y} - y)\,dy}{\displaystyle\int_0^1 (\sqrt{y} - y)\,dy}$$

$$= \frac{\frac{1}{2}\displaystyle\int_0^1 (y - y^2)\,dy}{\displaystyle\int_0^1 (\sqrt{y} - y)\,dy} = \frac{\frac{1}{12}}{\frac{1}{6}} = 0.5 \text{ m} \qquad\qquad Ans.$$

E X A M P L E   **9.7**

Locate the $\bar{y}$ centroid for the paraboloid of revolution, which is generated by revolving the shaded area shown in Fig. 9–15a about the y axis.

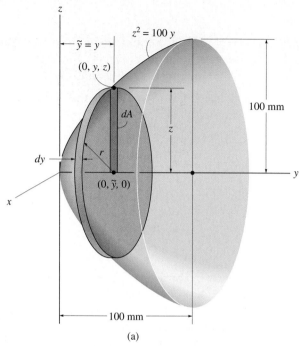

(a)

Fig. 9–15

Solution I

Differential Element. An element having the shape of a *thin disk* is chosen, Fig. 9–15a. This element has a thickness dy. In this "disk" method of analysis, the element of planar area, dA, is always taken *perpendicular* to the axis of revolution. Here the element intersects the generating curve at the *arbitrary point* $(0, y, z)$, and so its radius is $r = z$.

Area and Moment Arm. The volume of the element is $dV = (\pi z^2)\, dy$, and its centroid is located at $\tilde{y} = y$.

Integration Applying the second of Eqs. 9–5 and integrating with respect to y yields

$$\bar{y} = \frac{\displaystyle\int_V \tilde{y}\, dV}{\displaystyle\int_V dV} = \frac{\displaystyle\int_0^{100} y(\pi z^2)\, dy}{\displaystyle\int_0^{100} (\pi z^2)\, dy} = \frac{100\pi \displaystyle\int_0^{100} y^2\, dy}{100\pi \displaystyle\int_0^{100} y\, dy} = 66.7 \text{ mm} \quad Ans.$$

(b)

Solution II

Differential Element. As shown in Fig. 9–15b, the volume element can be chosen in the form of a *thin cylindrical shell*, where the shell's thickness is dz. In this "shell" method of analysis, the element of planar area, dA, is always taken *parallel* to the axis of revolution. Here the element intersects the generating curve at point $(0, y, z)$, and so the radius of the shell is $r = z$.

Area and Moment Arm. The volume of the element is $dV = 2\pi r\, dA = 2\pi z(100 - y)\, dz$, and its centroid is located at $\tilde{y} = y + (100 - y)/2 = (100 + y)/2$.

Integrations. Applying the second of Eqs. 9–5 and integrating with respect to z yields

$$\overline{y} = \frac{\displaystyle\int_V \tilde{y}\, dV}{\displaystyle\int_V dV} = \frac{\displaystyle\int_0^{100} [(100 + y)/2]2\pi z(100 - y)\, dz}{\displaystyle\int_0^{100} 2\pi z(100 - y)\, dz}$$

$$= \frac{\pi \displaystyle\int_0^{100} z(10^4 - 10^{-4}z^4)\, dz}{2\pi \displaystyle\int_0^{100} z(100 - 10^{-2}z^2)\, dz} = 66.7 \text{ mm} \qquad\qquad \textit{Ans.}$$

E X A M P L E **9.8**

Determine the location of the center of mass of the cylinder shown in Fig. 9–16a if its density varies directly with its distance from the base, i.e., $\rho = 200z$ kg/m³.

(a)

Solution

For reasons of material symmetry,

$$\bar{x} = \bar{y} = 0 \qquad \qquad Ans.$$

Differential Element. A disk element of radius 0.5 m and thickness dz is chosen for integration, Fig. 9–16a, since the *density of the entire element is constant* for a given value of z. The element is located along the z axis at the *arbitrary point* $(0, 0, z)$.

Volume and Moment Arm. The volume of the element is $dV = \pi(0.5)^2\, dz$, and its centroid is located at $\tilde{z} = z$.

Integrations. Using an equation similar to the third of Eqs. 9–4 and integrating with respect to z, noting that $\rho = 200z$, we have

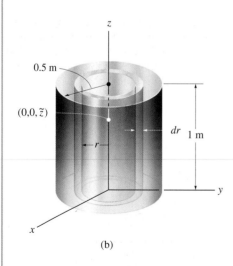

(b)

Fig. 9–16

$$\bar{z} = \frac{\displaystyle\int_v \tilde{z}\rho\, dV}{\displaystyle\int_v \rho\, dV} = \frac{\displaystyle\int_0^1 z(200z)\pi(0.5)^2\, dz}{\displaystyle\int_0^1 (200z)\pi(0.5)^2\, dz}$$

$$= \frac{\displaystyle\int_0^1 z^2\, dz}{\displaystyle\int_0^1 z\, dz} = 0.667 \text{ m} \qquad \qquad Ans.$$

Note: It is not possible to use a shell element for integration such as shown in Fig. 9–16b since the density of the material composing the shell would *vary* along the shell's height and hence the location of $\tilde{z}$ for the element cannot be specified.

PROBLEMS

9-1. Determine the distance $\bar{x}$ to the center of mass of the homogeneous rod bent into the shape shown. If the rod has a mass per unit length of 0.5 kg/m, determine the reactions at the fixed support O.

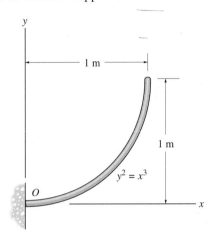

Prob. 9–1

9-3. Locate the center of mass of the homogeneous rod bent into the shape of a circular arc.

Prob. 9–3

9-2. Determine the location $(\bar{x}, \bar{y})$ of the centroid of the wire.

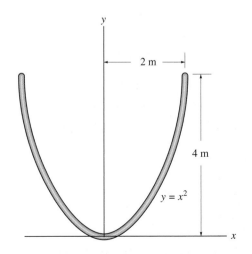

Prob. 9–2

*9-4. Locate the center of gravity $\bar{x}$ of the homogeneous rod bent in the form of a semicircular arc. The rod has a weight per unit length of 5 N/m. Also, determine the horizontal reaction at the smooth support B and the x and y components of reaction at the pin A.

Prob. 9–4

9-5. Determine the distance $\bar{x}$ to the center of gravity of the homogeneous rod bent into the parabolic shape. If the rod has a weight per unit length of 0.5 N/m, determine the reactions at the fixed support O.

9-6. Determine the distance $\bar{y}$ to the center of gravity of the homogeneous rod bent into the parabolic shape.

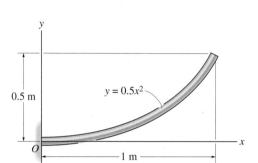

Probs. 9–5/6

9-7. Locate the centroid of the parabolic area.

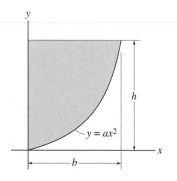

Prob. 9–7

*9-8.** Locate the centroid $(\bar{x}, \bar{y})$ of the shaded area.

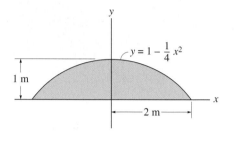

Prob. 9–8

9-9. Locate the centroid of the shaded area.

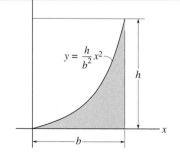

Prob. 9–9

9-10. Locate the centroid $\bar{x}$ of the shaded area.

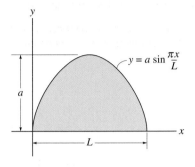

Prob. 9–10

9-11. Locate the centroid $\bar{x}$ of the shaded area.

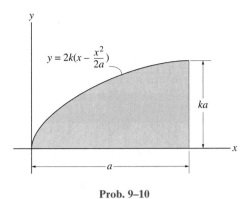

Prob. 9–11

***9-12.** Locate the centroid of the shaded area.

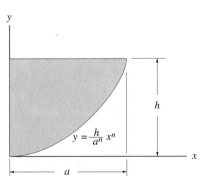

Prob. 9–12

9-13. The plate has a thickness of 2.5 cm and a specific weight of $\gamma = 80 \text{ kN/m}^3$. Determine the location of its center of gravity. Also, find the tension in each of the cords used to support it.

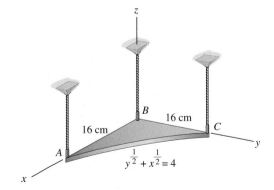

Prob. 9–13

9-14. Locate the centroid $\bar{y}$ of the shaded area.

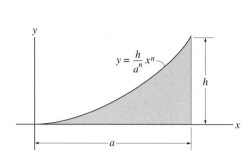

Prob. 9–14

9-15. Locate the centroid of the shaded area.

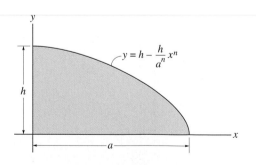

Prob. 9–15

*9-16. Locate the centroid of the shaded area bounded by the parabola and the line $y = a$.

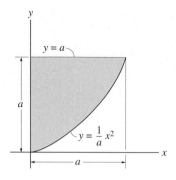

Prob. 9–16

9-17. Locate the centroid of the quarter elliptical area.

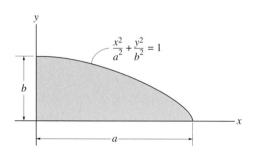

Prob. 9–17

9-18. Locate the centroid $(\bar{x}, \bar{y})$ of the shaded area.

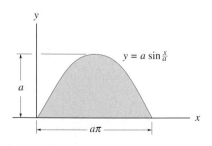

Prob. 9–18

9-19. Locate the centroid of the shaded area.

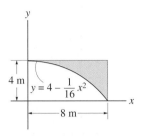

Prob. 9–19

*■**9-20.** Locate the centroid $\bar{x}$ of the shaded area. Solve the problem by evaluating the integrals using Simpson's rule.

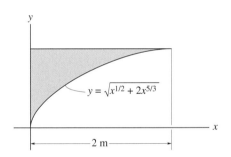

Prob. 9–20

■**9-21.** Locate the centroid $\bar{y}$ of the shaded area. Solve the problem by evaluating the integrals using Simpson's rule.

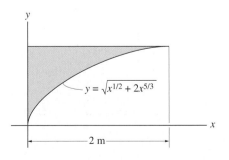

Prob. 9–21

9-22. The steel plate is 0.3 m thick and has a density of 7850 kg/m³. Determine the location of its center of mass. Also compute the reactions at the pin and roller support.

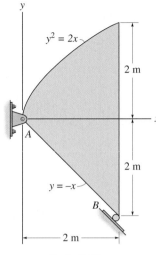

Prob. 9–22

9-23. Locate the centroid $\bar{x}$ of the shaded area.

*■**9-24.** Locate the centroid $\bar{y}$ of the shaded area.

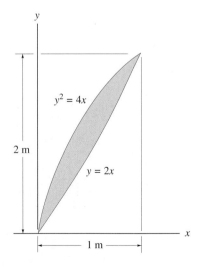

Probs. 9–23/24

9-25. Locate the centroid $\bar{x}$ of the shaded area.

9-26. Locate the centroid $\bar{y}$ of the shaded area.

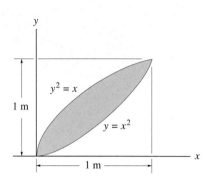

Probs. 9–25/26

9-29. Locate the centroid $\bar{x}$ of the shaded area.

9-30. Locate the centroid $\bar{y}$ of the shaded area.

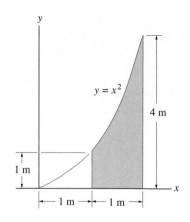

Probs. 9–29/30

9-27. Locate the centroid $\bar{x}$ of the shaded area.

*9-28. Locate the centroid $\bar{y}$ of the shaded area.

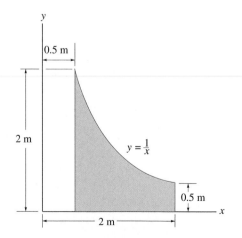

Probs. 9–27/28

9-31. Determine the location $\bar{r}$ of the centroid C of the cardioid, $r = a(1 - \cos \theta)$.

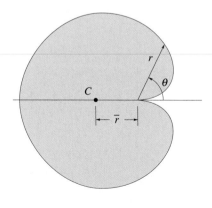

Prob. 9–31

*9-32. Locate the centroid of the ellipsoid of revolution.

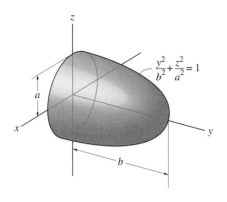

Prob. 9–32

9-34. Locate the centroid $\bar{z}$ of the hemisphere.

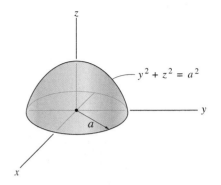

Prob. 9–34

9-33. Locate the center of gravity of the volume. The material is homogeneous.

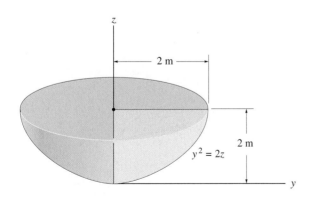

Prob. 9–33

9-35. Locate the centroid of the solid.

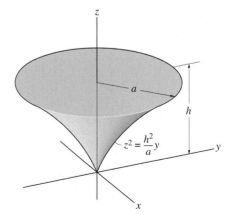

Prob. 9–35

*9-36. Locate the centroid of the quarter-cone.

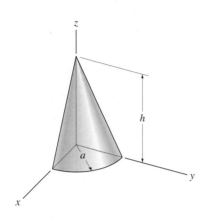

Prob. 9–36

9-38. Locate the centroid $\bar{z}$ of the right-elliptical cone.

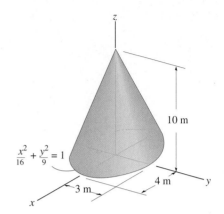

$$\frac{x^2}{16} + \frac{y^2}{9} = 1$$

10 m

3 m

4 m

Prob. 9–38

9-37. Locate the center of mass $\bar{x}$ of the hemisphere. The density of the material varies linearly from zero at the origin O to ρ_0 at the surface. *Suggestion:* Choose a hemispherical shell element for integration.

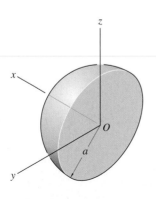

Prob. 9–37

9-39. Locate the centroid $\bar{y}$ of the paraboloid.

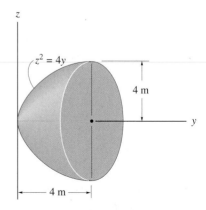

$z^2 = 4y$

4 m

4 m

Prob. 9–39

*9-40. The king's chamber of the Great Pyramid of Giza is located at its centroid. Assuming the pyramid to be a solid, prove that this point is at $\bar{z} = \frac{1}{4}h$, *Suggestion:* Use a rectangular differential plate element having a thickness dz and area $(2x)(2y)$.

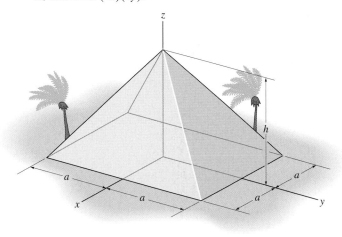

Prob. 9–40

9-41. Locate the centroid $\bar{z}$ of the frustum of the right-circular cone.

Prob. 9–41

9-42. The hemisphere of radius r is made from a stack of very thin plates such that the density varies with height $= kz$, where k is a constant. Determine its mass and the distance $\bar{z}$ to the center of mass G.

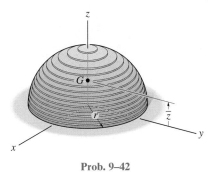

Prob. 9–42

9-43. Determine the location $\bar{z}$ of the centroid for the tetrahedron. *Suggestion:* Use a triangular "plate" element parallel to the x–y plane and of thickness dz.

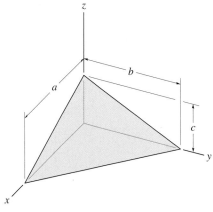

Prob. 9–43

9.3 Composite Bodies

A *composite body* consists of a series of connected "simpler" shaped bodies, which may be rectangular, triangular, semicircular, etc. Such a body can often be sectioned or divided into its composite parts and, provided the *weight* and location of the center of gravity of each of these parts are known, we can eliminate the need for integration to determine the center of gravity for the entire body. The method for doing this requires treating each composite part like a particle and following the

In order to determine the force required to tip over this concrete barrier it is first necessary to determine the location of its center of gravity.

procedure outlined in Sec. 9.1. Formulas analogous to Eqs. 9–1 result since we must account for a finite number of weights. Rewriting these formulas, we have

$$\overline{x} = \frac{\Sigma \widetilde{x} W}{\Sigma W} \qquad \overline{y} = \frac{\Sigma \widetilde{y} W}{\Sigma W} \qquad \overline{z} = \frac{\Sigma \widetilde{z} W}{\Sigma W} \qquad (9\text{–}8)$$

Here

$\overline{x}, \overline{y}, \overline{z}$ represent the coordinates of the center of gravity G of the composite body.

$\widetilde{x}, \widetilde{y}, \widetilde{z}$ represent the coordinates of the center of gravity of each composite part of the body.

ΣW is the sum of the weights of all the composite parts of the body, or simply the total weight of the body.

When the body has a *constant density or specific weight*, the center of gravity *coincides* with the centroid of the body. The centroid for composite lines, areas, and volumes can be found using relations analogous to Eqs. 9–8; however, the W's are replaced by L's, A's, and V's, respectively. Centroids for common shapes of lines, areas, shells, and volumes that often make up a composite body are given in the table on the inside back cover.

PROCEDURE FOR ANALYSIS

The location of the center of gravity of a body or the centroid of a composite geometrical object represented by a line, area, or volume can be determined using the following procedure.

Composite Parts.
- Using a sketch, divide the body or object into a finite number of composite parts that have simpler shapes.
- If a composite part has a *hole*, or a geometric region having no material, then consider the composite part without the hole and consider the hole as an *additional* composite part having *negative* weight or size.

Moment Arms.
- Establish the coordinate axes on the sketch and determine the coordinates $\widetilde{x}, \widetilde{y}, \widetilde{z}$ of the center of gravity or centroid of each part.

Summations.
- Determine $\overline{x}, \overline{y}, \overline{z}$ by applying the center of gravity equations, Eqs. 9–8, or the analogous centroid equations.
- If an object is *symmetrical* about an axis, the centroid of the object lies on this axis.

If desired, the calculations can be arranged in tabular form, as indicated in the following three examples.

EXAMPLE 9.9

Locate the centroid of the wire shown in Fig. 9–17a.

Solution

Composite Parts. The wire is divided into three segments as shown in Fig. 9–17b.

Moment Arms. The location of the centroid for each piece is determined and indicated in the figure. In particular, the centroid of segment ① is determined either by integration or by using the table on the inside back cover.

Summations. The calculations are tabulated as follows:

Segment	L (mm)	$\tilde{x}$ (mm)	$\tilde{y}$ (mm)	$\tilde{z}$ (mm)	$\tilde{x}L$ (mm²)	$\tilde{y}L$ (mm²)	$\tilde{z}L$ (mm²)
1	$\pi(60) = 188.5$	60	-38.2	0	11 310	-7200	0
2	40	0	20	0	0	800	0
3	20	0	40	-10	0	800	-200
	$\Sigma L = 248.5$				$\Sigma\tilde{x}L = 11\,310$	$\Sigma\tilde{y}L = -5600$	$\Sigma\tilde{z}L = -200$

Thus,

$$\overline{x} = \frac{\Sigma\tilde{x}L}{\Sigma L} = \frac{11310}{248.5} = 45.5 \text{ mm} \qquad Ans.$$

$$\overline{y} = \frac{\Sigma\tilde{y}L}{\Sigma L} = \frac{-5600}{248.5} = -22.5 \text{ mm} \qquad Ans.$$

$$\overline{z} = \frac{\Sigma\tilde{z}L}{\Sigma L} = \frac{-200}{248.5} = -0.805 \text{ mm} \qquad Ans.$$

(a)   (b)

Fig. 9–17

E X A M P L E **9.10**

Locate the centroid of the plate area shown in Fig. 9–18a.

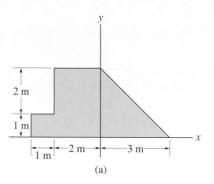

(a)

Fig. 9–18

Solution

Composite Parts. The plate is divided into three segments as shown in Fig. 9–18b. Here the area of the small rectangle ③ is considered "negative" since it must be subtracted from the larger one ②.

Moment Arms. The centroid of each segment is located as indicated in the figure. Note that the $\widetilde{x}$ coordinates of ② and ③ are *negative*.

Summations. Taking the data from Fig. 9–18b, the calculations are tabulated as follows:

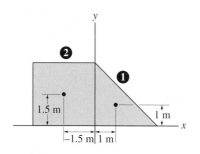

(b)

Segment	A (m^2)	$\widetilde{x}$ (m)	$\widetilde{y}$ (m)	$\widetilde{x}A$ (m^3)	$\widetilde{y}A$ (m^3)
1	$\frac{1}{2}(3)(3) = 4.5$	1	1	4.5	4.5
2	$(3)(3) = 9$	−1.5	1.5	−13.5	13.5
3	$-(2)(1) = -2$	−2.5	2	5	−4
	$\Sigma A = 11.5$			$\Sigma \widetilde{x}A = -4$	$\Sigma \widetilde{y}A = 14$

Thus,

$$\overline{x} = \frac{\Sigma \widetilde{x}A}{\Sigma A} = \frac{-4}{11.5} = -0.348 \text{ m} \qquad \textit{Ans.}$$

$$\overline{y} = \frac{\Sigma \widetilde{y}A}{\Sigma A} = \frac{14}{11.5} = 1.22 \text{ m} \qquad \textit{Ans.}$$

EXAMPLE 9.11

Locate the center of mass of the composite assembly shown in Fig. 9–19a. The conical frustum has a density of $\rho_c = 8 \text{ Mg/m}^3$, and the hemisphere has a density of $\rho_h = 4 \text{ Mg/m}^3$. There is a 25-mm radius cylindrical hole in the center.

Solution

Composite Parts. The assembly can be thought of as consisting of four segments as shown in Fig. 9–19b. For the calculations, ③ and ④ must be considered as "negative" volumes in order that the four segments, when added together, yield the total composite shape shown in Fig. 9–19a.

Moment Arm. Using the table on the inside back cover, the computations for the centroid $\tilde{z}$ of each piece are shown in the figure.

Summations. Because of *symmetry*, note that

$$\bar{x} = \bar{y} = 0 \qquad \textit{Ans.}$$

Since $W = mg$ and g is constant, the third of Eqs. 9–8 becomes $\bar{z} = \Sigma\tilde{z}m/\Sigma m$. The mass of each piece can be computed from $m = \rho V$ and used for the calculations. Also, $1 \text{ Mg/m}^3 = 10^{-6} \text{ kg/mm}^3$, so that

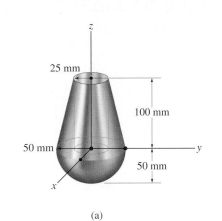

25 mm

100 mm

50 mm

50 mm

(a)

Fig. 9–19

Segment	m (kg)	$\tilde{z}$ (mm)	$\tilde{z}m$ (kg·mm)
1	$8(10^{-6})(\frac{1}{3})\pi(50)^2(200) = 4.189$	50	209.440
2	$4(10^{-6})(\frac{2}{3})\pi(50)^3 = 1.047$	-18.75	-19.635
3	$-8(10^{-6})(\frac{1}{3})\pi(25)^2(100) = -0.524$	$100 + 25 = 125$	-65.450
4	$-8(10^{-6})\pi(25)^2(100) = -1.571$	50	-78.540
	$\Sigma m = 3.141$		$\Sigma\tilde{z}m = 45.815$

Thus,

$$\tilde{z} = \frac{\Sigma\tilde{z}m}{\Sigma m} = \frac{45.815}{3.141} = 14.6 \text{ mm} \qquad \textit{Ans.}$$

200 mm

①

$\frac{200 \text{ mm}}{4} = 50 \text{ mm}$

50 mm

50 mm

②

$-\frac{3}{8}(50) = -18.75$ mm

③

100 mm

$\frac{100 \text{ mm}}{4} = 25$ mm

25 mm

100 mm

25 mm

100 mm

50 mm

④

(b)

PROBLEMS

*9-44. Locate the center of gravity G of the five particles with respect to the origin O.

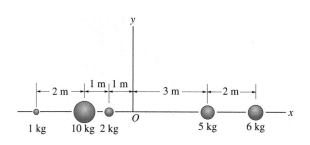

Prob. 9–44

9-45. Locate the center of mass $(\bar{x}, \bar{y})$ of the four particles.

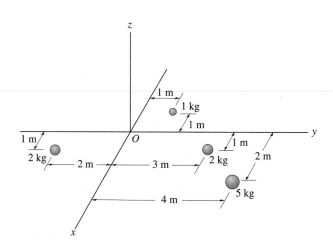

Prob. 9–45

9-46. Locate the centroid $(\bar{x}, \bar{y})$ of the uniform wire bent in the shape shown.

Prob. 9–46

9-47. The steel and aluminum plate assembly is bolted together and fastened to the wall. Each plate has a constant width in the z direction of 200 mm and thickness of 20 mm. If the density of A and B is $\rho_s = 7.85 \text{ Mg/m}^3$, and for C, $\rho_{al} = 2.71 \text{ Mg/m}^3$, determine the location $\bar{x}$ of the center of mass. Neglect the size of the bolts.

Prob. 9–47

***9-48.** The truss is made from five members, each having a length of 4 m and a mass of 7 kg/m. If the mass of the gusset plates at the joints and the thickness of the members can be neglected, determine the distance d to where the hoisting cable must be attached, so that the truss does not tip (rotate) when it is lifted.

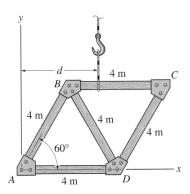

Prob. 9–48

9-49. Locate the centroid for the wire. Neglect the thickness of the material and slight bends at the corners.

Prob. 9–49

9-50. Locate the centroid ($\bar{x}$, $\bar{y}$) of the metal cross section. Neglect the thickness of the material and slight bends at the corners.

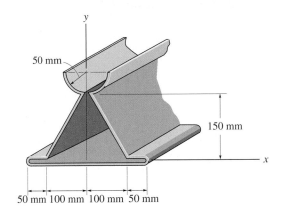

Prob. 9–50

9-51. The three members of the frame each have a weight per unit length of 4 kN/m. Locate the position ($\bar{x}$, $\bar{y}$) of the center of gravity. Neglect the size of the pins at the joints and the thickness of the members. Also, calculate the reactions at the fixed support A.

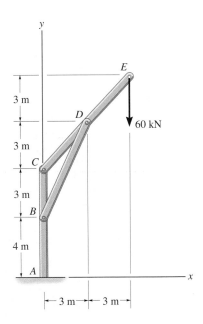

Prob. 9–51

*9-52. Each of the three members of the frame has a mass per unit length of 6 kg/m. Locate the position $(\bar{x}, \bar{y})$ of the center of gravity. Neglect the size of the pins at the joints and the thickness of the members. Also, calculate the reactions at the pin A and roller E.

Prob. 9–52

9-54. The gravity wall is made of concrete. Determine the location $(\bar{x}, \bar{y})$ of the center of gravity G for the wall.

Prob. 9–54

9-53. Determine the location $\bar{y}$ of the centroid of the beam's cross-sectional area. Neglect the size of the corner welds at A and B for the calculation.

Prob. 9–53

9-55. An aluminum strut has a cross section referred to as a deep hat. Locate the centroid $\bar{y}$ of its area. Each segment has a thickness of 10 mm.

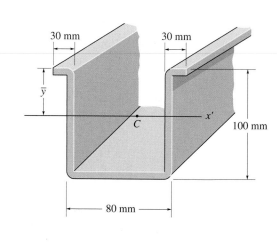

Prob. 9–55

*9-56. Locate the centroid $\bar{y}$ for the cross-sectional area of the angle.

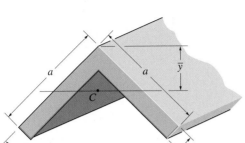

Prob. 9–56

9-58. Determine the location $(\bar{x}, \bar{y})$ of the centroid C of the area.

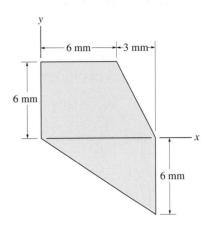

Prob. 9–58

9-57. Determine the location $\bar{y}$ of the centroidal axis $\bar{x}\,\bar{x}$ of the beam's cross-sectional area. Neglect the size of the corner welds at A and B for the calculation.

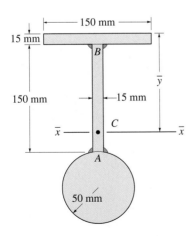

Prob. 9–57

9-59. Locate the centroid $(\bar{x}, \bar{y})$ for the angle's cross-sectional area.

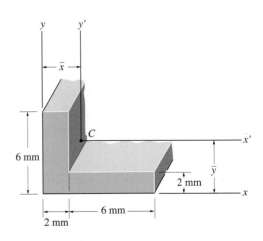

Prob. 9–59

*9-60. The wooden table is made from a square board having a weight of 75 N ($\approx$ 7.5 kg). Each of the legs weighs 10 N ($\approx$ 1 kg) and is 1 m long. Determine how high its center of gravity is from the floor. Also, what is the angle, measured from the horizontal, through which its top surface can be tilted on two of its legs before it begins to overturn? Neglect the thickness of each leg.

Prob. 9–60

9-61. Locate the centroid $\bar{y}$ of the cross-sectional area of the beam.

Prob. 9–61

9-62. Determine the location $\bar{x}$ of the centroid C of the shaded area which is part of a circle having a radius r.

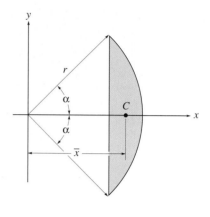

Prob. 9–62

9-63. Locate the centroid $\bar{y}$ of the channel's cross-sectional area.

Prob. 9–63

*9-64. Locate the centroid $\bar{y}$ of the cross-sectional area of the beam constructed from a channel and a plate. Assume all corners are square and neglect the size of the weld at A.

Prob. 9–64

9-66. Locate the centroid $\bar{y}$ of the concrete beam having the tapered cross section shown.

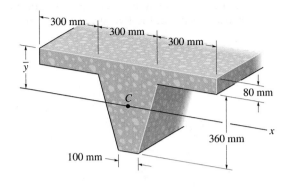

Prob. 9–66

9-65. Locate the centroid $(\bar{x}, \bar{y})$ of the member's cross-sectional area.

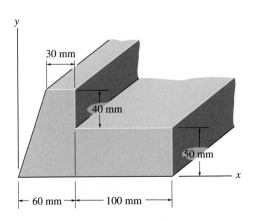

Prob. 9–65

9-67. Locate the centroid $\bar{y}$ of the beam's cross-section built up from a channel and a wide-flange beam.

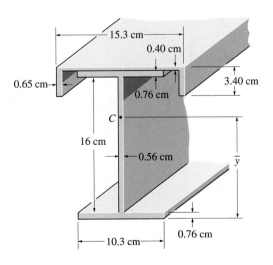

Prob. 9–67

*9-68. Locate the centroid $\bar{y}$ of the bulb-tee cross section.

Prob. 9–68

9-69. Determine the distance h to which a 100-mm-diameter hole must be bored into the base of the cone so that the center of mass of the resulting shape is located at $\bar{z} = 115$ mm. The material has a density of 8 Mg/m³.

9-70. Determine the distance $\bar{z}$ to the centroid of the shape which consists of a cone with a hole of height $h = 50$ mm bored into its base.

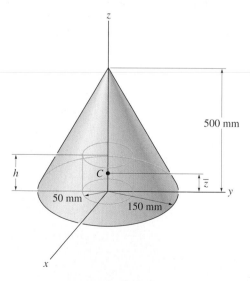

Probs. 9–69/70

9-71. The sheet metal part has the dimensions shown. Determine the location $(\bar{x}, \bar{y}, \bar{z})$ of its centroid.

*9-72. The sheet metal part has a weight per unit area of 0.01 N/cm² and is supported by the smooth rod and at C. If the cord is cut, the part will rotate about the y axis until it reaches equilibrium. Determine the equilibrium angle of tilt, measured downward from the negative x axis, that AD makes with the $-x$ axis.

Probs. 9–71/72

9-73. Determine the location $(\bar{x}, \bar{y})$ of the centroid C of the cross-sectional area for the structural member constructed from two equal-sized channels welded together as shown. Assume all corners are square. Neglect the size of the welds.

Prob. 9–73

9-74. Determine the location $(\bar{x}, \bar{y})$ of the center of gravity of the three-wheeler. The location of the center of gravity of each component and its weight are tabulated in the figure. If the three-wheeler is symmetrical with respect to the x–y plane, determine the normal reactions each of its wheels exerts on the ground.

1. Rear wheels	90 N
2. Mechanical components	425 N
3. Frame	600 N
4. Front wheel	40 N

Prob. 9–74

9-75. Major floor loadings in a shop are caused by the weights of the objects shown. Each force acts through its respective center of gravity G. Locate the center of gravity $(\bar{x}, \bar{y})$ of all these components.

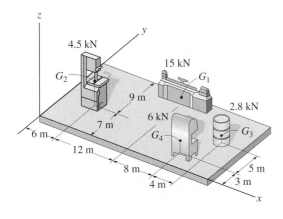

Prob. 9–75

*9-76. Locate the center of gravity of the two-block assembly. The specific weights of the materials A and B are $\gamma_A = 24$ kN/m^3 and $\gamma_B = 64$ kN/m^3, respectively.

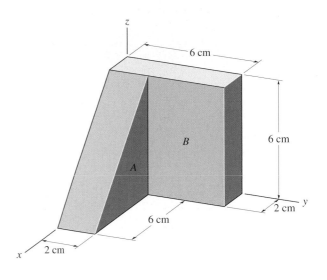

Prob. 9–76

9-77. The buoy is made from two homogeneous cones each having a radius of 1.5 m. If $h = 1.2$ m, find the distance $\bar{z}$ to the buoy's center of gravity G.

9-78. The buoy is made from two homogeneous cones each having a radius of 1.5 m. If it is required that the buoy's center of gravity G be located at $\bar{z} = 0.5$ m, determine the height h of the top cone.

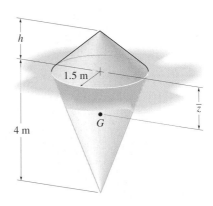

Probs. 9–77/78

9-79. Locate the centroid $\bar{z}$ of the top made from a hemisphere and a cone.

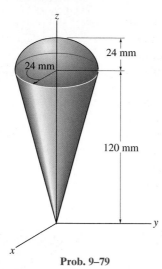

Prob. 9–79

*9-80. A triangular plate made of homogeneous material has a constant thickness which is very small. If it is folded over as shown, determine the location $\bar{y}$ of the plate's center of gravity G.

9-81. A triangular plate made of homogeneous material has a constant thickness which is very small. If it is folded over as shown, determine the location $\bar{z}$ of the plate's center of gravity G.

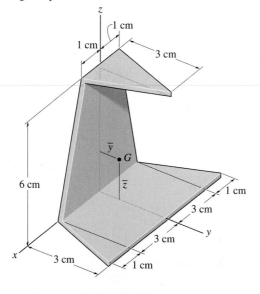

Probs. 9–80/81

9-82. Locate the center of mass $\bar{z}$ of the assembly. The material has a density of $\rho = 3 \text{ Mg/m}^3$. There is a 30-mm diameter hole bored through the center.

Prob. 9–82

9-83. The assembly consists of a 20-cm wooden dowel rod and a tight-fitting steel collar. Determine the distance $\bar{x}$ to its center of gravity if the specific weights of the materials are $\gamma_w = 24 \text{ kN/m}^3$ and $\gamma_{st} = 78 \text{ kN/m}^3$. The radii of the dowel and collar are shown.

Prob. 9–83

*9.4 Theorems of Pappus and Guldinus

The two *theorems of Pappus and Guldinus*, which were first developed by Pappus of Alexandria during the third century A.D. and then restated at a later time by the Swiss mathematician Paul Guldin or Guldinus (1577–1643), are used to find the surface area and volume of any object of revolution.

Fig. 9–20

A *surface area of revolution* is generated by revolving a *plane curve* about a nonintersecting fixed axis in the plane of the curve; whereas a *volume of revolution* is generated by revolving a *plane area* about a nonintersecting fixed axis in the plane of the area. For example, if the *line AB* shown in Fig. 9–20 is rotated about a fixed axis, it generates the *surface area* of a cone (less the area of the base); if the triangular *area ABC* shown in Fig. 9–21 is rotated about the axis, it generates the *volume* of a cone.

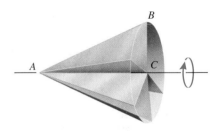

The statements and proofs of the theorems of Pappus and Guldinus follow. The proofs require that the generating curves and areas do *not* cross the axis about which they are rotated; otherwise, two sections on either side of the axis would generate areas or volumes having opposite signs and hence cancel each other.

Fig. 9–21

Surface Area. *The area of a surface of revolution equals the product of the length of the generating curve and the distance traveled by the centroid of the curve in generating the surface area.*

Proof. When a differential length dL of the curve shown in Fig. 9–22 is revolved about an axis through a distance $2\pi r$, it generates a ring having a surface area $dA = 2\pi r \, dL$. The entire surface area, generated by revolving the entire curve about the axis, is therefore $A = 2\pi \int_L r \, dL$. This equation may be simplified, however, by noting that the location r of the centroid for the line of total length L can be determined from an equation having the form of Eqs. 9–7, namely, $\int_L r \, dL = \bar{r} L$. Thus, the total surface area becomes $A = 2\pi \bar{r} L$. In general, though, if the line does not undergo a complete revolution, then,

$$A = \theta \bar{r} L \qquad (9\text{–}9)$$

Fig. 9–22

where A = surface area of revolution

θ = angle of revolution measured in radians, $\theta \leq 2\pi$

$\bar{r}$ = perpendicular distance from the axis of revolution to the centroid of the generating curve

L = length of the generating curve

Fig. 9–23

Volume. *The volume of a body of revolution equals the product of the generating area and the distance traveled by the centroid of the area in generating the volume.*

Proof. When the differential area dA shown in Fig. 9–23 is revolved about an axis through a distance $2\pi r$, it generates a ring having a volume $dV = 2\pi r\, dA$. The entire volume, generated by revolving A about the axis, is therefore $V = 2\pi \int_V r\, dA$. Here the integral can be eliminated by using an equation analogous to Eqs. 9–6, $\int_V r\, dA = \bar{r}A$, where $\bar{r}$ locates the centroid C of the generating area A. The volume becomes $V = 2\pi\bar{r}A$. In general, though,

$$\boxed{V = \theta\bar{r}A} \tag{9–10}$$

The surface area and the amount of water that can be stored in this water tank can be determined by using the theorems of Pappus and Guldinus.

where V = volume of revolution

θ = angle of revolution measured in radians, $\theta \le 2\pi$

$\bar{r}$ = perpendicular distance from the axis of revolution to the centroid of the generating area

A = generating area

Composite Shapes. We may also apply the above two theorems to lines or areas that may be composed of a series of composite parts. In this case the total surface area or volume generated is the addition of the surface areas or volumes generated by each of the composite parts. Since each part undergoes the *same* angle of revolution, θ, and the distance from the axis of revolution to the centroid of each composite part is $\tilde{r}$, then

$$A = \theta\Sigma(\tilde{r}L) \tag{9–11}$$

and

$$V = \theta\Sigma(\tilde{r}A) \tag{9–12}$$

The amount of roofing material used on this storage building can be estimated by using the theorem of Pappus and Guldinus to determine its surface area.

Application of the above theorems is illustrated numerically in the following example.

E X A M P L E   **9.12**

Show that the surface area of a sphere is $A = 4\pi R^2$ and its volume is $V = \frac{4}{3}\pi R^3$.

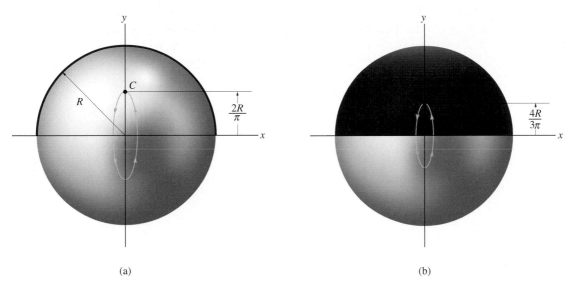

(a)   (b)

Fig. 9–24

Solution

Surface Area. The surface area of the sphere in Fig. 9–24a is generated by rotating a semicircular *arc* about the x axis. Using the table on the inside back cover, it is seen that the centroid of this arc is located at a distance $\bar{r} = 2R/\pi$ from the x axis of rotation. Since the centroid moves through an angle of $\theta = 2\pi$ rad in generating the sphere, then applying Eq. 9–9 we have

$$A = \theta \bar{r} L; \qquad A = 2\pi\left(\frac{2R}{\pi}\right)\pi R = 4\pi R^2 \qquad\qquad Ans.$$

Volume. The volume of the sphere is generated by rotating the semicircular *area* in Fig. 9–24b about the x axis. Using the table on the inside back cover to locate the centroid of the area, i.e., $\bar{r} = 4R/3\pi$, and applying Eq. 9–10, we have

$$V = \theta \bar{r} A; \qquad V = 2\pi\left(\frac{4R}{3\pi}\right)\left(\frac{1}{2}\pi R^2\right) = \frac{4}{3}\pi R^3 \qquad\qquad Ans.$$

PROBLEMS

*9-84. Using integration, determine both the area and the centroidal distance $\bar{x}$ of the shaded area. Then, using the second theorem of Pappus–Guldinus, determine the volume of the solid generated by revolving the area about the y axis.

9-86. Using integration, determine both the area and the distance $\bar{y}$ to the centroid of the shaded area. Then using the second theorem of Pappus–Guldinus, determine the volume of the solid generated by revolving the shaded area about the x axis.

Prob. 9–84

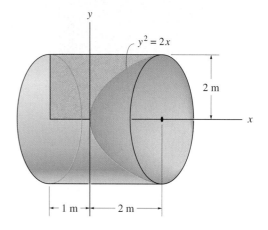

Prob. 9–86

9-85. The anchor ring is made of steel having a specific weight of $\gamma_{st} = 78$ kN/m³. Determine the surface area of the ring. The cross section is circular as shown.

9-87. A steel wheel has a diameter of 840 mm and a cross section as shown in the figure. Determine the total mass of the wheel if $\rho = 5$ Mg/m³.

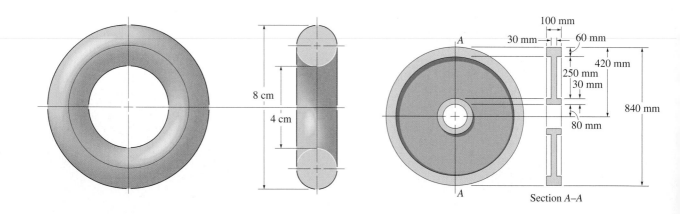

Prob. 9–85

Prob. 9–87

*9-88. The hopper is filled to its top with coal. Determine the volume of coal if the voids (air space) are 35 percent of the volume of the hopper.

1.5 m

4 m

1.2 m

0.2 m

Prob. 9–88

9-89. Sand is piled between two walls as shown. Assume the pile to be a quarter section of a cone and that 26 percent of this volume is voids (air space). Use the second theorem of Pappus–Guldinus to determine the volume of sand.

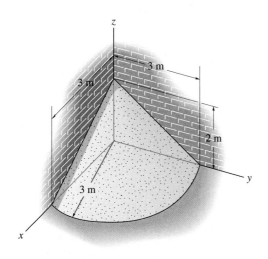

3 m

3 m

3 m

2 m

Prob. 9–89

9-90. The *rim* of a flywheel has the cross section *A-A* shown. Determine the volume of material needed for its construction.

300 mm

60 mm

20 mm

40 mm

20 mm

Section *A-A*

Prob. 9–90

9-91. The open tank is fabricated from a hemisphere and cylindrical shell. Determine the vertical reactions that each of the four symmetrically placed legs exerts on the floor if the tank contains water which is 6 m deep in the tank. The specific gravity of water is 10.4 kN/m^3. Neglect the weight of the tank.

*9-92. Determine the approximate amount of paint needed to cover the outside surface of the open tank. Assume that a litre of paint covers 8 m^2.

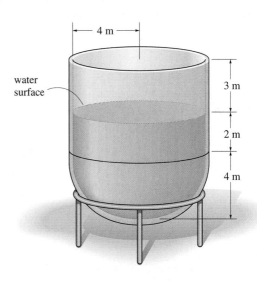

4 m

water surface

3 m

2 m

4 m

Probs. 9–91/92

9-93. Determine the volume of material needed to make the casting.

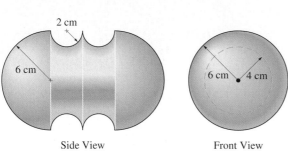

Side View   Front View

Prob. 9–93

9-94. A circular sea wall is made of concrete. Determine the total weight of the wall if the concrete has a specific weight of $\gamma_c = 24$ kN/m³.

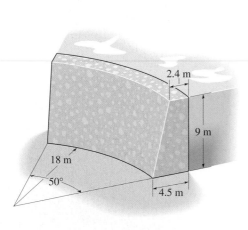

Prob. 9–94

9-95. Determine the outside surface area of the storage tank.

*9-96.** Determine the volume of the storage tank.

Probs. 9–95/96

9-97. The water-supply tank has a hemispherical bottom and cylindrical sides. Determine the weight of water in the tank when it is filled to the top at *C*. Take $\gamma_w = 10$ kN/m³.

9-98. Determine the amount of paint needed to paint the outside surface of the water-supply tank, which consists of a hemispherical bottom, cylindrical sides, and conical top. Each litre of paint can cover 6 m².

Probs. 9–97/98

9-99. The process tank is used to store liquids during manufacturing. Estimate both the volume of the tank and its surface area. The tank has a flat top and the plates from which the tank is made have negligible thickness.

Prob. 9–99

9-101. A V-belt has as inner radius of 6 cm, and a cross-sectional area as shown. Determine the volume of material used in making the V-belt.

Prob. 9–101

9-102. The full circular aluminum housing is used in an automotive brake system. The cross section is shown in the figure. Determine its weight if aluminum has a specific weight of 28 kN/m^3.

Prob. 9–102

*9-100.** Determine the height h to which liquid should be poured into the cup so that it contacts half the surface area on the inside of the cup. Neglect the cup's thickness for the calculation.

Prob. 9–100

9-103. Determine the height h to which liquid should be poured into the conical cup so that it contacts half the surface area on the inside of the cup.

100 mm

150 mm

h

Prob. 9–103

9-105. Determine the interior surface area of the brake piston. It consists of a full circular part. Its cross section is shown in the figure.

40 mm

60 mm

80 mm

20 mm

40 mm 30 mm 20 mm

Prob. 9–105

*9-104. Determine the surface area of the roof of the structure if it is formed by rotating the parabola about the y axis.

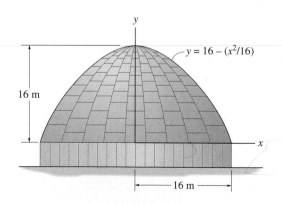

$y = 16 - (x^2/16)$

16 m

16 m

Prob. 9–104

*9.5 Resultant of a General Distributed Loading

In Sec. 4.10, we discussed the method used to simplify a distributed loading that is uniform along an axis of a rectangular surface. In this section we will generalize this method to include surfaces that have an arbitrary shape and are subjected to a variable load distribution. As a specific application, in Sec. 9.6 we will find the resultant loading acting on the surface of a body that is submerged in a fluid.

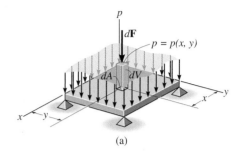

(a)

Pressure Distribution over a Surface.   Consider the flat plate shown in Fig. 9–25a, which is subjected to the loading function $p = p(x, y)$ Pa, where Pa (pascal) $= 1$ N/m^2. Knowing this function, we can determine the force dF acting on the differential area dA m^2 of the plate, located at the arbitrary point (x, y). This force magnitude is simply $dF = [p(x, y)$ N/m$^2](dA$ m$^2) = [p(x, y)\, dA]$ N. The entire loading on the plate is therefore represented as a system of *parallel forces* infinite in number and each acting on a separate differential area dA. This system will now be simplified to a single resultant force $\mathbf{F}_R$ acting through a unique point $(\bar{x}, \bar{y})$ on the plate, Fig. 9–25b.

(b)

Fig. 9–25

Magnitude of Resultant Force.   To determine the *magnitude* of $\mathbf{F}_R$, it is necessary to sum each of the differential forces dF acting over the plate's *entire surface area A*. This sum may be expressed mathematically as an integral:

$$F_R = \Sigma F; \qquad \boxed{F_R = \int_A p(x, y)\, dA = \int_V dV} \qquad (9\text{–}13)$$

Here $p(x, y)\, dA = dV$, the colored differential *volume element* shown in Fig. 9–25a. Therefore, the result indicates that the *magnitude of the resultant force is equal to the total volume under the distributed-loading diagram.*

Location of Resultant Force.   The location $(\bar{x}, \bar{y})$ of $\mathbf{F}_R$ is determined by setting the moments of $\mathbf{F}_R$ equal to the moments of all the forces dF about the respective y and x axes: From Figs. 9–25a and 9–25b, using Eq. 9–13, this results in

$$\boxed{\bar{x} = \frac{\displaystyle\int_A x p(x, y)\, dA}{\displaystyle\int_A p(x, y)\, dA} = \frac{\displaystyle\int_V x\, dV}{\displaystyle\int_V dV} \qquad \bar{y} = \frac{\displaystyle\int_A y p(x, y)\, dA}{\displaystyle\int_A p(x, y)\, dA} = \frac{\displaystyle\int_V y\, dV}{\displaystyle\int_V dV}} \qquad (9\text{–}14)$$

Hence, it can be seen that the *line of action of the resultant force passes through the geometric center or centroid of the volume under the distributed loading diagram.*

*9.6 Fluid Pressure

According to Pascal's law, a fluid at rest creates a pressure p at a point that is the *same* in *all* directions. The magnitude of p, measured as a force per unit area, depends on the specific weight γ or mass density ρ of the fluid and the depth z of the point from the fluid surface.* The relationship can be expressed mathematically as

$$p = \gamma z = \rho g z \qquad (9\text{--}15)$$

where g is the acceleration due to gravity. Equation 9–15 is valid only for fluids that are assumed *incompressible*, as in the case of most liquids. Gases are compressible fluids, and since their density changes significantly with both pressure and temperature, Eq. 9–15 cannot be used.

To illustrate how Eq. 9–15 is applied, consider the submerged plate shown in Fig. 9–26. Three points on the plate have been specified. Since point B is at depth z_1 from the liquid surface, the *pressure* at this point has a magnitude $p_1 = \gamma z_1$. Likewise, points C and D are both at depth z_2; hence, $p_2 = \gamma z_2$. In all cases, the pressure acts *normal* to the surface area dA located at the specified point. Using Eq. 9-15 and the results of Sec. 9.5, it is possible to determine the resultant force caused by a liquid pressure distribution and specify its location on the surface of a submerged plate. Three different shapes of plates will now be considered.

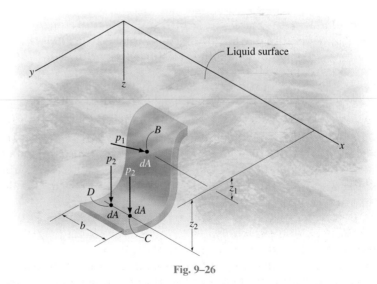

Fig. 9–26

*In particular, for water $\gamma = \rho g = 9810 \ \text{N/m}^3$ since $\rho = 1000 \ \text{kg/m}^3$ and $g = 9.81 \ \text{m/s}^2$.

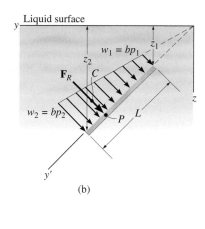

(a)

(b)

Fig. 9–27

Flat Plate of Constant Width.   A flat rectangular plate of constant width, which is submerged in a liquid having a specific weight γ, is shown in Fig. 9–27a. The plane of the plate makes an angle with the horizontal, such that its top edge is located at a depth z_1 from the liquid surface and its bottom edge is located at a depth z_2. Since pressure varies linearly with depth, Eq. 9–15, the distribution of pressure over the plate's surface is represented by a trapezoidal volume having an intensity of $p_1 = \gamma z_1$ at depth z_1 and $p_2 = \gamma z_2$ at depth z_2. As noted in Sec. 9.5, the magnitude of the *resultant force* $\mathbf{F}_R$ is equal to the *volume* of this loading diagram and $\mathbf{F}_R$ has a *line of action* that passes through the volume's centroid C. Hence, $\mathbf{F}_R$ does *not* act at the centroid of the plate; rather, it acts at point P, called the *center of pressure*.

Since the plate has a *constant width*, the loading distribution may also be viewed in two dimensions, Fig. 9–27b. Here the loading intensity is measured as force/length and varies linearly from $w_1 = bp_1 = b\gamma z_1$ to $w_2 = bp_2 = b\gamma z_2$. The magnitude of $\mathbf{F}_R$ in this case equals the trapezoidal *area*, and $\mathbf{F}_R$ has a *line of action* that passes through the area's *centroid* C. For numerical applications, the area and location of the centroid for a trapezoid are tabulated on the inside back cover.

y

(a)

Fig. 9–28

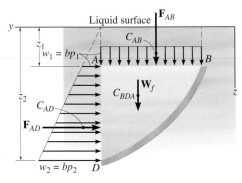

Fig. 9–29

Curved Plate of Constant Width.

When the submerged plate is curved, the pressure acting normal to the plate continually changes direction, and therefore calculation of the magnitude of $\mathbf{F}_R$ and its location P is more difficult than for a flat plate. Three- and two-dimensional views of the loading distribution are shown in Figs. 9–28a and 9–28b, respectively. Here integration can be used to determine both F_R and the location of the centroid C or center of pressure P.

A simpler method exists, however, for calculating the magnitude of $\mathbf{F}_R$ and its location along a curved (or flat) plate having a *constant width*. This method requires separate calculations for the horizontal and vertical *components* of $\mathbf{F}_R$. For example, the distributed loading acting on the curved plate DB in Fig. 9–28b can be represented by the *equivalent loading* shown in Fig. 9–29. Here the plate supports the weight of liquid W_f contained within the block BDA. This force has a magnitude $W_f = (\gamma b)(\text{area}_{BDA})$ and acts through the centroid of BDA. In addition, there are the pressure distributions caused by the liquid acting along the vertical and horizontal sides of the block. Along the vertical side AD, the force $\mathbf{F}_{AD}$ has a magnitude that equals the area under the trapezoid and acts through the centroid C_{AD} of this area. The distributed loading along the horizontal side AB is constant since all points lying in this plane are at the same depth from the surface of the liquid. The magnitude of $\mathbf{F}_{AB}$ is simply the area of the rectangle. This force acts through the area's centroid C_{AB} or the midpoint of AB. Summing the three forces in Fig. 9–29 yields $\mathbf{F}_R = \Sigma\mathbf{F} = \mathbf{F}_{AD} + \mathbf{F}_{AB} + \mathbf{W}_f$, which is shown in Fig. 9–28. Finally, the location of the center of pressure P on the plate is determined by applying the equation $M_{R_O} = \Sigma M_O$, which states that the moment of the resultant force about a convenient reference point O, such as D or B, in Fig. 9–28, is equal to the sum of the moments of the three forces in Fig. 9–29 about the same point.

Flat Plate of Variable Width. The pressure distribution acting on the surface of a submerged plate having a variable width is shown in Fig. 9–30. The resultant force of this loading equals the volume described by the plate area as its base and linearly varying pressure distribution as its altitude. The shaded element shown in Fig. 9–30 may be used if integration is chosen to determine this volume. The element consists of a rectangular strip of area $dA = x\,dy'$ located at a depth z below the liquid surface. Since a uniform pressure $p = \gamma z$ (force/area) acts on dA, the magnitude of the differential force dF is equal to $dF = dV = p\,dA = \gamma z(x\,dy')$. Integrating over the entire volume yields Eq. 9–13; i.e.,

$$F_R = \int_A p\,dA = \int_V dV = V$$

The resultant force of the water and its location on the elliptical back plate of this tank truck must be determined by integration.

From Eq. 9–14, the centroid of V defines the point through which $\mathbf{F}_R$ acts. The center of pressure, which lies on the surface of the plate just below C, has coordinates $P(\overline{x}, \overline{y}')$ defined by the equations

$$\overline{x} = \frac{\displaystyle\int_V \widetilde{x}\,dV}{\displaystyle\int_V dV} \qquad \overline{y}' = \frac{\displaystyle\int_V \widetilde{y}'\,dV}{\displaystyle\int_V dV}$$

This point should *not* be mistaken for the centroid of the plate's *area*.

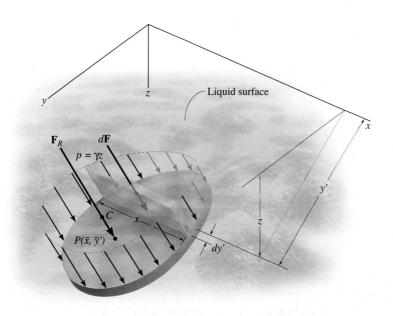

Fig. 9–30

E X A M P L E 9.13

(a)

Determine the magnitude and location of the resultant hydrostatic force acting on the submerged rectangular plate AB shown in Fig. 9–31a. The plate has a width of 1.5 m; $\rho_w = 1000 \text{ kg/m}^3$.

Solution

The water pressures at depths A and B are

$$p_A = \rho_w g z_A = (1000 \text{ kg/m}^3)(9.81 \text{ m/s}^2)(2 \text{ m}) = 19.62 \text{ kPa}$$

$$p_B = \rho_w g z_B = (1000 \text{ kg/m}^3)(9.81 \text{ m/s}^2)(5 \text{ m}) = 49.05 \text{ kPa}$$

Since the plate has a constant width, the distributed loading can be viewed in two dimensions as shown in Fig. 9–31b. The intensities of the load at A and B are

$$w_A = b p_A = (1.5 \text{ m})(19.62 \text{ kPa}) = 29.43 \text{ kN/m}$$

$$w_B = b p_B = (1.5 \text{ m})(49.05 \text{ kPa}) = 73.58 \text{ kN/m}$$

(b)

From the table on the inside back cover, the magnitude of the resultant force $\mathbf{F}_R$ created by the distributed load is

$$F_R = \text{area of trapezoid}$$
$$= \tfrac{1}{2}(3)(29.4 + 73.6) = 154.5 \text{ kN} \qquad \textit{Ans.}$$

This force acts through the centroid of the area,

$$h = \frac{1}{3}\left(\frac{2(29.43) + 73.58}{29.43 + 73.58}\right)(3) = 1.29 \text{ m} \qquad \textit{Ans.}$$

measured upward from B, Fig. 9–31b.

The same results can be obtained by considering two components of $\mathbf{F}_R$ defined by the triangle and rectangle shown in Fig. 9–31c. Each force acts through its associated centroid and has a magnitude of

$$F_{Re} = (29.43 \text{ kN/m})(3 \text{ m}) = 88.3 \text{ kN}$$

$$F_t = \tfrac{1}{2}(44.15 \text{ kN/m})(3 \text{ m}) = 66.2 \text{ kN}$$

Hence,

$$F_R = F_{Re} + F_t = 88.3 + 66.2 = 154.5 \text{ kN} \qquad \textit{Ans.}$$

The location of $\mathbf{F}_R$ is determined by summing moments about B, Fig. 9–31b and c, i.e.,

$$\zeta + (M_R)_B = \Sigma M_B; \qquad (154.5)h = 88.3(1.5) + 66.2(1)$$
$$h = 1.29 \text{ m} \qquad \textit{Ans.}$$

(c)

Fig. 9–31

EXAMPLE 9.14

Determine the magnitude of the resultant hydrostatic force acting on the surface of a seawall shaped in the form of a parabola as shown in Fig. 9–32a. The wall is 5 m long; $\rho_w = 1020$ kg/m^3.

(a)

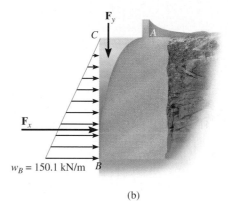

(b)

Fig. 9–32

Solution

The horizontal and vertical components of the resultant force will be calculated, Fig. 9–32b. Since

$$p_B = \rho_w g z_B = (1020 \text{ kg/m}^3)(9.81 \text{ m/s}^2)(3 \text{ m}) = 30.02 \text{ kPa}$$

then

$$w_B = b p_B = 5 \text{ m}(30.02 \text{ kPa}) = 150.1 \text{ kN/m}$$

Thus,

$$F_x = \tfrac{1}{2}(3 \text{ m})(150.1 \text{ kN/m}) = 225.1 \text{ kN}$$

The area of the parabolic sector ABC can be determined using the table on the inside back cover. Hence, the weight of water within this region is

$$F_y = (\rho_w g b)(\text{area}_{ABC})$$
$$= (1020 \text{ kg/m}^3)(9.81 \text{ m/s}^2)(5 \text{ m})[\tfrac{1}{3}(1 \text{ m})(3 \text{ m})] = 50.0 \text{ kN}$$

The resultant force is therefore

$$F_R = \sqrt{F_x^2 + F_y^2} = \sqrt{(225.1)^2 + (50.0)^2}$$
$$= 231 \text{ kN} \qquad\qquad Ans.$$

E X A M P L E **9.15**

Determine the magnitude and location of the resultant force acting on the triangular end plates of the water trough shown in Fig. 9–33a; $\rho_w = 1000 \text{ kg/m}^3$.

(a)

Solution

The pressure distribution acting on the end plate E is shown in Fig. 9–33b. The magnitude of the resultant force $\mathbf{F}$ is equal to the volume of this loading distribution. We will solve the problem by integration. Choosing the differential volume element shown in the figure, we have

$$dF = dV = p\,dA = \rho_w g z(2x\,dz) = 19\,620zx\,dz$$

The equation of line AB is

$$x = 0.5(1 - z)$$

Hence, substituting and integrating with respect to z from $z = 0$ to $z = 1$ m yields

$$F = V = \int_V dV = \int_0^1 (19\,620)z[0.5(1 - z)]\,dz$$

$$= 9810 \int_0^1 (z - z^2)\,dz = 1635 \text{ N} = 1.64 \text{ kN} \qquad \textit{Ans.}$$

This resultant passes through the centroid of the volume. Because of symmetry,

$$\bar{x} = 0 \qquad \textit{Ans.}$$

Since $\tilde{z} = z$ for the volume element in Fig. 9–33b, then

$$\bar{z} = \frac{\int_V \tilde{z}\,dV}{\int_V dV} = \frac{\int_0^1 z(19\,620)z[0.5(1 - z)]\,dz}{1635} = \frac{9810 \int_0^1 (z^2 - z^3)\,dz}{1635}$$

$$= 0.5 \text{ m} \qquad \textit{Ans.}$$

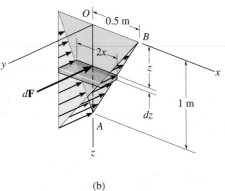

(b)

Fig. 9–33

PROBLEMS

9-106. Determine the magnitude of the resultant hydrostatic force acting on the dam and its location, measured from the top surface of the water. The width of the dam is 8 m; $\rho_w = 1.0$ Mg/m³.

Prob. 9–106

9-107. The tank is filled with water to a depth of $d = 4$ m. Determine the resultant force the water exerts on side A and side B of the tank. If oil instead of water is placed in the tank, to what depth d should it reach so that it creates the same resultant forces? $\rho_o = 900$ kg/m³ and $\rho_w = 1000$ kg/m³.

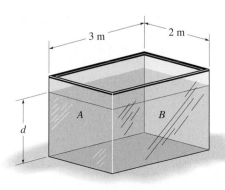

Prob. 9–107

*9-108. When the tide water A subsides, the tide gate automatically swings open to drain the marsh B. For the condition of high tide shown, determine the horizontal reactions developed at the hinge C and stop block D. The length of the gate is 6 m and its height is 4 m. $\rho_w = 1.0$ Mg/m³.

Prob. 9–108

9-109. The concrete "gravity" dam is held in place by its own weight. If the density of concrete is $\rho_c = 2.5$ Mg/m³, and water has a density of $\rho_w = 1.0$ Mg/m³, determine the smallest dimension d that will prevent the dam from overturning about its end A.

Prob. 9–109

9-110. The concrete dam is designed so that its face AB has a gradual slope into the water as shown. Because of this, the frictional force at the base BD of the dam is increased due to the hydrostatic force of the water acting on the dam. Calculate the hydrostatic force acting on the face AB of the dam. The dam is 60 m wide. $\gamma_w = 10$ kN/m³.

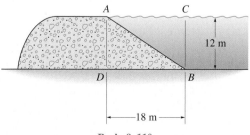

Prob. 9–110

9-111. The semicircular tunnel passes under a river which is 9 m deep. Determine the vertical resultant hydrostatic force acting per meter of length along the length of the tunnel. The tunnel is 6 m wide; $\rho_w = 1.0$ Mg/m^3.

9 m

6 m

Prob. 9–111

*9-112. The tank is used to store a liquid having a specific weight of 13 kN/m^3. If it is filled to the top, determine the magnitude of force the liquid exerts on each of its two sides $ABDC$ and $BDFE$.

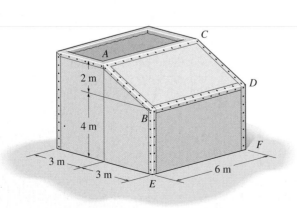

Prob. 9–112

9-113. Determine the resultant horizontal and vertical force components that the water exerts on the side of the dam. The dam is 25 m long and $\gamma_A = 10.4$ kN/m^3.

$y = 0.25\, x^2$

y

25 m

10 m

x

Prob. 9–113

9-114. The gate AB is 8 m wide. Determine the horizontal and vertical components of force acting on the pin at B and the vertical reaction at the smooth support A. $\rho_w = 1.0$ Mg/m^3.

5 m

B

4 m

A

3 m

Prob. 9–114

9-115. The storage tank contains oil having a specific weight of $\gamma_O = 9$ kN/m³. If the tank is 6 m wide, calculate the resultant force acting on the inclined side BC of the tank, caused by the oil, and specify its location along BC, measured from B. Also compute the total resultant force acting on the bottom of the tank.

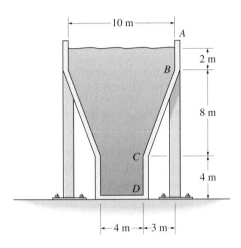

Prob. 9–115

*9-116. The arched surface AB is shaped in the form of a quarter circle. If it is 8 m long, determine the horizontal and vertical components of the resultant force caused by the water acting on the surface. $\rho_w = 1.0$ Mg/m³.

Prob. 9–116

9-117. The rectangular bin is filled with coal, which creates a pressure distribution along wall A that varies as shown, i.e., $p = 4z^3$ kN/m², where z is measured in metre. Determine the resultant force created by the coal, and specify its location measured from the top surface of the coal.

Prob. 9–117

9-118. The semicircular drainage pipe is filled with water. Determine the resultant horizontal and vertical force components that the water exerts on the side AB of the pipe per foot of pipe length; $\gamma_w = 10.4$ kN/m³.

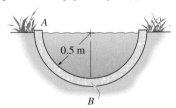

Prob. 9–118

9-119. The pressure loading on the plate is described by the function $p = 10[6/(x + 1) + 8]$ N/m². Determine the magnitude of the resultant force and the coordinates $(\bar{x}, \bar{y})$ of the point where the line of action of the force intersects the plate.

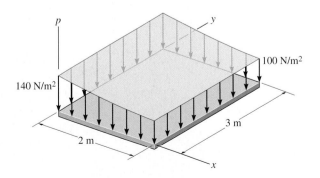

Prob. 9–119

*9-120. The tank is filled to the top ($y = 0.5$ m) with water having a density of $\rho_w = 1.0$ Mg/m³. Determine the resultant force of the water pressure acting on the flat end plate C of the tank, and its location, measured from the top of the tank.

9-122. The loading acting on a square plate is represented by a parabolic pressure distribution. Determine the magnitude of the resultant force and the coordinates ($\bar{x}, \bar{y}$) of the point where the line of action of the force intersects the plate. Also, what are the reactions at the rollers B and C and the ball-and-socket joint A? Neglect the weight of the plate.

Prob. 9–120

Prob. 9–122

9-121. The wind blows uniformly on the front surface of the metal building with a pressure of 30 kN/m². Determine the resultant force it exerts on the surface and the position of this resultant.

9-123. The tank is filled with a liquid which has a density of 900 kg/m³. Determine the resultant force that it exerts on the elliptical end plate, and the location of the center of pressure, measured from the x axis.

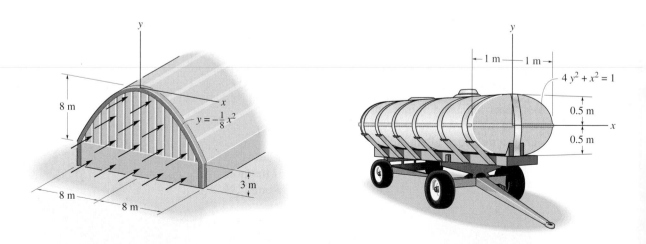

Prob. 9–121

Prob. 9–123

CHAPTER REVIEW

- ***Center of Gravity and Centroid.*** The *center of gravity* represents a point where the weight of the body can be considered concentrated. The distance $\bar{s}$ to this point can be determined from a balance of moments. This requires that the moment of the weight of all the particles of the body about some point must equal the moment of the entire body about the point, $\bar{s}W = \Sigma \tilde{s}W$. The *centroid* is the location of the geometric center for the body. It is determined in a similar manner, using a moment balance of geometric elements such as line, area, or volume segments. For bodies having a continuous shape, moments are summed (integrated) using differential elements. If the body is a composite of several shapes, each having a known location for its center of gravity or centroid, then the location is determined from a discrete summation using its composite parts.

- ***Theorems of Pappus and Guldinus.*** These theorems can be used to determine the surface area and volume of a body of revolution. The *surface area* equals the product of the length of the generating curve and the distance traveled by the centroid of the curve needed to generate the area $A = \theta \bar{r} L$. The *volume* of the body equals the product of the generating area and the distance traveled by the centroid of this area needed to generate the volume, $V = \theta \bar{r} A$.

- ***Fluid Pressure.*** The pressure developed by a liquid at a point on a submerged surface depends upon the depth of the point and the density of the liquid in accordance with Pascal's law, $p = \rho g h = \gamma h$. This pressure will create a *linear distribution* of loading on a flat vertical or inclined surface. If the surface is horizontal, then the loading will be *uniform*. In any case, the resultants of these loadings can be determined by finding the volume or area under the loading curve. The line of action of the resultant force passes through the centroid of the loading diagram.

REVIEW PROBLEMS

*9-124. A circular V-belt has an inner radius of 600 mm and a cross-sectional area as shown. Determine the volume of material required to make the belt.

9-125. A circular V-belt has an inner radius of 600 mm and a cross-sectional area as shown. Determine the surface area of the belt.

Probs. 9–124/125

9-126. Locate the centroid $\bar{y}$ of the beam's cross-sectional area.

Prob. 9–126

9-127. Locate the centroid of the solid.

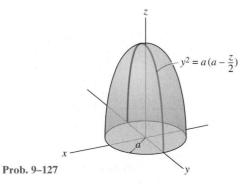

Prob. 9–127

*9-128. Determine the magnitude of the resultant hydrostatic force acting per metre of length on the sea wall; $\gamma_w = 10$ kN/m³.

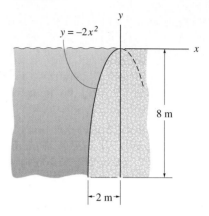

Prob. 9–128

9-129. The tank and compressor have a mass of 15 kg and mass center at G_T, and the motor has a mass of 70 kg and a mass center at G_M. Determine the angle of tilt, θ, of the tank so that the unit will be on the verge of tipping over.

Prob. 9–129

9-130. The thin-walled channel and stiffener have the cross section shown. If the material has a constant thickness, determine the location $\bar{y}$ of its centroid. The dimensions are indicated to the center of each segment.

Prob. 9–130

9-131. Locate the center of gravity of the homogeneous rod. The rod has a weight of 30 N/m. Also, compute the x, y, z components of reaction at the fixed support A.

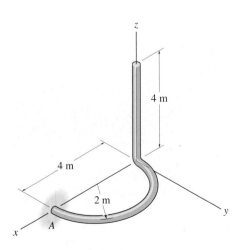

Prob. 9–131

*9-132. The rectangular bin is filled with coal, which creates a pressure distribution along wall A that varies as shown, i.e., $p = z^{1/3}$ kN/m² where z is in metre. Compute the resultant force created by the coal, and its location, measured from the top surface of the coal.

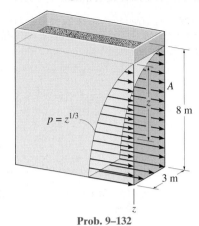

Prob. 9–132

9-133. The load over the plate varies linearly along the sides of the plate such that $p = \frac{2}{3}[x(4 - y)]$ kPa. Determine the resultant force and its position $(\bar{x}, \bar{y})$ on the plate.

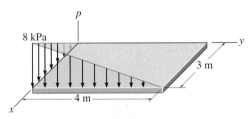

Prob. 9–133

9-134. The pressure loading on the plate is described by the function $p = \{-240/(x + 1) + 340\}$ Pa. Determine the magnitude of the resultant force and coordinates of the point where the line of action of the force intersects the plate.

Prob. 9–134

The design of a structural member, such as a beam or column, requires calculation of its cross-sectional moment of inertia. In this chapter, we will discuss how this is done.

CHAPTER

10

Moments of Inertia

CHAPTER OBJECTIVES

- To develop a method for determining the moment of inertia for an area.

- To introduce the product of inertia and show how to determine the maximum and minimum moments of inertia of an area.

- To discuss the mass moment of inertia.

10.1   Definition of Moments of Inertia for Areas

In the last chapter, we determined the centroid for an area by considering the first moment of the area about an axis; that is, for the computation we had to evaluate an integral of the form $\int x \, dA$. An integral of the second moment of an area, such as $\int x^2 \, dA$, is referred to as the *moment of inertia* for the area. The terminology "moment of inertia" as used here is actually a misnomer; however, it has been adopted because of the similarity with integrals of the same form related to mass.

499

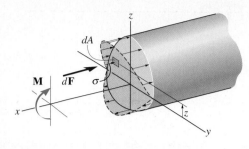

Fig. 10–1

The moment of inertia of an area orginates whenever one relates the normal stress σ (sigma), or force per unit area, acting on the transverse cross section of an elastic beam, to the applied external moment **M**, which causes bending of the beam. From the theory of mechanics of materials, it can be shown that the stress within the beam varies linearly with its distance from an axis passing through the centroid C of the beam's cross-sectional area; i.e., $\sigma = kz$, Fig. 10–1. The magnitude of force acting on the area element dA, shown in the figure, is therefore $dF = \sigma\,dA = kz\,dA$. Since this force is located a distance z from the y axis, the moment of $d\mathbf{F}$ about the y axis is $dM = dFz = kz^2\,dA$. The resulting moment of the entire stress distribution is equal to the applied moment **M**; hence, $M = k\int z^2\,dA$. Here the integral represents the moment of inertia of the area about the y axis. Since integrals of this form often arise in formulas used in mechanics of materials, structural mechanics, fluid mechanics, and machine design, the engineer should become familiar with the methods used for their computation.

Moment of Inertia. Consider the area A, shown in Fig. 10–2, which lies in the x–y plane. By definition, the moments of inertia of the differential planar area dA about the x and y axes are $dI_x = y^2\,dA$ and $dI_y = x^2\,dA$, respectively. For the entire area the *moments of inertia* are determined by integration; i.e.,

$$
\begin{aligned}
I_x &= \int_A y^2\,dA \\
I_y &= \int_A x^2\,dA
\end{aligned}
\qquad (10\text{–}1)
$$

We can also formulate the second moment of dA about the pole O or z axis, Fig. 10–2. This is referred to as the polar moment of inertia, $dJ_O = r^2\,dA$. Here r is the perpendicular distance from the pole (z axis) to the element dA. For the entire area the *polar moment of inertia* is

$$
J_O = \int_A r^2\,dA = I_x + I_y
\qquad (10\text{–}2)
$$

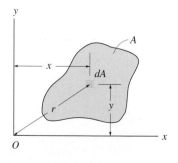

Fig. 10–2

The relationship between J_O and I_x, I_y is possible since $r^2 = x^2 + y^2$, Fig. 10–2.

From the above formulations it is seen that I_x, I_y, and J_O will *always* be *positive* since they involve the product of distance squared and area. Furthermore, the units for moment of inertia involve length raised to the fourth power, e.g., m^4, mm^4.

10.2 Parallel-Axis Theorem for an Area

If the moment of inertia for an area is known about an axis passing through its centroid, which is often the case, it is convenient to determine the moment of inertia of the area about a corresponding parallel axis using the *parallel-axis theorem*. To derive this theorem, consider finding the moment of inertia of the shaded area shown in Fig. 10–3 about the x axis. In this case, a differential element dA is located at an arbitrary distance y' from the *centroidal* x' axis, whereas the *fixed distance* between the parallel x and x' axes is defined as d_y. Since the moment of inertia of dA about the x axis is $dI_x = (y' + d_y)^2 \, dA$, then for the entire area,

$$I_x = \int_A (y' + d_y)^2 \, dA$$

$$= \int_A y'^2 \, dA + 2d_y \int_A y' \, dA + d_y^2 \int_A dA$$

The first integral represents the moment of inertia of the area about the centroidal axis, $\bar{I}_{x'}$. The second integral is zero since the x' axis passes through the area's centroid C; i.e., $\int y' \, dA = \bar{y} \int dA = 0$ since $\bar{y} = 0$. Realizing that the third integral represents the total area A, the final result is therefore

$$\boxed{I_x = \bar{I}_{x'} + Ad_y^2} \tag{10–3}$$

A similar expression can be written for I_y; i.e.,

$$\boxed{I_y = \bar{I}_{y'} + Ad_x^2} \tag{10–4}$$

And finally, for the polar moment of inertia about an axis perpendicular to the x–y plane and passing through the pole O (z axis), Fig. 10–3, we have

$$\boxed{J_O = \bar{J}_C + Ad^2} \tag{10–5}$$

The form of each of these three equations states that *the moment of inertia of an area about an axis is equal to the moment of inertia of the area about a parallel axis passing through the area's centroid plus the product of the area and the square of the perpendicular distance between the axes.*

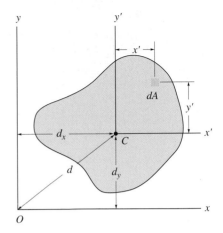

Fig. 10–3

10.3 Radius of Gyration of an Area

The *radius of gyration* of a planar area has units of length and is a quantity that is often used for the design of columns in structural mechanics. Provided the areas and moments of inertia are *known*, the radii of gyration are determined from the formulas

$$k_x = \sqrt{\frac{I_x}{A}} \qquad k_y = \sqrt{\frac{I_y}{A}} \qquad k_O = \sqrt{\frac{J_O}{A}} \tag{10–6}$$

The form of these equations is easily remembered since it is similar to that for finding the moment of inertia of a differential area about an axis. For example, $I_x = k_x^2 A$; whereas for a differential area, $dI_x = y^2 \, dA$.

10.4 Moments of Inertia for an Area by Integration

When the boundaries for a planar area are expressed by mathematical functions, Eqs. 10–1 may be integrated to determine the moments of inertia for the area. If the element of area chosen for integration has a differential size in two directions as shown in Fig. 10–2, a double integration must be performed to evaluate the moment of inertia. Most often, however, it is easier to perform only a single integration by choosing an element having a differential size or thickness in only one direction.

PROCEDURE FOR ANALYSIS

- If a single integration is performed to determine the moment of inertia of an area about an axis, it will be necessary to specify the differential element dA.
- Most often this element will be rectangular, such that it will have a finite length and differential width.
- The element should be located so that it intersects the boundary of the area at the *arbitrary point* (x, y). There are two possible ways to orient the element with respect to the axis about which the moment of inertia is to be determined.

Case 1

- The *length* of the element can be oriented *parallel* to the axis. This situation occurs when the rectangular element shown in Fig. 10–4 is used to determine I_y for the area. Direct application of Eq. 10–1, i.e., $I_y = \int x^2 \, dA$, can be made in this case since the element has an infinitesimal thickness dx and therefore *all parts* of the element lie at the *same* moment-arm distance x from the y axis.*

Case 2

- The *length* of the element can be oriented *perpendicular* to the axis. Here Eq. 10–1 *does not apply* since all parts of the element will *not* lie at the same moment-arm distance from the axis. For example, if the rectangular element in Fig. 10–4 is used for determining I_x for the area, it will first be necessary to calculate the moment of inertia of the *element* about a horizontal axis passing through the element's centroid and then determine the moment of inertia of the *element* about the x axis by using the parallel-axis theorem. Integration of this result will yield I_x.

*In the case of the element $dA = dx \, dy$, Fig. 10–2, the moment arms y and x are appropriate for the formulation of I_x and I_y (Eq. 10–1) since the *entire* element, because of its infinitesimal size, lies at the specified y and x perpendicular distances from the x and y axes.

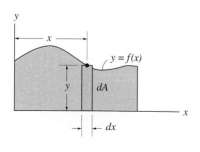

Fig. 10–4

E X A M P L E   **10.1**

Determine the moment of inertia for the rectangular area shown in Fig. 10–5 with respect to (a) the centroidal x' axis, (b) the axis x_b passing through the base of the rectangle, and (c) the pole or z' axis perpendicular to the $x'-y'$ plane and passing through the centroid C.

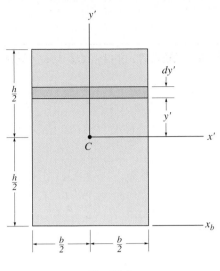

Fig. 10–5

Solution (Case 1)

Part (a).   The differential element shown in Fig. 10–5 is chosen for integration. Because of its location and orientation, the *entire element* is at a distance y' from the x' axis. Here it is necessary to integrate from $y' = -h/2$ to $y' = h/2$. Since $dA = b\,dy'$, then

$$\bar{I}_{x'} = \int_A y'^2\,dA = \int_{-h/2}^{h/2} y'^2(b\,dy') = b\int_{-h/2}^{h/2} y'^2\,dy$$

$$= \frac{1}{12}bh^3 \qquad\qquad\qquad\qquad Ans.$$

Part (b).   The moment of inertia about an axis passing through the base of the rectangle can be obtained by using the result of part (a) and applying the parallel-axis theorem, Eq. 10–3.

$$I_{x_b} = \bar{I}_{x'} + Ad_y^2$$

$$= \frac{1}{12}bh^3 + bh\left(\frac{h}{2}\right)^2 = \frac{1}{3}bh^3 \qquad\qquad Ans.$$

Part (c).   To obtain the polar moment of inertia about point C, we must first obtain $\bar{I}_{y'}$, which may be found by interchanging the dimensions b and h in the result of part (a), i.e.,

$$\bar{I}_{y'} = \frac{1}{12}hb^3$$

Using Eq. 10–2, the polar moment of inertia about C is therefore

$$\bar{J}_C = \bar{I}_{x'} + \bar{I}_{y'} = \frac{1}{12}bh(h^2 + b^2) \qquad\qquad Ans.$$

E X A M P L E 10.2

Determine the moment of inertia of the shaded area shown in Fig. 10–6a about the x axis.

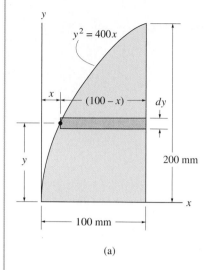

(a)

Solution I (Case 1)
A differential element of area that is *parallel* to the x axis, as shown in Fig. 10–6a, is chosen for integration. Since the element has a thickness dy and intersects the curve at the *arbitrary point* (x, y), the area is $dA = (100 - x)\,dy$. Furthermore, all parts of the element lie at the same distance y from the x axis. Hence, integrating with respect to y, from $y = 0$ to $y = 200$ mm, yields

$$I_x = \int_A y^2\,dA = \int_A y^2(100 - x)\,dy$$

$$= \int_0^{200} y^2\left(100 - \frac{y^2}{400}\right) dy = 100\int_0^{200} y^2\,dy - \frac{1}{400}\int_0^{200} y^4\,dy$$

$$= 107(10^6)\ \text{mm}^4 \qquad\qquad\qquad Ans.$$

(b)

Fig. 10–6

Solution II (Case 2)
A differential element *parallel* to the y axis, as shown in Fig. 10–6b, is chosen for integration. It intersects the curve at the *arbitrary point* (x, y). In this case, all parts of the element do *not* lie at the same distance from the x axis, and therefore the parallel-axis theorem must be used to determine the *moment of inertia of the element* with respect to this axis. For a rectangle having a base b and height h, the moment of inertia about its centroidal axis has been determined in part (a) of Example 10.1. There it was found that $\bar{I}_{x'} = \frac{1}{12}bh^3$. For the differential element shown in Fig. 10–6b, $b = dx$ and $h = y$, and thus $d\bar{I}_{x'} = \frac{1}{12}dx\,y^3$. Since the centroid of the element is at $\tilde{y} = y/2$ from the x axis, the moment of inertia of the element about this axis is

$$dI_x = d\bar{I}_{x'} + dA\,\tilde{y}^2 = \frac{1}{12}dx\,y^3 + y\,dx\left(\frac{y}{2}\right)^2 = \frac{1}{3}y^3\,dx$$

This result can also be concluded from part (b) of Example 10.1. Integrating with respect to x, from $x = 0$ to $x = 100$ mm, yields

$$I_x = \int dI_x = \int_A \frac{1}{3}y^3\,dx = \int_0^{100} \frac{1}{3}(400x)^{3/2}\,dx$$

$$= 107(10^6)\ \text{mm}^4 \qquad\qquad\qquad Ans.$$

E X A M P L E **10.3**

Determine the moment of inertia with respect to the x axis of the circular area shown in Fig. 10–7a.

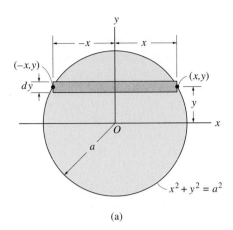

(a)

Solution I *(Case 1)*

Using the differential element shown in Fig. 10–7a, since $dA = 2x\, dy$, we have

$$I_x = \int_A y^2\, dA = \int_A y^2(2x)\, dy$$

$$= \int_{-a}^{a} y^2(2\sqrt{a^2 - y^2})\, dy = \frac{\pi a^4}{4} \qquad \textit{Ans.}$$

Solution II *(Case 2)*

When the differential element is chosen as shown in Fig. 10–7b, the centroid for the element happens to lie on the x axis, and so, applying Eq. 10–3, noting that $d_y = 0$ and for a rectangle $\bar{I}_{x'} = \frac{1}{12}bh^3$, we have

$$dI_x = \frac{1}{12}dx\,(2y)^3$$

$$= \frac{2}{3}y^3\, dx$$

Integrating with respect to x yields

$$I_x = \int_{-a}^{a} \frac{2}{3}(a^2 - x^2)^{3/2}\, dx = \frac{\pi a^4}{4} \qquad \textit{Ans.}$$

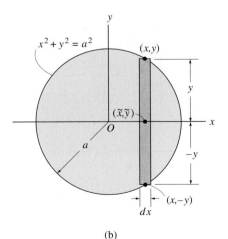

(b)

Fig. 10–7

E X A M P L E **10.4**

Determine the moment of inertia of the shaded area shown in Fig. 10–8a about the x axis.

Solution I *(Case 1)*

The differential element parallel to the x axis is chosen for integration, Fig. 10–8a. The element intersects the curve at the *arbitrary points* (x_2, y) and (x_1, y). Consequently, its area is $dA = (x_1 - x_2)\, dy$. Since all parts of the element lie at the same distance y from the x axis, we have

$$I_x = \int_A y^2\, dA = \int_0^1 y^2 (x_1 - x_2)\, dy = \int_0^1 y^2 (\sqrt{y} - y)\, dy$$

$$I_x = \frac{2}{7} y^{7/2} - \frac{1}{4} y^4 \Big|_0^1 = 0.0357 \text{ m}^4 \qquad \text{Ans.}$$

Solution II *(Case 2)*

The differential element parallel to the y axis is shown in Fig. 10–8b. It intersects the curves at the *arbitrary points* (x, y_2) and (x, y_1). Since all parts of its entirety do *not* lie at the same distance from the x axis, we must first use the parallel-axis theorem to find the *element's* moment of mertia about the x axis, using $\bar{I}_{x'} = \frac{1}{12} bh^3$, then integrate this result to determine I_x. Thus,

$$dI_x = d\bar{I}_{x'} + dA\, \tilde{y}^2 = \frac{1}{12} dx (y_2 - y_1)^3$$

$$+ (y_2 - y_1)\, dx \left(y_1 + \frac{y_2 - y_1}{2} \right)^2$$

$$= \frac{1}{3} (y_2^3 - y_1^3)\, dx = \frac{1}{3} (x^3 - x^6)\, dx$$

$$I_x = \frac{1}{3} \int_0^1 (x^3 - x^6)\, dx = \frac{1}{12} x^4 - \frac{1}{21} x^7 \Big|_0^1 = 0.0357 \text{ m}^4 \quad \text{Ans.}$$

By comparison, Solution I requires much less computation. Therefore, if an integral using a particular element appears difficult to evaluate, try solving the problem using an element oriented in the other direction.

(a)

(b)

Fig. 10–8

PROBLEMS

10-1. Determine the moment of inertia of the shaded area about the x axis.

10-2. Determine the moment of inertia of the shaded area about the y axis.

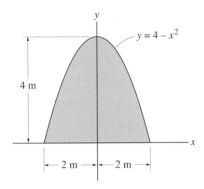

Probs. 10–1/2

10-3. Determine the moment of inertia of the area about the x axis. Solve the problem in two ways, using rectangular differential elements: (a) having a thickness dx and (b) having a thickness of dy.

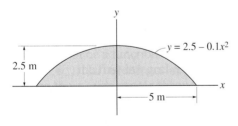

Prob. 10–3

*10-4. Determine the moment of inertia of the area about the x axis. Solve the problem in two ways, using rectangular differential elements: (a) having a thickness of dx, and (b) having a thickness of dy.

10-5. Determine the moment of inertia of the area about the y axis. Solve the problem in two ways, using rectangular differential elements: (a) having a thickness of dx, and (b) having a thickness of dy.

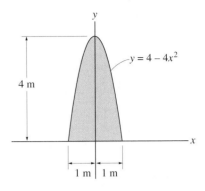

Probs. 10–4/5

10-6. Determine the moment of inertia of the shaded area about the x axis.

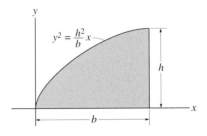

Prob. 10–6

10-7. Determine the moment of inertia of the shaded area about the x axis.

*10-8. Determine the moment of inertia of the shaded area about the y axis.

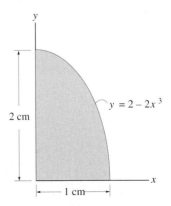

Probs. 10–7/8

10-9. Determine the moment of inertia of the shaded area about the x axis.

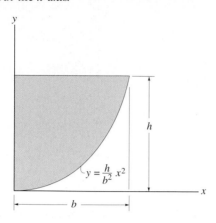

Prob. 10–9

10-10. Determine the moment of inertia of the shaded area about the y axis.

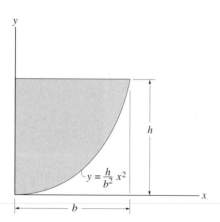

Prob. 10–10

10-11. Determine the moment of inertia of the shaded area about the x axis.

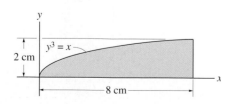

Prob. 10–11

*10-12. Determine the moment of inertia of the shaded area about the x axis.

10-13. Determine the moment of inertia of the shaded area about the y axis.

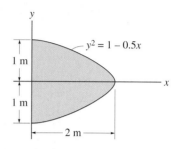

Probs. 10–12/13

10-14. Determine the moment of inertia of the shaded area about the x axis.

10-15. Determine the moment of inertia of the shaded area about the y axis.

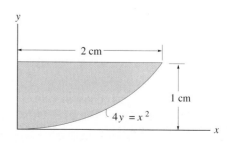

Probs. 10–14/15

*■**10-16.** Determine the moment of inertia of the area about the y axis. Use Simpson's rule to evaluate the integral.

■**10-17.** Determine the moment of inertia of the area about the x axis. Use Simpson's rule to evaluate the integral.

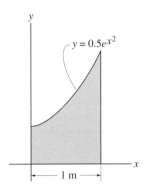

$y = 0.5e^{x^2}$

1 m

Probs. 10–16/17

10-18. Determine the moment of inertia of the shaded area about the x axis.

10-19. Determine the moment of inertia of the shaded area about the y axis.

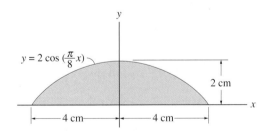

$y = 2\cos\left(\frac{\pi}{8}x\right)$

2 cm

4 cm

4 cm

Probs. 10–18/19

***10-20.** Determine the moment of inertia of the shaded area about the x axis.

10-21. Determine the moment of inertia of the shaded area about the y axis.

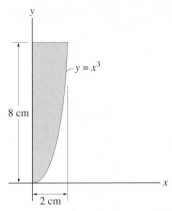

$y = x^3$

8 cm

2 cm

Probs. 10–20/21

10-22. Determine the moment of inertia of the shaded area about the x axis.

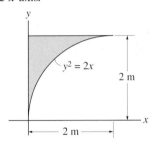

$y^2 = 2x$

2 m

2 m

Prob. 10–22

■**10-23.** Determine the moment of inertia of the shaded area about the y axis. Use Simpson's rule to evaluate the integral.

*■**10-24.** Determine the moment of inertia of the shaded area about the x axis. Use Simpson's rule to evaluate the integral.

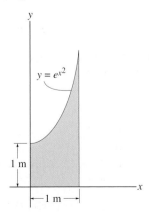

$y = e^{x^2}$

1 m

1 m

Probs. 10–23/24

10.5 Moments of Inertia for Composite Areas

A composite area consists of a series of connected "simpler" parts or shapes, such as semicircles, rectangles, and triangles. Provided the moment of inertia of each of these parts is known or can be determined about a common axis, then the moment of inertia of the composite area equals the *algebraic sum* of the moments of inertia of all its parts.

Structural members have various cross-sectional shapes, and it is necessary to calculate their moments of inertia in order to determine the stress in these members.

PROCEDURE FOR ANALYSIS

The moment of inertia of a composite area about a reference axis can be determined using the following procedure.

Composite Parts.

- Using a sketch, divide the area into its composite parts and indicate the perpendicular distance from the *centroid* of each part to the reference axis.

Parallel-Axis Theorem.

- The moment of inertia of each part should be determined about its centroidal axis, which is parallel to the reference axis. For the calculation use the table given on the inside back cover.

- If the centroidal axis does not coincide with the reference axis, the parallel-axis theorem, $I = \bar{I} + Ad^2$, should be used to determine the moment of inertia of the part about the reference axis.

Summation.

- The moment of inertia of the entire area about the reference axis is determined by summing the results of its composite parts.

- If a composite part has a "hole," its moment of inertia is found by "subtracting" the moment of inertia for the hole from the moment of inertia of the entire part including the hole.

EXAMPLE **10.5**

Compute the moment of inertia of the composite area shown in Fig. 10–9a about the x axis.

(a)   (b)

Fig. 10–9

Solution

Composite Parts. The composite area is obtained by *subtracting* the circle from the rectangle as shown in Fig. 10–9b. The centroid of each area is located in the figure.

Parallel-Axis Theorem. The moments of inertia about the x axis are determined using the parallel-axis theorem and the data in the table on the inside back cover.

Circle

$$I_x = \bar{I}_{x'} + Ad_y^2$$

$$= \frac{1}{4}\pi(25)^4 + \pi(25)^2(75)^2 = 11.4(10^6) \text{ mm}^4$$

Rectangle

$$I_x = \bar{I}_{x'} + Ad_y^2$$

$$= \frac{1}{12}(100)(150)^3 + (100)(150)(75)^2 = 112.5(10^6) \text{ mm}^4$$

Summation. The moment of inertia for the composite area is thus

$$I_x = -11.4(10^6) + 112.5(10^6)$$

$$= 101(10^6) \text{ mm}^4 \qquad\qquad Ans.$$

EXAMPLE 10.6

(a)

(b)

Fig. 10–10

Determine the moments of inertia of the beam's cross-sectional area shown in Fig. 10–10a about the x and y centroidal axes.

Solution

Composite Parts. The cross section can be considered as three composite rectangular areas A, B, and D shown in Fig. 10–10b. For the calculation, the centroid of each of these rectangles is located in the figure.

Parallel-Axis Theorem. From the table on the inside back cover, or Example 10.1, the moment of inertia of a rectangle about its centroidal axis is $\bar{I} = \frac{1}{12}bh^3$. Hence, using the parallel-axis theorem for rectangles A and D, the calculations are as follows:

Rectangle A

$$I_x = \bar{I}_{x'} + Ad_y^2 = \frac{1}{12}(100)(300)^3 + (100)(300)(200)^2$$
$$= 1.425(10^9)\ mm^4$$

$$I_y = \bar{I}_{y'} + Ad_x^2 = \frac{1}{12}(300)(100)^3 + (100)(300)(250)^2$$
$$= 1.90(10^9)\ mm^4$$

Rectangle B

$$I_x = \frac{1}{12}(600)(100)^3 = 0.05(10^9)\ mm^4$$

$$I_y = \frac{1}{12}(100)(600)^3 = 1.80(10^9)\ mm^4$$

Rectangle D

$$I_x = \bar{I}_{x'} + Ad_y^2 = \frac{1}{12}(100)(300)^3 + (100)(300)(200)^2$$
$$= 1.425(10^9)\ mm^4$$

$$I_y = \bar{I}_{y'} + Ad_x^2 = \frac{1}{12}(300)(100)^3 + (100)(300)(250)^2$$
$$= 1.90(10^9)\ mm^4$$

Summation. The moments of inertia for the entire cross section are thus

$$I_x = 1.425(10^9) + 0.05(10^9) + 1.425(10^9)$$
$$= 2.90(10^9)\ mm^4 \qquad \textit{Ans.}$$
$$I_y = 1.90(10^9) + 1.80(10^9) + 1.90(10^9)$$
$$= 5.60(10^9)\ mm^4 \qquad \textit{Ans.}$$

PROBLEMS

10-25. The polar moment of inertia of the area is $\bar{J}_C = 23$ cm^4 about the z axis passing through the centroid C. If the moment of inertia about the y'axis is 5 cm^4, and the moment of inertia about the x axis is 40 cm^4, determine the area A.

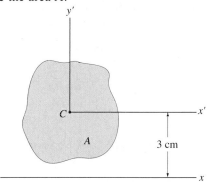

Prob. 10–25

10-26. The polar moment of inertia of the area is $\bar{J}_C = 548(10^6)$ mm^4, about the z' axis passing through the centroid C. The moment of inertia about the y' axis is $383(10^6)$ mm^4, and the moment of inertia about the x axis is $856(10^6)$ mm^4. Determine the area A.

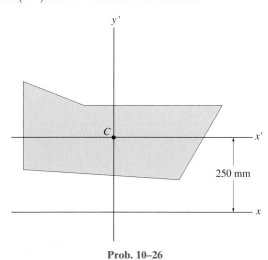

Prob. 10–26

10-27. The beam is constructed from the two channels and two cover plates. If each channel has a cross-sectional area of $A_c = 11.8$ cm^2 and a moment of inertia about a horizontal axis passing through its own centroid, C_c, of $(\bar{I}_{\bar{x}})_{C_c} = 349$ cm^4, determine the moment of inertia of the beam about the x axis.

*10-28. The beam is constructed from the two channels and two cover plates. If each channel has a cross-sectional area of $A_c = 11.8$ cm^2 and a moment of inertia about a vertical axis passing through its own centroid, C_c, of $(\bar{I}_{\bar{y}})_{C_c} = 9.23$ cm^4, determine the moment of inertia of the beam about the y axis.

Probs. 10–27/28

10-29. Determine the moment of inertia of the beam's cross-sectional area with respect to the x' centroidal axis. Neglect the size of all the rivet heads. R, for the calculation. Handbook values for the area, moment of inertia, and location of the centroid C of one of the angles are listed in the figure.

$(I_a)_{x_a'} = 1.32(10^6)$ mm^4

$A_a = 1.36(10^3)$ mm^2

Prob. 10–29

10-30. Locate the centroid $\bar{y}$ of the cross-sectional area for the angle. Then find the moment of inertia $\bar{I}_{x'}$ about the x' centroidal axis.

10-31. Locate the centroid $\bar{x}$ of the cross-sectional area for the angle. Then find the moment of inertia $\bar{I}_{y'}$ about the y' centroidal axis.

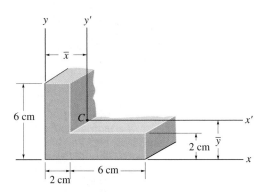

Probs. 10–30/31

10-34. Determine the moments of inertia of the shaded area about the x and y axes.

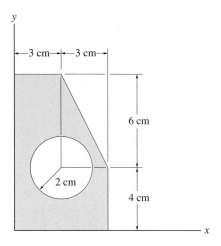

Prob. 10–34

***10-32.** Determine the distance $\bar{x}$ to the centroid of the beam's cross-sectional area: then find the moment of inertia about the y' axis.

10-33. Determine the moment of inertia of the beam's cross-sectional area about the x' axis.

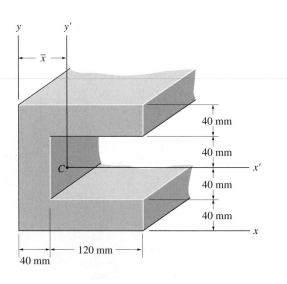

Probs. 10–32/33

10-35. Determine the moment of inertia of the beam's cross-sectional area about the x' axis. Neglect the size of the corner welds at A and B for the calculation, $\bar{y} = 154.4$ mm.

Prob. 10–35

*10-36. Compute the moments of inertia I_x and I_y for the beam's cross-sectional area about the x and y axes.

10-37. Determine the distance $\bar{y}$ to the centroid C of the beam's cross-sectional area and then compute the moment of inertia $\bar{I}_{x'}$ about the x' axis.

10-38. Determine the distance $\bar{x}$ to the centroid C of the beam's cross-sectional area and then compute the moment of inertia $\bar{I}_{y'}$ about the y' axis.

Probs. 10–36/37/38

10-39. Locate the centroid $\bar{y}$ of the cross section and determine the moment of inertia of the section about the x' axis.

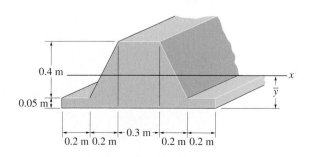

Prob. 10–39

*10-40. Determine $\bar{y}$, which locates the centroidal axis x' for the cross-sectional area of the T-beam, and then find the moments of inertia $\bar{I}_{x'}$ and $\bar{I}_{y'}$.

Prob. 10–40

10-41. Determine the distance $\bar{y}$ to the centroid for the beam's cross-sectional area; then determine the moment of inertia about the x' axis.

10-42. Determine the moment of inertia of the beam's cross-sectional area about the y axis.

Probs. 10–41/42

10-43. Determine the moment of inertia I_x of the shaded area about the x axis.

*10-44. Determine the moment of inertia I_y of the shaded area about the y axis.

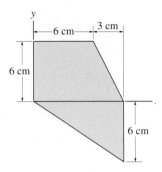

Probs. 10–43/44

10-45. Locate the centroid $\bar{y}$ of the channel's cross-sectional area, and then determine the moment of inertia with respect to the x' axis passing through the centroid.

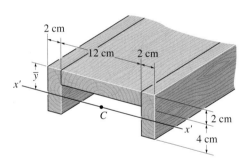

Prob. 10–45

10-46. Determine the moments of inertia I_x and I_y of the shaded area.

Prob. 10–46

10-47. Determine the moment of inertia of the parallelogram about the x' axis, which passes through the centroid C of the area.

*10-48. Determine the moment of inertia of the parallelogram about the y' axis, which passes through the centroid C of the area.

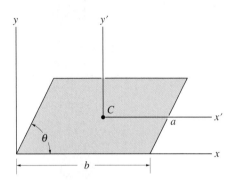

Probs. 10–47/48

10-49. An aluminum strut has a cross section referred to as a deep hat. Determine the location $\bar{y}$ of the centroid of its area and the moment of inertia of the area about the x' axis. Each segment has a thickness of 10 mm.

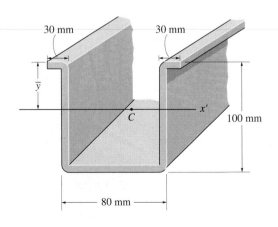

Prob. 10–49

10-50. Determine the moment of inertia of the beam's cross-sectional area with respect to the x' axis passing through the centroid C of the cross section. Neglect the size of the corner welds at A and B for the calculation, $\bar{y} = 104.3$ mm.

Prob. 10–50

10-51. Determine the location $\bar{y}$ of the centroid of the channel's cross-sectional area and then calculate the moment of inertia of the area about this axis.

Prob. 10–51

*10-52. Determine the radius of gyration k_x for the column's cross-sectional area.

Prob. 10–52

10-53. Determine the moments of inertia of the triangular area about the x' and y' axes, which pass through the centroid C of the area.

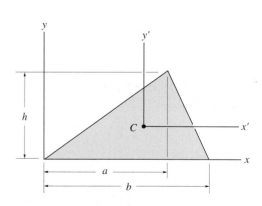

Prob. 10–53

*10.6   Product of Inertia for an Area

In general, the moment of inertia for an area is different for every axis about which it is computed. In some applications of structural or mechanical design it is necessary to know the orientation of those axes which give, respectively, the maximum and minimum moments of inertia for the area. The method for determining this is discussed in Sec. 10.7. To use this method, however, one must first compute the product of inertia for the area as well as its moments of inertia for given x, y axes.

The product of inertia for an element of area dA located at point (x, y), Fig. 10–11, is defined as $dI_{xy} = xy\, dA$. Thus, for the entire area A, the *product of inertia* is

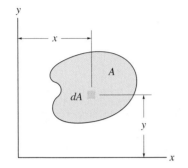

$$I_{xy} = \int_A xy\, dA \qquad (10\text{--}7)$$

Fig. 10–11

If the element of area chosen has a differential size in two directions, as shown in Fig. 10–11, a double integration must be performed to evaluate I_{xy}. Most often, however, it is easier to choose an element having a differential size or thickness in only one direction in which case the evaluation requires only a single integration (see Example 10.7).

Like the moment of inertia, the product of inertia has units of length raised to the fourth power, e.g., m^4, mm^4. However, since x or y may be a negative quantity, while the element of area is always positive, the product of inertia may be positive, negative, or zero, depending on the location and orientation of the coordinate axes. For example, the product of inertia I_{xy} for an area will be *zero* if either the x or y axis is an axis of *symmetry* for the area. To show this, consider the shaded area in Fig. 10–12, where for every element dA located at point (x, y) there is a corresponding element dA located at $(x, -y)$. Since the products of inertia for these elements are, respectively, $xy\, dA$ and $-xy\, dA$, the algebraic sum or integration of all the elements that are chosen in this way will cancel each other. Consequently, the product of inertia for the total area becomes zero. It also follows from the definition of I_{xy} that the "sign" of this quantity depends on the quadrant where the area is located. As shown in Fig. 10–13, if the area is rotated from one quadrant to another, the sign of I_{xy} will change.

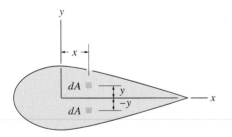

Fig. 10–12

Parallel-Axis Theorem.   Consider the shaded area shown in Fig. 10–14, where x' and y' represent a set of axes passing through the *centroid* of the area, and x and y represent a corresponding set of parallel

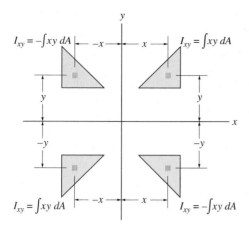

$I_{xy} = -\int xy \, dA$ $I_{xy} = \int xy \, dA$

$I_{xy} = \int xy \, dA$ $I_{xy} = -\int xy \, dA$

Fig. 10–13

axes. Since the product of inertia of dA with respect to the x and y axes is $dI_{xy} = (x' + d_x)(y' + d_y) \, dA$, then for the entire area,

$$I_{xy} = \int_A (x' + d_x)(y' + d_y) \, dA$$

$$= \int_A x'y' \, dA + d_x \int_A y' \, dA + d_y \int_A x' \, dA + d_x d_y \int_A dA$$

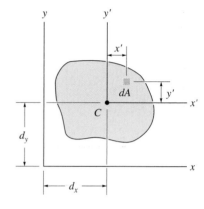

Fig. 10–14

The first term on the right represents the product of inertia of the area with respect to the centroidal axis, $\bar{I}_{x'y'}$. The integrals in the second and third terms are zero since the moments of the area are taken about the centroidal axis. Realizing that the fourth integral represents the total area A, the final result is therefore

$$\boxed{I_{xy} = \bar{I}_{x'y'} + Ad_x d_y} \tag{10–8}$$

The similarity between this equation and the parallel-axis theorem for moments of inertia should be noted. In particular, it is important that the *algebraic signs* for d_x and d_y be maintained when applying Eq. 10–8. As illustrated in Example 10.8, the parallel-axis theorem finds important application in determining the product of inertia of a *composite area* with respect to a set of x, y axes.

E X A M P L E 10.7

Determine the product of inertia I_{xy} of the triangle shown in Fig. 10–15a.

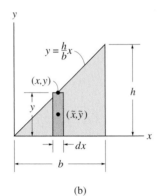

(a)

Solution I

A differential element that has a thickness dx, Fig. 10–15b, has an area $dA = y\,dx$. The product of inertia of the element about the x, y axes is determined using the parallel-axis theorem.

$$dI_{xy} = d\bar{I}_{x'y'} + dA\,\tilde{x}\,\tilde{y}$$

where $(\tilde{x}, \tilde{y})$ locates the *centroid* of the element or the origin of the x', y' axes. Since $d\bar{I}_{x'y'} = 0$, due to symmetry, and $\tilde{x} = x$, $\tilde{y} = y/2$, then

$$dI_{xy} = 0 + (y\,dx)x\left(\frac{y}{2}\right) = \left(\frac{h}{b}x\,dx\right)x\left(\frac{h}{2b}x\right)$$

$$= \frac{h^2}{2b^2}x^3\,dx$$

Integrating with respect to x from $x = 0$ to $x = b$ yields

$$I_{xy} = \frac{h^2}{2b^2}\int_0^b x^3\,dx = \frac{b^2h^2}{8} \qquad \textit{Ans.}$$

Solution II

The differential element that has a thickness dy, Fig. 10–15c, and area $dA = (b - x)\,dy$ can also be used. The *centroid* is located at point $\tilde{x} = x + (b - x)/2 = (b + x)/2$, $\tilde{y} = y$, so the product of inertia of the element becomes

$$dI_{xy} = d\bar{I}_{x'y'} + dA\,\tilde{x}\,\tilde{y}$$

$$= 0 + (b - x)\,dy\left(\frac{b + x}{2}\right)y$$

$$= \left(b - \frac{b}{h}y\right)dy\left[\frac{b + (b/h)y}{2}\right]y = \frac{1}{2}y\left(b^2 - \frac{b^2}{h^2}y^2\right)dy$$

Integrating with respect to y from $y = 0$ to $y = h$ yields

$$I_{xy} = \frac{1}{2}\int_0^h y\left(b^2 - \frac{b^2}{h^2}y^2\right)dy = \frac{b^2h^2}{8} \qquad \textit{Ans.}$$

(b)

(c)

Fig. 10–15

EXAMPLE **10.8**

Compute the product of inertia of the beam's cross-sectional area, shown in Fig. 10–16a, about the x and y centroidal axes.

(a)

(b)

Fig. 10–16

Solution

As in Example 10.6, the cross section can be considered as three composite rectangular areas A, B, and D, Fig. 10–16b. The coordinates for the centroid of each of these rectangles are shown in the figure. Due to symmetry, the product of inertia of *each rectangle* is *zero* about each set of x', y' axes that passes through the rectangle's centroid. Hence, application of the parallel-axis theorem to each of the rectangles yields

Rectangle A

$$I_{xy} = \bar{I}_{x'y'} + Ad_xd_y$$
$$= 0 + (300)(100)(-250)(200)$$
$$= -1.50(10^9) \text{ mm}^4$$

Rectangle B

$$I_{xy} = \bar{I}_{x'y'} + Ad_xd_y$$
$$= 0 + 0$$
$$= 0$$

Rectangle D

$$I_{xy} = \bar{I}_{x'y'} + Ad_xd_y$$
$$= 0 + (300)(100)(250)(-200)$$
$$= -1.50(10^9) \text{ mm}^4$$

The product of inertia for the entire cross section is therefore

$$I_{xy} = -1.50(10^9) + 0 - 1.50(10^9) = -3.00(10^9) \text{ mm}^4 \quad Ans.$$

*10.7 Moments of Inertia for an Area About Inclined Axes

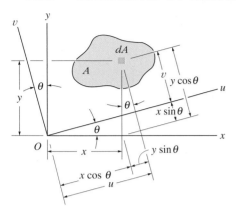

Fig. 10–17

In structural and mechanical design, it is sometimes necessary to calculate the moments and product of inertia I_u, I_v, and I_{uv} for an area with respect to a set of inclined u and v axes when the values for θ, I_x, I_y, and I_{xy} are *known*. To do this we will use *transformation equations* which relate the x, y and u, v coordinates. From Fig. 10–17, these equations are

$$u = x \cos \theta + y \sin \theta$$
$$v = y \cos \theta - x \sin \theta$$

Using these equations, the moments and product of inertia of dA about the u and v axes become

$$dI_u = v^2 \, dA = (y \cos \theta - x \sin \theta)^2 \, dA$$
$$dI_v = u^2 \, dA = (x \cos \theta + y \sin \theta)^2 \, dA$$
$$dI_{uv} = uv \, dA = (x \cos \theta + y \sin \theta)(y \cos \theta - x \sin \theta) \, dA$$

Expanding each expression and integrating, realizing that $I_x = \int y^2 \, dA$, $I_y = \int x^2 \, dA$, and $I_{xy} = \int xy \, dA$, we obtain

$$I_u = I_x \cos^2 \theta + I_y \sin^2 \theta - 2I_{xy} \sin \theta \cos \theta$$
$$I_v = I_x \sin^2 \theta + I_y \cos^2 \theta + 2I_{xy} \sin \theta \cos \theta$$
$$I_{uv} = I_x \sin \theta \cos \theta - I_y \sin \theta \cos \theta + I_{xy}(\cos^2 \theta - \sin^2 \theta)$$

These equations may be simplified by using the trigonometric identities $\sin 2\theta = 2 \sin \theta \cos \theta$ and $\cos 2\theta = \cos^2 \theta - \sin^2 \theta$, in which case

$$I_u = \frac{I_x + I_y}{2} + \frac{I_x - I_y}{2} \cos 2\theta - I_{xy} \sin 2\theta$$
$$I_v = \frac{I_x + I_y}{2} - \frac{I_x - I_y}{2} \cos 2\theta + I_{xy} \sin 2\theta$$
$$I_{uv} = \frac{I_x - I_y}{2} \sin 2\theta + I_{xy} \cos 2\theta$$

(10–9)

If the first and second equations are added together, we can show that the polar moment of inertia about the z axis passing through point O is *independent* of the orientation of the u and v axes; i.e.,

$$J_O = I_u + I_v = I_x + I_y$$

Principal Moments of Inertia. Equations 10–9 show that I_u, I_v, and I_{uv} depend on the angle of inclination, θ, of the u, v axes. We will now determine the orientation of these axes about which the moments of inertia for the area, I_u and I_v, are maximum and minimum. This particular set of axes is called the *principal axes* of the area, and the corresponding moments of inertia with respect to these axes are called the *principal moments of inertia*. In general, there is a set of principal axes for every chosen origin O. For the structural and mechanical design of a member, the origin O is generally located at the cross-sectional area's centroid.

The angle $\theta = \theta_p$, which defines the orientation of the principal axes for the area, may be found by differentiating the first of Eqs. 10–9 with respect to θ and setting the result equal to zero. Thus,

$$\frac{dI_u}{d\theta} = -2\left(\frac{I_x - I_y}{2}\right)\sin 2\theta - 2I_{xy}\cos 2\theta = 0$$

Therefore, at $\theta = \theta_p$,

$$\tan 2\theta_p = \frac{-I_{xy}}{(I_x - I_y)/2} \qquad (10\text{–}10)$$

This equation has two roots, θ_{p_1} and θ_{p_2}, which are 90° apart and so specify the inclination of the principal axes. In order to substitute them into Eq. 10–9, we must first find the sine and cosine of $2\theta_{p_1}$ and $2\theta_{p_2}$. This can be done using the triangles shown in Fig. 10–18, which are based on Eq. 10–10.

For θ_{p_1},

$$\sin 2\theta_{p_1} = -I_{xy} \bigg/ \sqrt{\left(\frac{I_x - I_y}{2}\right)^2 + I_{xy}^2}$$

$$\cos 2\theta_{p_1} = \left(\frac{I_x - I_y}{2}\right) \bigg/ \sqrt{\left(\frac{I_x - I_y}{2}\right)^2 + I_{xy}^2}$$

For θ_{p_2},

$$\sin 2\theta_{p_2} = I_{xy} \bigg/ \sqrt{\left(\frac{I_x - I_y}{2}\right)^2 + I_{xy}^2}$$

$$\cos 2\theta_{p_2} = -\left(\frac{I_x - I_y}{2}\right) \bigg/ \sqrt{\left(\frac{I_x - I_y}{2}\right)^2 + I_{xy}^2}$$

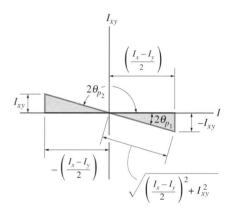

Fig. 10–18

Substituting these two sets of trigonometric relations into the first or second of Eqs. 10–9 and simplifying, we obtain

$$I_{\substack{\max \\ \min}} = \frac{I_x + I_y}{2} \pm \sqrt{\left(\frac{I_x - I_y}{2}\right)^2 + I_{xy}^2} \qquad (10\text{–}11)$$

Depending on the sign chosen, this result gives the maximum or minimum moment of inertia for the area. Furthermore, if the above trigonometric relations for θ_{p_1} and θ_{p_2} are substituted into the third of Eqs. 10–9, it can be shown that $I_{uv} = 0$; that is, the *product of inertia with respect to the principal axes is zero*. Since it was indicated in Sec. 10.6 that the product of inertia is zero with respect to any symmetrical axis, it therefore follows that *any symmetrical axis represents a principal axis of inertia for the area*.

E X A M P L E **10.9**

100 mm

400 mm

100 mm

C ———— *x*

400 mm

100 mm

600 mm

(a)

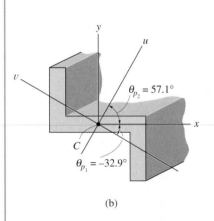

y

u

v

$\theta_{p_2} = 57.1°$

C ———— *x*

$\theta_{p_1} = -32.9°$

(b)

Fig. 10–19

Determine the principal moments of inertia for the beam's cross-sectional area shown in Fig. 10–19a with respect to an axis passing through the centroid.

Solution

The moments and product of inertia of the cross section with respect to the *x*, *y* axes have been computed in Examples 10.6 and 10.8. The results are

$$I_x = 2.90(10^9) \text{ mm}^4 \quad I_y = 5.60(10^9) \text{ mm}^4 \quad I_{xy} = -3.00(10^9) \text{ mm}^4$$

Using Eq. 10–10, the angles of inclination of the principal axes *u* and *v* are

$$\tan 2\theta_p = \frac{-I_{xy}}{(I_x - I_y)/2} = \frac{3.00(10^9)}{[2.90(10^9) - 5.60(10^9)]/2} = -2.22$$

$$2\theta_{p_1} = -65.8° \quad \text{and} \quad 2\theta_{p_2} = 114.2°$$

Thus, as shown in Fig. 10–19b,

$$\theta_{p_1} = -32.9° \quad \text{and} \quad \theta_{p_2} = 57.1°$$

The principal moments of inertia with respect to the *u* and *v* axes are determined from Eq. 10–11. Hence,

$$I_{\min}^{\max} = \frac{I_x + I_y}{2} \pm \sqrt{\left(\frac{I_x - I_y}{2}\right)^2 + I_{xy}^2}$$

$$= \frac{2.90(10^9) + 5.60(10^9)}{2}$$

$$\pm \sqrt{\left[\frac{2.90(10^9) - 5.60(10^9)}{2}\right]^2 + [-3.00(10^9)]^2}$$

$$I_{\min}^{\max} = 4.25(10^9) \pm 3.29(10^9)$$

or

$$I_{\max} = 7.54(10^9) \text{ mm}^4 \quad I_{\min} = 0.960(10^9) \text{ mm}^4 \quad \textit{Ans.}$$

Specifically, the maximum moment of inertia, $I_{\max} = 7.54(10^9) \text{ mm}^4$, occurs with respect to the selected *u* axis since *by inspection* most of the cross-sectional area is farthest away from this axis. Or, stated in another manner, $I_{\max}$ occurs about the *u* axis since it is located within $\pm 45°$ of the *y* axis, which has the largest value of I ($I_y > I_x$). Also, this may be concluded by substituting the data with $\theta = 57.1°$ into the first of Eqs. 10–9.

*10.8 Mohr's Circle for Moments of Inertia

Equations 10–9 to 10–11 have a graphical solution that is convenient to use and generally easy to remember. Squaring the first and third of Eqs. 10–9 and adding, it is found that

$$\left(I_u - \frac{I_x + I_y}{2} \right)^2 + I_{uv}^2 = \left(\frac{I_x - I_y}{2} \right)^2 + I_{xy}^2 \qquad (10\text{–}12)$$

In a given problem, I_u and I_{uv} are *variables*, and I_x, I_y, and I_{xy} are *known constants*. Thus, Eq. 10–12 may be written in compact form as

$$(I_u - a)^2 + I_{uv}^2 = R^2$$

When this equation is plotted on a set of axes that represent the respective moment of inertia and the product of inertia, Fig. 10–20, the resulting graph represents a *circle* of radius

$$R = \sqrt{ \left(\frac{I_x - I_y}{2} \right)^2 + I_{xy}^2 }$$

having its center located at point $(a, 0)$, where $a = (I_x + I_y)/2$. The circle so constructed is called *Mohr's circle*, named after the German engineer Otto Mohr (1835–1918).

(a)

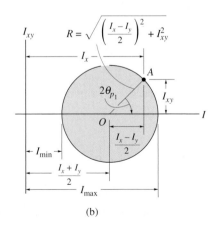

(b)

Fig. 10–20

PROCEDURE FOR ANALYSIS

The main purpose in using Mohr's circle here is to have a convenient means for transforming I_x, I_y, and I_{xy} into the principal moments of inertia. The following procedure provides a method for doing this.

Determine I_x, I_y, and I_{xy}.

- Establish the x, y axes for the area, with the origin located at the point P of interest, and determine I_x, I_y, and I_{xy}, Fig. 10–20a.

Construct the Circle.

- Construct a rectangular coordinate system such that the abscissa represents the moment of inertia I, and the ordinate represents the product of inertia I_{xy}, Fig. 10–20b.

- Determine the center of the circle, O, which is located at a distance $(I_x + I_y)/2$ from the origin, and plot the reference point A having coordinates (I_x, I_{xy}). By definition, I_x is always positive, whereas I_{xy} will be either positive or negative.

- Connect the reference point A with the center of the circle and determine the distance OA by trigonometry. This distance represents the radius of the circle, Fig. 10–20b. Finally, draw the circle.

Principal Moments of Inertia.

- The points where the circle intersects the abscissa give the values of the principal moments of inertia I_{min} and I_{max}. Notice that the *product of inertia will be zero at these points*, Fig. 10–20b.

Principal Axes.

- To find the direction of the major principal axis, determine by trigonometry the angle $2\theta_{p_1}$, *measured from the radius OA to the positive I axis*, Fig. 10–20b. This angle represents *twice* the angle from the x axis of the area in question to the axis of maximum moment of inertia I_{max}, Fig. 10–20a. Both the angle on the circle, $2\theta_{p_1}$, and the angle to the axis on the area, θ_{p_1}, *must be measured in the same sense*, as shown in Fig. 10–20. The axis for minimum moment of inertia I_{min} is perpendicular to the axis for I_{max}.

Using trigonometry, the above procedure may be verified to be in accordance with the equations developed in Sec. 10.7.

E X A M P L E 10.10

Using Mohr's circle, determine the principal moments of inertia for the beam's cross-sectional area shown in Fig. 10–21a, with respect to an axis passing through the centroid.

(a)

(b)

Solution

Determine I_x, I_y, I_{xy}. The moment of inertia and the product of inertia have been determined in Examples 10.6 and 10.8 with respect to the x, y axes shown in Fig. 10–21a. The results are $I_x = 2.90(10^9)$ mm^4, $I_y = 5.60(10^9)$ mm^4, and $I_{xy} = -3.00(10^9)$ mm^4.

Construct the Circle. The I and I_{xy} axes are shown in Fig. 10–21b. The center of the circle, O, lies at a distance $(I_x + I_y)/2 = (2.90 + 5.60)/2 = 4.25$ from the origin. When the reference point $A(2.90, -3.00)$ is connected to point O, the radius OA is determined from the triangle OBA using the Pythagorean theorem.

$$OA = \sqrt{(1.35)^2 + (-3.00)^2} = 3.29$$

The circle is constructed in Fig. 10–21c.

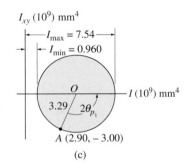

(c)

Principal Moments of Inertia. The circle intersects the I axis at points $(7.54, 0)$ and $(0.960, 0)$. Hence,

$$I_{max} = 7.54(10^9) \text{ mm}^4 \qquad Ans.$$
$$I_{min} = 0.960(10^9) \text{ mm}^4 \qquad Ans.$$

Principal Axes. As shown in Fig. 10–21c, the angle $2\theta_{p_1}$ is determined from the circle by measuring counterclockwise from OA to the direction of the *positive I* axis. Hence,

$$2\theta_{p_1} = 180° - \sin^{-1}\left(\frac{|BA|}{|OA|}\right) = 180° - \sin^{-1}\left(\frac{3.00}{3.29}\right) = 114.2°$$

The principal axis for $I_{max} = 7.54(10^9)$ mm^4 is therefore oriented at an angle $\theta_{p_1} = 57.1°$, measured *counterclockwise*, from the *positive x* axis to the *positive u* axis. The v axis is perpendicular to this axis. The results are shown in Fig. 10–21d.

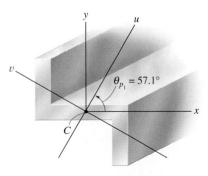

Fig. 10–21

PROBLEMS

10-54. Determine the product of inertia of the shaded portion of the parabola with respect to the x and y axes.

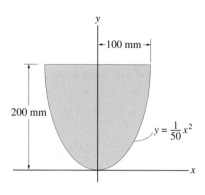

Prob. 10–54

10-55. Determine the product of inertia of the shaded area with respect to the x and y axes.

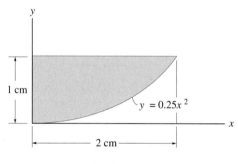

Prob. 10–55

*10-56. Determine the product of inertia of the shaded area of the ellipse with respect to the x and y axes.

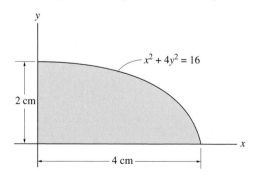

Prob. 10–56

10-57. Determine the product of inertia of the parabolic area with respect to the x and y axes.

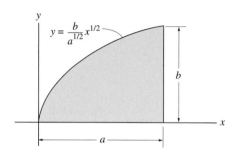

Prob. 10–57

10-58. Determine the product of inertia of the shaded area with respect to the x and y axes.

Prob. 10–58

10-59. Determine the product of inertia of the shaded area with respect to the x and y axes.

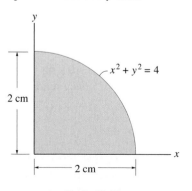

Prob. 10–59

*10-60. Determine the product of inertia of the shaded area with respect to the x and y axes.

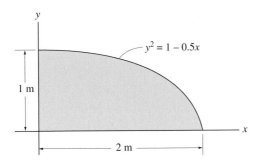

$y^2 = 1 - 0.5x$

1 m

2 m

Prob. 10–60

10-61. Determine the product of inertia of the shaded area with respect to the x and y axes.

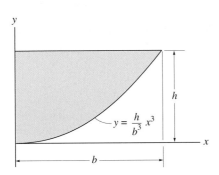

h

$y = \dfrac{h}{b^3} x^3$

b

Prob. 10–61

10-62. Determine the product of inertia of the shaded area with respect to the x and y axes.

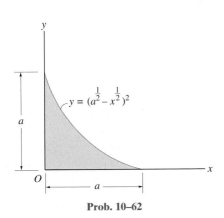

$y = (a^{\frac{1}{2}} - x^{\frac{1}{2}})^2$

a

O

a

Prob. 10–62

10-63. Determine the product of inertia of the shaded area with respect to the x and y axes.

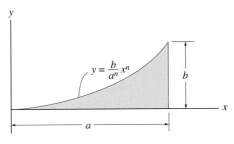

$y = \dfrac{b}{a^n} x^n$

b

a

Prob. 10–63

*10-64. Determine the product of inertia of the shaded area with respect to the x and y axes.

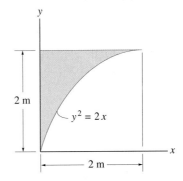

2 m

$y^2 = 2x$

2 m

Prob. 10–64

■10-65. Determine the product of inertia of the shaded area with respect to the x and y axes. Use Simpson's rule to evaluate the integral.

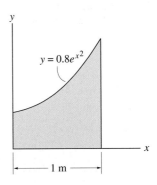

$y = 0.8e^{x^2}$

1 m

Prob. 10–65

10-66. Determine the product of inertia of the thin strip of area with respect to the x and y axes. The strip is oriented at an angle θ from the x axis. Assume that $t \ll l$.

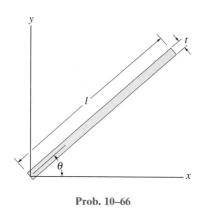

Prob. 10–66

10-67. Determine the product of inertia of the beam's cross-sectional area with respect to the x and y axes that have their origin located at the centroid C.

*10-68. Determine the product of inertia of the beam's cross-sectional area with respect to the x and y axes.

Prob. 10–68

10-69. Determine the product of inertia of the cross-sectional area with respect to the x and y axes that have their origin located at the centroid C.

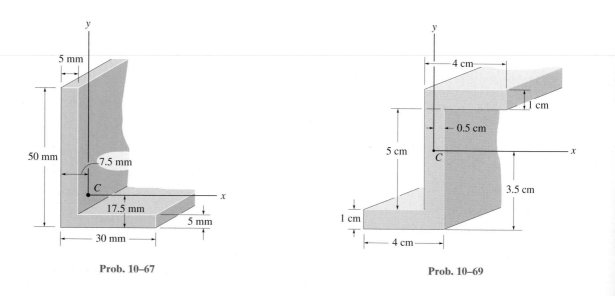

Prob. 10–67

Prob. 10–69

10-70. Determine the product of inertia of the parallelogram with respect to the *x* and *y* axes.

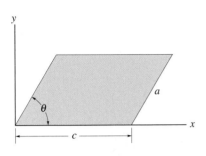

Prob. 10–70

10-71. Determine the product of inertia of the cross-sectional area with respect to the *x* and *y* axes.

Prob. 10–71

*10-72. Determine the product of inertia of the beam's cross-sectional area with respect to the *x* and *y* axes that have their origin located at the centroid *C*.

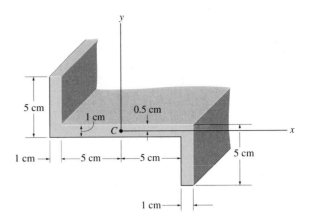

Prob. 10–72

10-73. Determine the product of inertia for the angle with respect to the *x* and *y* axes passing through the centroid *C*. Assume all corners to be square.

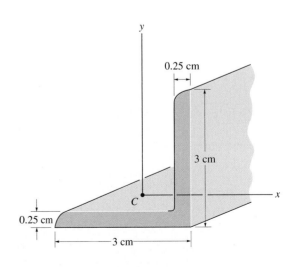

Prob. 10–73

10-74. Determine the product of inertia for the beam's cross-sectional area with respect to the u and v axes.

Prob. 10–74

10-75. Determine the moments of inertia I_u and I_v of the cross-sectional area.

Prob. 10–75

*10-76. Determine the distance $\bar{y}$ to the centroid of the area and then calculate the moments of inertia I_u and I_v of the channel's cross-sectional area. The u and v axes have their origin at the centroid C. For the calculation, assume all corners to be square.

Prob. 10–76

10-77. Determine the moments of inertia of the shaded area with respect to the u and v axes.

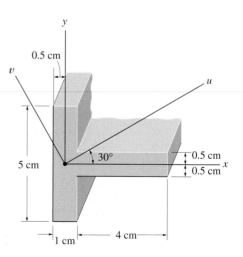

Prob. 10–77

10-78. Determine the directions of the principal axes with origin located at point O, and the principal moments of inertia for the rectangular area about these axes.

Prob. 10–78

10-79. Determine the moments of inertia I_u, I_v and the product of inertia I_{uv} of the beam's cross-sectional area. Take $\theta = 45°$.

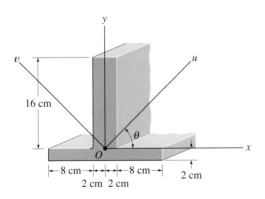

Prob. 10–79

*****10-80.** Determine the directions of the principal axes with origin located at point O, and the principal moments of inertia of the area about these axes.

Prob. 10–80

10-81. Determine the principal moments of inertia of the beam's cross-sectional area about the principal axes that have their origin located at the centroid C. Use the equations developed in Section 10.7. For the calculation, assume all corners to be square.

Prob. 10–81

10-82. Determine the principal moments of inertia for the angle's cross-sectional area with respect to a set of principal axes that have their origin located at the centroid C. Use the equation developed in Section 10.7. For the calculation, assume all corners to be square.

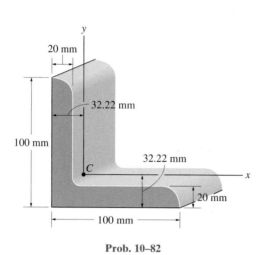

Prob. 10–82

10-83. The area of the cross section of an airplane wing has the following properties about the x and y axes passing through the centroid C: $\bar{I}_x = 450$ cm⁴, $\bar{I}_y = 1730$ cm⁴, $\bar{I}_{xy} = 138$ cm⁴. Determine the orientation of the principal axes and the principal moments of inertia.

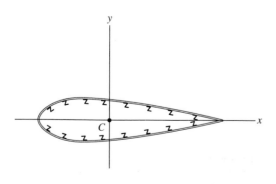

Prob. 10–83

*10-84. Determine the moments of inertia I_u and I_v of the shaded area.

Prob. 10–84

10-85. Solve Prob. 10-78 using Mohr's circle.

10-86. Solve Prob. 10-81 using Mohr's circle.

10-87. Solve Prob. 10-82 using Mohr's circle.

*10-88. Solve Prob. 10-80 using Mohr's circle.

10-89. Solve Prob. 10-83 using Mohr's circle.

10.9 Mass Moment of Inertia

The mass moment of inertia of a body is a property that measures the resistance of the body to angular acceleration. Since it is used in dynamics to study rotational motion, methods for its calculation will now be discussed.

We define the *mass moment of inertia* as the integral of the "second moment" about an axis of all the elements of mass dm which compose the body.* For example, consider the rigid body shown in Fig. 10–22. The body's moment of inertia about the z axis is

$$I = \int_m r^2 \, dm \qquad (10\text{–}13)$$

Here the "moment arm" r is the perpendicular distance from the axis to the arbitrary element dm. Since the formulation involves r, the value of I is *unique* for each axis z about which it is computed. However, the axis which is generally chosen for analysis passes through the body's mass center G. The moment of inertia computed about this axis will be defined as I_G. Realize that because r is squared in Eq. 10–13, the mass moment of inertia is always a *positive quantity*. Common unit used for its measurement is $kg \cdot m^2$.

Fig. 10–22

---

*Another property of the body which measures the symmetry of the body's mass with respect to a coordinate system is the mass product of inertia. This property most often applies to the three-dimensional motion of a body and is discussed in *Engineering Mechanics: Dynamics* (Chapter 21).

(a)

(b)

(c)

Fig. 10–23

If the body consists of material having a variable density, $\rho = \rho(x, y, z)$, the elemental mass dm of the body may be expressed in terms of its density and volume as $dm = \rho \, dV$. Substituting dm into Eq. 10–13, the body's moment of inertia is then computed using *volume elements* for integration; i.e.

$$I = \int_V r^2 \rho \, dV \qquad (10\text{–}14)$$

In the special case of ρ being a *constant*, this term may be factored out of the integral, and the integration is then purely a function of geometry:

$$I = \rho \int_V r^2 \, dV \qquad (10\text{–}15)$$

When the elemental volume chosen for integration has differential sizes in all three directions, e.g., $dV = dx \, dy \, dz$, Fig. 10–23a, the moment of inertia of the body must be determined using "triple integration." The integration process can, however, be simplified to a *single integration* provided the chosen elemental volume has a differential size or thickness in only *one direction*. Shell or disk elements are often used for this purpose.

PROCEDURE FOR ANALYSIS

For integration, we will consider only symmetric bodies having surfaces which are generated by revolving a curve about an axis. An example of such a body which is generated about the z axis is shown in Fig. 10–23.

Shell Element

- If a *shell element* having a height z, radius y, and thickness dy is chosen for integration, Fig. 10–23b, then the volume $dV = (2\pi y)(z) \, dy$.

- This element may be used in Eq. 10–14 or 10–15 for determining the moment of inertia I_z of the body about the z axis since the *entire element*, due to its "thinness," lies at the *same* perpendicular distance $r = y$ from the z axis (see Example 10.11).

Disk Element

- If a disk element having a radius y and a thickness dz is chosen for integration, Fig. 10–23c, then the volume $dV = (\pi y^2) \, dz$.

- In this case the element is *finite* in the radial direction, and consequently its parts *do not* all lie at the *same radial distance* r from the z axis. As a result, Eqs. 10–14 or 10–15 *cannot* be used to determine I_z. Instead, to perform the integration using this element, it is first necessary to determine the moment of inertia *of the element* about the z axis and then integrate this result (see Example 10.12).

E X A M P L E   **10.11**

Determine the mass moment of inertia of the cylinder shown in Fig. 10–24a about the z axis. The density of the material, ρ, is constant.

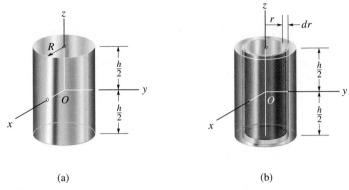

(a)   (b)

Fig. 10–24

Solution

Shell Element.   This problem may be solved using the *shell element* in Fig. 10–24b and single integration. The volume of the element is $dV = (2\pi r)(h)\,dr$, so that its mass is $dm = \rho\,dV = \rho(2\pi h r\,dr)$. Since the *entire element* lies at the same distance r from the z axis, the moment of inertia *of the element* is

$$dI_z = r^2\,dm = \rho 2\pi h r^3\,dr$$

Integrating over the entire region of the cylinder yields

$$I_z = \int_m r^2\,dm = \rho 2\pi h \int_0^R r^3\,dr = \frac{\rho\pi}{2}R^4 h$$

The mass of the cylinder is

$$m = \int_m dm = \rho 2\pi h \int_0^R r\,dr = \rho\pi h R^2$$

so that

$$I_z = \frac{1}{2}mR^2 \qquad\qquad Ans.$$

E X A M P L E **10.12**

A solid is formed by revolving the shaded area shown in Fig. 10–25*a* about the *y* axis. If the density of the material is 5 Mg/m³, determine the mass moment of inertia about the *y* axis.

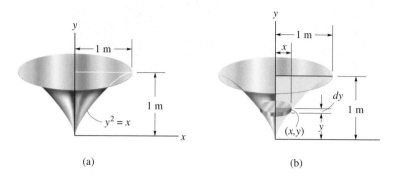

(a)   (b)

Fig. 10–25

Solution

Disk Element.   The moment of inertia will be determined using a *disk element*, as shown in Fig. 10–25*b*. Here the element intersects the curve at the arbitrary point (x, y) and has a mass

$$dm = \rho \, dV = \rho(\pi x^2) \, dy$$

Although all portions of the element are *not* located at the same distance from the *y* axis, it is still possible to determine the moment of inertia dI_y *of the element* about the *y* axis. In Example 10.11 it was shown that the moment of inertia of a cylinder about its longitudinal axis is $I = \frac{1}{2}mR^2$, where m and R are the mass and radius of the cylinder. Since the height of the cylinder is not involved in this formula, we can also use it for a disk. Thus, for the disk element in Fig. 10–25*b*, we have

$$dI_y = \frac{1}{2}(dm)x^2 = \frac{1}{2}[\rho(\pi x^2) \, dy]x^2$$

Substituting $x = y^2$, $\rho = 5$ Mg/m³, and integrating with respect to y, from $y = 0$ to $y = 1$ m, yields the moment of inertia for the entire solid:

$$I_y = \frac{5\pi}{2} \int_0^1 x^4 \, dy = \frac{5\pi}{2} \int_0^1 y^8 \, dy = 0.873 \text{ Mg} \cdot \text{m}^2 = 873 \text{ kg} \cdot \text{m}^2 \quad \textit{Ans.}$$

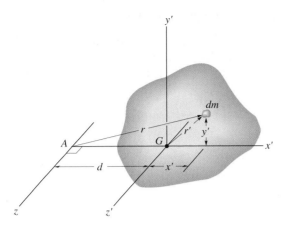

Fig. 10–26

Parallel-Axis Theorem.   If the moment of inertia of the body about an axis passing through the body's mass center is known, then the moment of inertia about any other *parallel axis* may be determined by using the *parallel-axis theorem.* This theorem can be derived by considering the body shown in Fig. 10–26. The z' axis passes through the mass center G, whereas the corresponding *parallel z axis* lies at a constant distance d away. Selecting the differential element of mass dm which is located at point (x', y') and using the Pythagorean theorem, $r^2 = (d + x')^2 + y'^2$, we can express the moment of inertia of the body about the z axis as

$$I = \int_m r^2 \, dm = \int_m [(d + x')^2 + y'^2] \, dm$$

$$= \int_m (x'^2 + y'^2) \, dm + 2d \int_m x' \, dm + d^2 \int_m dm$$

Since $r'^2 = x'^2 + y'^2$, the first integral represents I_G. The second integral equals *zero*, since the z' axis passes through the body's mass center, i.e., $\int x' \, dm = \bar{x} \int dm = 0$ since $\bar{x} = 0$. Finally, the third integral represents the total mass m of the body. Hence, the moment of inertia about the z axis can be written as

$$\boxed{I = I_G + md^2}$$

$$(10\text{–}16)$$

where

I_G = moment of inertia about the z' axis passing through the mass center G

m = mass of the body

d = perpendicular distance between the parallel axes

Radius of Gyration.   Occasionally, the moment of inertia of a body about a specified axis is reported in handbooks using the *radius of gyration, k*. This value has units of length, and when it and the body's mass *m* are known, the moment of inertia is determined from the equation

$$I = mk^2 \quad \text{or} \quad k = \sqrt{\frac{I}{m}} \qquad (10\text{–}17)$$

Note the *similarity* between the definition of k in this formula and r in the equation $dI = r^2\, dm$, which defines the moment of inertia of an elemental mass dm of the body about an axis.

Composite Bodies.   If a body is constructed from a number of simple shapes such as disks, spheres, and rods, the moment of inertia of the body about any axis z can be determined by adding algebraically the moments of inertia of all the composite shapes computed about the z axis. Algebraic addition is necessary since a composite part must be considered as a negative quantity if it has already been included within another part—for example, a "hole" subtracted from a solid plate. The parallel-axis theorem is needed for the calculations if the center of mass of each composite part does not lie on the z axis. For the calculation, then, $I = \Sigma(I_G + md^2)$, where I_G for each of the composite parts is computed by integration or can be determined from a table, such as the one given on the inside back cover of this book.

This flywheel, which operates a metal cutter, has a large moment of inertia about its center. Once it begins rotating it is difficult to stop it and therefore a uniform motion can be effectively transferred to the cutting blade.

E X A M P L E 10.13

If the plate shown in Fig. 10–27a has a density of 8000 kg/m³ and a thickness of 10 mm, determine its mass moment of inertia about an axis directed perpendicular to the page and passing through point O.

(a)   (b)

Fig. 10–27

Solution

The plate consists of two composite parts, the 250-mm-radius disk *minus* a 125-mm-radius disk, Fig. 10–27b. The moment of inertia about O can be determined by computing the moment of inertia of each of these parts about O and then *algebraically* adding the results. The computations are performed by using the parallel-axis theorem in conjunction with the data listed in the table on the inside back cover.

Disk. The moment of inertia of a disk about an axis perpendicular to the plane of the disk is $I_G = \frac{1}{2}mr^2$. The mass center of the disk is located at a distance of 0.25 m from point O. Thus,

$$m_d = \rho_d V_d = 8000 \text{ kg/m}^3[\pi(0.25 \text{ m})^2(0.01 \text{ m})] = 15.71 \text{ kg}$$

$$(I_O)_d = \frac{1}{2}m_d r_d^2 + m_d d^2$$

$$= \frac{1}{2}(15.71 \text{ kg})(0.25 \text{ m})^2 + (15.71 \text{ kg})(0.25 \text{ m})^2$$

$$= 1.473 \text{ kg} \cdot \text{m}^2$$

Hole. For the 125-mm-radius disk (hole), we have

$$m_h = \rho_h V_h = 8000 \text{ kg/m}^3[\pi(0.125 \text{ m})^2(0.01 \text{ m})] = 3.93 \text{ kg}$$

$$(I_O)_h = \frac{1}{2}m_h r_h^2 + m_h d^2$$

$$= \frac{1}{2}(3.93 \text{ kg})(0.125 \text{ m})^2 + (3.93 \text{ kg})(0.25 \text{ m})^2$$

$$= 0.276 \text{ kg} \cdot \text{m}^2$$

The moment of inertia of the plate about point O is therefore

$$I_O = (I_O)_d - (I_O)_h$$

$$= 1.473 \text{ kg} \cdot \text{m}^2 - 0.276 \text{ kg} \cdot \text{m}^2$$

$$= 1.20 \text{ kg} \cdot \text{m}^2 \qquad\qquad\qquad Ans.$$

E X A M P L E 10.14

Fig. 10–28

The pendulum in Fig. 10–28 consists of two thin rods each having a mass of 100 kg. Determine the pendulum's mass moment of inertia about an axis passing through (a) the pin at O, and (b) the mass center G of the pendulum.

Solution

Part (a). Using the table on the inside back cover, the moment of inertia of rod OA about an axis perpendicular to the page and passing through the end point O of the rod is $I_O = \frac{1}{3}ml^2$. Hence,

$$(I_{OA})_O = \frac{1}{3}ml^2 = \frac{1}{3}(100 \text{ kg})(3 \text{ m})^2 = 300 \text{ kg} \cdot \text{m}^2$$

This same value may be computed using $I_G = \frac{1}{12}ml^2$ and the parallel-axis theorem; i.e.,

$$(I_{OA})_O = \frac{1}{12}ml^2 + md^2 = \frac{1}{12}(100 \text{ kg})(3 \text{ m})^2 + (100 \text{ kg})(1.5 \text{ m})^2$$
$$= 300 \text{ kg} \cdot \text{m}^2$$

For rod BC we have

$$(I_{BC})_O = \frac{1}{12}ml^2 + md^2 = \frac{1}{12}(100 \text{ kg})(3 \text{ m})^2 + (100 \text{ kg})(3 \text{ m})^2$$
$$= 975 \text{ kg} \cdot \text{m}^2$$

The moment of inertia of the pendulum about O is therefore

$$I_O = 300 + 975 = 1275 \text{ kg} \cdot \text{m}^2 \qquad \textit{Ans.}$$

Part (b). The mass center G will be located relative to the pin at O. Assuming this distance to be $\bar{y}$, Fig. 10–28, and using the formula for determining the mass center, we have

$$\bar{y} = \frac{\Sigma \tilde{y}m}{\Sigma m} = \frac{1.5 \text{ m} (100 \text{ kg}) + 3 \text{ m} (100 \text{ kg})}{100 \text{ kg} + 100 \text{ kg}} = 2.25 \text{ m}$$

The moment of inertia I_G may be computed in the same manner as I_O, which requires successive applications of the parallel-axis theorem in order to transfer the moments of inertia of rods OA and BC to G. A more direct solution, however, involves applying the parallel-axis theorem using the result for I_O determined above; i.e.,

$$I_O = I_G + md^2; \quad 1275 \text{ kg} \cdot \text{m}^2 = I_G + (200 \text{ kg})(2.25 \text{ m})^2$$

$$I_G = 262.5 \text{ kg} \cdot \text{m}^2 \qquad \textit{Ans.}$$

PROBLEMS

10-90. Determine the moment of inertia. I_y for the slender rod. The rod's density ρ and cross-sectional area A are constant. Express the result in terms of the rod's total mass m.

Prob. 10–90

10-91. Determine the moment of inertia of the thin ring about the z axis. The ring has a mass m.

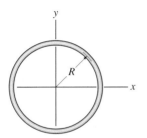

Prob. 10–91

*10-92. Determine the moment of inertia I_x of the right circular cone and express the result in terms of the total mass m of the cone. The cone has a constant density ρ.

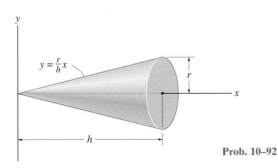

Prob. 10–92

10-93. Determine the moment of inertia I_x of the sphere and express the result in terms of the total mass m of the sphere. The sphere has a constant density ρ.

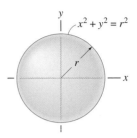

Prob. 10–93

10-94. Determine the radius of gyration k_x of the paraboloid. The density of the material is $\rho = 5 \text{ Mg/m}^3$.

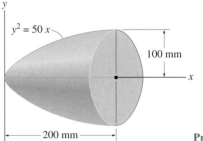

Prob. 10–94

10-95. Determine the moment of inertia of the semi-ellipsoid with respect to the x axis and express the result in terms of the mass m of the semiellipsoid. The material has a constant density ρ.

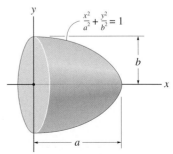

Prob. 10–95

*10-96. Determine the radius of gyration k_x of the body. The density of the material is $\rho = 6000 \text{ kg/m}^3$.

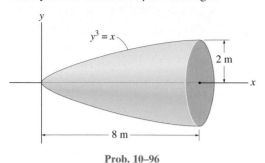

Prob. 10–96

10-97. Determine the moment of inertia of the ellipsoid with respect to the x axis and express the result in terms of the mass m of the ellipsoid. The material has a constant density ρ.

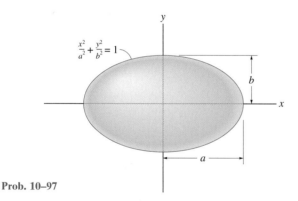

Prob. 10–97

10-98. Determine the moment of inertia of the homogenous triangular prism with respect to the y axis. Express the result in terms of the mass m of the prism. *Hint:* For integration, use thin plate elements parallel to the x-y plane having a thickness of dz.

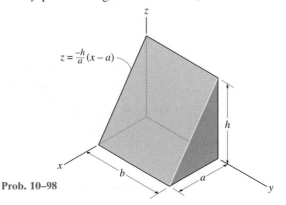

Prob. 10–98

10-99. The concrete shape is formed by rotating the shaded area about the y axis. Determine the moment of inertia I_y. The density of concrete is $\rho = 2400 \text{ kg/m}^3$.

Prob. 10–99

*10-100. Determine the moment of inertia of the wire triangle about an axis perpendicular to the page and passing through point O. Also, locate the mass center G and determine the moment of inertia about an axis perpendicular to the page and passing through point G. The wire has a mass of 0.3 kg/m. Neglect the size of the ring at O.

Prob. 10–100

10-101. Determine the moment of inertia I_z of the frustum of the cone which has a conical depression. The material has a density of 200 kg/m^3.

Prob. 10–101

10-102. Determine the moment of inertia of the wheel about the x axis that passes through the center of mass G. The material has a density of $\rho = 1500 \text{ kg/m}^3$.

10-103. Determine the moment of inertia of the wheel about the x' axis that passes through point O. The material has a density of $\rho = 1500 \text{ kg/m}^3$.

10-105. The slender rods have a mass of 4 kg/m. Determine the moment of inertia of the assembly about an axis perpendicular to the page and passing through point A.

Prob. 10–105

Probs. 10–102/103

****10-104.** The pendulum consists of a disk having a mass of 6 kg and slender rods AB and DC which have a mass of 2 kg/m. Determine the length L of DC so that the center of the mass is at the bearing O. What is the moment of inertia of the assembly about an axis perpendicular to the page and passing through point O?

10-106. Determine the moment of inertia I_z of the frustrum of the cone which has a conical depression. The material has a density of 200 kg/m³.

Prob. 10–106

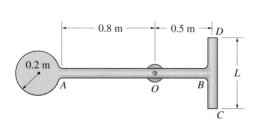

Prob. 10–104

10-107. The slender rods have a mass of 3 kg/m. Determine the moment of inertia of the assembly about an axis perpendicular to the page and passing through point A.

Prob. 10–107

*10-108. The pendulum consists of a plate having a mass of 60 kg and a slender rod having a mass of 20 kg. Determine the radius of gyration of the pendulum about an axis perpendicular to the page and passing through point O.

Prob. 10–108

10-109. Determine the moment of inertia of the overhung crank about the x axis. The material is steel having a density of $\rho = 7.85$ Mg/m³.

10-110. Determine the moment of inertia of the overhung crank about the x' axis. The material is steel having a density of $\rho = 7.85$ Mg/m³.

Probs. 10–109/110

10-111. Determine the location of $\bar{y}$ of the center of mass G of the assembly and then calculate the moment of inertia about an axis perpendicular to the page and passing through G. The block has a mass of 3 kg and the mass of the semicylinder is 5 kg.

Prob. 10–111

*10-112. The pendulum consists of two slender rods AB and OC which have a mass of 3 kg/m. The thin plate has a mass of 12 kg/m². Determine the location $\bar{y}$ of the center of mass G of the pendulum, then calculate the moment of inertia of the pendulum about an axis perpendicular to the page and passing through G.

Prob. 10–112

CHAPTER REVIEW

- **Area Moment of Inertia.**   The *area moment of inertia* represents the second moment of the area about an axis, $I = \int r^2 \, dA$. It is frequently used in formulas related to strength and stability of structural members or mechanical elements. If the area shape is irregular, then a differential element must be selected and integration over the entire area must be performed. Tabular values of the moment of inertia of common shapes about their *centroidal axis* are available. To determine the moment of inertia of these shapes about some *other axis*, the parallel-axis theorem must be used, $I = \bar{I} + Ad^2$. If an area is a composite of these shapes, then its moment of inertia is equal to the sum of the moments of inertia of each of its parts.

- **Product of Inertia.**   The *product of inertia* of an area is used to determine the location of an axis about which the moment of inertia for the area is a maximum or minimum. This property is determined from $I_{xy} = \int xy \, dA$, where the integration is performed over the entire area. If the product of inertia for an area is known about its centroidal x', y' axes, then its value can be determined about any x, y axes using the parallel-axis theorem for the product of inertia, $I_{xy} = \bar{I}_{x'y'} + A \, d_x d_y$.

- **Principal Moments of Inertia.**   Provided the moments of inertia I_x and I_y, and the product of inertia I_{xy} are known, then formulas, or Mohr's circle, can be used to determine the maximum and minimum or *principal moments of inertia* for the area, as well as finding the orientation of the principal axes of inertia.

- **Mass Moment of Inertia.**   The *mass moment of inertia* is a property of a body that measures its resistance to a change in its rotation. It is defined as the second moment of the mass elements of the body about an axis, $I = \int r^2 \, dm$. For bodies having axial symmetry, it can be determined by integration, using either disk or shell elements. The mass moment of inertia of a composite body is determined by using tabular values of its composite shapes along with the parallel-axis theorem, $I = \bar{I} + md^2$.

REVIEW PROBLEMS

10-113. Determine the area moment of inertia of the beam's cross-sectional area about the x axis which passes through the centroid C.

10-114. Determine the area moment of inertia of the beam's cross-sectional area about the y axis which passes through the centroid C.

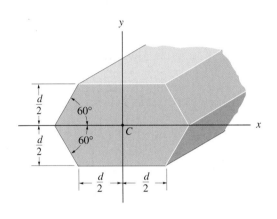

Probs. 10–113/114

*10-116. Determine the area moments of inertia I_x and I_y of the shaded area.

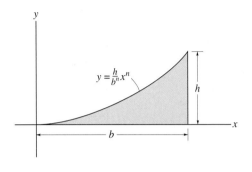

Prob. 10–116

10-115. Determine the mass moment of inertia I_x of the body and express the result in terms of the total mass m of the body. The density is constant.

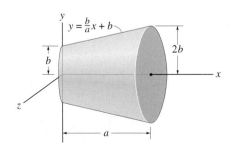

Prob. 10–115

10-117. Determine the area moments of inertia I_u and I_v and the product of inertia I_{uv} for the semicircular area.

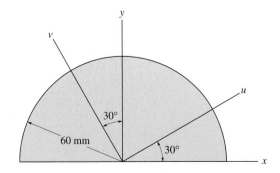

Prob. 10–117

10-118. Determine the area moment of inertia of the shaded area about the y axis.

10-119. Determine the area moment of inertia of the shaded area about the x axis.

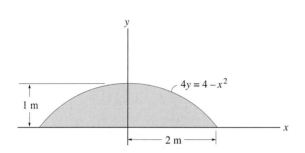

Probs. 10–118/119

10-121. Determine the area moment of inertia of the triangular area about (a) the x axis, and (b) the centroidal x' axis.

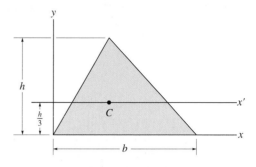

Prob. 10–121

*10-120. Determine the area moment of inertia of the area about the x axis. Then, using the parallel-axis theorem, find the area moment of inertia about the x' axis that passes through the centroid C of the area. $\bar{y} = 120$ mm.

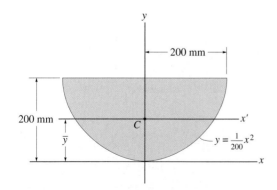

Prob. 10–120

10-122. Determine the product of inertia of the shaded area with respect to the x and y axes.

Prob. 10–122

Equilibrium and stability of this articulated crane boom as a function of its position can be analyzed using methods based on work and energy, which are explained in this chapter.

<table>
<tr><td>

CHAPTER

11

</td><td>

Virtual Work

</td></tr>
</table>

CHAPTER OBJECTIVES

- To introduce the principle of virtual work and show how it applies to determining the equilibrium configuration of a series of pin-connected members.

- To establish the potential energy function and use the potential-energy method to investigate the type of equilibrium or stability of a rigid body or configuration.

11.1 Definition of Work and Virtual Work

Work of a Force. In mechanics a force **F** does work only when it undergoes a displacement in the direction of the force. For example, consider the force **F** in Fig. 11–1, which is located on the path s specified by the position vector **r**. If the force moves along the path to a new position $\mathbf{r}' = \mathbf{r} + d\mathbf{r}$, the displacement is $d\mathbf{r}$, and therefore the work dU is a *scalar quantity* defined by the dot product

$$dU = \mathbf{F} \cdot d\mathbf{r}$$

Fig. 11–1

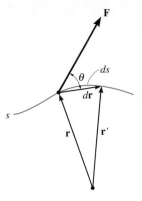

Fig. 11–1

Because $d\mathbf{r}$ is infinitesimal, the magnitude of $d\mathbf{r}$ can be represented by ds, the differential arc segment along the path. If the angle between the tails of $d\mathbf{r}$ and $\mathbf{F}$ is θ, Fig. 11–1, then by definition of the dot product, the above equation may also be written as

$$dU = F\, ds \cos \theta$$

Work expressed by this equation may be interpreted in one of two ways: either as the product of F and the component of displacement in the direction of the force, i.e., $ds \cos \theta$; or as the product of ds and the component of force in the direction of displacement, i.e., $F \cos \theta$. Note that if $0° \leq \theta < 90°$, then the force component and the displacement have the *same sense*, so that the work is *positive*; whereas if $90° < \theta \leq 180°$, these vectors have an *opposite sense*, and therefore the work is *negative*. Also, $dU = 0$ if the force is *perpendicular* to displacement, since $\cos 90° = 0$, or if the force is applied at a *fixed point*, in which case the displacement $ds = 0$.

The basic unit for work combines the units of force and displacement. In the SI system a *joule* (J) is equivalent to the work done by a force of 1 newton which moves 1 meter in the direction of the force $(1\,\text{J} = 1\,\text{N} \cdot \text{m})$. The moment of a force has the same combination of units; however, the concepts of moment and work are in no way related. A moment is a vector quantity, whereas work is a scalar.

(a)

Translation

(b)

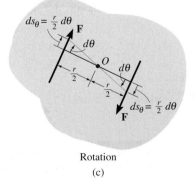

Rotation

(c)

Fig. 11–2

Work of a Couple. The two forces of a couple do work when the couple *rotates* about an axis perpendicular to the plane of the couple. To show this, consider the body in Fig. 11–2a, which is subjected to a couple whose moment has a magnitude $M = Fr$. Any general differential displacement of the body can be considered as a combination of a translation and rotation. When the body *translates* such that the *component of displacement* along the line of action of each force is ds_t, clearly the "positive" work of one force ($F\,ds_t$) *cancels* the "negative" work of the other ($-F\,ds_t$), Fig. 11–2b. Consider now a differential *rotation* $d\theta$ of the body about an axis perpendicular to the plane of the couple, which intersects the plane at point O, Fig. 11–2c. (For the derivation, any other point in the plane may also be considered.) As shown, each force undergoes a displacement $ds_\theta = (r/2)\,d\theta$ in the direction of the force; hence, the work of both forces is

$$dU = F\left(\frac{r}{2}d\theta\right) + F\left(\frac{r}{2}d\theta\right) = (Fr)\,d\theta$$

or

$$dU = M\,d\theta$$

The resultant work is *positive* when the sense of $\mathbf{M}$ is the *same* as that of $d\boldsymbol{\theta}$, and negative when they have an opposite sense. As in the case of the moment vector, the *direction and sense* of $d\boldsymbol{\theta}$ are defined by the right-hand rule, where the fingers of the right hand follow the rotation or "curl" and the thumb indicates the direction of $d\boldsymbol{\theta}$. Hence, the line of action of $d\boldsymbol{\theta}$ will be *parallel* to the line of action of $\mathbf{M}$ if movement of the body occurs in the *same plane*. If the body rotates in space, however, the *component* of $d\boldsymbol{\theta}$ in the direction of $\mathbf{M}$ is required. Thus, in general, the work done by a couple is defined by the dot product, $dU = \mathbf{M} \cdot d\boldsymbol{\theta}$.

Virtual Work. The definitions of the work of a force and a couple have been presented in terms of *actual movements* expressed by differential displacements having magnitudes of ds and $d\theta$. Consider now an *imaginary* or *virtual movement*, which indicates a displacement or rotation that is *assumed* and *does not actually exist*. These movements are first-order differential quantities and will be denoted by the symbols δs and $\delta\theta$ (delta s and delta θ), respectively. The *virtual work* done by a force undergoing a virtual displacement δs is

$$\boxed{\delta U = F\cos\theta\,\delta s} \tag{11–1}$$

Similarly, when a couple undergoes a virtual rotation $\delta\theta$ in the plane of the couple forces, the *virtual work* is

$$\boxed{\delta U = M\,\delta\theta} \tag{11–2}$$

11.2 Principle of Virtual Work for a Particle and a Rigid Body

Fig. 11–3

Fig. 11–4

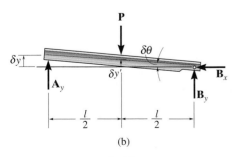

(a)

(b)

Fig. 11–5

Particle. If the particle in Fig. 11–3 undergoes an imaginary or virtual displacement $\delta\mathbf{r}$, then the virtual work (δU) done by the force system becomes

$$\delta U = \Sigma\mathbf{F}\cdot\delta\mathbf{r}$$
$$= (\Sigma F_x\mathbf{i} + \Sigma F_y\mathbf{j} + \Sigma F_z\mathbf{k})\cdot(\delta x\mathbf{i} + \delta y\mathbf{j} + \delta z\mathbf{k})$$
$$= \Sigma F_x\,\delta x + \Sigma F_y\,\delta y + \Sigma F_z\,\delta z$$

For equilibrium $\Sigma F_x = 0$, $\Sigma F_y = 0$, $\Sigma F_z = 0$, and so the virtual work must also be zero, i.e.,

$$\delta U = 0$$

In other words, we can write three independent virtual work equations corresponding to the three equations of equilibrium.

For example, consider the free body diagram of the ball which rests on the floor, Fig. 11–4. If we "imagine" the ball to be displaced downwards a virtual amount δy, then the weight does positive virtual work, $W\delta y$, and the normal force does negative virtual work, $-N\delta y$. For equilibrium the total virtual work must be zero, so that $\delta U = W\delta y - N\delta y = (W - N)\delta y = 0$. Since $\delta y \neq 0$, then $N = W$ as required.

Rigid Body. In a similar manner, we can also write a set of three virtual work equations ($\delta U = 0$) for a rigid body subjected to a coplanar force system. If these equations involve separate virtual translations in the x and y directions and a virtual rotation about an axis perpendicular to the x–y plane and passing through an arbitrary point O, then it can be shown that they will correspond to the three equilibrium equations, $\Sigma F_x = 0$, $\Sigma F_y = 0$, and $\Sigma M_O = 0$. When writing these equations, it is *not necessary* to include the work done by the *internal forces* acting within the body since a rigid body *does not deform* when subjected to an external loading, and furthermore, when the body moves through a virtual displacement, the internal forces occur in equal but opposite collinear pairs, so that the corresponding work done by each pair of forces *cancels*.

To demonstrate an application, consider the simply supported beam in Fig. 11–5a. When the beam is given a virtual rotation $\delta\theta$ about point B, Fig. 11–5b, the only forces that do work are $\mathbf{P}$ and $\mathbf{A}_y$. Since $\delta y = l\delta\theta$ and $\delta y' = (l/2)\delta\theta$, the virtual work equation for this case is $\delta U = A_y(l\delta\theta) - P(l/2)\delta\theta = (A_y - P/2)l\delta\theta = 0$. Since $\delta\theta \neq 0$, then $A_y = P/2$. Excluding $\delta\theta$, notice that the terms in parentheses actually represent moment equilibrium about point B.

As in the case of a particle, no added advantage is gained by solving rigid-body equilibrium problems using the principle of virtual work. This is because for each application of the virtual-work equation the virtual displacement, common to every term, factors out, leaving an equation that could have been obtained in a more *direct manner* by simply applying the equations of equilibrium.

11.3 Principle of Virtual Work for a System of Connected Rigid Bodies

The method of virtual work is most suitable for solving equilibrium problems that involve a system of several *connected* rigid bodies such as the ones shown in Fig. 11–6. Before we can apply the principle of virtual work to these systems, however, we must first specify the number of degrees of freedom for a system and establish coordinates that define the position of the system.

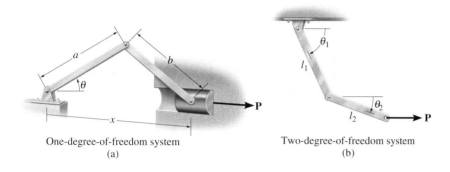

One-degree-of-freedom system
(a)

Two-degree-of-freedom system
(b)

Fig. 11–6

Degrees of Freedom.   A system of connected bodies takes on a unique shape that can be specified provided we know the position of a number of specific points on the system. These positions are defined using *independent coordinates q*, which are measured from fixed reference points. For every coordinate established, the system will have a *degree of freedom* for displacement along the coordinate axis such that it is consistent with the constraining action of the supports. Thus, an *n*-degree-of-freedom system requires *n* independent coordinates q_n to specify the location of all its members. For example, the link and sliding-block arrangement shown in Fig. 11–6a is an example of a one-degree-of-freedom system. The independent coordinate $q = \theta$ may be used to specify the location of the two connecting links and the block. The coordinate x could also be used as the independent coordinate. However, since the block is constrained to move within the slot, x is not independent of θ; rather, it can be related to θ using the cosine law, $b^2 = a^2 + x^2 - 2ax \cos \theta$. The double-link arrangement, shown in Fig. 11–6b, is an example of a two-degrees-of-freedom system. To specify the location of each link, the coordinate angles θ_1 and θ_2 must be known since a rotation of one link is independent of a rotation of the other.

During operation the scissors lift has one degree of freedom. Without dismembering the mechanism, the hydraulic force required to provide the lift can be determined *directly* by using the principle of virtual work.

Principle of Virtual Work.   The principle of virtual work for a system of rigid bodies whose connections are *frictionless* may be stated as follows: *A system of connected rigid bodies is in equilibrium provided the virtual work done by all the external forces and couples acting on the system is zero for each independent virtual displacement of the system.* Mathematically, this may be expressed as

$$\delta U = 0 \qquad\qquad (11\text{–}3)$$

where δU represents the virtual work of all the external forces (and couples) acting on the system during any independent virtual displacement.

As stated above, if a system has n degrees of freedom it takes n independent coordinates q_n to completely specify the location of the system. Hence, for the system it is possible to write n independent virtual-work equations, one for every virtual displacement taken along each of the independent coordinate axes, while the remaining $n - 1$ independent coordinates are held *fixed.**

IMPORTANT POINTS

- A force does work when it moves through a displacement in the direction of the force. A couple moment does work when it moves through a collinear rotation. Specifically, positive work is done when the force or couple moment and its displacement have the same sense of direction.

- The principle of virtual work is generally used to determine the equilibrium configuration for a series of multiply-connected members.

- A virtual displacement is imaginary, i.e., does not really happen. It is a differential that is given in the positive direction of the position coordinate.

- Forces or couple moments that do not virtually displace do no virtual work.

---

*This method of applying the principle of virtual work is sometimes called the *method of virtual displacements* since a virtual displacement is applied, resulting in the calculation of a real force. Although it is not to be used here, realize that we can also apply the principle of virtual work as a method of virtual forces. This method is often used to determine the displacements of points on deformable bodies. See R. C. Hibbeler, *Mechanics of Materials*, SI edition, Prentice Hall, Inc., 2004.

PROCEDURE FOR ANALYSIS

The equation of virtual work can be used to solve problems involving a system of frictionless connected rigid bodies having a single degree of freedom by using the following procedure.

Free-Body Diagram.

- Draw the free-body diagram of the entire system of connected bodies and define the *independent coordinate q*.

- Sketch the "deflected position" of the system on the free-body diagram when the system undergoes a positive virtual displacement δq.

Virtual Displacements.

- Indicate *position coordinates s_i*, measured from a *fixed point* on the free-body diagram to each of the *i* number of "active" forces and couples, i.e., those that do work.

- Each coordinate axis should be parallel to the line of action of the "active" force to which it is directed, so that the virtual work along the coordinate axis can be calculated.

- Relate each of the position coordinates s_i to the independent coordinate *q*; then *differentiate* these expressions in order to express the virtual displacements δs_i in terms of δq.

Virtual Work Equation.

- Write the *virtual-work equation* for the system assuming that, whether possible or not, all the position coordinates s_i undergo *positive* virtual displacements δs_i.

- Using the relations for δs_i, express the work of *each* "active" force and couple in the equation in terms of the *single* independent virtual displacement δq.

- Factor out this common displacement from all the terms and solve for the unknown force, couple, or equilibrium position, *q*.

- If the system contains *n* degrees of freedom, *n* independent coordinates q_n must be specified. Follow the above procedure and let *only one* of the independent coordinates undergo a virtual displacement, while the remaining $n - 1$ coordinates are *held fixed*. In this way, *n* virtual-work equations can be written, one for each independent coordinate.

EXAMPLE 11.1

Determine the angle θ for equilibrium of the two-member linkage shown in Fig. 11–7a. Each member has a mass of 10 kg.

Solution

Free-Body Diagram. The system has only one degree of freedom since the location of both links may be specified by the single independent coordinate $(q =) \theta$. As shown on the free-body diagram in Fig. 11–7b, when θ undergoes a *positive* (clockwise) virtual rotation $\delta\theta$, only the active forces, **F** and the two 98.1-N weights, do work. (The reactive forces $\mathbf{D}_x$ and $\mathbf{D}_y$ are fixed, and $\mathbf{B}_y$ does not move along its line of action.)

Virtual Displacements. If the origin of coordinates is established at the *fixed* pin support D, the location of **F** and **W** may be specified by the *position coordinates* x_B and y_w, as shown in the figure. In order to determine the work, note that these coordinates are parallel to the lines of action of their associated forces.

Expressing the position coordinates in terms of the independent coordinate θ and taking the derivatives yields

$$x_B = 2(1 \cos \theta) \text{ m} \qquad \delta x_B = -2 \sin \theta \, \delta\theta \text{ m} \qquad (1)$$

$$y_w = \tfrac{1}{2}(1 \sin \theta) \text{ m} \qquad \delta y_w = 0.5 \cos \theta \, \delta\theta \text{ m} \qquad (2)$$

It is seen by the *signs* of these equations, and indicated in Fig. 11–7b, that an *increase* in θ (i.e., $\delta\theta$) causes a *decrease* in x_B and an *increase* in y_w.

Virtual-Work Equation. If the virtual displacements δx_B and δy_w were *both positive*, then the forces **W** and **F** would do positive work since the forces and their corresponding displacements would have the same sense. Hence, the virtual-work equation for the displacement $\delta\theta$ is

$$\delta U = 0; \qquad W \, \delta y_w + W \, \delta y_w + F \, \delta x_B = 0 \qquad (3)$$

Substituting Eqs. 1 and 2 into Eq. 3 in order to relate the virtual displacements to the common virtual displacement $\delta\theta$ yields

$$98.1(0.5 \cos \theta \, \delta\theta) + 98.1(0.5 \cos \theta \, \delta\theta) + 25(-2 \sin \theta \, \delta\theta) = 0$$

Notice that the "negative work" done by **F** (force in the opposite sense to displacement) has been *accounted for* in the above equation by the "negative sign" of Eq. 1. Factoring out the *common displacement* $\delta\theta$ and solving for θ, noting that $\delta\theta \neq 0$, yields

$$(98.1 \cos \theta - 50 \sin \theta) \, \delta\theta = 0$$

$$\theta = \tan^{-1}\frac{98.1}{50} = 63.0° \qquad \qquad Ans.$$

If this problem had been solved using the equations of equilibrium, it would have been necessary to dismember the links and apply three scalar equations to *each* link. The principle of virtual work, by means of calculus, has eliminated this task so that the answer is obtained directly.

(a)

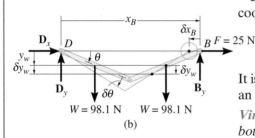

(b)

Fig. 11–7

E X A M P L E 11.2

Determine the angle θ required to maintain equilibrium of the mechanism in Fig. 11–8a. Neglect the weight of the links. The spring is unstretched when $\theta = 0°$, and it maintains a horizontal position due to the roller.

Solution

Free-Body Diagram. The mechanism has one degree of freedom, and therefore the location of each member may be specified using the independent coordinate θ. When θ undergoes a *positive* virtual displacement $\delta\theta$, as shown on the free-body diagram in Fig. 11–8b, links AB and EC rotate by the same amount since they have the same length, and link BC only translates. Since a couple moment does work *only* when it rotates, the work done by $\mathbf{M}_2$ is zero. The reactive forces at A and E do no work. Why?

Fig. 11–8

Virtual Displacements. The position coordinates x_B and x_D are *parallel* to the lines of action of $\mathbf{P}$ and $\mathbf{F}_s$, and these coordinates locate the forces with respect to the *fixed points* A and E. From Fig. 11–8b,

$$x_B = 0.4 \sin\theta \text{ m}$$
$$x_D = 0.2 \sin\theta \text{ m}$$

Thus,

$$\delta x_B = 0.4 \cos\theta \; \delta\theta \text{ m}$$
$$\delta x_D = 0.2 \cos\theta \; \delta\theta \text{ m}$$

Virtual-Work Equation. For *positive* virtual displacements, $\mathbf{F}_s$ is opposite to δx_D and hence does negative work. Thus,

$$\delta U = 0; \qquad M_1 \, \delta\theta + P \, \delta x_B - F_s \, \delta x_D = 0$$

Relating each of the virtual displacements to the *common* virtual displacement $\delta\theta$ yields

$$0.5 \, \delta\theta + 2(0.4 \cos\theta \; \delta\theta) - F_s(0.2 \cos\theta \; \delta\theta) = 0$$
$$(0.5 + 0.8 \cos\theta - 0.2 F_s \cos\theta) \, \delta\theta = 0 \qquad (1)$$

For the arbitrary angle θ, the spring is stretched a distance of $x_D = (0.2 \sin\theta) \text{ m}$; and therefore, $F_s = 60 \text{ N/m}(0.2 \sin\theta) \text{ m} = (12 \sin\theta) \text{ N}$. Substituting into Eq. 1 and noting that $\delta\theta \neq 0$, we have

$$0.5 + 0.8 \cos\theta - 0.2(12 \sin\theta) \cos\theta = 0$$

Since $\sin 2\theta = 2 \sin\theta \cos\theta$, then

$$1 = 2.4 \sin 2\theta - 1.6 \cos\theta$$

Solving for θ by trial and error yields

$$\theta = 36.3° \qquad\qquad\qquad Ans.$$

E X A M P L E **11.3**

(a)

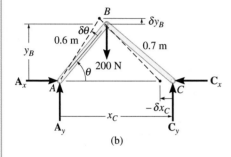

(b)

Fig. 11–9

Determine the horizontal force C_x that the pin at C must exert on BC in order to hold the mechanism shown in Fig. 11–9a in equilibrium when $\theta = 45°$. Neglect the weight of the members.

Solution

Free-Body Diagram. The reaction $\mathbf{C}_x$ can be obtained by *releasing* the pin constraint at C in the x direction and allowing the frame to be displaced in this direction. The system then has only one degree of freedom, defined by the independent coordinate θ, Fig. 11–9b. When θ undergoes a *positive* virtual displacement $\delta\theta$, only $\mathbf{C}_x$ and the 200-N force do work.

Virtual Displacements. Forces $\mathbf{C}_x$ and 200 N are located from the fixed origin A using position coordinates y_B and x_C. From Fig. 11–9b, x_C can be related to θ by the "law of cosines." Hence,

$$(0.7)^2 = (0.6)^2 + x_C^2 - 2(0.6)x_C \cos\theta \tag{1}$$

$$0 = 0 + 2x_C \, \delta x_C - 1.2 \, \delta x_C \cos\theta + 1.2x_C \sin\theta \, \delta\theta$$

$$\delta x_C = \frac{1.2x_C \sin\theta}{1.2 \cos\theta - 2x_C}\delta\theta \tag{2}$$

Also,

$$y_B = 0.6 \sin\theta$$

$$\delta y_B = 0.6 \cos\theta \, \delta\theta \tag{3}$$

Virtual-Work Equation. When y_B and x_C undergo *positive* virtual displacements δy_B and δx_C, $\mathbf{C}_x$ and 200 N do *negative work* since they both act in the opposite sense to $\delta\mathbf{y}_B$ and $\delta\mathbf{x}_C$. Hence,

$$\delta U = 0; \qquad -200 \, \delta y_B - C_x \, \delta x_C = 0$$

Substituting Eqs. 2 and 3 into this equation, factoring out $\delta\theta$, and solving for C_x yields

$$-200(0.6 \cos\theta \, \delta\theta) - C_x \frac{1.2x_C \sin\theta}{1.2 \cos\theta - 2x_C}\delta\theta = 0$$

$$C_x = \frac{-120 \cos\theta \, (1.2 \cos\theta - 2x_C)}{1.2x_C \sin\theta} \tag{4}$$

At the required equilibrium position $\theta = 45°$, the corresponding value of x_C can be found by using Eq. 1, in which case

$$x_C^2 - 1.2 \cos 45° \, x_C - 0.13 = 0$$

Solving for the positive root yields

$$x_C = 0.981 \text{ m}$$

Thus, from Eq. 4,

$$C_x = 114 \text{ N} \qquad\qquad Ans.$$

EXAMPLE 11.4

Determine the equilibrium position of the two-bar linkage shown in Fig. 11–10a. Neglect the weight of the links.

Solution

The system has two degrees of freedom since the *independent coordinates* θ_1 and θ_2 must be known to locate the position of both links. The position coordinate x_B, measured from the fixed point O, is used to specify the location of **P**, Fig. 11–10b and c.

If θ_1 is held *fixed* and θ_2 varies by an amount $\delta\theta_2$, as shown in Fig. 11–10b, the virtual-work equation becomes

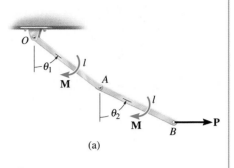

$$[\delta U = 0]_{\theta_2}; \qquad P(\delta x_B)_{\theta_2} - M\,\delta\theta_2 = 0 \qquad (1)$$

Here P and M represent the magnitudes of the applied force and couple moment acting on link AB.

When θ_2 is held *fixed* and θ_1 varies by an amount $\delta\theta_1$, as shown in Fig. 11–10c, then AB translates and the virtual-work equation becomes

$$[\delta U = 0]_{\theta_1}; \qquad P(\delta x_B)_{\theta_1} - M\,\delta\theta_1 = 0 \qquad (2)$$

The *position coordinate* x_B may be related to the independent coordinates θ_1 and θ_2 by the equation

$$x_B = l\sin\theta_1 + l\sin\theta_2 \qquad (3)$$

To obtain the variation δx_B in terms of $\delta\theta_2$, it is necessary to take the *partial derivative* of x_B with respect to θ_2 since x_B is a function of both θ_1 and θ_2. Hence,

$$\frac{\partial x_B}{\partial\theta_2} = l\cos\theta_2 \qquad (\delta x_B)_{\theta_2} = l\cos\theta_2\,\delta\theta_2$$

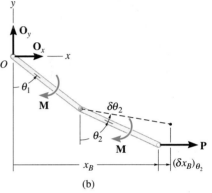

(b)

Substituting into Eq. 1, we have

$$(Pl\cos\theta_2 - M)\,\delta\theta_2 = 0$$

Since $\delta\theta_2 \neq 0$, then

$$\theta_2 = \cos^{-1}\left(\frac{M}{Pl}\right) \qquad \textit{Ans.}$$

Using Eq. 3 to obtain the variation of x_B with θ_1 yields

$$\frac{\partial x_B}{\partial\theta_1} = l\cos\theta_1 \qquad (\delta x_B)_{\theta_1} = l\cos\theta_1\,\delta\theta_1$$

Substituting into Eq. 2, we have

$$(Pl\cos\theta_1 - M)\,\delta\theta_1 = 0$$

Since $\delta\theta_1 \neq 0$, then

$$\theta_1 = \cos^{-1}\left(\frac{M}{Pl}\right) \qquad \textit{Ans.}$$

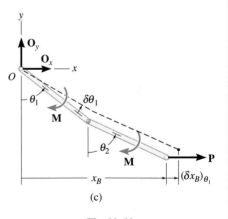

(c)

Fig. 11–10

PROBLEMS

11-1. Use the method of virtual work to determine the tensions in cable *AC*. The lamp weighs 100 N ($\approx$ 10 kg).

Prob. 11–1

11-2. The uniform rod *OA* has a weight of 100 N ($\approx$ 10 kg). When the rod is in vertical position, $\theta = 0°$, the spring is unstretched. Determine the angle θ for equilibrium if the end of the spring wraps around the periphery of the disk as the disk turns.

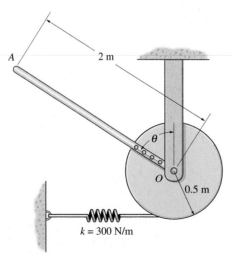

Prob. 11–2

11-3. Determine the force **F** acting on the cord which is required to maintain equilibrium of the horizontal 10-kg bar *AB*. *Hint:* Express the total *constant vertical length l* of the cord in terms of position coordinates s_1 and s_2. The derivative of this equation yields a relationship between δ_1 and δ_2.

Prob. 11–3

*■**11-4.** Each member of the pin-connected mechanism has a mass of 8 kg. If the spring is unstretched when $\theta = 0°$, determine the angle θ for equilibrium. Set $k = 2500$ N/m and $M = 50$ N·m.

11-5. Each member of the pin-connected mechanism has a mass of 8 kg. If the spring is unstretched when $\theta = 0°$, determine the required stiffness k so that the mechanism is in equilibrium when $\theta = 30°$. Set **M** = 0.

Probs. 11–4/5

11-6. The crankshaft is subjected to a torque of $M = 50\,\text{N}\cdot\text{m}$. Determine the horizontal compressive force F applied to the piston for equilibrium when $\theta = 60°$.

11-7. The crankshaft is subjected to a torque of $M = 50\,\text{N}\cdot\text{m}$. Determine the horizontal compressive force F and plot the result of F (ordinate) versus θ (abscissa) for $0° \le \theta \le 90°$.

11-9. Determine the angles θ for equilibrium of the 4-kN (≈ 400-kg) disk using the principle of the virtual work. Neglect the weight of the rod. The spring is unstretched when $\theta = 0°$ and always remains in the vertical position due to the roller guide.

Probs. 11–6/7

Prob. 11–9

***11-8.** Determine the force developed in the spring required to keep the 100-N (≈ 10-kg) uniform rod AB in equilibrium when $\theta = 35°$.

11-10. If each of the three links of the mechanism has a weight of 200 N (≈ 20 kg), determine the angle θ for equilibrium of the spring, which, due to the roller guide, always remains horizontal and is unstretched when $\theta = 0°$.

Prob. 11–8

Prob. 11–10

11-11. When $\theta = 20°$, the 5-kN ($\approx$ 500-kg) uniform block compresses the two vertical springs 4 m. If the uniform links AB and CD each weigh 1 kN ($\approx$ 100 kg), determine the magnitude of the applied couple moments **M** needed to maintain equilibrium when $\theta = 20°$.

Prob. 11–11

*11-12.** The spring is unstretched when $\theta = 0°$. If $P = 8$ kN, determine the angle θ for equilibrium. Due to the roller guide, the spring always remains vertical. Neglect the weight of the links.

Prob. 11–12

11-13. The thin rod of weight W rest against the smooth wall and floor. Determine the magnitude of force **P** needed to hold it in equilibrium for a given angle θ.

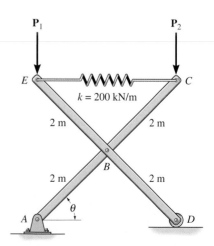

Prob. 11–13

■**11-14.** The 4-m members of the mechanism are pin-connected at their centers. If vertical forces $P_1 = P_2 = 30$ kN act at C and E as shown, determine the angle θ for equilibrium. The spring is unstretched when $\theta = 45°$. Neglect the weight of the members.

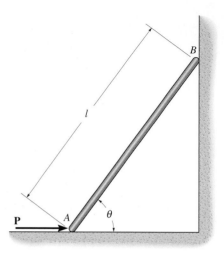

Prob. 11–14

11-15. The spring has an unstretched length of 0.3 m. Determine the angle θ for equilibrium if the uniform links each have a mass of 5 kg.

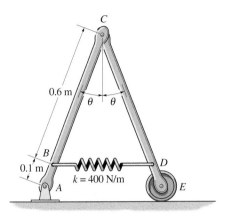

Prob. 11–15

11-17. The machine shown is used for forming metal plates. It consists of two toggles ABC and DEF, which are operated by hydraulic cylinder BE. The toggles push the moveable bar FC forward, pressing the plate p into the cavity. If the force which the plate exerts on the head is $P = 8\,\text{kN}$, determine the force **F** in the hydraulic cylinder when $\theta = 30°$.

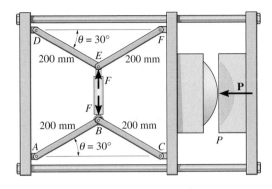

Prob. 11–17

***11-16.** Determine the force F needed to lift the block having a weight of 100 N ($\approx$ 10 kg). *Hint:* Note that the coordinates s_A and s_B can be related to the *constant* vertical length l of the cord.

Prob. 11–16

11-18. The vent plate is supported at B by a pin. If it weighs 150 N ($\approx$ 15 kg) and has a center of gravity at G, determine the stiffness k of the spring so that the plate remains in equilibrium at $\theta = 30°$. The spring is unstretched when $\theta = 0°$.

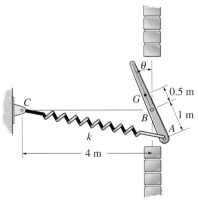

Prob. 11–18

11-19. The scissors jack supports a load **P**. Determine the axial force in the screw necessary for equilibrium when the jack is in the position θ. Each of the four links has a length L and is pin-connected at its center. Points B and D can move horizontally.

11-21. The piston C moves vertically between the two smooth walls. If the spring has a stiffness of $k = 1.5$ kN/m and is unstretched when $\theta = 0°$, determine the couple **M** that must be applied to link AB to hold the mechanism in equilibrium; $\theta = 30°$.

Prob. 11–19

Prob. 11–21

*11-20. Determine the mass of A and B required to hold the 400-g desk lamp in balance for any angles θ and ϕ. Neglect the weight of the mechanism and the size of the lamp.

11-22. The crankshaft is subjected to a torque of $M = 50$ N · cm. Determine the vertical compressive force **F** applied to the piston for equilibrium when $\theta = 60°$.

Prob. 11–20

Prob. 11–22

11-23. The assembly is used for exercise. It consist of four pin-connected bars, each of length L, and a spring of stiffness k and unstretched length a ($<2L$). If horizontal forces $\mathbf{P}$ and $-\mathbf{P}$ are applied to the handles so that θ is slowly decreased, determine the angle θ at which the magnitude of $\mathbf{P}$ becomes a maximum.

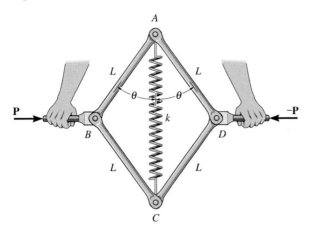

Prob. 11–23

11-25. Rods AB and BC have centers of mass located at their midpoints. If all contacting surfaces are smooth and BC has a mass of 100 kg, determine the appropriate mass of AB required for equilibrium.

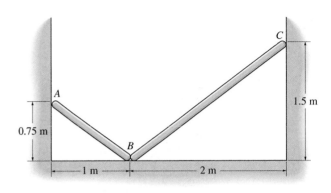

Prob. 11–25

*11-24. Determine the weight W of the crate if the angle $\theta = 45°$. The springs are unstretched when $\theta = 60°$. Neglect the weights of the members.

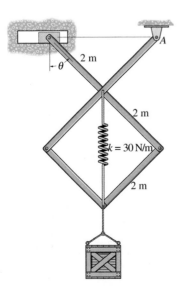

Prob. 11–24

*11.4 Conservative Forces

The work done by a force when it undergoes a *differential displacement* has been defined as $dU = F \cos \theta \, ds$, Fig. 11–1. If the force is displaced over a path that has a *finite length s*, the work is determined by integrating over the path: i.e.,

$$U = \int_s F \cos \theta \, ds$$

To evaluate the integral, it is necessary to obtain a relationship between F and the component of displacement $ds \cos \theta$. In some instances, however, the work done by a force will be *independent* of its path and, instead, will depend only on the initial and final locations of the force along the path. A force that has this property is called a *conservative force*.

Weight. Consider the body in Fig. 11–11, which is initially at P'. If the body is moved *down* along the *arbitrary path A* to the second position, then, for a given displacement ds along the path, the displacement component in the direction of **W** has a magnitude of $dy = ds \cos \theta$, as shown. Since both the force and displacement are in the same direction, the work is positive; hence,

$$U = \int_s W \cos \theta \, ds = \int_0^y W \, dy$$

or

$$U = Wy$$

In a similar manner, the work done by the weight when the body moves up a distance y back to P', along the arbitrary path A', is

$$U = -Wy$$

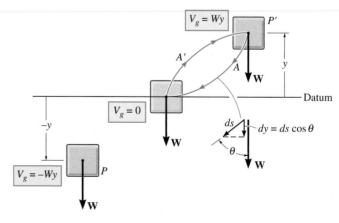

Fig. 11–11

Why is the work negative?

The weight of a body is therefore a conservative force since the work done by the weight depends *only* on the body's *vertical displacement* and is independent of the path along which the body moves.

Elastic Spring. The force developed by an elastic spring ($F_s = ks$) is also a conservative force. If the spring is attached to a body and the body is displaced along *any path*, such that it causes the spring to elongate or compress from a position s_1 to a further position s_2, the work will be negative since the spring exerts a force $\mathbf{F}_s$ *on the body* that is opposite to the body's displacement ds, Fig. 11–12. For either extension or compression, the work is independent of the path and is simply

$$U = \int_{s_1}^{s_2} F_s \, ds = \int_{s_1}^{s_2} (-ks) \, ds$$

$$= -\left(\tfrac{1}{2}ks_2^2 - \tfrac{1}{2}ks_1^2\right)$$

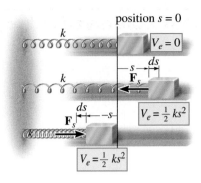

Fig. 11–12

Friction. In contrast to a conservative force, consider the force of *friction* exerted on a sliding body by a fixed surface. The work done by the frictional force depends on the path; the longer the path, the greater the work. Consequently, frictional forces are *nonconservative*, and the work done is dissipated from the body in the form of heat.

*11.5 Potential Energy

When a conservative force acts on a body, it gives the body the capacity to do work. This capacity, measured as *potential energy*, depends on the location of the body.

Gravitational Potential Energy. If a body is located a distance y *above* a fixed horizontal reference or datum, Fig. 11–11, the weight of the body has *positive* gravitational potential energy V_g since $\mathbf{W}$ has the capacity of doing positive work when the body is moved back down to the datum. Likewise, if the body is located a distance y *below* the datum, V_g is *negative* since the weight does negative work when the body is moved back up to the datum. At the datum, $V_g = 0$.

Measuring y as *positive upward*, the gravitational potential energy of the body's weight $\mathbf{W}$ is thus

$$V_g = Wy \qquad (11\text{–}4)$$

Datum

(a)

$F_s = ky_{eq}$

(b)

Fig. 11–13

Elastic Potential Energy. The elastic potential energy V_e that a spring produces on an attached body, when the spring is elongated or compressed from an undeformed position ($s = 0$) to a final position s, is

$$V_e = \tfrac{1}{2}ks^2 \tag{11–5}$$

Here V_e is *always positive* since in the deformed position the spring has the capacity of doing *positive work* in *returning* the body back to the spring's undeformed position, Fig. 11–12.

Potential Function. In the general case, if a body is subjected to *both* gravitational and elastic forces, the *potential energy or potential function* V of the body can be expressed as the algebraic sum

$$V = V_g + V_e \tag{11–6}$$

where measurement of V depends on the location of the body with respect to a selected datum in accordance with Eqs. 11–4 and 11–5.

In general, if a system of frictionless connected rigid bodies has a *single degree of freedom* such that its position from the datum is defined by the independent coordinate q, then the potential function for the system can be expressed as $V = V(q)$. The work done by all the conservative forces acting on the system in moving it from q_1 to q_2 is measured by the *difference* in V; i.e.,

$$U_{1-2} = V(q_1) - V(q_2) \tag{11–7}$$

For example, the potential function for a system consisting of a block of weight $\mathbf{W}$ supported by a spring, Fig. 11–13a, can be expressed in terms of its independent coordinate $(q =) y$, measured from a fixed datum located at the unstretched length of the spring; we have

$$V = V_g + V_e$$
$$= -Wy + \tfrac{1}{2}ky^2 \tag{11–8}$$

If the block moves from y_1 to a farther downward position y_2, then the work of $\mathbf{W}$ and $\mathbf{F}_s$ is

$$U_{1-2} = V(y_1) - V(y_2) = -W[y_1 - y_2] + \tfrac{1}{2}ky_1^2 - \tfrac{1}{2}ky_2^2$$

*11.6 Potential-Energy Criterion for Equilibrium

System Having One Degree of Freedom. When the displacement of a frictionless connected system is *infinitesimal*, i.e., from q to $q + dq$, Eq. 11–7 becomes

$$dU = V(q) - V(q + dq)$$

or

$$dU = -dV$$

Furthermore, if the system undergoes a *virtual displacement* δq, rather than an actual displacement dq, then $\delta U = -\delta V$. For equilibrium, the principle of virtual work requires that $\delta U = 0$, and therefore, provided the potential function for the system is known, this also requires that $\delta V = 0$. We can also express this requirement as

$$\boxed{\frac{dV}{dq} = 0} \tag{11-9}$$

Hence, *when a frictionless connected system of rigid bodies is in equilibrium, the first variation or change in V is zero.* This change is determined by taking the *first derivative* of the potential function and setting it equal to zero. For example, using Eq. 11–8 to determine the equilibrium position for the spring and block in Fig. 11–13a, we have

$$\frac{dV}{dy} = -W + ky = 0$$

Hence, the equilibrium position $y = y_{eq}$ is

$$y_{eq} = \frac{W}{k}$$

Of course, the *same result* is obtained by applying $\Sigma F_y = 0$ to the forces acting on the free-body diagram of the block, Fig. 11–13b.

System Having *n* Degrees of Freedom. When the system of connected bodies has n degrees of freedom, the total potential energy stored in the system will be a function of n independent coordinates q_n, i.e., $V = V(q_1, q_2, \ldots, q_n)$. In order to apply the equilibrium criterion $\delta V = 0$, it is necessary to determine the change in potential energy δV by using the "chain rule" of differential calculus; i.e.,

$$\delta V = \frac{\partial V}{\partial q_1} \delta q_1 + \frac{\partial V}{\partial q_2} \delta q_2 + \cdots + \frac{\partial V}{\partial q_n} \delta q_n = 0$$

Since the virtual displacements $\delta q_1, \delta q_2, \ldots, \delta q_n$ are independent of one another, the equation is satisfied provided

$$\frac{\partial V}{\partial q_1} = 0, \quad \frac{\partial V}{\partial q_2} = 0, \quad \ldots, \quad \frac{\partial V}{\partial q_n} = 0$$

Hence *it is possible to write n independent equations for a system having n degrees of freedom.*

*11.7 Stability of Equilibrium

Stable equilibrium   Neutral equilibrium

Unstable equilibrium

Fig. 11–14

Once the equilibrium configuration for a body or system of connected bodies is defined, it is sometimes important to investigate the "type" of equilibrium or the stability of the configuration. For example, consider the position of a ball resting at a point on each of the three paths shown in Fig. 11–14. Each situation represents an equilibrium state for the ball. When the ball is at A, it is said to be in *stable equilibrium* because if it is given a small displacement up the hill, it will always *return* to its original, lowest, position. At A, its total potential energy is a *minimum*. When the ball is at B, it is in *neutral equilibrium*. A small displacement to either the left or right of B will not alter this condition. The ball *remains* in equilibrium in the displaced position, and therefore its potential energy is *constant*. When the ball is at C, it is in *unstable equilibrium*. Here a small displacement will cause the ball's potential energy to be *decreased*, and so it will roll farther *away* from its original, highest position. At C, the potential energy of the ball is a *maximum*.

Types of Equilibrium. The example just presented illustrates that one of three types of equilibrium positions can be specified for a body or system of connected bodies.

1. *Stable equilibrium* occurs when a small displacement of the system causes the system to return to its original position. In this case the original potential energy of the system is a minimum.

2. *Neutral equilibrium* occurs when a small displacement of the system causes the system to remain in its displaced state. In this case the potential energy of the system remains constant.

3. *Unstable equilibrium* occurs when a small displacement of the system causes the system to move farther away from its original position. In this case the original potential energy of the system is a maximum.

During high winds and when going around a curve, these sugar-cane trucks can become unstable and tip over since their center of gravity is high off the road when they are fully loaded.

System Having One Degree of Freedom. For *equilibrium* of a system having a single degree of freedom, defined by the independent coordinate q, it has been shown that the first derivative of the potential function for the system must be equal to zero; i.e., $dV/dq = 0$. If the potential function $V = V(q)$ is plotted, Fig. 11–15, the first derivative (equilibrium position) is represented as the slope dV/dq, which is zero when the function is maximum, minimum, or an inflection point.

If the *stability* of a body is to be investigated, it is necessary to determine the *second derivative* of V and evaluate it at the equilibrium position $q = q_{eq}$. As shown in Fig. 11–15a, if $V = V(q)$ is a *minimum*, then

$$\frac{dV}{dq} = 0, \quad \frac{d^2V}{dq^2} > 0 \qquad \text{stable equilibrium} \qquad (11\text{–}10)$$

If $V = V(q)$ is a *maximum*, Fig. 11–15b, then

$$\frac{dV}{dq} = 0, \quad \frac{d^2V}{dq^2} < 0 \qquad \text{unstable equilibrium} \qquad (11\text{–}11)$$

If the second derivative is zero, it will be necessary to investigate *higher-order* derivatives to determine the stability. In particular, stable equilibrium will occur if the order of the lowest remaining nonzero derivative is *even* and the sign of this nonzero derivative is positive when it is evaluated at $q = q_{eq}$; otherwise, it is unstable.

If the system is in neutral equilibrium, Fig. 11–15c, it is required that

$$\frac{dV}{dq} = \frac{d^2V}{dq^2} = \frac{d^3V}{dq^3} = \ldots = 0 \qquad \text{neutral equilibrium} \qquad (11\text{–}12)$$

since then V must be constant at and around the "neighborhood" of q_{eq}.

System Having Two Degrees of Freedom. A criterion for investigating stability becomes increasingly complex as the number of degrees of freedom for the system increases. For a system having two degrees of freedom, defined by independent coordinates (q_1, q_2), it may be verified (using the calculus of functions of two variables) that equilibrium and stability occur at a point (q_{1eq}, q_{2eq}) when

$$\frac{\partial V}{\partial q_1} = \frac{\partial V}{\partial q_2} = 0$$

$$\left[\left(\frac{\partial^2 V}{\partial q_1\, \partial q_2} \right)^2 - \left(\frac{\partial^2 V}{\partial q_1^2} \right)\left(\frac{\partial^2 V}{\partial q_2^2} \right) \right] < 0$$

$$\frac{\partial^2 V}{\partial q_1^2} > 0 \quad \text{or} \quad \frac{\partial^2 V}{\partial q_2^2} > 0$$

Stable equilibrium

(a)

Unstable equilibrium

(b)

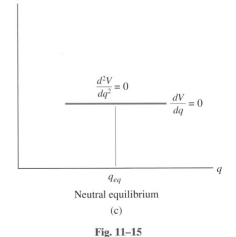

Neutral equilibrium

(c)

Fig. 11–15

Both equilibrium and instability occur when

$$\frac{\partial V}{\partial q_1} = \frac{\partial V}{\partial q_2} = 0$$

$$\left[\left(\frac{\partial^2 V}{\partial q_1\, \partial q_2} \right)^2 - \left(\frac{\partial^2 V}{\partial q_1^2} \right) \left(\frac{\partial^2 V}{\partial q_2^2} \right) \right] < 0$$

$$\frac{\partial^2 V}{\partial q_1^2} < 0 \quad \text{or} \quad \frac{\partial^2 V}{\partial q_2^2} < 0$$

PROCEDURE FOR ANALYSIS

Using potential-energy methods, the equilibrium positions and the stability of a body or a system of connected bodies having a single degree of freedom can be obtained by applying the following procedure.

Potential Function.

- Sketch the system so that it is located at some *arbitrary position* specified by the independent coordinate q.
- Establish a horizontal *datum* through a *fixed point*[*] and express the *gravitational potential energy* V_g in terms of the weight W of each member and its vertical distance y from the datum, $V_g = Wy$.
- Express the elastic potential energy V_e of the system in terms of the stretch or compression, s, of any connecting spring and the spring's stiffness k, $V_e = \frac{1}{2}ks^2$.
- Formulate the potential function $V = V_g + V_e$ and express the *position coordinates* y and s in terms of the independent coordinate q.

Equilibrium Position.

- The equilibrium position is determined by taking the first derivative of V and setting it equal to zero, $\delta V = 0$.

Stability.

- Stability at the equilibrium position is determined by evaluating the second or higher-order derivatives of V.
- If the second derivative is greater than zero, the body is stable, if all derivatives are equal to zero the body is in neutral equilibrium, and if the second derivative is less than zero, the body is unstable.

[*]The location of the datum is *arbitrary* since only the *changes* or differentials of V are required for investigation of the equilibrium position and its stability.

EXAMPLE 11.5

The uniform link shown in Fig. 11–16a has a mass of 10 kg. The spring is unstretched when $\theta = 0°$. Determine the angle θ for equilibrium and investigate the stability at the equilibrium position.

Solution

Potential Function. The datum is established at the top of the link when the *spring is unstretched*, Fig. 11–16b. When the link is located at the arbitrary position θ, the spring increases its potential energy by stretching and the weight decreases its potential energy. Hence,

$$V = V_e + V_g = \frac{1}{2}ks^2 - W\left(s + \frac{l}{2}\cos\theta\right)$$

Since $l = s + l\cos\theta$ or $s = l(1 - \cos\theta)$, then

$$V = \frac{1}{2}kl^2(1 - \cos\theta)^2 - \frac{Wl}{2}(2 - \cos\theta)$$

Equilibrium Position. The first derivative of V gives

$$\frac{dV}{d\theta} = kl^2(1 - \cos\theta)\sin\theta - \frac{Wl}{2}\sin\theta = 0$$

or

$$l\left[kl(1 - \cos\theta) - \frac{W}{2}\right]\sin\theta = 0$$

This equation is satisfied provided

$$\sin\theta = 0 \qquad \theta = 0° \qquad \textit{Ans.}$$

$$\theta = \cos^{-1}\left(1 - \frac{W}{2kl}\right) = \cos^{-1}\left[1 - \frac{10(9.81)}{2(200)(0.6)}\right] = 53.8° \qquad \textit{Ans.}$$

Stability. Determining the second derivative of V gives

$$\frac{d^2V}{d\theta^2} = kl^2(1 - \cos\theta)\cos\theta + kl^2\sin\theta\sin\theta - \frac{Wl}{2}\cos\theta$$

$$= kl^2(\cos\theta - \cos 2\theta) - \frac{Wl}{2}\cos\theta$$

Substituting values for the constants, with $\theta = 0°$ and $\theta = 53.8°$, yields

$$\left.\frac{d^2V}{d\theta^2}\right|_{\theta=0°} = 200(0.6)^2(\cos 0° - \cos 0°) - \frac{10(9.81)(0.6)}{2}\cos 0°$$

$$= -29.4 < 0 \qquad \text{(unstable equilibrium at } \theta = 0°) \qquad \textit{Ans.}$$

$$\left.\frac{d^2V}{d\theta^2}\right|_{\theta=53.8°} = 200(0.6)^2(\cos 53.8° - \cos 107.6°) - \frac{10(9.81)(0.6)}{2}\cos 53.8°$$

$$= 46.9 > 0 \qquad \text{(stable equilibrium at } \theta = 53.8°) \qquad \textit{Ans.}$$

(a)

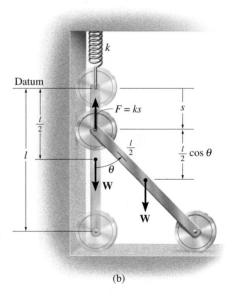

(b)

Fig. 11–16

E X A M P L E 11.6

(a)

(b)

Fig. 11–17

Determine the mass m of the block required for equilibrium of the uniform 10-kg rod shown in Fig. 11–17a when $\theta = 20°$. Investigate the stability at the equilibrium position.

Solution

Potential Function. The datum is established through point A, Fig. 11–17b. When $\theta = 0°$, the block is assumed to be suspended $(y_W)_1$ below the datum. Hence, in the position θ,

$$V = V_e + V_g = 98.1\left(\frac{1.5 \sin \theta}{2}\right) - m(9.81)(\Delta y) \qquad (1)$$

The distance $\Delta y = (y_W)_2 - (y_W)_1$ may be related to the independent coordinate θ by measuring the difference in cord lengths $B'C$ and BC. Since

$$B'C = \sqrt{(1.5)^2 + (1.2)^2} = 1.92$$
$$BC = \sqrt{(1.5 \cos \theta)^2 + (1.2 - 1.5 \sin \theta)^2} = \sqrt{3.69 - 3.60 \sin \theta}$$

then

$$\Delta y = B'C - BC = 1.92 - \sqrt{3.69 - 3.60 \sin \theta}$$

Substituting the above result into Eq. 1 yields

$$V = 98.1\left(\frac{1.5 \sin \theta}{2}\right) - m(9.81)(1.92 - \sqrt{3.69 - 3.60 \sin \theta}) \qquad (2)$$

Equilibrium Position.

$$\frac{dV}{d\theta} = 73.6 \cos \theta - \left[\frac{m(9.81)}{2}\right]\left(\frac{3.60 \cos \theta}{\sqrt{3.69 - 3.60 \sin \theta}}\right) = 0$$

$$\left.\frac{dV}{d\theta}\right|_{\theta=20°} = 69.14 - 10.58m = 0$$

$$m = \frac{69.14}{10.58} = 6.53 \text{ kg} \qquad\qquad Ans.$$

Stability. Taking the second derivative of Eq. 2, we obtain

$$\frac{d^2V}{d\theta^2} = -73.6 \sin \theta - \left[\frac{m(9.81)}{2}\right]\left(\frac{-1}{2}\right)\frac{-(3.60 \cos \theta)^2}{(3.69 - 3.60 \sin \theta)^{3/2}}$$

$$- \left[\frac{m(9.81)}{2}\right]\left(\frac{-3.60 \sin \theta}{\sqrt{3.69 - 3.60 \sin \theta}}\right)$$

For the equilibrium position $\theta = 20°$, with $m = 6.53$ kg, then

$$\frac{d^2V}{d\theta^2} = -47.6 < 0 \qquad \text{(unstable equilibrium at } \theta = 20°) \quad Ans.$$

E X A M P L E   11.7

The homogeneous block having a mass m rests on the top surface of the cylinder, Fig. 11–18a. Show that this is a condition of unstable equilibrium if $h > 2R$.

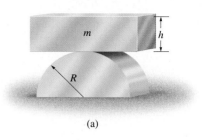

(a)

Solution

Potential Function.   The datum is established at the base of the cylinder, Fig. 11–18b. If the block is displaced by an amount θ from the equilibrium position, the potential function may be written in the form

$$V = V_e + V_g$$
$$= 0 + mgy$$

From Fig. 11–18b,

$$y = \left(R + \frac{h}{2} \right) \cos \theta + R\theta \sin \theta$$

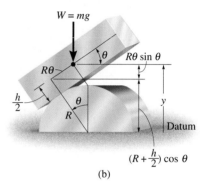

(b)

Fig. 11–18

Thus,

$$V = mg\left[\left(R + \frac{h}{2} \right) \cos \theta + R\theta \sin \theta \right]$$

Equilibrium Position.

$$\frac{dV}{d\theta} = mg\left[-\left(R + \frac{h}{2} \right) \sin \theta + R \sin \theta + R\theta \cos \theta \right] = 0$$

$$= mg\left(-\frac{h}{2} \sin \theta + R\theta \cos \theta \right) = 0$$

Obviously, $\theta = 0°$ is the equilibrium position that satisfies this equation.

Stability.   Taking the second derivative of V yields

$$\frac{d^2V}{d\theta^2} = mg\left(-\frac{h}{2} \cos \theta + R \cos \theta - R\theta \sin \theta \right)$$

At $\theta = 0°$,

$$\frac{d^2V}{d\theta^2}\bigg|_{\theta=0°} = -mg\left(\frac{h}{2} - R \right)$$

Since all the constants are positive, the block is in unstable equilibrium if $h > 2R$, for then $d^2V/d\theta^2 < 0$.

PROBLEMS

11-26. If the potential function for a conservative one-degree-of-freedom system is $V = (8x^3 - 2x^2 - 10)$ J, where x is given in meters, determine the positions for equilibrium and investigate the stability at each of these positions.

11-27. If the potential function for a conservative one-degree-of-freedom system is $V = (12 \sin 2\theta + 15 \cos \theta)$ J, where $0° < \theta < 180°$, determine the positions for equilibrium and investigate the stability at each of these positions.

*11-28. If the potential function for a conservative one-degree-of-freedom system is $V = (10 \cos 2\theta + 25 \sin \theta)$ J, where $0° < \theta < 180°$, determine the positions for equilibrium and investigate the stability at each of these positions.

11-29. If the potential function for a conservative two-degree-of-freedom system is $V = (9y^2 + 18x^2)$ J, where x and y are given in meters, determine the equilibrium position and investigate the stability at this position.

11-30. The spring of the scale has an unstretched length of a. Determine the angle θ for equilibrium when a weight W is supported on the platform. Neglect the weight of the members. What value W would be required to keep the scale in neutral equilibrium when $\theta = 0°$?

11-31. The two bars each have a weight of 80 N (≈ 8 kg). Determine the required stiffness k of the spring so that the two bars are in equilibrium when $\theta = 30°$. The spring has an unstretched length of 1 m.

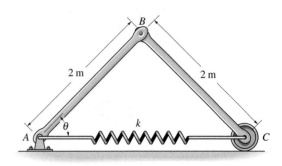

Prob. 11–31

*11-32. The two bars each have a weight of 80 N (≈ 8 kg). Determine the angle θ for the equilibrium and investigate the stability at the equilibrium position. The spring has an unstretched length of 1 m.

Prob. 11–30

Prob. 11–32

11-33. The truck has a mass of 20 Mg and a mass center at G. Determine the steepest grade θ along which it can park without overturning and investigate the stability in this position.

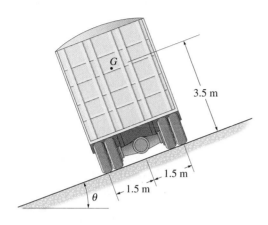

Prob. 11–33

11-35. The cylinder is made of two materials such that it has a mass of m and a center of gravity at point G. Show that when G lies above the centroid C of the cylinder, the equilibrium is unstable.

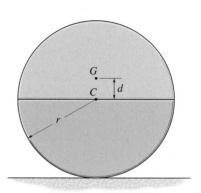

Prob. 11–35

11-34. The bar supports a weight of $W = 5000$ N (≈ 500 kg) at its end. If the springs are originally unstretched when the bar is vertical, determine the required stiffness $k_1 = k_2 = k$ of the springs so that the bar is in neutral equilibrium when it is vertical.

Prob. 11–34

*****11-36.** Determine the angle θ for equilibrium and investigate the stability at this position. The bars each have a mass of 3 kg and the suspended block D has a mass of 7 kg. Cord DC has a total length of 1 m.

Prob. 11–36

11-37. The cup has a hemispherical bottom and a mass m. Determine the position h of the center of mass G so that the cup is in neutral equilibrium.

Prob. 11–37

11-38. If each of the three links of the mechanism has a weight W, determine the angle θ for equilibrium. The spring, which always remains vertical, is unstretched when $\theta = 0°$.

Prob. 11–38

11-39. If the uniform rod OA has a mass of 12 kg, determine the mass m that will hold the rod in equilibrium when $\theta = 30°$. Point C is coincident with B when OA is horizontal. Neglect the size of the pulley at B.

Prob. 11–39

*11-40. The uniform right circular cone having a mass m is suspended from the cord as shown. Determine the angle θ at which it hangs from the wall for equilibrium. Is the cone in stable equilibrium?

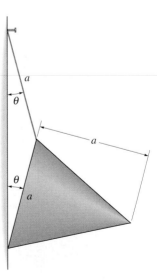

Prob. 11–40

11-41. The homogeneous cylinder has a conical cavity cut into its base as shown. Determine the depth d of the cavity so that the cylinder balances on the pivot and remains in neutral equilibrium.

Prob. 11–41

11-42. A homogeneous block rests on top of the cylindrical surface. Derive the relationship between the radius of the cylinder, r, and the dimension of the block, b, for stable equilibrium. *Hint*: Establish the potential energy function for a small angle θ, i.e., approximate $\sin \theta \approx \theta$, and $\cos \theta \approx 1 - \theta^2/2$.

Prob. 11–42

11-43. The homogeneous cone has a conical cavity cut into it as shown. Determine the depth of d of the cavity in terms of h so that the cone balances on the pivot and remains in neutral equilibrium.

Prob. 11–43

*11-44. The triangular block of weight W rests on the smooth corners which are a distance a apart. If the block has three equal sides of length d, determine the angle θ for equilibrium.

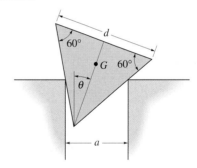

Prob. 11–44

11-45. Two uniform bars, each having a weight W, are pin-connected at their ends. If they are placed over a smooth cylindrical surface, show that the angle θ for equilibrium must satisfy the equation $\cos \theta / \sin^3 \theta = a/2r$.

Prob. 11–45

CHAPTER REVIEW

- *Principle of Virtual Work.* The forces on a body will do *virtual work* when the body undergoes an *imaginary* differential displacement or rotation. For equilibrium, the sum of the virtual work done by all the forces acting on the body must be equal to zero for any virtual displacement. This is referred to as the *principle of virtual work*, and it is useful for finding the equilibrium configuration for a mechanism or a reactive force acting on a series of connected members. If this system has one degree of freedom, then its position can be specified by one independent coordinate q. To apply the principle of virtual work, it is first necessary to use *position coordinates* to locate all the forces and moments on the mechanism that will do work when the mechanism undergoes a virtual movement δq. The coordinates are related to the independent coordinate q and then these expressions are differentiated in order to relate the *virtual* coordinate displacements to δq. Finally, the equation of virtual work is written for the mechanism in terms of the common displacement δq, and then it is set equal to zero. By factoring δq out of the equation, it is then possible to determine either the unknown force or couple moment, or the equilibrium position q.

- *Potential Energy Criterion for Equilibrium.* When a system is subjected only to conservative forces, such as weight or spring forces, then the equilibrium configuration can be determined using the *potential energy function V* for the system. This function is established by expressing the weight and spring potential energy for the system in terms of the independent coordinate q. Once it is formulated, its first derivative is set equal to zero, $dV/dq = 0$. The solution yields the equilibrium position q_{eq} for the system. The stability of the system can be investigated by taking the second derivative of V. If this is evaluated at q_{eq} and $d^2V/dq^2 > 0$, then *stable equilibrium* occurs. If $d^2V/dq^2 < 0$, then *unstable equilibrium* occurs. And if all higher derivatives are zero, then the system is in *neutral equilibrium*.

REVIEW PROBLEMS

11-46. The uniform links AB and BC each weigh 2 N (≈ 0.2 kg) and the cylinder weighs 20 N (≈ 2 kg). Determine the horizontal force **P** required to hold the mechanism in the position when $\theta = 45°$. The spring has an unstretched length of 6 cm.

Prob. 11–46

11-47. The spring attached to the mechanism has an unstretched length when $\theta = 90°$. Determine the position θ for equilibrium and investigate the stability of the mechanism at this position. Disk A is pin-connected to the frame at B and has a weight of 20 N (≈ 2 kg). Neglect the weight of the bars.

Prob. 11–47

*11-48. The toggle joint is subjected to the load **P**. Determine the compressive force F it creates on the cylinder at A as a function of θ.

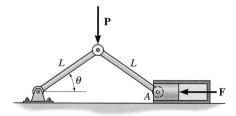

Prob. 11–48

11-49. The uniform beam AB weighs 100 N (≈ 10 kg). If both springs DE and BC are unstretched when $\theta = 90°$, determine the angle θ for equilibrium using the principle of potential energy. Investigate the stability at the equilibrium position. Both springs always act in the horizontal position because of the roller guides at C and E.

Prob. 11–49

11-50. The uniform bar AB weighs 10 N (≈ 1 kg). If the attached spring is unstretched when $\theta = 90°$, use the method of virtual work and determine the angle θ for equilibrium. Note that the spring always remains in the vertical position due to the roller guide.

11-51. Solve Prob. 11-50 using the principle of potential energy. Investigate the stability of the bar when it is in the equilibrium position.

Probs. 11–50/51

*11-52. The punch press consists of the ram R, connecting rod AB, and a flywheel. If a torque of $M = 50$ N · m is applied to the flywheel, determine the force **F** applied at the ram to hold the rod in the position $\theta = 60°$.

Prob. 11–52

Mathematical Expressions

Quadratic Formula

If $ax^2 + bx + c = 0$, then $x = \dfrac{-b \pm \sqrt{b^2 - 4ac}}{2a}$

Hyperbolic Functions

$\sinh x = \dfrac{e^x - e^{-x}}{2}$, $\cosh x = \dfrac{e^x + e^{-x}}{2}$, $\tanh x = \dfrac{\sinh x}{\cosh x}$

Trigonometric Identities

$\sin \theta = \dfrac{A}{C}$, $\csc \theta = \dfrac{C}{A}$

$\cos \theta = \dfrac{B}{C}$, $\sec \theta = \dfrac{C}{B}$

$\tan \theta = \dfrac{A}{B}$, $\cot \theta = \dfrac{B}{A}$

$\sin^2 \theta + \cos^2 \theta = 1$

$\sin(\theta \pm \phi) = \sin \theta \cos \phi \pm \cos \theta \sin \phi$

$\sin 2\theta = 2 \sin \theta \cos \theta$

$\cos(\theta \pm \phi) = \cos \theta \cos \phi \mp \sin \theta \sin \phi$

$\cos 2\theta = \cos^2 \theta - \sin^2 \theta$

$\cos \theta = \pm\sqrt{\dfrac{1 + \cos 2\theta}{2}}$, $\sin \theta = \pm\sqrt{\dfrac{1 - \cos 2\theta}{2}}$

$\tan \theta = \dfrac{\sin \theta}{\cos \theta}$

$1 + \tan^2 \theta = \sec^2 \theta \qquad 1 + \cot^2 \theta = \csc^2 \theta$

Power-Series Expansions

$\sin x = x - \dfrac{x^3}{3!} + \cdots, \quad \cos x = 1 - \dfrac{x^2}{2!} + \cdots$

$\sinh x = x + \dfrac{x^3}{3!} + \cdots, \quad \cosh x = 1 + \dfrac{x^2}{2!} + \cdots$

Derivatives

$\dfrac{d}{dx}(u^n) = nu^{n-1}\dfrac{du}{dx}$ $\qquad \dfrac{d}{dx}(\sin u) = \cos u \dfrac{du}{dx}$

$\dfrac{d}{dx}(uv) = u\dfrac{dv}{dx} + v\dfrac{du}{dx}$ $\qquad \dfrac{d}{dx}(\cos u) = -\sin u \dfrac{du}{dx}$

$\dfrac{d}{dx}\left(\dfrac{u}{v}\right) = \dfrac{v\dfrac{du}{dx} - u\dfrac{dv}{dx}}{v^2}$ $\qquad \dfrac{d}{dx}(\tan u) = \sec^2 u \dfrac{du}{dx}$

$\dfrac{d}{dx}(\cot u) = -\csc^2 u \dfrac{du}{dx}$ $\qquad \dfrac{d}{dx}(\sinh u) = \cosh u \dfrac{du}{dx}$

$\dfrac{d}{dx}(\sec u) = \tan u \sec u \dfrac{du}{dx}$ $\dfrac{d}{dx}(\cosh u) = \sinh u \dfrac{du}{dx}$

$\dfrac{d}{dx}(\csc u) = -\csc u \cot u \dfrac{du}{dx}$

Integrals

$$\int x^n \, dx = \frac{x^{n+1}}{n+1} + C, \, n \neq -1$$

$$\int \frac{dx}{a+bx} = \frac{1}{b} \ln(a+bx) + C$$

$$\int \frac{dx}{a+bx^2} = \frac{1}{2\sqrt{-ba}} \ln\left[\frac{a+x\sqrt{-ab}}{a-x\sqrt{-ab}}\right] + C,$$
$$ab < 0$$

$$\int \frac{x \, dx}{a+bx^2} = \frac{1}{2b} \ln(bx^2 + a) + C$$

$$\int \frac{x^2 \, dx}{a+bx^2} = \frac{x}{b} - \frac{a}{b\sqrt{ab}} \tan^{-1}\frac{x\sqrt{ab}}{a} + C, \, ab > 0$$

$$\int \sqrt{a+bx} \, dx = \frac{2}{3b} \sqrt{(a+bx)^3} + C$$

$$\int x\sqrt{a+bx} \, dx = \frac{-2(2a-3bx)\sqrt{(a+bx)^3}}{15b^2} + C$$

$$\int x^2\sqrt{a+bx} \, dx =$$
$$\frac{2(8a^2 - 12abx + 15b^2x^2)\sqrt{(a+bx)^3}}{105b^3} + C$$

$$\int \sqrt{a^2 - x^2} \, dx = \frac{1}{2}\left[x\sqrt{a^2-x^2} + a^2 \sin^{-1}\frac{x}{a}\right] + C,$$
$$a > 0$$

$$\int x\sqrt{a^2-x^2} \, dx = -\frac{1}{3}\sqrt{(a^2-x^2)^3} + C$$

$$\int x^2\sqrt{a^2-x^2} \, dx = -\frac{x}{4}\sqrt{(a^2-x^2)^3}$$
$$+ \frac{a^2}{8}\left(x\sqrt{a^2-x^2} + a^2 \sin^{-1}\frac{x}{a}\right) + C, \, a > 0$$

$$\int \sqrt{x^2 \pm a^2} \, dx =$$
$$\frac{1}{2}\left[x\sqrt{x^2 \pm a^2} \pm a^2 \ln(x + \sqrt{x^2 \pm a^2})\right] + C$$

$$\int x\sqrt{x^2 \pm a^2} \, dx = \frac{1}{3}\sqrt{(x^2 \pm a^2)^3} + C$$

$$\int x^2\sqrt{x^2 \pm a^2} \, dx = \frac{x}{4}\sqrt{(x^2 \pm a^2)^3}$$
$$\mp \frac{a^2}{8} x \sqrt{x^2 \pm a^2} - \frac{a^4}{8} \ln(x + \sqrt{x^2 \pm a^2}) + C$$

$$\int \frac{dx}{\sqrt{a+bx}} = \frac{2\sqrt{a+bx}}{b} + C$$

$$\int \frac{x \, dx}{\sqrt{x^2 \pm a^2}} = \sqrt{x^2 \pm a^2} + C$$

$$\int \frac{dx}{\sqrt{a+bx+cx^2}} = \frac{1}{\sqrt{c}} \ln\left[\sqrt{a+bx+cx^2} + \right.$$
$$\left. x\sqrt{c} + \frac{b}{2\sqrt{c}}\right] + C, \, c > 0$$
$$= \frac{1}{\sqrt{-c}} \sin^{-1}\left(\frac{-2cx - b}{\sqrt{b^2 - 4ac}}\right) + C, \, c < 0$$

$$\int \sin x \, dx = -\cos x + C$$

$$\int \cos x \, dx = \sin x + C$$

$$\int x \cos(ax) \, dx = \frac{1}{a^2} \cos(ax) + \frac{x}{a} \sin(ax) + C$$

$$\int x^2 \cos(ax) \, dx = \frac{2x}{a^2} \cos(ax) + \frac{a^2x^2 - 2}{a^3} \sin(ax) + C$$

$$\int e^{ax} \, dx = \frac{1}{a} e^{ax} + C$$

$$\int x e^{ax} \, dx = \frac{e^{ax}}{a^2}(ax - 1) + C$$

$$\int \sinh x \, dx = \cosh x + C$$

$$\int \cosh x \, dx = \sinh x + C$$

APPENDIX

B

Numerical and Computer Analysis

Occasionally the application of the laws of mechanics will lead to a system of equations for which a closed-form solution is difficult or impossible to obtain. When confronted with this situation, engineers will often use a numerical method which in most cases can be programmed on a microcomputer or "programmable" pocket calculator. Here we will briefly present a computer program for solving a set of linear algebraic equations and three numerical methods which can be used to solve an algebraic or transcendental equation, evaluate a definite integral, and solve an ordinary differential equation. Application of each method will be explained by example, and an associated computer program written in Microsoft BASIC, which is designed to run on most personal computers, is provided.* A text on numerical analysis should be consulted for further discussion regarding a check of the accuracy of each method and the inherent errors that can develop from the methods.

B.1 Linear Algebraic Equations

Application of the equations of static equilibrium or the equations of motion sometimes requires solving a set of linear algebraic equations. The computer program listed in Fig. B–1 can be used for this purpose. It is based on the method of a Gaussian elimination and can solve at

---

*Similar types of programs can be written or purchased for programmable pocket calculators.

```
1  PRINT"Linear system of equations":PRINT
2  DIM A(10,11)
3  INPUT"Input number of equations : ",N
4  PRINT
5  PRINT"A  coefficients"
6  FOR I = 1 TO N
7  FOR J = 1 TO N
8  PRINT "A(";I;",";J;
9  INPUT")=",A(I,J)
10 NEXT J
11 NEXT I
12 PRINT
13 PRINT"B  coefficients"
14 FOR I = 1 TO N
15 PRINT "B(";I;
16 INPUT")=",A(I,N+1)
17 NEXT I
18 GOSUB 25
19 PRINT
```

```
20 PRINT"Unknowns"
21 FOR I = 1 TO N
22 PRINT "X(";I;")=";A(I,N+1)
23 NEXT I
24 END
25 REM Subroutine Guassian
26 FOR M=1 TO N
27 NP=M
28 BG=ABS(A(M,M))
29 FOR I = M TO N
30 IF ABS(A(I,M))<=BG THEN 33
31 BG=ABS(A(I,M))
32 NP=I
33 NEXT I
34 IF NP=M THEN 40
35 FOR I = M TO N+1
36 TE=A(M,I)
37 A(M,I)=A(NP,I)
38 A(NP,I)=TE
```

```
39 NEXT I
40 FOR I = M+1 TO N
41 FC=A(I,M)/A(M,M)
42 FOR J = M+1 TO N+1
43 A(I,J)=A(I,J)-FC*A(M,J)
44 NEXT J
45 NEXT I
46 NEXT M
47 A(N,N+1)=A(N,N+1)/A(N,N)
48 FOR I = N-1 TO 1 STEP -1
49 SM=0
50 FOR J=I+1 TO N
51 SM=SM+A(I,J)*A(J,N+1)
52 NEXT J
53 A(I,N+1)=(A(I,N+1)-SM)/A(I,I)
54 NEXT I
55 RETURN
```

Fig. B–1

most 10 equations with 10 unknowns. To do so, the equations should first be written in the following general format:

$$A_{11}x_1 + A_{12}x_2 + \cdots + A_{1n}x_n = B_1$$
$$A_{21}x_1 + A_{22}x_2 + \cdots + A_{2n}x_n = B_2$$
$$\vdots$$
$$A_{n1}x_1 + A_{n2}x_2 + \cdots + A_{nn}x_n = B_n$$

The "A" and "B" coefficients are "called" for when running the program. The output presents the unknowns $x_1, \ldots, x_n$.

EXAMPLE B.1

Solve the two equations

$$3x_1 + x_2 = 4$$
$$2x_1 - x_2 = 10$$

Solution

When the program begins to run, it first calls for the number of equations (2); then the A coefficients in the sequence $A_{11} = 3$, $A_{12} = 1$, $A_{21} = 2$, $A_{22} = -1$; and finally the B coefficients $B_1 = 4$, $B_2 = 10$. The output appears as

Unknowns

$$X(1) = 2.8 \qquad \textit{Ans.}$$
$$X(2) = -4.4 \qquad \textit{Ans.}$$

B.2 Simpson's Rule

Simpson's rule is a numerical method that can be used to determine the area under a curve given as a graph or as an explicit function $y = f(x)$. Likewise, it can be used to compute the value of a definite integral which involves the function $y = f(x)$. To do so, the area must be subdivided into an *even number* of strips or intervals having a width h. The curve between three consecutive ordinates is approximated by a parabola, and the entire area or definite integral is then determined from the formula

$$\int_{x_0}^{x_n} f(x) \, dx \simeq \frac{h}{3} [y_0 + 4(y_1 + y_3 + \cdots + y_{n-1})$$

$$+ 2(y_2 + y_4 + \cdots + y_{n-2}) + y_n] \quad \text{(B–1)}$$

The computer program for this equation is given in Fig. B–2. For its use, we must first specify the function (on line 6 of the program). The upper and lower limits of the integral and the number of intervals are called for when the program is executed. The value of the integral is then given as the output.

```
1 PRINT"Simpson's rule":PRINT
2 PRINT" To execute this program :":PRINT
3 PRINT"    1- Modify right-hand side of the equation given below,
4 PRINT"       then press RETURN key"
5 PRINT"    2- Type  RUN 6":PRINT:EDIT 6
6 DEF FNF(X)=LOG(X)
7 PRINT:INPUT" Enter Lower Limit = ",A
8 INPUT" Enter Upper Limit = ",B
9 INPUT" Enter Number (even) of Intervals = ",N%
10 H=(B-A)/N%:AR=FNF(A):X=A+H
11 FOR J%=2 TO N%
12 K=2*(2-J%+2*INT(J%/2))
13 AR=AR+K*FNF(X)
14 X=X+H:NEXT J%
15 AR=H*(AR+FNF(B))/3
16 PRINT" Integral = ",AR
17 END
```

Fig. B–2

E X A M P L E   B.2

Evaluate the definite integral

$$\int_2^5 \ln x \, dx$$

Solution

The interval $x_0 = 2$ to $x_6 = 5$ will be divided into six equal parts ($n = 6$), each having a width $h = (5 - 2)/6 = 0.5$. We then compute $y = f(x) = \ln x$ at each point of subdivision.

n	x_n	y_n
0	2	0.693
1	2.5	0.916
2	3	1.099
3	3.5	1.253
4	4	1.386
5	4.5	1.504
6	5	1.609

Thus, Eq. B–1 becomes

$$\int_2^5 \ln x \, dx \simeq \frac{0.5}{3}[0.693 + 4(0.916 + 1.253 + 1.504)$$

$$+ \, 2(1.099 + 1.386) + 1.609]$$

$$\simeq 3.66 \qquad\qquad\qquad Ans.$$

This answer is equivalent to the exact answer to three significant figures. Obviously, accuracy to a greater number of significant figures can be improved by selecting a smaller interval h (or larger n).

Using the computer program, we first specify the function $\ln x$, line 6 in Fig. B–2. During execution, the program input requires the upper and lower limits 2 and 5, and the number of intervals $n = 6$. The output appears as

$$\text{Integral} = 3.66082 \qquad\qquad Ans.$$

B.3 The Secant Method

The secant method is used to find the real roots of an algebraic or transcendental equation $f(x) = 0$. The method derives its name from the fact that the formula used is established from the slope of the secant line to the graph $y = f(x)$. This slope is $[f(x_n) - f(x_{n-1})]/(x_n - x_{n-1})$, and the secant formula is

$$x_{n+1} = x_n - f(x_n)\left[\frac{x_n - x_{n-1}}{f(x_n) - f(x_{n-1})}\right] \qquad (B-2)$$

For application it is necessary to provide two initial guesses, x_0 and x_1, and thereby evaluate x_2 from Eq. B–2 ($n = 1$). One then proceeds to reapply Eq. B–2 with x_1 and the calculated value of x_2 and obtain x_3 ($n = 2$), etc., until the value $x_{n+1} \simeq x_n$. One can see this will occur if x_n is approaching the root of the function $f(x) = 0$, since the correction term on the right of Eq. B–2 will tend toward zero. In particular, the larger the slope, the smaller the correction to x_n, and the faster the root will be found. On the other hand, if the slope is very small in the neighborhood of the root, the method leads to large corrections for x_n, and convergence to the root is slow and may even lead to a failure to find it. In such cases other numerical techniques must be used for solution.

A computer program based on Eq. B–2 is listed in Fig. B–3. We must first specify the function on line 7 of the program. When the program is executed, two initial guesses, x_0 and x_1, must be entered in order to approximate the solution. The output specifies the value of the root. If it cannot be determined, this is so stated.

```
1 PRINT"Secant method":PRINT
2 PRINT" To execute this program :":PRINT
3 PRINT"    1) Modify right hand side of the equation given below,"
4 PRINT"       then press RETURN key."
5 PRINT"    2) Type  RUN 7"
6 PRINT:EDIT 7
7 DEF FNF(X)=.5*SIN(X)-2*COS(X)+1.3
8 INPUT"Enter point #1 =",X
9 INPUT"Enter point #2 =",X1
10 IF X=X1 THEN 14
11 EP=.00001:TL=2E-20
12 FP=(FNF(X1)-FNF(X))/(X1-X)
13 IF ABS(FP)>TL THEN 15
14 PRINT"Root can not be found.":END
15 DX=FNF(X1)/FP
16 IF ABS(DX)>EP THEN 19
17 PRINT "Root = ";X1;"      Function evaluated at this root = ";FNF(X1)
18 END
19 X=X1:X1=X1-DX
20 GOTO 12
```

Fig. B–3

E X A M P L E B.3

Determine the root of the equation

$$f(x) = 0.5 \sin x - 2 \cos x + 1.30 = 0$$

Solution

Guesses of the initial roots will be $x_0 = 45°$ and $x_1 = 30°$. Applying Eq. B–2,

$$x_2 = 30° - (-0.1821)\frac{(30° - 45°)}{(-0.1821 - 0.2393)} = 36.48°$$

Using this value in Eq. B–2, along with $x_1 = 30°$, we have

$$x_3 = 36.48° - (-0.0108)\frac{36.48° - 30°}{(-0.0108 + 0.1821)} = 36.89°$$

Repeating the process with this value and $x_2 = 36.48°$ yields

$$x_4 = 36.89° - (0.0005)\left[\frac{36.89° - 36.48°}{(0.0005 + 0.0108)}\right] = 36.87°$$

Thus $x = 36.9°$ is appropriate to three significant figures.

If the problem is solved using the computer program, first we specify the function, line 7 in Fig. B–3. During execution, the first and second guesses must be entered in radians. Choosing these to be 0.8 rad and 0.5 rad, the result appears as

Root = 0.6435022

Function evaluated at this root = 1.66893E−06

This result converted from radians to degrees is therefore

$$x = 36.9° \qquad\qquad \textit{Ans.}$$

Review for the Fundamentals of Engineering Examination

The Fundamentals of Engineering (FE) exam is given semiannually by the National Council of Engineering Examiners (NCEE) and is one of the requirements for obtaining a Professional Engineering License. A portion of this exam contains problems in statics, and this appendix provides a review of the subject matter most often asked on this exam. Before solving any of the problems, you should review the sections indicated in each chapter in order to become familiar with the boldfaced definitions and the procedures used to solve the various types of problems. Also, review the example problems in these sections.

The following problems are arranged in the same sequence as the topics in each chapter. Besides helping as a preparation for the FE exam, these problems also provide additional examples for general practice of the subject matter. Solutions to *all the problems* are given at the back of this appendix.

Chapter 2—Review All Sections

C-1. Two forces act on the hook. Determine the magnitude of the resultant force.

Prob. C–1

C-2. The force $F = 450$ N acts on the frame. Resolve this force into components acting along members AB and AC, and determine the magnitude of each component.

Prob. C–2

C-3. Determine the magnitude and direction of the resultant force.

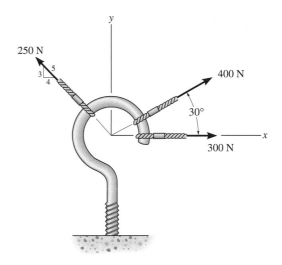

Prob. C–3

C-4. If $\mathbf{F} = \{30\mathbf{i} + 50\mathbf{j} - 45\mathbf{k}\}$ N, determine the magnitude and coordinate direction angles of the force.

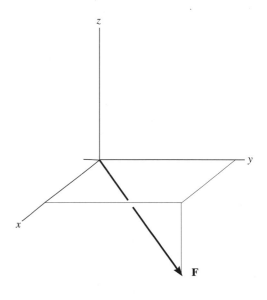

Prob. C–4

C-5. The force has a component of 20 N directed along the $-y$ axis as shown. Represent the force **F** as a Cartesian vector.

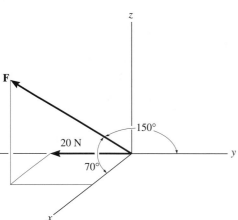

Prob. C–5

C-6. The force acts on the beam as shown. Determine its coordinate direction angles.

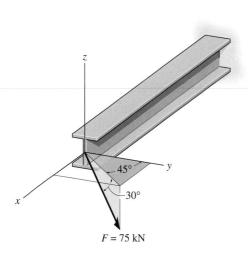

Prob. C–6

C-7. The cables supporting the antenna are subjected to the forces shown. Represent each force as a Cartesian vector.

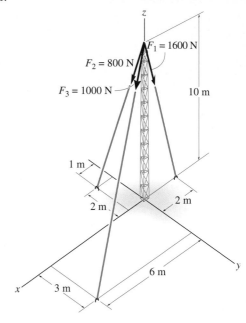

Prob. C–7

C-8. Determine the angle θ between the two cords.

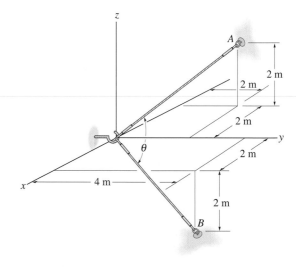

Prob. C–8

C-9. Determine the component the of projection of the force **F** along the pipe *AB*.

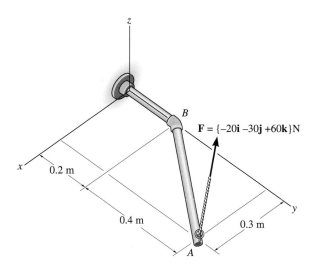

Prob. C–9

Chapter 3—Review Sections 3.1–3.3

C-10. The crate at *D* has a weight of 550 N (≈ 55 kg). Determine the force in each supporting cable.

Prob. C–10

C-11. The beam has a weight of 700 N (≈ 70 kg). Determine the shortest cable *ABC* that can be used to lift it if the maximum force the cable can sustain is 1500 N.

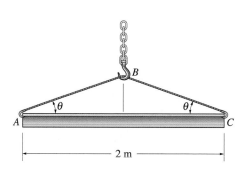

Prob. C–11

C-12. The block has a mass of 5 kg and rests on the smooth plane. Determine the unstretched length of the spring.

Prob. C–12

C-13. The post can be removed by a vertical force of 400 N. Determine the force P that must be applied to the cord in order to pull the post out of the ground.

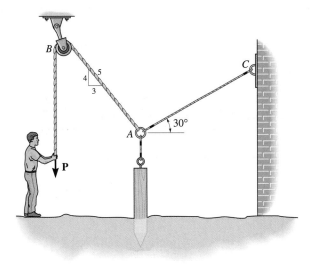

Prob. C–13

Chapter 4—Review All Sections

C-14. Determine the moment of the force about point O.

Prob. C–14

C-15. Determine the moment of the force about point O. Neglect the thickness of the member.

Prob. C–15

C-16. Determine the moment of the force about point O.

Prob. C–16

C-17. Determine the moment of the force about point A. Express the result as a Cartesian vector.

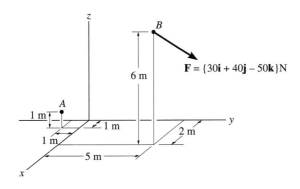

$$\mathbf{F} = \{30\mathbf{i} + 40\mathbf{j} - 50\mathbf{k}\}\text{N}$$

Prob. C–17

C-18. Determine the moment of the force about point A. Express the result as a Cartesian vector.

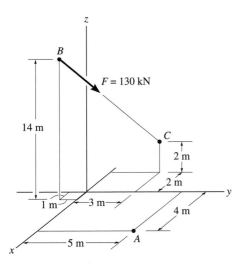

$F = 130$ kN

Prob. C–18

C-19. Determine the resultant couple moment acting on the beam.

Prob. C–19

C-20. Determine the resultant couple moment acting on the triangular plate.

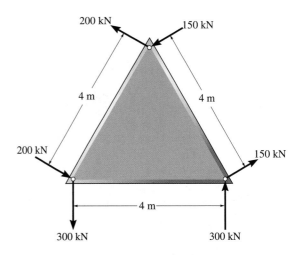

Prob. C–20

C-21. Replace the loading shown by an equivalent resultant force and couple-moment system at point A.

Prob. C–21

C-22. Replace the loading shown by an equivalent resultant force and couple-moment system at point A.

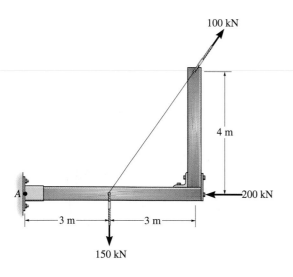

Prob. C–22

C-23. Replace the loading shown by an equivalent single resultant force and specify where the force acts, measured from point O.

Prob. C–23

C-24. Replace the loading shown by an equivalent single resultant force and specify the x and y coordinates of its line of action.

Prob. C–24

C-25. Replace the loading shown by an equivalent single resultant force and specify the x and y coordinates of its line of action.

C-27. Determine the resultant force and specify where it acts on the beam measured from A.

Prob. C–25

Prob. C–27

C-26. Determine the resultant force and specify where it acts on the beam measured from A.

C-28. Determine the resultant force and specify where it acts on the beam measured from A.

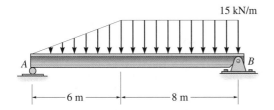

Prob. C–26

Prob. C–28

Chapter 5—Review Sections 5.1–5.6

C-29. Determine the horizontal and vertical components of reaction at the supports. Neglect the thickness of the beam.

Prob. C–29

C-31. Determine the components of reaction at the fixed support A. Neglect the thickness of the beam.

Prob. C–31

C-30. Determine the horizontal and vertical components of reaction at the supports.

Prob. C–30

C-32. Determine the tension in the cable and the horizontal and vertical components of reaction at the pin A. Neglect the size of the pulley.

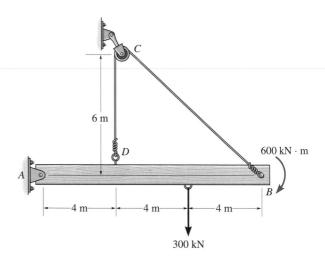

Prob. C–32

C-33. The uniform plate has a weight of 500 N ($\approx$ 50 kg). Determine the tension in each of the supporting cables.

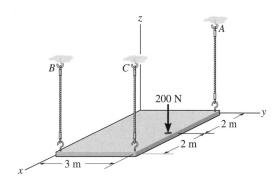

Prob. C–33

C-35. Determine the force in members AE and DC. State if the members are in tension or compression.

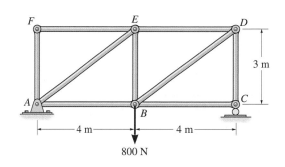

Prob. C–35

Chapter 6—Review Sections 6.1–6.4, 6.6

C-34. Determine the force in each member of the truss. State if the members are in tension or compression.

C-36. Determine the force in members BC, CF, and FE. State if the members are in tension or compression.

Prob. C–34

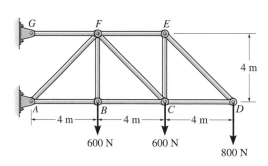

Prob. C–36

C-37. Determine the force in members *GF, FC,* and *CD*. State if the members are in tension or compression.

C-39. Determine the horizontal and vertical components of reaction at pin *C*.

Prob. C–37

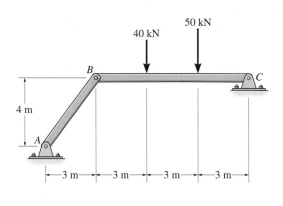

Prob. C–39

C-38. Determine the force *P* needed to hold the 60-N (≈ 6-kg) weight in equilibrium.

C-40. Determine the horizontal and vertical components of reaction at pin *C*.

Prob. C–38

Prob. C–40

C-41. Determine the normal force that the 1000-N ($\approx$ 100-kg) plate A exerts on the 300-N ($\approx$ 30-kg) plate B.

Prob. C–41

C-42. Determine the force P needed to lift the load. Also, determine the proper placement x of the hook for equilibrium. Neglect the weight of the beam.

Prob. C–42

Chapter 7—Review Section 7.1

C-43. Determine the internal normal force, shear force, and moment acting in the beam at point B.

Prob. C–43

C-44. Determine the internal normal force, shear force, and moment acting in the beam at point B, which is located just to the left of the 800-N force.

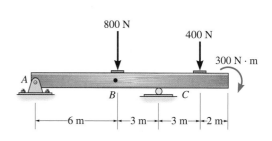

Prob. C–44

C-45. Determine the internal normal force, shear force, and moment acting in the beam at point B.

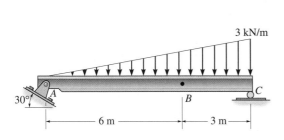

Prob. C–45

C-47. Determine the vertical force P needed to rotate the 200-N ($\approx$ 20-kg) spool. The coefficient of static friction at all contacting surfaces is $\mu_s = 0.4$.

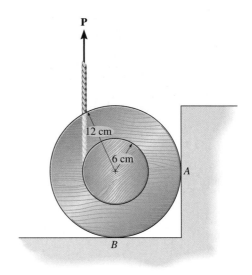

Prob. C–47

Chapter 8—Review Sections 8.1–8.2

C-46. Determine the force P needed to move the 100-N ($\approx$ 10-kg) block. The coefficient of static friction is $\mu_s = 0.3$, and the coefficient of kinetic friction is $\mu_k = 0.25$. Neglect tipping.

C-48. Block A has a weight of 30 N ($\approx$ 3 kg) and block B weighs 50 N ($\approx$ 5 kg). If the coefficient of static friction is $\mu_s = 0.4$ between all contacting surfaces, determine the frictional force at each surface.

Prob. C–46

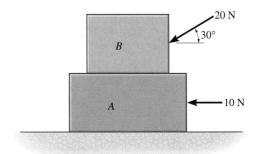

Prob. C–48

C-49. Determine the force P necessary to move the 250-N ($\approx$ 25-kg) crate which has a center of gravity at G. The coefficient of static friction at the floor is $\mu_s = 0.4$.

Prob. C–49

C-50. The filing cabinet A has a mass of 60 kg and center of mass at G. It rests on a 10-kg plank. Determine the smallest force P needed to move it. The coefficient of static friction between the cabinet A and the plank B is $\mu_s = 0.4$, and between the plank and the floor $\mu_s = 0.3$.

Prob. C–50

Chapter 9—Review Sections 9.1–9.3
(Integration is covered in the mathematics portion of the exam.)

C-51. Determine the location $(\bar{x}, \bar{y})$ of the centroid of the area.

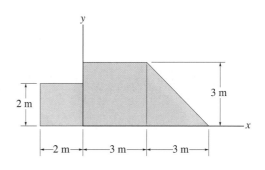

Prob. C–51

C-52. Determine the location $(\bar{x}, \bar{y})$ of the centroid of the area.

Prob. C–52

Chapter 10—Review Sections 10.1–10.5
(Integration is covered in the mathematics portion of the exam.)

C-53. Determine the moment of inertia of the cross-sectional area of the channel with respect to the *y* axis.

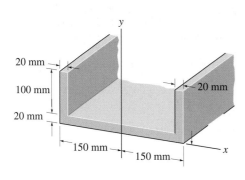

Prob. C–53

C-54. Determine the moment of inertia of the area with respect to the *x* axis.

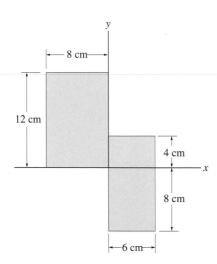

Prob. C–54

C-55. Determine the moment of inertia of the cross-sectional area of the T-beam with respect to the *x'* axis passing through the centroid of the cross section.

Prob. C–55

Partial Solutions and Answers

C-1. $F_R = \sqrt{200^2 + 500^2 - 2(200)(500) \cos 140°}$
$= 666$ N **Ans.**

C-2. $\dfrac{F_{AB}}{\sin 105°} = \dfrac{450}{\sin 30°}$
$= 869$ N **Ans.**

$\dfrac{F_{AC}}{\sin 45°} = \dfrac{450}{\sin 30°}$

$F_{AC} = 636$ N **Ans.**

C-3. $F_{Rx} = 300 + 400 \cos 30° - 250\left(\dfrac{4}{5}\right) = 446.4$ N

$F_{Ry} = 400 \sin 30° + 250\left(\dfrac{3}{5}\right) = 350$ N

$F_R = \sqrt{(446.4)^2 + 350^2} = 567$ N **Ans.**

$\theta = \tan^{-1}\dfrac{350}{446.4} = 38.1°$ ⦪ **Ans.**

C-4. $F = \sqrt{30^2 + 50^2 + (-45)^2} = 73.7$ N **Ans.**

$\alpha = \cos^{-1}\left(\dfrac{30}{73.7}\right) = 66.0°$ **Ans.**

$\beta = \cos^{-1}\left(\dfrac{50}{73.7}\right) = 47.2°$ **Ans.**

$\gamma = \cos^{-1}\left(\dfrac{-45}{73.7}\right) = 128°$ **Ans.**

C-5. $F_y = -20$

$\dfrac{F_y}{|F|} = \cos \beta$

$|F| = \left|\dfrac{-20}{\cos 150°}\right| = 23.09$ N

$\cos \gamma = \sqrt{1 - \cos^2 70° - \cos^2 150°}$
$\gamma = 68.61°$ (From Fig. $\gamma < 90°$)
$F = 23.09 \cos 70°\mathbf{i} + 23.09 \cos 150°\mathbf{j}$
$\quad + 23.09 \cos 68.61°\mathbf{k}$
$= \{7.90\mathbf{i} - 20\mathbf{j} + 8.42\mathbf{k}\}$ N **Ans.**

C-6. $F_x = 75 \cos 30° \sin 45° = 45.93$
$F_y = 75 \cos 30° \cos 45° = 45.93$
$F_z = -75 \sin 30° = -37.5$

$\alpha = \cos^{-1}\left(\dfrac{45.93}{75}\right) = 52.2°$ **Ans.**

$\beta = \cos^{-1}\left(\dfrac{45.93}{75}\right) = 52.2°$ **Ans.**

$\gamma = \cos^{-1}\left(\dfrac{-37.5}{75}\right) = 120°$ **Ans.**

C-7. $F_1 = 1600$ N$\left(-\dfrac{2}{10.2}\mathbf{i} - \dfrac{10}{10.2}\mathbf{k}\right)$
$= \{-314\mathbf{i} - 1570\mathbf{k}\}$ N **Ans.**

$F_2 = 800$ N$\left(\dfrac{1}{10.25}\mathbf{i} - \dfrac{2}{10.25}\mathbf{j} - \dfrac{10}{10.25}\mathbf{k}\right)$
$= \{78.1\mathbf{i} - 156\mathbf{j} - 781\mathbf{k}\}$ N **Ans.**

$F_3 = 1000$ N$\left(\dfrac{6}{12.04}\mathbf{i} + \dfrac{3}{12.04}\mathbf{j} - \dfrac{10}{12.04}\mathbf{k}\right)$
$= \{498\mathbf{i} + 249\mathbf{j} - 830\mathbf{k}\}$ N **Ans.**

C-8. $\mathbf{r}_{OA} = \{-2\mathbf{i} + 2\mathbf{j} + 2\mathbf{k}\}$ m
$\mathbf{r}_{OB} = \{2\mathbf{i} + 4\mathbf{j} - 2\mathbf{k}\}$ m

$\cos \theta = \dfrac{\mathbf{r}_{OA} \cdot \mathbf{r}_{OB}}{|r_{OA}||r_{OB}|}$

$\dfrac{(-2\mathbf{i} + 2\mathbf{j} + 2\mathbf{k}) \cdot (2\mathbf{i} + 4\mathbf{j} - 2\mathbf{k})}{\sqrt{12}\sqrt{24}} = 0$

$\theta = 90°$ **Ans.**

C-9. $|F_{AB}| = \mathbf{F} \cdot \mathbf{u}_{AB}$

$= (-20\mathbf{i} - 30\mathbf{j} + 60\mathbf{k}) \cdot \left(-\dfrac{0.3}{0.5}\mathbf{i} - \dfrac{0.4}{0.5}\mathbf{j}\right) = 36$ **Ans.**

C-10. $\overset{+}{\rightarrow}\Sigma F_x = 0; \dfrac{4}{5}F_{AC} - F_{AB} \cos 30° = 0$

$+\uparrow\Sigma F_y = 0; \dfrac{3}{5}F_{AC} + F_{AB} \sin 30° - 550 = 0$

$F_{AB} = 478$ N **Ans.**, $F_{AC} = 518$ N **Ans.**

C-11. $+\uparrow\Sigma F_y = 0; -2(1500) \sin \theta + 700 = 0$
$\theta = 13.5°$

$L_{ABC} = 2\left(\dfrac{1\text{ m}}{\cos 13.5°}\right) = 2.06$ m

C-12. $+\nearrow\Sigma F_x = 0; \dfrac{4}{5}(F_{sp}) - 5(9.81) \sin 45° = 0$

$F_{sp} = 43.35$ N

$F_{sp} = k(l - l_0); 43.35 = 200(0.5 - l_0)$

$l_0 = 0.283$ m **Ans.**

C-13. At A:

$\overset{+}{\leftarrow}\Sigma F_x = 0; \dfrac{3}{5}P - T_{AC} \cos 30° = 0$

$+\uparrow\Sigma F_y = 0; \dfrac{4}{5}P + T_{AC} \sin 30° - 400 = 0$

$P = 349$ N **Ans.**, $T_{AC} = 242$ N **Ans.**

C-14. $\downarrow+M_O = 600 \sin 50° (5) + 600 \cos 50° (0.5)$
$= 2.49$ kN · m **Ans.**

C-15. $\curvearrowleft + M_O = 50 \sin 60° (0.1 + 0.2 \cos 45° + 0.1)$
$\qquad - 50 \cos 60° (0.2 \sin 45°)$
$\qquad = 11.2 \text{ N} \cdot \text{m} \quad \textit{Ans.}$

C-16. $\curvearrowright + M_O = 500 \sin 45° (3 + 3 \cos 45°)$
$\qquad - 500 \cos 45° (3 \sin 45°)$
$\qquad = 1.06 \text{ kN} \cdot \text{m} \quad \textit{Ans.}$

C-17. $\mathbf{M}_A = \mathbf{r}_{AB} \times \mathbf{F} = \begin{vmatrix} \mathbf{i} & \mathbf{j} & \mathbf{k} \\ 1 & 6 & 5 \\ 30 & 40 & -50 \end{vmatrix}$
$\qquad = \{-500\mathbf{i} + 200\mathbf{j} - 140\mathbf{k}\} \text{ N} \cdot \text{m} \quad \textit{Ans.}$

C-18. $\mathbf{F} = 130 \text{ kN}\left(-\dfrac{3}{13}\mathbf{i} + \dfrac{4}{13}\mathbf{j} - \dfrac{12}{13}\mathbf{k}\right)$
$\qquad = \{-30\mathbf{i} + 40\mathbf{j} - 120\mathbf{k}\} \text{ kN}$

$\qquad \mathbf{M}_A = \mathbf{r}_{AB} \times \mathbf{F} = \begin{vmatrix} \mathbf{i} & \mathbf{j} & \mathbf{k} \\ -3 & -6 & 14 \\ -30 & 40 & -120 \end{vmatrix}$
$\qquad = \{160\mathbf{i} - 780\mathbf{j} - 300\mathbf{k}\} \text{ kN} \cdot \text{m} \quad \textit{Ans.}$

C-19. $\curvearrowleft + M_{C_R} = \Sigma M_A = 400(3) - 400(5) + 300(5)$
$\qquad + 200(0.2) = 740 \text{ N} \cdot \text{m} \quad \textit{Ans.}$

Also,
$\curvearrowleft + M_{C_R} = 300(5) - 400(2) + 200(0.2)$
$\qquad = 740 \text{ N} \cdot \text{m} \quad \textit{Ans.}$

C-20. $\curvearrowright + M_{C_R} = 300(4) + 200(4) + 150(4)$
$\qquad = 2600 \text{ kN} \cdot \text{m} \quad \textit{Ans.}$

C-21. $\xrightarrow{+} F_{Rx} = \Sigma F_x; \quad F_{Rx} = \dfrac{4}{5}(50) = 40 \text{ N}$

$+\downarrow F_{Ry} = \Sigma F_y; \quad F_{Ry} = 40 + 30 + \dfrac{3}{5}(50)$
$\qquad\qquad\qquad\qquad = 100 \text{ N}$
$\qquad F_R = \sqrt{(40)^2 + (100)^2} = 108 \text{ N} \quad \textit{Ans.}$

$\theta = \tan^{-1}\left(\dfrac{100}{40}\right) = 68.2° \quad \textit{Ans.}$

$+\downarrow M_{A_R} = \Sigma M_A; \; M_{A_R} = 30(3) + \dfrac{3}{5}(50)(6) + 200$
$\qquad\qquad = 470 \text{ N} \cdot \text{m} \quad \textit{Ans.}$

C-22. $\xleftarrow{+} F_{Rx} = \Sigma F_x; \; F_{Rx} = 200 - \dfrac{3}{5}(100) = 140 \text{ kN}$

$+\downarrow F_{Ry} = \Sigma F_y; \; F_{Ry} = 150 - \dfrac{4}{5}(100) = 70 \text{ kN}$
$\qquad F_R = \sqrt{140^2 + 70^2} = 157 \text{ kN} \quad \textit{Ans.}$
$\qquad \theta = \tan^{-1}\left(\dfrac{70}{140}\right) = 26.6° \quad \textit{Ans.}$

$+\downarrow M_{A_R} = \Sigma M_A; \; M_{A_R} = \dfrac{3}{5}(100)(4) - \dfrac{4}{5}(100)(6) + 150(3)$
$\qquad\qquad M_{R_A} = 210 \text{ kN} \cdot \text{m} \quad \textit{Ans.}$

C-23. $+\downarrow F_R = \Sigma F_y; \quad F_R = 500 + 250 + 500$
$\qquad\qquad\qquad\qquad = 1250 \text{ kN} \quad \textit{Ans.}$
$+\downarrow F_R x = \Sigma M_O; \qquad 1250(x) = 500(3) + 250(6) + 500(9)$
$\qquad\qquad\qquad\qquad m = 6 \text{ m} \quad \textit{Ans.}$

C-24. $+\downarrow F_R = \Sigma F_z; \quad F_R = 400 + 500 - 100$
$\qquad\qquad\qquad\qquad = 800 \text{ N} \quad \textit{Ans.}$
$M_{Rx} = \Sigma M_x; \quad -800y = -400(4) - 500(4)$
$\qquad\qquad\qquad\qquad y = 4.50 \text{ m} \quad \textit{Ans.}$
$M_{Ry} = \Sigma M_y; \quad 800x = 500(4) - 100(3)$
$\qquad\qquad\qquad\qquad x = 2.125 \text{ m} \quad \textit{Ans.}$

C-25. $+\downarrow F_R = \Sigma F_y; \quad F_R = 200 + 200 + 100 + 100$
$\qquad\qquad\qquad\qquad = 600 \text{ N} \quad \textit{Ans.}$
$M_{Rx} = \Sigma M_x; \quad -600y = 200(1) + 200(1)$
$\qquad\qquad\qquad\qquad + 100(3) - 100(3)$
$\qquad\qquad\qquad\qquad y = -0.667 \text{ m} \quad \textit{Ans.}$
$M_{Ry} = \Sigma M_y; \quad 600x = 100(3) + 100(3)$
$\qquad\qquad\qquad\qquad + 200(2) - 200(3)$
$\qquad\qquad\qquad\qquad x = 0.667 \text{ m} \quad \textit{Ans.}$

C-26. $F_R = \dfrac{1}{2}(6)(15) + 8(15) = 165 \text{ kN} \quad \textit{Ans.}$
$+\downarrow M_{A_R} = \Sigma M_A;$
$\qquad 165d = \left[\dfrac{1}{2}(6)(15)\right](4) + [8(15)](10)$
$\qquad\qquad d = 8.36 \text{ m} \quad \textit{Ans.}$

C-27. $F_R = \displaystyle\int w(x) \, dx = \int_0^4 2.5x^3 \, dx = 160 \text{ N} \quad \textit{Ans.}$
$+\downarrow M_{A_R} = \Sigma M_A;$
$\qquad x = \dfrac{\displaystyle\int xw(x)\,dx}{\displaystyle\int w(x)\,dx} = \dfrac{\displaystyle\int_0^4 2.5x^4\,dx}{160} = 3.20 \text{ m} \quad \textit{Ans.}$

C-28. $+\downarrow F_R = \Sigma F_y; \quad F_R = \dfrac{1}{2}(5)(6) + (15)(6) + 50$
$\qquad\qquad\qquad\qquad = 155 \text{ kN} \quad \textit{Ans.}$
$+\downarrow M_{A_R} = \Sigma M_A;$
$\qquad 155d = \left[\dfrac{1}{2}(5)(6)\right](4) + [15(6)](3) + 50(9)$
$\qquad\qquad d = 5.03 \text{ m} \quad \textit{Ans.}$

C-29. $\xrightarrow{+} \Sigma F_x = 0; \quad -A_x + 50\left(\dfrac{3}{5}\right) = 0$
$\qquad\qquad\qquad\qquad A_x = 30 \text{ kN} \quad \textit{Ans.}$
$+\curvearrowleft \Sigma M_A = 0; \quad B_y(10) - 50\left(\dfrac{4}{5}\right)(5) - 60 = 0$
$\qquad\qquad\qquad\qquad B_y = 26 \text{ kN} \quad \textit{Ans.}$

$+\uparrow\Sigma F_y = 0;\quad A_y + 26 - 50\left(\dfrac{4}{5}\right) = 0$

$A_y = 14$ kN *Ans.*

C-30. $\xrightarrow{+}\Sigma F_x = 0;\ -A_x + 400 = 0;\ A_x = 400$ N *Ans.*

$\zeta + \Sigma M_A = 0;\quad B_y(4) - 400(0.5) - 500(2) = 0$

$B_y = 300$ N *Ans.*

$+\uparrow\Sigma F_y = 0;\quad A_y + 300 - 500 = 0$

$A_y = 200$ N *Ans.*

C-31. $\xrightarrow{+}\Sigma F_x = 0;\ -A_x + 400\cos 30° = 0$

$A_x = 346$ N *Ans.*

$+\uparrow\Sigma F_y = 0;\ A_y - 200 - 200 - 200 - 400\sin 30° = 0$

$A_y = 800$ N *Ans.*

$\zeta + \Sigma M_A = 0; M_A - 200(2.5) - 200(3.5) - 200(4.5)$

$-400\sin 30°(4.5) - 400\cos 30°(3\sin 60°) = 0$

$M_A = 3.90$ kN·m *Ans.*

C-32. $+\gamma\Sigma M_A = 0;\ T(4) + \dfrac{3}{5}T(12) - 300(8) - 600 = 0$

$T = 267.9 = 268$ kN *Ans.*

$\xrightarrow{+}\Sigma F_x = 0;\ A_x - \left(\dfrac{4}{5}\right)(267.9) = 0$

$A_x = 214$ kN *Ans.*

$+\uparrow\Sigma F_y = 0;\ A_y + 267.9 + \left(\dfrac{3}{5}\right)(267.9) - 300 = 0$

$A_y = -129$ kN *Ans.*

C-33. $\Sigma F_z = 0;\ T_A + T_B + T_C - 200 - 500 = 0$

$\Sigma M_x = 0;\ T_A(3) + T_C(3) - 500(1.5) - 200(3) = 0$

$\Sigma M_y = 0;\ -T_B(4) - T_C(4) + 500(2) + 200(2) = 0$

$T_A = 350$ N, $T_B = 250$ N, $T_C = 100$ N *Ans.*

C-34. Joint D:

$+\uparrow\Sigma F_y = 0;\ \dfrac{3}{5}F_{CD} - 300 = 0;\ F_{CD} = 500$ N (T) *Ans.*

$\xrightarrow{+}\Sigma F_x = 0;\ -F_{AD} + \dfrac{4}{5}(500) = 0;$

$F_{AD} = 400$ N (C) *Ans.*

Joint C:

$+\searrow\Sigma F_y = 0;\ F_{CA} = 0$ *Ans.*

$+\nearrow\Sigma F_x = 0;\ F_{CB} - 500 = 0;$

$F_{CB} = 500$ N (T) *Ans.*

Joint A:

$+\uparrow\Sigma F_y = 0;\ F_{AB} = 0$ *Ans.*

C-35. $Ax = 0,\ Ay = Cy = 400$ N

Joint A:

$+\uparrow\Sigma F_y = 0;\ -\dfrac{3}{5}F_{AE} + 400 = 0;\ F_{AE} = 667$ N (C) *Ans.*

Joint C:

$+\uparrow\Sigma F_y = 0;\ -F_{DC} + 400 = 0;\ F_{DC} = 400$ N (C) *Ans.*

C-36. Section truss through *FE, FC, BC*. Use the right segment.

$+\uparrow\Sigma F_y = 0;\ F_{CF}\sin 45° - 600 - 800 = 0$

$F_{CE} = 1980$ N (T) *Ans.*

$+\gamma\Sigma M_C = 0;\ F_{FE}(4) - 800(4) = 0$

$F_{FE} = 800$ N (T) *Ans.*

$\downarrow + \Sigma M_F = 0;\ F_{BC}(4) - 600(4) - 800(8) = 0$

$F_{BC} = 2200$ N (C) *Ans.*

C-37. Section truss through *GF, FC, DC*. Use the top segment.

$+\gamma\Sigma M_C = 0;\ F_{GF}(4) - (7)(3) - (10)(6) = 0$

$F_{GF} = 20.25$ kN (T) *Ans.*

$\xrightarrow{+}\Sigma F_x = 0;\ -\dfrac{4}{5}F_{FC} + 7 + 10 = 0$

$F_{FC} = 21.25$ kN (C) *Ans.*

$\zeta + \Sigma M_F = 0;\ F_{CD}(4) - (10)(3) = 0$

$F_{CD} = 7.5$ kN (C) *Ans.*

C-38. $+\uparrow\Sigma F_y = 0;\ 3P - 60 = 0$

$P = 20$ N *Ans.*

C-39. $+\gamma\Sigma M_C = 0;\ -\left(\dfrac{4}{5}\right)(F_{AB})(9) + (40)(6) + (50)(3) = 0$

$F_{AB} = 54.167$ kN

$\xrightarrow{+}\Sigma F_x = 0;\ -C_x + \dfrac{3}{5}(54.167) = 0$

$C_x = 32.5$ kN *Ans.*

$+\uparrow\Sigma F_y = 0;\ C_y + \dfrac{4}{5}(54.167) - 40 - 50 = 0$

$C_y = 46.7$ kN *Ans.*

C-40. $+\gamma\Sigma M_C = 0;\ F_{AB}\cos 45°(1) - F_{AB}\sin 45°(3) + 800 + 400(2) = 0$

$F_{AB} = 1131.37$ N

$\xrightarrow{+}\Sigma F_x = 0;\ -C_x + 1131.37\cos 45° = 0$

$C_x = 800$ N *Ans.*

$+\uparrow\Sigma F_y = 0;\ -C_y + 1131.37\sin 45° - 400 = 0$

$C_y = 400$ N *Ans.*

C-41. Plate A:

$+\uparrow\Sigma F_y = 0;\ 2T + N_{AB} - 1000 = 0$

Plate B:

$+\uparrow\Sigma F_y = 0;\ 2T + N_{AB} - 300 = 0$

$T = 325$ N, $N_{AB} = 350$ N *Ans.*

C-42. Pulley C:

$+\uparrow\Sigma F_y = 0;\ T - 2P = 0;\ T = 2P$

Beam:

$+\uparrow\Sigma F_y = 0;\ 2P + P - 6 = 0$

$P = 2$ kN *Ans.*

$+\gamma\Sigma M_A = 0;\ 2(1) - 6(x) = 0$

$x = 0.333$ m *Ans.*

C-43. $A_y = 8.75$ kN. Use segment AB:

$\xrightarrow{+} \Sigma F_x = 0;$ $\qquad N_B = 0$ **Ans.**

$+\uparrow \Sigma F_y = 0;$ $\qquad 8.75 - 3(1.5) - V_B = 0$

$\qquad\qquad V_B = 4.25$ kN **Ans.**

$+\uparrow \Sigma M_B = 0;$ $\quad M_B + 3(1.5)(0.75) - 8.75(1.5) = 0$

$\qquad\qquad M_B = 9.75$ kN · m **Ans.**

C-44. $A_y = 0, A_y = 100$ N. Use segment AB.

$\xrightarrow{+} \Sigma F_x = 0;$ $\qquad N_B = 0$ **Ans.**

$+\uparrow \Sigma F_y = 0;$ $\qquad 100 - V_B = 0$

$\qquad\qquad V_B = 100$ N **Ans.**

$+\uparrow \Sigma M_B = 0;$ $\qquad M_B - 100(6) = 0$

$\qquad\qquad M_B = 600$ N · m **Ans.**

C-45. $A_x = 0, A_y = 4.5$ kN, $w_B = 2$ kN/m. Use segment AB.

$\xrightarrow{+} \Sigma F_x = 0;$ $\qquad N_B = 0$ **Ans.**

$+\uparrow \Sigma F_y = 0;$ $\qquad 4.5 - \dfrac{1}{2}(6)(2) + V_B = 0$

$\qquad\qquad V_B = 1.5$ kN **Ans.**

$+\uparrow \Sigma M_B = 0;$ $\quad M_B + \left[\dfrac{1}{2}(6)(2)\right](2) - 4.5(6) = 0$

$\qquad\qquad M_B = 15$ kN · m **Ans.**

C-46. $+\uparrow \Sigma F_y = 0;$ $\quad N_b - P \sin 30° - 100 = 0$

$\xrightarrow{+} \Sigma F_x = 0;$ $\quad -P \cos 30° + 0.3 N_b = 0$

$\qquad\qquad P = 41.9$ N **Ans.**

C-47. $\xrightarrow{+} \Sigma F_x = 0;$ $\quad 0.4 N_B - N_A = 0$

$+\uparrow \Sigma M_B = 0;$ $\quad 0.4 N_A(12) + N_A(12) - P(6) = 0$

$+\uparrow \Sigma F_y = 0;$ $\quad P + 0.4 N_A + N_B - 200 = 0$

$\qquad\qquad P = 98.2$ N **Ans.**

C-48. Block B:

$+\uparrow \Sigma F_y = 0;$ $\quad N_B - 20 \sin 30° - 50 = 0$

$\qquad\qquad N_B = 60$ N

$\xrightarrow{+} \Sigma F_x = 0;$ $\quad F_B - 20 \cos 30° = 0$

$\qquad\qquad F_B = 17.3$ N ($< 0.4(60$ N)) **Ans.**

Blocks A and B:

$+\uparrow \Sigma F_y = 0;$ $\quad N_A - 30 - 50 - 20 \sin 30° = 0$

$\qquad\qquad N_A = 90$ N

$\xrightarrow{+} \Sigma F_x = 0;$ $\quad F_A - 20 \cos 30° - 10 = 0$

$\qquad\qquad F_A = 27.3$ N ($< 0.4(90$ N)) **Ans.**

C-49. If slipping occurs:

$+\uparrow \Sigma F_y = 0;$ $\quad N_C = 250$ N $= 0$

$\qquad\qquad N_C = 250$ N

$\xrightarrow{+} \Sigma F_x = 0;$ $\quad P - 0.4(250) = 0$

$\qquad\qquad P = 100$ N

If tipping occurs:

$\downarrow+\Sigma M_A = 0;$ $\quad -P(0.45) = 250(0.15) = 0$

$\qquad\qquad P = 83.3$ N **Ans.**

C-50. P for A to slip on B:

$+\uparrow \Sigma F_y = 0;$ $\quad N_A - 60(9.81) = 0$

$\qquad\qquad N_A = 588.6$ N

$\xrightarrow{+} \Sigma F_x = 0;$ $\quad 0.4(588.6) - P = 0$

$\qquad\qquad P = 235$ N

P for B to slip:

$+\uparrow \Sigma F_y = 0;$ $\quad N_B - 60(9.81) - 10(9.81) = 0$

$\qquad\qquad N_B = 686.7$ N

$\xrightarrow{+} \Sigma F_x = 0;$ $\quad 0.3(686.7) - P = 0$

$\qquad\qquad P = 206$ N

P to tip A:

$\downarrow+\Sigma M_C = 0;$ $\quad P(1.3) - 60(9.81)(0.2) = 0$

$\qquad\qquad P = 90.6$ N **Ans.**

C-51. $\bar{x} = \dfrac{\Sigma \tilde{x} A}{\Sigma A} =$

$\dfrac{(-1)(2)(2) + 1.5(3)(3) + 4\left(\dfrac{1}{2}\right)(3)(3)}{2(2) + 3(3) + \dfrac{1}{2}(3)(3)} = 1.57$ m **Ans.**

$\bar{y} = \dfrac{\Sigma \tilde{y} A}{\Sigma A} =$

$\dfrac{1(2)(2) + 1.5(3)(3) + 1\left(\dfrac{1}{2}\right)(3)(3)}{2(2) + 3(3) + \dfrac{1}{2}(3)(3)} = 1.26$ m **Ans.**

C-52. $\bar{x} = 0$ (symmetry) **Ans.**

$\bar{y} = \dfrac{\Sigma \tilde{y} A}{\Sigma A} = \dfrac{4(1(8)) + 9(6)(2)}{1(8) + 6(2)} = 7$ cm **Ans.**

C-53. $I_y = \dfrac{1}{12}(120)(300)^3 - \dfrac{1}{12}(100)(260)^3$

$= 124 (10^6)$ mm⁴ **Ans.**

C-54. $I = \Sigma(\bar{I} + Ad^2) = \left[\dfrac{1}{12}(8)(12)^3 + (8)(12)(6)^2\right]$

$+ \left[\dfrac{1}{12}(6)(12)^3 + (6)(12)(-2)^2\right] = 5760$ cm⁴ **Ans.**

C-55. $\bar{x} = \dfrac{\Sigma \tilde{x} A}{\Sigma A} = \dfrac{4(8)(2) + 9(2)(8)}{8(2) + 2(8)} = 6.5$ cm

$\bar{I}_{x'} = \Sigma(\bar{I} + Ad^2) = \left[\dfrac{1}{12}(2)(8)^3 + (8)(2)(6.5 - 4)^2\right]$

$+ \left[\dfrac{1}{12}(8)(2)^3 + 2(8)(9 - 6.5)^2\right] = 291$ cm⁴ **Ans.**

Answers to Selected Problems

Chapter 1

1–1. **a)** 4.66 m, **b)** 55.6 s, **c)** 4.56 kN, **d)** 2.77 Mg

1–2. **a)** 0.000431 kg = 0.431 g,
b) $35.3(10^3)$ N = 35.3 kN,
c) 0.00532 km = 5.32 m

1–4. **a)** $(430 \text{ kg})^2 = 0.185 \text{ Mg}^2$,
b) $(0.002 \text{ mg})^2 = 4 \text{ }\mu\text{g}^2$,
c) $(230 \text{ m})^3 = 0.0122 \text{ km}^3$

1–6. **a)** $W = 98.1$ N,
b) $W = 4.90$ mN,
c) $W = 44.1$ kN

1–7. **a)** $(354 \text{ mg})(45 \text{ km})/0.0356 \text{ kN} = 0.447 \text{ kg}\cdot\text{m/N}$,
b) $(0.00453 \text{ Mg})(201 \text{ ms}) = 0.911 \text{ kg}\cdot\text{s}$,
c) 435 MN/23.2 mm = 18.8 GN/m

1–9. **a)** $m = \dfrac{W}{g} = 2.04$ g,

b) $m = \dfrac{W}{g} = 15.3$ Mg,

c) $m = \dfrac{W}{g} = 6.12$ Gg,

1–10. $F = 7.41(10^{-6})$ N = 7.41 μN

Chapter 2

2–1. $F_R = 867$ N, $\phi = 108°$

2–2. **a)** $F_R = 111$ N, **b)** $F_R{}' = 143$ N

2–3. $F_R = 393$ N, $\phi = 353°$

2–5. $F_{1u} = 205$ N, $F_{1v} = 160$ N

2–6. $F_{2u} = 376$ N, $F_{2v} = 482$ N

2–7. $F_R = 10.8$ kN, $\phi = 3.16°$

2–9. $F_{AB} = 448$ N, $F_{AC} = 366$ N

2–10. $F_{AB} = 314$ N, $F_{AC} = 256$ N

2–11. $F_a = 30.6$ N, $F_b = 26.9$ N

2–13. $F_{AB} = 485$ N

2–14. $T = 744$ N, $\theta = 23.8°$

2–15. $\theta = 53.5°$, $F_{AB} = 621$ N

2–17. **a)** $F_y = 163.3$ N, $F_n = -223$ N,
b) $F_t = 59.8$ N, $F_x = 163.3$ N

2–18. $\theta = 18.6°$, $F = 319$ N

2–19. $\phi = \theta = 70.5°$

2–21. $F_B = 325$ N, $F_A = 893$: $\vartheta = 70.0°$

2–22. $F_R = 19.2$ N, $\theta = 2.37°$

2–23. $F_R = 19.2$ N, $\theta = 2.37°$ ⦨θ

2–25. $F_A = 3.66$ kN, $F_B = 7.07$ kN

2–26. $F_B = 5.00$ kN, $F_A = 8.66$ kN, $\theta = 60.0°$

2–27. $F_A = 439$ N, $F_B = 311$ N

2–29. $\theta = 10.9°$, $F_{min} = 235$ N

2–30. $F = 97.4$ N, $\theta = 16.2°$

2–31. $F_x = 514$ N, $F_y = -613$ N

2–33. $F = 11.3$ kN

2–34. $F_R = 546$ N, $\theta = 253°$

2–35. $\theta = 37.0°$, $F_1 = 889$ N

2–37. $\theta = 29.1°$, $F_1 = 275$ N

2–38. $F_R = 1.03$ kN, $\theta = 87.9°$

2–39. $\mathbf{F}_1 = \{-15.0\mathbf{i} - 26.0\mathbf{j}\}$ kN,
$\mathbf{F}_2 = \{-10.0\mathbf{i} + 24.0\mathbf{j}\}$ kN

2–41. $F_R = 867$ N, $\theta = 108°$

2–42. $F_R = 19.2$ N, $\theta = 2.37°$ ⦨θ

2–43. $\theta = 68.6°$, $F_B = 960$ N

2–45. $F_{1x} = 141$ N, $F_{1y} = 141$ N, $F_{2x} = -130$ N,
$F_{2y} = 75$ N

2–46. $F_R = 217$ N, $\theta = 87.0°$

2–47. $F_{1x} = -2$ kN, $F_{1y} = 0$, $F_{2x} = 3200$ N,
$F_{2y} = -2400$ N, $F_{3x} = 1800$ N, $F_{3y} = 2400$ N,
$F_{4x} = -3000$ N, $F_{4y} = 0$, $F_R = 0$

2–49. $\theta = 54.3°$, $F_A = 686$ N

2–50. $F_R = 1.23$ kN, $\theta = 6.08°$

2–51. $\mathbf{F}_1 = \{90\mathbf{i} - 120\mathbf{j}\}$ N, $\mathbf{F}_2 = \{-275\mathbf{j}\}$ N,
$\mathbf{F}_3 = \{-37.5\mathbf{i} - 65.0\mathbf{j}\}$ N, $F_R = 463$ N

2–53. $F = 5.96$ kN, $F_R = 2.33$ kN

2–54. $\theta = 67.0°$, $F_1 = 434$ N

2–55. $\theta = 117°$, $F_3 = 1.12 F_1$

2–57. $F_R = 161$ kN, $\theta = 38.3°$

2–58. $F = 2.03$ kN, $F_R = 7.87$ kN

2–59. $F_1 = 87.7$ N, $\alpha_1 = 46.9°$, $\beta_1 = 125°$, $\gamma_1 = 62.9°$,
$F_2 = 98.6$ N, $\alpha_2 = 114°$, $\beta_2 = 150°$, $\gamma_2 = 72.3°$

2–61. $\mathbf{F} = \{13.7\mathbf{i} + 37.6\mathbf{j} + 30.0\mathbf{k}\}$ N,
$F = 50$ N, $\alpha = 74.1°$, $\beta = 41.3°$, $\gamma = 53.1°$

2–62. $F_R = 39.4$ N, $\alpha = 52.8°$, $\beta = 141°$, $\gamma = 99.5°$

2–63. $\beta = 90°$, $\mathbf{F} = \{-30\mathbf{i} - 52.0\mathbf{k}\}$ N

2–65. $\mathbf{F}_1 = \{53.1\mathbf{i} - 44.5\mathbf{j} + 40\mathbf{k}\}$ N, $\alpha_1 = 48.4°$,
$\beta_1 = 124°$, $\gamma_1 = 60°$, $\mathbf{F}_2 = \{-130\mathbf{k}\}$ N,
$\alpha_2 = 90°$, $\beta_2 = 90°$, $\gamma_2 = 180°$

2–66. $\alpha_1 = 45.6°$, $\beta_1 = 53.1°$, $\gamma_1 = 66.4°$

2–67. $\alpha_1 = 90°$, $\beta_1 = 53.1°$, $\gamma_1 = 66.4°$

2–69. $\mathbf{F}_1 = \{176\mathbf{j} - 605\mathbf{k}\}$ N,
$\mathbf{F}_2 = \{125\mathbf{i} - 177\mathbf{j} + 125\mathbf{k}\}$ N,
$\mathbf{F}_R = \{125\mathbf{i} - 0.377\mathbf{j} - 480\mathbf{k}\}$ N,
$F_R = 496$ N, $\alpha = 75.4°$, $\beta = 90.0°$, $\gamma = 165°$

2–70. $F_R = 369$ N, $\alpha = 19.5°$, $\beta = 78.3°$, $\gamma = 105°$

2–71. $F_2 = 66.4$ N, $\alpha = 59.8°$, $\beta = 107°$, $\gamma = 144°$

2–73. $\mathbf{F}_1 = \{86.5\mathbf{i} + 186\mathbf{j} - 143\mathbf{k}\}$ N,
$\mathbf{F}_2 = \{-200\mathbf{i} + 283\mathbf{j} + 200\mathbf{k}\}$ N,
$\mathbf{F}_R = \{-113\mathbf{i} + 468\mathbf{j} + 56.6\mathbf{k}\}$ N,
$F_R = 485$ N, $\alpha = 104°$, $\beta = 15.1°$, $\gamma = 83.3°$

2–74. $F_x = 1.28$ kN, $F_y = 2.60$ kN, $F_z = 0.776$ kN

2–75. $F = 2.02$ kN, $F_y = 0.523$ kN

2–77. $F_3 = 166$ N, $\alpha = 97.5°$, $\beta = 63.7°$, $\gamma = 27.5°$

2–78. $\alpha_{F_1} = 36.9°$, $\beta_{F_1} = 90.0°$, $\gamma_{F_1} = 53.1°$,
$\alpha_R = 69.3°$, $\beta_R = 52.2°$, $\gamma_R = 45.0°$

2–79. $F_x = 40$ N, $F_y = 40$ N, $F_z = 56.6$ N

2–81. $\alpha = 69.6°$, $\beta = 116°$, $\gamma = 34.4°$

2–82. $\mathbf{r}_{AB} = \{2\mathbf{i} - 7\mathbf{j} - 5\mathbf{k}\}$ m, $r_{AB} = 8.83$ m,
$\alpha = 76.9°$, $\beta = 142°$, $\gamma = 124°$

2–83. $\mathbf{r} = \{2\mathbf{i} + 3\mathbf{j} + 6\mathbf{k}\}$ m, $r_{AB} = 7$ m,
$\alpha = 73.4°$, $\beta = 64.6°$, $\gamma = 31.0°$

2–85. $\mathbf{r} = \{-1.175\mathbf{i} + 1.965\mathbf{j} + 1.855\mathbf{k}\}$ m, $r = 2.945$ m,
$\alpha = 113°$, $\beta = 48.2°$, $\gamma = 51.0°$

2–86. $\mathbf{F} = F\mathbf{u}_{AB} = \{404\mathbf{i} + 276\mathbf{j} - 101\mathbf{k}\}$ kN,
$\alpha = 36.0°$, $\beta = 56.5°$, $\gamma = 102°$

2–87. $r_{AB} = 2.11$ m

2–89. $r_{AB} = 4.39$ m,
$$\mathbf{u}_{AB} = -\frac{2}{4.39}\mathbf{i} - \frac{2.5}{4.39}\mathbf{j} + \frac{3}{4.39}\mathbf{k}$$
$\mathbf{F} = F\mathbf{u}_{AB} = \{-154.9\mathbf{i} - 193.62\mathbf{j} + 232.35\mathbf{k}\}$ N

2–90. $r_{AB} = 467$ mm

2–91. $r_{AD} = 1.50$ m, $r_{BD} = 1.50$ m, $r_{CD} = 1.73$ m

2–93. $\mathbf{F} = \{452\mathbf{i} + 370\mathbf{j} - 136\mathbf{k}\}$ kN, $\alpha = 41.1°$,
$\beta = 51.9°$, $\gamma = 103°$

2–94. $F_R = 316$ N, $\alpha = 60.1°$, $\beta = 74.6°$, $\gamma = 146°$

2–95. $\mathbf{F}_A = F_A\mathbf{u}_{AB} = \{285\mathbf{j} - 93.0\mathbf{k}\}$ N,
$\mathbf{F}_C = F_C\mathbf{u}_{CD} = \{159\mathbf{i} + 183\mathbf{j} - 59.7\mathbf{k}\}$ N

2–97. $\mathbf{F}_{AB} = F_{AB}\mathbf{u}_{AB} = \{302.04\mathbf{i} - 174.36\mathbf{j} - 488.2\mathbf{k}\}$ N,
$\mathbf{F}_{BC} = F_{BC}\mathbf{u}_{BC} = \{107.08\mathbf{i} + 133.84\mathbf{j} - 361.4\mathbf{k}\}$ N,
$F_R = 943.86$ N, $\alpha = 64.3°$, $\beta = 92.5°$, $\gamma = 154°$

2–98. $\mathbf{F}_A = F_A\mathbf{u}_{AC} = \{-43.5\mathbf{i} + 174\mathbf{j} - 174\mathbf{k}\}$ N,
$\mathbf{F}_B = F_B\mathbf{u}_{BD} = \{53.2\mathbf{i} - 79.8\mathbf{j} - 146\mathbf{k}\}$ N

2–99. $\mathbf{F}_1 = \{-26.2\mathbf{i} - 41.9\mathbf{j} + 62.9\mathbf{k}\}$ kN,
$\mathbf{F}_2 = \{13.4\mathbf{i} - 26.7\mathbf{j} - 40.1\mathbf{k}\}$ kN,
$F_R = 73.5$ kN, $\alpha = 100°$, $\beta = 159°$, $\gamma = 71.9°$

2–101. $\mathbf{F} = F\mathbf{u}_{AB} = \{66.9\mathbf{i} + 116.1\mathbf{j} + 267.9\mathbf{k}\}$ N

2–102. $\mathbf{F} = F\mathbf{u}_{AB} = \{-66\mathbf{i} - 37.3\mathbf{j} + 92.9\mathbf{k}\}$ N

2–103. $x = 3.82$ m, $y = 2.12$ m, $z = 1.88$ m

2–104. $x = 4.33$ m, $y = 0.948$ m

2–105. $\mathbf{F}_{EA} = \{12\mathbf{i} - 8\mathbf{j} - 24\mathbf{k}\}$ kN,
$\mathbf{F}_{EB} = \{12\mathbf{i} + 8\mathbf{j} - 24\mathbf{k}\}$ kN,
$\mathbf{F}_{EC} = \{-12\mathbf{i} + 8\mathbf{j} - 24\mathbf{k}\}$ kN,
$\mathbf{F}_{ED} = \{-12\mathbf{i} - 8\mathbf{j} - 24\mathbf{k}\}$ kN, $\mathbf{F}_R = \{-96\mathbf{k}\}$ kN

2–106. $F_R = 1.50$ kN, $\alpha = 77.6°$, $\beta = 90.6°$, $\gamma = 168°$

2–107. $\mathbf{F} = F\mathbf{u}_{AB} = \{1430\mathbf{i} + 2480\mathbf{j} - 2010\mathbf{k}\}$ N

2–109. Since the component of $(\mathbf{B} + \mathbf{D})$ is equal to the sum of the components of $\mathbf{B}$ and $\mathbf{D}$, then $\mathbf{A} \cdot (\mathbf{B} + \mathbf{D}) = \mathbf{A} \cdot \mathbf{B} + \mathbf{A} \cdot \mathbf{D}$.

2–110. $\theta = 121°$

2–111. $\theta = 109°$

2–113. $\theta = 70.5°$

2–114. $F_1 = \mathbf{F} \cdot \mathbf{u}_{AB} = 19.4$ N, $F_2 = 53.4$ N

2–115. $\theta = 74.2°$

2–117. $F_\parallel = 99.1$ N, $F_\perp = 592$ N

2–118. $F_\parallel = 82.4$ N, $F_\perp = 594$ N

2–119. $F_1 = \mathbf{F} \cdot \mathbf{u}_{AO} = 333$ N, $F_2 = 373$ N

2–121. Proj. $F = 31.1$ N

2–122. $\theta = 70.5°$

2–123. $\phi = 65.8°$

2–125. The magnitude is $(\mathbf{F}_1)_{F_1} = 5.44$ N

2–126. $\theta = 100°$

2–127. $\theta = 34.2°$

2–129. $\theta = 82.0°$

2–130. The magnitude is $(\mathbf{F}_1)_{F_2} = 50.6$ N

2–131. $\theta = 97.3°$

2–133. $F_3 = 428$ N, $\alpha = 88.3°$, $\beta = 20.6°$, $\gamma = 69.5°$

2–134. $F_3 = 250$ N, $\alpha = 87.0°$, $\beta = 143°$, $\gamma = 53.1°$

2–135. $F_{BA} = 215$ N, $\theta = 52.7°$

2–137. $\phi = \dfrac{\theta}{2}$, $F_R = 2F \cos\left(\dfrac{\theta}{2}\right)$

2–138. $\theta = 74.0°$, $\phi = 33.9°$

2–139. Proj $F_{AB} = 70.5$ N, Proj $F_{AC} = 65.1$ N

2–141. $\theta = 60°$, $P = 200$ N, $T = 346.4$ N

Chapter 3

3–1. $F_1 = 435$ N, $F_2 = 171$ N

3–2. $\theta = 31.8°$, $F = 4.94$ kN

3–3. $\theta = 12.9°$, $F_1 = 552$ N

3–5. $F_1 = 1.83$ kN, $F_2 = 9.60$ kN

3–6. $\theta = 4.69°$, $F_1 = 4.31$ kN

3–7. $F_{BC} = 2.99$ kN, $F_{AB} = 3.78$ kN

3–9. $\theta = 34.2°$

3–10. Thus, the smallest angle is $\theta = 11.5°$

3–11. $F = 1.13$ mN

3–13. $x_{AC} = 0.793$ m, $x_{AB} = 0.467$ m

3–14. $m = 12.8$ kg

3–15. $F = 158$ N

3–17. $W = 383.2$ N

3–18. $\theta = 78.7°$, $F_{CD} = 625$ N

3–19. $\theta = 78.7°$, $W = 255$ N

3–21. $d = 2.42$ m

3–22. $\theta = 60°$, $T_{AB} = 34.6$ N

3–23. $\theta = 60°$, $W = 46.2$ N

3–25. $s = 0.533$ m

3–26. $W = 6$ N

3–27. $F_{AC} = F_{AB} = F = \{2.45 \csc \theta\}$ kN, $l = 1.72$ m

3–29. $l = 265.2$ mm

3–30. $T = 212.1$ N

3–31. $\theta = 35.0°$

3–33. $W_E = 91.65$ N

3–34. $l = 0.8$ m

3–35. $F_{BD} = 171$ N, $F_{BC} = 145$ N

3–37. $\theta = 43.0°$

3–38. $y = 6.59$ m

3–39. $m_B = 3.58$ kg, $N = 19.7$ N

3–41. $F_1 = 608$ N, $\alpha = 79.2°$, $\beta = 16.4°$, $\gamma = 77.8°$

3–42. $F_1 = 800$ N, $F_2 = 147$ N, $F_3 = 564$ N

3–43. $F_1 = 5.60$ kN, $F_2 = 8.55$ kN, $F_3 = 9.44$ kN

3–45. $F_{AD} = 1.20$ kN, $F_{AC} = 0.40$ kN, $F_{AB} = 0.80$ kN

3–46. $F_{AC} = 130$ N, $F_{AD} = 510$ N, $F = 1.06$ kN

3–47. $S_{OB} = 327$ mm, $S_{OA} = 218$ mm

3–49. $F_{AB} = 0.980$ kN, $F_{AC} = 0.463$ kN, $F_{AD} = 1.55$ kN

3–50. $F_{AO} = 319$ N, $F_{AB} = 110$ N, $F_{AC} = 85.8$ N

3–51. $F_{AC} = 80.8$ N, $F_{AB} = 104$ N, $W = 138$ N

3–53. $F_{AE} = F_{AD} = 1.42$ kN $F_{AB} = 1.32$ kN

3–54. $F_{AB} = F_{AC} = 16.6$ kN, $F_{AD} = 55.2$ kN

3–55. $F_B = 19.2$ kN, $F_C = 10.4$ kN, $F_D = 6.32$ kN

3–57. $F_{AB} = 520$ N, $F_{AC} = F_{AD} = 260$ N, $d = 3.61$ m

3–58. $F_{AB} = 179.5$ N, $F_{AC} = F_{AD} = 127.0$ N

3–59. $W = 1337$ N

3–61. $F_{AB} = 469$ N, $F_{AC} = F_{AD} = 331$ N

3–62. $x = 0.190$ m, $y = 0.0123$ m

3–63. $F_{AD} = 7.10$ N, $F_{AC} = 4.57$ N, $F_{AB} = 7.35$ N

3–65. $F_{OB} = 120$ N, $F_{OC} = 150$ N, $F_{OD} = 480$ N

3–66. $F_A = 173.2$ N, $F_B = 286.5$ N

3–67. $F = 40.8$ N

3–69. $T_{AC} = 368.15$ N < 2000 N

Yes, Romeo can climb up the rope.

$T_{AC} = 708$ N < 2000 N

Yes, Romeo and Juliet can climb down.

3–70. $F_1 = 8.26$ kN, $F_2 = 3.84$ kN, $F_3 = 12.2$ kN

3–71. $\theta = 90°$, $F_{AB} = 800$ N, $\theta = 120°$, $F_{AB} = 800$ N

3–73. $W = 480$ N

3–74. $F_{CD} = 625$ N, $F_{CA} = 198$ N

3–75. $F_1 = 0$, $F_2 = 311$ N, $F_3 = 238$ N

Chapter 4

4–3. If $\mathbf{A} \cdot (\mathbf{B} \times \mathbf{C}) = 0$, then the volume equals zero, so that $\mathbf{A}$, $\mathbf{B}$, and $\mathbf{C}$ are coplanar.

4–5. $\curvearrowleft+$ $M_P = 2.37$ kN $\cdot$ m $\curvearrowright$

4–6. $\curvearrowleft+$ $M_O = 2.88$ kN $\cdot$ m $\curvearrowleft$

4–7. $\curvearrowleft+$ $M_P = 3.15$ kN $\cdot$ m $\curvearrowleft$

4–9. $\curvearrowleft+$ $M_P = 3.15$ kN $\cdot$ m $\curvearrowright$

4–10. $\curvearrowleft+$ $(M_{F_1})_O = 24.1$ N $\cdot$ m $\curvearrowleft$,
$\curvearrowleft+$ $(M_{F_2})_O = 14.5$ N $\cdot$ m $\curvearrowleft$

4–11. $M_O = 2.42$ kN $\cdot$ m $\curvearrowleft$

4–13. $\curvearrowleft+$ $(M_{F_1})_B = 825$ N $\cdot$ m $\curvearrowleft$,
$\curvearrowleft+$ $(M_{F_2})_B = 400$ N $\cdot$ m $\curvearrowleft$,
$\curvearrowleft+$ $(M_{F_3})_B = 8$ N $\cdot$ m $\curvearrowleft$

4–14. $\downarrow+ M_B = 108.8 \text{ N} \cdot \text{m} \uparrow$, $\downarrow+ M_C = 173.2 \text{ N} \cdot \text{m} \uparrow$

4–15. $\downarrow+ M_A = 237.5 \text{ N} \cdot \text{m} \uparrow$

4–17. $M_O = 28.1 \text{ N} \cdot \text{m} \uparrow$, $\theta = 88.6°$,
$(M_A)_{max} = 32.0 \text{ N} \cdot \text{m} \uparrow$

4–18. **a)** $\uparrow+ (M_A)_{max} = 412 \text{ N} \cdot \text{m}$, $\theta = 76.0°$
b) $\uparrow+ (M_A)_{min} = 0$, $\theta = 166°$

4–19. $\downarrow+ M_O = 120 \text{ N} \cdot \text{m} \downarrow$, $\downarrow+ M_O = 520 \text{ N} \cdot \text{m} \downarrow$

4–21. **a)** $M_A = 13.0 \text{ N} \cdot \text{m}$ **b)** $F = 35.2 \text{ N}$

4–22. $\downarrow+ (M_{F_1})_A = 433 \text{ N} \cdot \text{m} \downarrow$,
$\downarrow+ (M_{F_2})_A = 1.30 \text{ kN} \cdot \text{m} \downarrow$,
$\downarrow+ (M_{F_3})_A = 800 \text{ N} \cdot \text{m} \downarrow$

4–23. $\theta = 8.05°$

4–25. $F_A = 115.5 \text{ N}$

4–26. $\uparrow+ (M_O)_{max} = 80 \text{ kN} \cdot \text{m}$, $x = 24.0 \text{ m}$

4–27. $\uparrow+ (M_O)_{max} = 80 \text{ kN} \cdot \text{m}$, $\theta = 33.6°$, $\theta = 33.6°$

4–29. $\downarrow+ M_A = 1200 \sin \theta + 800 \cos \theta$

4–30. $\downarrow+ M_A = 0.418 \text{ N} \cdot \text{m} \downarrow$,
$\downarrow+ M_B = 4.92 \text{ N} \cdot \text{m} \downarrow$

4–31. $\downarrow+ M_A = \{1.18 \cos \theta (7.5 + x)\} \text{ kN} \cdot \text{m} \downarrow$,
$\downarrow+ (M_A)_{max} = 14.7 \text{ kN} \cdot \text{m} \downarrow$

4–33. $F = 4 \text{ kN}$

4–34. $\mathbf{M}_O = \{260\mathbf{i} + 180\mathbf{j} + 510\mathbf{k}\} \text{ N} \cdot \text{m}$

4–35. $\mathbf{M}_O = \{440\mathbf{i} + 220\mathbf{j} + 990\mathbf{k}\} \text{ N} \cdot \text{m}$

4–37. $\mathbf{M}_P = \{-116\mathbf{i} + 16\mathbf{j} - 135\mathbf{k}\} \text{ kN} \cdot \text{m}$

4–38. $\mathbf{M}_O = \{-128\mathbf{i} + 128\mathbf{j} - 257\mathbf{k}\} \text{ N} \cdot \text{m}$

4–39. $\mathbf{M}_B = \{-37.6\mathbf{i} + 90.7\mathbf{j} - 155\mathbf{k}\} \text{ N} \cdot \text{m}$

4–41. $\mathbf{M}_C = \{-3.54\mathbf{i} - 12.8\mathbf{j} - 22.2\mathbf{k}\} \text{ N} \cdot \text{m}$

4–42. $\mathbf{M}_A = \{-16.0\mathbf{i} - 32.1\mathbf{k}\} \text{ N} \cdot \text{m}$
$\mathbf{M}_A = \{-16.0\mathbf{i} - 32.1\mathbf{k}\} \text{ N} \cdot \text{m}$

4–43. $F_{AB} = 185.6 \text{ N}$

4–45. $\mathbf{M}_B = \{10.6\mathbf{i} + 13.1\mathbf{j} + 29.2\mathbf{k}\} \text{ N} \cdot \text{m}$

4–46. $\mathbf{M}_O = \{373\mathbf{i} - 99.9\mathbf{j} + 173\mathbf{k}\} \text{ N} \cdot \text{m}$

4–47. $\mathbf{M}_R = \{-1.90\mathbf{i} + 6.00\mathbf{j}\} \text{ kN} \cdot \text{m}$

4–49. $d = 1.15 \text{ m}$

4–50. $\theta = 0°, 90°, 180°$

4–51. $(\mathbf{M}_{Oa})_P = \{218\mathbf{j} + 163\mathbf{k}\} \text{ N} \cdot \text{m}$

4–53. $(\mathbf{M}_R)_{Oa} = \{26.1\mathbf{i} - 15.1\mathbf{j}\} \text{ N} \cdot \text{m}$

4–54. **a)** $(M_{AB})_1 = 72.0 \text{ N} \cdot \text{m}$, $(M_{AB})_2 = (M_{AB})_3 = 0$
b) $(M_{AB})_2 = (M_{AB})_3 = 0$, $(M_{AB})_1 = 72.0 \text{ N} \cdot \text{m}$

4–55. $M_x = 66.6 \text{ N} \cdot \text{m}$

4–57. $M_y = 0.277 \text{ N} \cdot \text{m}$

4–58. $\mathbf{M}_y = \{-66.67\mathbf{j}\} \text{ N} \cdot \text{m}$

4–59. $M_x = 15.0 \text{ N} \cdot \text{m}$, $M_y = 4.00 \text{ N} \cdot \text{m}$,
$M_z = 36.0 \text{ N} \cdot \text{m}$

4–61. $M_x = 3.75 \text{ N} \cdot \text{m}$

4–62. $M_z = 109 \text{ N} \cdot \text{cm}$

4–63. $|M_{CA}| = 226 \text{ N} \cdot \text{m}$

4–65. $P = 33.83 \text{ N}$

4–66. $M_y = 936 \text{ N} \cdot \text{m}$

4–67. $\mathbf{M}_z = \{35.4\mathbf{k}\} \text{ N} \cdot \text{m}$

4–69. $M_C = 18.3 \text{ kN} \cdot \text{m} \uparrow$

4–70. $M_C = 100 \text{ N} \cdot \text{m} \uparrow$

4–71. $M_C = 17.6 \text{ kN} \cdot \text{m} \uparrow$

4–73. $F = 133 \text{ N}$, $F = 800 \text{ N}$

4–74. $T = 0.909 \text{ kN}$

4–75. $N = 26.0 \text{ N}$

4–77. $M = 1200 \text{ N} \cdot \text{m}$, $R_B = 2300 \text{ N}$

4–78. $F = 762 \text{ N}$

4–79. $\mathbf{M}_C = \{-5\mathbf{i} + 8.75\mathbf{j}\} \text{ N} \cdot \text{m}$

4–81. $d = 1.54 \text{ m}$

4–82. $F = 167 \text{ N}$
Resultant couple can act anywhere.

4–83. $d = 2.03 \text{ m}$

4–85. $\mathbf{M}_C = \{12.6\mathbf{k}\} \text{ N} \cdot \text{m}$, $M_C = 12.6 \text{ N} \cdot \text{m}$

4–86. $\mathbf{M}_C = \{-360\mathbf{i} + 380\mathbf{j} + 320\mathbf{k}\} \text{ N} \cdot \text{m}$

4–87. $\mathbf{M}_C = \{-411\mathbf{i} - 257\mathbf{j} - 651\mathbf{k}\} \text{ kN} \cdot \text{m}$

4–89. $(M_R)_{x'} = 5.45 \text{ kN} \cdot \text{m}$, $(M_R)_{y'} = 31.46 \text{ kN} \cdot \text{m}$

4–90. $\mathbf{M}_C = \{7.01\mathbf{i} + 42.1\mathbf{j}\} \text{ N} \cdot \text{m}$

4–91. $F = 35.1 \text{ N}$

4–93. $\mathbf{M}_R = \{11.0\mathbf{i} - 49.0\mathbf{j} - 40.0\mathbf{k}\} \text{ N} \cdot \text{m}$
$M_R = 64.2 \text{ N} \cdot \text{m}$, $\alpha = 80.1°$, $\beta = 140°$, $\gamma = 129°$

4–94. $M_R = 59.9 \text{ N} \cdot \text{m}$, $\alpha = 99.0°$, $\beta = 106°$, $\gamma = 18.3°$

4–95. $\alpha = 155°$, $\beta = 115°$, $\gamma = 90°$

4–97. $d = 342 \text{ mm}$

4–98. $F_O = 375 \text{ N}$, $M_O = 100 \text{ N} \cdot \text{m} \downarrow$

4–99. $F_P = 375 \text{ N}$, $M_P = 737 \text{ N} \cdot \text{m} \uparrow$

4–101. $F_R = 178 \text{ N}$, $\theta = 73.0°$, $M_{R_P} = 2.68 \text{ N} \cdot \text{m} \uparrow$

4–102. $F_R = 274 \text{ kN}$, $\theta = 5.24°$, $M_O = 4609 \text{ kN} \cdot \text{m} \uparrow$

4–103. $F_R = 274 \text{ kN}$, $\theta = 5.24°$, $M_P = 5476 \text{ kN} \cdot \text{m} \uparrow$

4–105. $F_R = 6.57 \text{ N}$, $\theta = 57.4°$, $M_{R_P} = 31.0 \text{ N} \cdot \text{m} \uparrow$

4–106. $F_R = 2.10 \text{ kN}$, $\theta = 81.6°$, $M_O = 10.6 \text{ kN} \cdot \text{m} \downarrow$

4–107. $F_R = 2.10 \text{ kN}$, $\theta = 81.6°$, $M_P = 16.8 \text{ kN} \cdot \text{m} \downarrow$

4–109. $F_R = 375$ N $\uparrow$, $x = 2.47$ m

4–110. $F_R = 5.93$ kN, $\theta = 77.8°$,
$M_{R_A} = -34.8$ kN $\cdot$ m $= 34.8$ kN $\cdot$ m $\downarrow$

4–111. $F_R = 5.93$ kN, $\theta = 77.8°$, $M_{R_B} = 11.6$ kN $\cdot$ m $\uparrow$

4–113. $F = 798$ N, $\theta = 67.9°$, $x = 3.28$ m

4–114. $F = 922$ N, $\theta = 77.5°$, $x = 0.36$ m

4–115. $F = 1302$ N, $\theta = 84.5°$, $x = 7.36$ m

4–117. $F_2 = 25.9$ N, $\theta = 18.1°$, $F_1 = 68.1$ N

4–118. $F_R = 49$ kN $\downarrow$, $M_{R_A} = 142$ kN $\cdot$ m $\uparrow$

4–119. $F_R = 49$ kN $\downarrow$, $d = 2.9$ m

4–121. $F_R = 991$ N, $\theta = 63.0°$, $x = 2.64$ m

4–122. $F_R = 65.9$ N, $\theta = 49.8°$, $d = 0.21$ m

4–123. $F_R = 65.9$ N, $\theta = 49.8°$, $d = 0.46$ m

4–125. $\mathbf{F}_R = \{8\mathbf{i} + 6\mathbf{j} + 8\mathbf{k}\}$ kN,
$\mathbf{M}_{RO} = \{-10\mathbf{i} + 18\mathbf{j} - 56\mathbf{k}\}$ kN $\cdot$ m

4–126. $\mathbf{F}_R = \{8\mathbf{i} + 6\mathbf{j} + 8\mathbf{k}\}$ kN,
$\mathbf{M}_{RP} = \{-46\mathbf{i} + 66\mathbf{j} - 56\mathbf{k}\}$ kN $\cdot$ m

4–127. $\mathbf{F}_R = \{8\mathbf{i} + 6\mathbf{j} + 8\mathbf{k}\}$ kN,
$\mathbf{M}_{RQ} = \{-10\mathbf{i} - 30\mathbf{j} - 20\mathbf{k}\}$ kN $\cdot$ m
$\mathbf{F}_R = \{-40\mathbf{j} - 40\mathbf{k}\}$ N,
$\mathbf{M}_{RA} = \{-12\mathbf{j} + 12\mathbf{k}\}$ N $\cdot$ m

4–129. $\mathbf{F}_R = \{-28.3\mathbf{j} - 68.3\mathbf{k}\}$ N,
$\mathbf{M}_{RA} = \{-20.5\mathbf{j} + 8.49\mathbf{k}\}$ N $\cdot$ m

4–130. $\mathbf{F}_R = \{400\mathbf{i} + 300\mathbf{j} - 650\mathbf{k}\}$ N,
$\mathbf{M}_{RA} = \{-3100\mathbf{i} + 4800\mathbf{j}\}$ N $\cdot$ m

4–131. $\mathbf{F}_R = \{0.232\mathbf{i} + 5.06\mathbf{j} + 12.4\mathbf{k}\}$ kN,
$\mathbf{M}_{R_O} = \{36.0\mathbf{i} - 26.1\mathbf{j} + 12.2\mathbf{k}\}$ kN $\cdot$ m

4–133. $F_R = 140$ kN $\downarrow$, $y = 7.14$ m, $x = 5.71$ m

4–134. $F_R = 140$ kN $\downarrow$, $x = 6.43$ m, $y = 7.29$ m

4–135. $\mathbf{F}_R = \{141\mathbf{i} + 100\mathbf{j} + 159\mathbf{k}\}$ N,
$\mathbf{M}_{R_O} = \{122\mathbf{i} - 183\mathbf{k}\}$ N $\cdot$ m

4–137. $F_R = 990$ N, $M_R = 3.07$ kN $\cdot$ m,
$x = 1.16$ m, $y = 2.06$ m

4–138. $F_R = 108$ kN,
$M_R = -62.4$ kN $\cdot$ m, $z = 0.87$ m, $y = 0.04$ m

4–139. $F_{RO} = 70$ N $\downarrow$, $x = 0.107$ m

4–141. $F_R = 51.0$ kN $\downarrow$, $M_{R_O} = 914$ kN $\cdot$ m $\downarrow$

4–142. $F_R = 51.0$ kN $\downarrow$, $d = 17.9$ m

4–143. $F_R = 32.5$ kN, $\theta = 67.2°$, $x = 1.286$ m

4–145. $F_R = 60$ kN $\downarrow$, $x = 3.89$ m

4–146. $F_R = 1.10$ kN $\downarrow$, $M_{R_O} = 3.10$ kN $\cdot$ m $\downarrow$

4–147. $d = 1.50$ m, $w = 175$ N/m

4–149. $F_{RO} = 13$ kN $\uparrow$, $d = 3.76$ m

4–150. $b = 1.5$ m, $a = 2.92$ m

4–151. $F_R = 35.5$ kN $\downarrow$, $x = 0.16$ m

4–152. $F_R = 1.35$ kN, $\theta = 42.0°$, $y = 0.1$ m

4–153. $F_R = 1.35$ kN, $\theta = 42.0°$, $x = 0.556$ m

4–154. $F_R = 95.6$ kN $\rightarrow$, $M_{R_O} = 349$ kN $\cdot$ m $\downarrow$

4–155. $F_R = 107$ kN $\leftarrow$, $h = 1.60$ m

4–157. $F_R = 3.60$ kN $\downarrow$, $M_{R_O} = 19.4$ kN $\cdot$ m $\downarrow$

4–158. $F_R = 597.3$ kN $\uparrow$, $\bar{x} = 4.85$ m

4–159. $F_R = 1.87$ kN $\uparrow$, $\bar{x} = 3.66$ m

4–161. $\mathbf{u}_{OA} = 0.3293\mathbf{i} + 0.7683\mathbf{j} + 0.5488\mathbf{k}$
$\alpha = 70.8°$, $\beta = 39.8°$, $\gamma = 56.7°$
$\mathbf{u}_{AO} = -0.3293\mathbf{i} - 0.7683\mathbf{j} - 0.5488\mathbf{k}$
$\alpha = 109°$, $\beta = 140°$, $\gamma = 123°$

4–162. $\mathbf{M}_O = \{298\mathbf{i} + 15.1\mathbf{j} - 200\mathbf{k}\}$ N $\cdot$ cm

4–163. $P = 94.8$ N

4–164. $\mathbf{M}_A = \{-59.7\mathbf{i} - 159\mathbf{k}\}$ N $\cdot$ m

4–165. $\mathbf{M}_{a-a} = 59.7$ N $\cdot$ m
The negative sign indicates that $\mathbf{M}_{a-a}$ is directed toward negative x axis. $\mathbf{M}_{a-a} = 59.7$ N $\cdot$ m

4–166. $\mathbf{M}_{CR} = \{63.6\mathbf{i} - 170\mathbf{j} + 264\mathbf{k}\}$ N $\cdot$ m

4–167. $\mathbf{F}_R = \{14.3\mathbf{i} + 21.4\mathbf{j} - 42.9\mathbf{k}\}$ kN
$\mathbf{M}_A = \{-192.9\mathbf{i} + 42.86\mathbf{j} - 42.86\mathbf{k}\}$ kN $\cdot$ m

4–169. $\mathbf{M}_O = \{1.06\mathbf{i} + 1.06\mathbf{j} - 4.03\mathbf{k}\}$ N $\cdot$ m,
$\alpha = 75.7°$, $\beta = 75.7°$, $\gamma = 160°$

4–170. $\mathbf{M}_{RP} = \{-26\mathbf{i} + 357\mathbf{j} + 127\mathbf{k}\}$ N $\cdot$ m

Chapter 5

5–1. W is the effect of gravity (weight) on the paper roll. N_A and N_B are the smooth blade reactions on the paper roll.

5–3. W is the effect of gravity (weight) on the dumpster. A_y and A_x are the pin A reactions on the dumpster. F_{BC} is the hydraulic cylinder BC reaction on the dumpster.

5–5. C_y and C_x are the pin C reactions on the truss. T_{AB} is the cable AB tension on the truss. 3 kN and 4 kN force are the effect of external applied forces on the truss.

5–6. W is the effect of gravity (weight) on the boom. A_y and A_x are the pin A reactions on the boom. T_{BC} is the cable BC force reactions on the boom. 5000-N force is the suspended load reaction on the boom.

5–11. $N_B = 245$ N, $N_A = 425$ N

5–13. $T_{AB} = 5.89$ kN, $C_x = 5.11$ kN, $C_y = 4.05$ kN

5–14. $T_{BC} = 44.2$ kN, $A_x = 40.8$ kN, $A_y = 24.6$ kN

5–15. $N_B = 2.14$ kN, $A_x = 1.29$ kN, $A_y = 1.49$ kN

5–17. $N_C = 493$ N, $N_B = 554$ N, $N_A = 247$ N

5–18. $B_y = 642$ N, $A_x = 192$ N, $A_y = 180$ N

5–19. $C_y = 586$ N,
$$F_A = \sqrt{(103.528)^2 + (400)^2} = 413 \text{ N}$$

5–21. $F_A = 120$ N, $F_B = 144.9$ N, $F_C = 37.5$ N

5–22. $F_H = 288.9$ N, $T_B = 328.9$ N

5–23. $F_{CD} = 975$ N, $A_x = 487.4$ N, $A_y = 155.8$ N

5–25. $(N_A)_r = 397.7$ N, $(N_A)_s = 396.5$ N

5–26. $N_B = 10.5$ N, $A_x = 42.0$ N, $A_y = 10.5$ N

5–27. $W_B = 314.29$ N

5–29. $F_B = 6.37765$ N = 6.38 N,
$A_x = 3.19$ N, $A_y = 2.48$ N

5–30. $F_{BC} = 574$ N, $A_x = 1.08$ kN, $A_y = 637$ N

5–31. $A_x = 1462$ N, $F_B = 1.66$ kN

5–33. $D_x = 0$, $D_y = 1.65$ kN, $M_D = 1.40$ kN · m,
$(M_D)_{max} = 3.00$ kN · m

5–34. $x = 2.5$ m, $A_x = N_B = 4.17$ kN, $A_y = 5.00$ kN
$x = 1$ m, $A_x = N_B = 1.67$ kN, $A_y = 5.00$ kN

5–35. $F_B = 105$ N

5–37. $F = 311$ kN, $A_x = 460$ kN, $A_y = 7.85$ kN

5–38. $h = 4.731$ m

5–39. $N_A = 408.1$ N, $F_B = 251.2$ N

5–41. $B_x = 989$ N, $A_x = 989$ N, $B_y = 186$ N

5–42. $w_1 = 413$ kN/m, $w_2 = 407$ kN/m

5–43. $T = 5$ kN, $T_{BC} = 16.4$ kN, $F_A = 20.6$ kN

5–45. $R_A = 204.67$ kN, $R_B = 625.33$ kN

5–46. $C_x = 1333.3$ N, $C_y = 2888.9$ N

5–47. $N_A = 8.76$ kN, $N_B = 4.64$ kN, $W = 18.96$ kN

5–49. $F_2 = 724$ N, $F_1 = 1.45$ kN, $F_A = 1.75$ kN

5–50. $d = \dfrac{3a}{4}$

5–51. $N_B = 2.11$ N, $F_A = 2.81$ N

5–53. $k = 336.8$ N/m

5–54. $R_C = 255.6$ N, $R_B = 47.8$ N, $R_A = 104$ N

5–55. $\alpha = \tan^{-1}\left(\dfrac{0.05333}{3}\right) = 1.02°$

5–57. $d = \dfrac{a}{\cos^3 \theta}$

5–58. $\theta = \tan^{-1}\dfrac{b}{a}$

5–59. $\alpha = 10.4°$

5–61. $\dfrac{2h}{s} = \dfrac{h^2 + s^2 - l^2}{2hs} \quad h\sqrt{\dfrac{s^2 - l^2}{3}}$

5–62. $\theta = 27.1°$

5–63. $T = 1.84$ kN, $F = 6.18$ kN

5–65. $T_B = 11$ kN, $T_C = 5.5$ kN, $T_A = 5.5$ kN

5–66. $W = 3000$ N, $x = 2.6$ m, $y = 2.63$ m

5–67. $F_A = 663$ N, $F_C = 569$ N, $F_B = 449$ N

5–69. $N_C = 289$ N, $N_A = 213$ N, $N_B = 332$ N

5–70. $A_x = 0$, $A_y = 1500$ kN, $A_z = 750$ kN, $T = 919$ kN

5–71. $F = 1.31$ kN, $A_x = 0$, $A_y = 1.31$ kN, $A_z = 653$ N

5–73. $P = 245$ N, $A_y = 0$, $B_z = 245$ N, $A_z = 245$ N
$B_x = 368$ N, $A_x = 123$ N

5–74. $O_x = 0$, $O_y = -8.49$ kN, $O_z = 8.0$ kN
$(M_O)_x = 94.8$ kN, $(M_O)_y = 0$, $(M_O)_z = 0$

5–75. $F_{BC} = 0$, $A_y = 0$, $A_z = 800$ N,
$(M_A)_x = 480$ N · m, $(M_A)_y = 0$, $(M_A)_z = 0$

5–77. $T = 58.0$ N, $C_z = 87.0$ N, $C_y = 28.8$ N,
$D_x = 0$, $D_y = 79.2$ N, $D_z = 58.0$ N

5–78. $T = 58.0$ N, $C_z = 77.6$ N, $C_y = 24.9$ N,
$D_x = 0$, $D_y = 68.5$ N, $D_z = 32.1$ N

5–79. $A_x = 633$ N, $A_y = -141$ N, $B_x = -721$ N,
$B_z = 895$ N, $C_y = 200$ N, $C_z = -506$ N

5–81. $A_z = 6110$ N, $B_z = 4667$ N, $C_z = 3223$ N

5–82. $T_{DE} = 32.1$ kN, $T_{BC} = 42.9$ kN, $A_x = 3.57$ kN,
$A_y = 50$ kN, $(M_A)_x = 0$, $(M_A)_y = -17.9$ kN · m

5–83. $T_B = 16.7$ kN, $A_x = 0$, $A_y = 5.00$ kN,
$A_z = 16.7$ kN

5–85. $T_B = 25$ kN, $B_y = 25$ kN,
$A_x = 25$ kN, $A_y = -25$ kN, $A_z = 50$ kN

5–86. $B_z = 25.0$ N, $P = 62.5$ N, $B_x = 22.3$ N,
$A_x = 84.8$ N, $B_y = 0$, $A_z = 25.0$ N

5–87. $F_{AC} = F_{BC} = 6.13$ kN, $F_{DE} = 19.6$ kN

5–89. $T_{BC} = 131$ kN, $T_{BD} = 510$ kN, $A_x = 0$,
$A_y = 0$, $A_z = 589$ kN
Also, note that BA is a two-force member, so that
$A_x = A_y = 0$.

5–90. $F_{BC} = 205$ N, $F_{ED} = 629$ N
$A_x = 32.4$ N, $A_y = 107$ N, $A_z = 1.28$ kN

5–91. $F_{CD} = 1.02$ kN, $A_z = -208$ N, $B_z = -139$ N,
$A_y = 573$ N, $B_y = 382$ N

5–93. $F = 354$ N

5–94. $N_A = 8.00$ kN, $B_x = 5.20$ kN, $B_y = 5.00$ kN

5–95. $N_B = 400$ N, $F_A = 721$ N

5–97. $N_B = 957$ N, $A_x = 0$

5–98. $A_x = 0, A_y = 0, A_z = B_z = C_z = 5.33$ kN

5–99. $A_x = 0$, $A_y = -200$ N, $A_z = 150$ N,
$(M_A)_x = -100$ N $\cdot$ m, $(M_A)_y = 0$,
$(M_A)_z = -500$ N $\cdot$ m

Chapter 6

6–1. Joint B: $F_{BA} = 286$ kN (T), $F_{BC} = 808$ kN (T)
Joint C: $F_{CA} = 571$ kN (C), $C_y = 571$ kN
Note: The support reactions A_x and A_y can be determined by analyzing Joint A using the results obtained above.

6–2. Joint B: $F_{BA} = 286$ kN (T), $F_{BC} = 384$ kN (T)
Joint C: $F_{CA} = 271$ kN (C), $C_y = 271.43$ kN
Note: The support reactions A_x and A_y can be determined by analyzing Joint A using the results obtained above.

6–3. Joint A: $F_{AD} = 849$ kN (C), $F_{AB} = 600$ kN (T)
Joint B: $F_{BD} = 400$ kN (C), $F_{BC} = 600$ kN (T)
Joint D: $F_{DC} = 1.41$ MN (T), $F_{DE} = 1.60$ MN (C)

6–5. Joint A: $F_{AE} = 8.94$ kN (C), $F_{AB} = 8.00$ kN (T)
Joint B: $F_{BC} = 8.00$ kN (T), $F_{BE} = 8.00$ kN (C)
Joint E: $F_{EC} = 8.94$ kN (T), $F_{ED} = 17.9$ kN (C)
Joint D: $F_{DC} = 8.00$ kN (T), $D_x = 16.0$ kN
Note: The support reactions C_x and C_y can be determined by analyzing Joint C using the results obtained above.

6–6. Joint A: $F_{AE} = 372$ N (C), $F_{AB} = 332$ N (T)
Joint B: $F_{BC} = 332$ N (T), $F_{BE} = 196$ N (C)
Joint E: $F_{EC} = 558$ N (T), $F_{ED} = 929$ N (C)
Joint D: $F_{DC} = 582$ N (T)

6–7. Joint B: $F_{BC} = 3$ kN (C), $F_{BA} = 8$ kN (C)
Joint A: $F_{AC} = 1.46$ kN (C), $F_{AF} = 4.17$ kN (T)
Joint C: $F_{CD} = 4.17$ kN (C), $F_{CF} = 3.12$ kN (C)

Joint E: $F_{EF} = 0$, $F_{ED} = 13.1$ kN (C)
Joint D: $F_{DF} = 5.21$ kN (T)

6–9. Joint C: $F_{CB} = 8.00$ kN (T), $F_{CD} = 6.93$ kN (C)
Joint D: $F_{DE} = 6.93$ kN (C), $F_{DB} = 4.00$ kN (T)
Joint B: $F_{BE} = 4.00$ kN (C), $F_{BA} = 12.0$ kN (T)
Note: The support reactions at support A and E can be determined by analyzing Joints A and E respectively using the results obtained above.

6–10. Joint A: $F_{AG} = 47.1$ kN (C), $F_{AB} = 33.3$ kN (T)
Joint B: $F_{BG} = 0$, $F_{BC} = 33.3$ kN (T)
Joint D: $F_{DE} = 94.3$ kN (C), $F_{DC} = 66.7$ kN (T)
Joint E: $F_{EC} = 66.7$ kN (T), $F_{EG} = 66.7$ kN (C)
Joint C: $F_{CG} = 47.1$ kN (T)

6–11. Joint A: $F_{AG} = 117.9$ kN C), $F_{AB} = 83.3$ kN (T)
Joint B: $F_{BC} = 83.3$ kN T), $F_{BG} = 50$ kN (T)
Joint D: $F_{DE} = 165$ kN (C), $F_{DC} = 116.7$ kN (T)
Joint E: $F_{EC} = 116.7$ kN (T), $F_{EG} = 116.7$ kN (C)
Joint C: $F_{CG} = 47.1$ kN (T)

6–13. $F_{GB} = 30$ kN (T), Joint A: $F_{AF} = 20$ kN (C),
$F_{AB} = 22.4$ kN (C)
Joint B: $F_{BF} = 20$ kN (T), $F_{BC} = 20$ kN (T)
Joint F: $F_{FC} = 28.3$ kN (C), $F_{FE} = 0$
Joint E: $F_{ED} = 0$, $F_{EC} = 20.0$ kN (T)
Joint D: $F_{DC} = 0$

6–14. Joint A: $F_{AB} = 33$ kN (C), $F_{AF} = 7.93$ kN (T)
Joint B: $F_{BF} = 23.3$ kN (T), $F_{BC} = 23.3$ kN (C)
Joint F: $F_{FC} = 4.71$ kN (C), $F_{FE} = 11.3$ kN (T)
Joint E: $F_{EC} = 30$ kN (T), $F_{ED} = 11.3$ kN (T)
Joint C: $F_{CD} = 37.7$ kN (C)

6–15. Joint A: $F_{AB} = 37.7$ kN (C), $F_{AF} = 19$ kN (T)
Joint B: $F_{BF} = 26.7$ kN (T), $F_{BC} = 26.7$ kN (C)
Joint F: $F_{FC} = 18.9$ kN (T), $F_{FE} = 5.67$ kN (T)
Joint E: $F_{ED} = 5.67$ kN (T), $F_{EC} = 0$
Joint C: $F_{CD} = 18.9$ kN (C)

6–17. $P = 5.20$ kN

6–18. Joint C: $F_{CB} = 400$ kN (C), $F_{CD} = 693$ kN (C)
Joint B: $F_{BD} = 667$ kN (T), $F_{BA} = 1133$ kN (C)
Member AB is a two-force member and exerts only a vertical force along AB at A.

6–19. Joint C: $F_{CD} = 3.61$ kN (C), $F_{CB} = 3$ kN (T)
Joint B: $F_{BA} = 3$ kN (T), $F_{BD} = 3$ kN (C)
Joint D: $F_{DA} = 2.70$ kN (T), $F_{DE} = 6.31$ kN (C)

6–21. Joint B: $F_{BA} = P \csc 2\theta$ (C), $F_{BC} = P \cot 2\theta$ (C)
Joint C:
$F_{CA} = (\cot \theta \cos \theta - \sin \theta + 2 \cos \theta)P$ (T),
$F_{CD} = (\cot 2\theta + 1)P$ (C)
Joint D:
$F_{DA} = (\cot 2\theta + 1)(\cot 2\theta)P$ (C)

6–22. $P_{max} = 732$ N

6–23. Joint D: $F_{DE} = 16.3$ kN (C), $F_{DC} = 8.40$ kN (T)
Joint E: $F_{EA} = 8.85$ kN (C), $F_{EC} = 6.20$ kN (C)
Joint C: $F_{CF} = 8.77$ kN (T), $F_{CB} = 2.20$ kN (T)
Joint B: $F_{BA} = 3.11$ kN (T), $F_{BF} = 6.20$ kN (C)
Joint F: $F_{FA} = 6.20$ kN (T)

6–25. Joint A: $F_{AB} = 7.5$ kN (T), $F_{AE} = 4.5$ kN (C)
Joint E: $F_{ED} = 4.5$ kN (C), $F_{EB} = 8$ kN (T)
Joint B: $F_{BD} = 19.8$ kN (C), $F_{BC} = 18.5$ kN (T)
C_y is zero because BC is a two-force member.

6–26. Joint A: $F_{AB} = 196$ N (T), $F_{AE} = 118$ N (C)
Joint E: $F_{ED} = 118$ N (C), $F_{EB} = 216$ N (T)
Joint B: $F_{BD} = 1.04$ kN (C), $F_{BC} = 857$ N (T)

6–27. $F_{CB} = F_{CD} = 0$ Joint A: $F_{AB} = 2.40P$ (C),
$F_{AF} = 2.00P$ (T)
Joint B: $F_{BF} = 1.86P$ (T), $F_{BD} = 0.373P$ (C)

6–29. $127° \le \theta \le 196°$, $336° \le \theta \le 347°$

6–30. $F_{HG} = 29.0$ kN (C), $F_{BC} = 20.5$ kN (T),
$F_{HC} = 12.0$ kN (T)

6–31. $F_{GF} = 29.0$ kN (C), $F_{CD} = 23.5$ kN (T),
$F_{CF} = 7.78$ kN (T)

6–33. $F_{KJ} = 13.3$ kN (T), $F_{BC} = 14.9$ kN (C),
$F_{CK} = 0$

6–34. $F_{KJ} = 112.5$ kN (T), $F_{CD} = 93.75$ kN (C),
$F_{CJ} = 31.25$ kN (C), $F_{DJ} = 0$

6–35. $F_{JI} = 7.5$ kN (T), $F_{EI} = 25$ kN (C)

6–37. $F_{FG} = 8.08$ kN (T), $F_{CD} = 8.47$ kN (C),
$F_{CF} = 0.770$ kN (T)

6–38. $F_{GF} = 67.1$ kN (C), $F_{GB} = 67.1$ kN (T)

6–39. $F_{BG} = -200\sqrt{L^2 + 9}$,
$F_{BC} = -200L$, $F_{HG} = 400L$

6–41. $AB, BC, CD, DE, HI,$ and GI are zero-force members.
Joint E: $F_{JE} = 9.38$ kN (C), $F_{GF} = 5.625$ kN (T)

6–42. $F_{BC} = 10.4$ kN (C), $F_{HG} = 9.16$ kN (T),
$F_{HC} = 2.24$ kN (T)

6–43. $F_{CD} = 11.2$ kN (C),
Joint G: $F_{CF} = 3.21$ kN (T)
$F_{GH} = 9.155$ kN (T), $F_{CG} = 6.80$ kN (C)

6–45. $F_{GJ} = 20$ kN (C)

6–46. $F_{GC} = 10$ kN (T)

6–47. $F_{GF} = 1.78$ kN (T), $F_{CD} = 2.23$ kN (C),
$F_{CF} = 0$

6–49. $F_{EF} = P$ (C),
$F_{CB} = 1.12P$ (T), $F_{BE} = 0.5P$ (T)

6–50. $F_{AB} = P$ (T),
$F_{EF} = P$ (C), $F_{BF} = 1.41P$ (C)

6–51. *Method of Joints:* By inspection, members $BN, NC,$
DO, OC, HJ, LE and JG are zero-force members.
$F_{CD} = 5.625$ kN (T), $F_{CM} = 2.00$ kN (T)

6–53. $F_{KJ} = 30.7$ kN (C), $F_{CD} = 30.7$ kN (T)
$F_{ND} = 1.67$ kN (T), $F_{NJ} = 1.67$ kN (C)

6–54. $F_{JI} = 21.3$ kN (C), $F_{DE} = 21.3$ kN (C)

6–55. $F_{AD} = 300$ N (C), $F_{BD} = 450$ N (C),
$F_{CD} = 568$ N (C)

6–57. $F_{DC} = F_{DA} = 2.59$ kN (C), $F_{DB} = 3.85$ kN (C)
$F_{BC} = F_{BA} = 0.890$ kN (T), $F_{AC} = 0.616$ kN (T)

6–58. Joint F: $F_{BF} = 0$ Joint B: $F_{BC} = 0$
$F_{BE} = 500$ kN (T), $F_{AB} = 300$ kN (C)
Joint A: $F_{AC} = 583$ kN (T), $F_{AD} = 333$ kN (T),
$F_{AE} = 667$ kN (C)
Joint E: $F_{EF} = 300$ kN (C),
Joint C: $F_{CD} = 300$ kN (C), $F_{CF} = 300$ kN (C)
Joint F: $F_{DF} = 424$ kN (T)

6–59. Joint F: $F_{BF} = 0$
Joint B: $F_{BC} = 0$, $F_{BE} = 500$ kN (T),
$F_{AB} = 300$ kN (C)
Joint A: $F_{AC} = 972$ kN (T), $F_{AD} = 0$
$F_{AE} = 367$ kN (C)
Joint E: $F_{DE} = 0$, $F_{EF} = 300$ kN (C)
Joint C: $F_{CD} = 500$ kN (C), $F_{CF} = 300$ kN (C)
Joint F: $F_{DF} = 424$ kN (T)

6–61. $F_{BC} = F_{BD} = 1.342 = 1.34$ kN (C)
Joint A: $F_{AB} = 2.4$ kN (C),
$F_{AG} = F_{AE} = 1.01$ kN (T)
Joint B: $F_{BG} = 1.80$ kN (T), $F_{BE} = 1.80$ kN (T)

6–62. Joint C: $F_{BC} = 1.15$ kN (C)
Joint D: $F_{DF} = 4.16$ kN (C)
Joint B: $F_{BE} = 4.16$ kN (T)

6–63. Joint C: $F_{CF} = 0$, $F_{CD} = 2.31$ kN (T)

Joint D: $F_{DF} = 4.163$ kN (C), $F_{ED} = 3.46$ kN (T)

Joint B: $F_{BE} = 4.163$ kN (T), $F_{AB} = 3.46$ kN (C)

6–65. Joint C: $F_{BC} = 0$, $F_{CD} = 0$, $F_{CF} = 8$ kN (C)

Joint B: $F_{BD} = 0$, $F_{BA} = 6$ kN (C)

Joint D: $F_{AD} = 0$, $F_{DF} = 0$, $F_{DE} = 9$ kN (C)

Joint E: $F_{EF} = 0$, $F_{EA} = 0$

Joint A: $F_{AF} = 0$

6–66. **a)** $P = 25.0$ N, **b)** $P = 33.3$ N, **c)** $P' = 33.33$ N,

$P = 11.1$ N

6–67. $F_P = 59.4$ N, $F_A = 852$ N

6–69. $R_E = 177$ N, $R_A = 128$ N

6–70. $P = 40.0$ N, $x = 240$ mm

6–71. $P = 21.8$ N, At A, $R_A = 2P = 43.6$ N,

At B, $R_B = 2P = 43.6$ N, At C, $R_C = 6P = 131$ N

6–73. $A_y = 9.59$ kN, $B_y = 8.54$ kN,

$C_y = 2.93$ kN, $C_x = 9.20$ kN

6–74. $P = 743$ N

6–75. $A_y = 300$ N, $A_x = 300$ N

For pin C,

$C_x = F_{BC} \sin 45° = 300$ N,

$C_y = F_{BC} \cos 45° = 300$ N

6–77. $B_y = 1.33$ kN, $B_x = 5.00$ kN

For pin A and C, $A_x = C_x = 5.00$ kN,

$A_y = C_y = 6.67$ kN

From FBD **(b)**, $M_D = 10.0$ kN · m,

$D_y = 8.00$ kN, $D_x = 0$

6–78. $C_x = 7.5$ kN, $C_y = 10$ kN

6–79. $A_x = 4.20$ kN, $B_x = 4.20$ kN, $A_y = 4.00$ kN,

$B_y = 3.20$ kN, $C_x = 3.40$ kN, $C_y = 4.00$ kN

6–81. $T = 100$ N, $\theta = 14.6°$

6–82. $x = 3.30$ m

6–83. Pulley E: $T = 350$ N, Member ABC: $A_y = 700$ N,

$A_x = 1.88$ kN

At D: $D_x = 1.70$ kN, $D_y = 1.70$ kN

6–85. $A_x = 80$ kN, $A_y = 80$ kN, $B_y = 1.33$ kN,

$B_x = 336$ kN, $C_x = 416$ kN, $C_y = 53.3$ kN

6–86. From FBD **(a)**, $F_{AB} = 9.23$ kN, $C_x = 2.17$ kN,

$C_y = 7.01$ kN

From FBD **(b)**, $D_x = 0$, $D_y = 1.96$ kN,

$M_D = 2.66$ kN · m

6–87. $C_y = 350$ N, $C_x = 166.7$ N

$B_x = 666.7$ N, $B_y = 150$ N

6–89. $C_x = D_x = 1600$ N, $C_y = D_y = 1067$ N,

$B_y = 266.7$ N, $B_x = 800$ N, $E_x = 0$,

$E_y = 266.7$ N, $A_x = 1600$ N

6–90. $F_E = 3.64F$

6–91. $A_y = 657$ N, $C_y = 229$ N, $C_x = 0$,

$B_x = 0$, $B_y = 429$ N

6–93. $m = 366$ kg, $F_A = 2.93$ kN

6–94. $F_{AD} = 1796.3$ N, $M = 376.3$ N · m

6–95. $P = 168.85$ N

6–97. From FBD **(a)**, $A_y = 34.0$ N, $A_x = 0$

From **(b)**, $C_y = 6.54$ kN, $C_x = 0$

From **(c)**, $x = 292$ mm, $B_y = 1.06$ N, $B_x = 0$

6–98. $F_{DE} = 1.07$ kN

6–99. From FBD **(b)**, $C_y = 1.33$ kN, $B_y = 549$ N,

From FBD **(a)**, $C_x = 2.98$ kN, $A_y = 235$ N,

$A_x = 2.98$ kN, $B_x = 2.98$ kN

6–101. $F = 47.12$ N

6–102. From FBD **(a)**, $F_{AC} = 6.28$ N

FBD **(b)**, $F_{AD} = 8.58$ kN

6–103. $W_C = 0.812W$

6–105. From FBD **(b)**, $B_y = 940$ kN, $A_y = 360$ kN

From FBD **(c)**, $M_C = 7.80$ MN · m

$C_y = 900$ kN, $C_x = 250$ kN

6–106. BCE: $N_C = 20$ kN, $B_x = 34$ kN, $B_y = 62$ kN,

ACD: $A_x = 34$ kN, $A_y = 12$ kN, $M_A = 336$ kN · m

6–107. $E_x = 6.79$ kN, $E_y = 1.55$ kN

At point D, $D_x = 981$ N, $D_y = 981$ N

6–109. $\theta = \sin^{-1}\left(\dfrac{8W}{kL}\right)$

6–110. Case 1: $B_y = 31.67$ kN > 27.5 kN   (**N.G.!**)

Case 2: $C_y = 21.25$ kN < 27.5 kN   (**O.K.!**)

$x = 5.21$ m

Case 3: $B_y = 22.69$ kN < 27.5 kN   (**O.K.!**),

$A_y = 24.81$ kN < 27.5 kN   (**O.K.!**)

0.525 m $\le x \le 5.21$ m

6–111. $F_E = 36.73$ N, $F_C = 22.04$ N, $F_B = 61.22$ N

$F_D = 20.8$ N, $F_F = 14.7$ N, $F_A = 24.5$ N

6–113. **a)** $F = 750$ N, $N_C = 1500$ N

b) $F = 375$ N, $N_C = 375$ N

6–114. **a)** $F = 900$ N, $N_C = 1650$ N

b) $F = 450$ N, $N_C = 300$ N

6–115. $M = 1.70$ N · m

6–117. $F_{AB} = 981$ N, $F_E = 2.64$ kN, $F_{CD} = 16.3$ kN,

$F_F = 14.0$ kN

6–118. $x = 87.5$ mm

6–119. $P = \dfrac{F}{A} = 24$ N/mm^2

6–121. $T_{AI} = T_{EF} = 288$ kN, $F_H = F_G = 399$ kN

6–122. $F_{CA} = 12.9$ kN, $F_{AB} = 11.9$ kN, $F_{AD} = 2.39$ kN

6–123. $W_1 = 3$ kN, $W_2 = 21$ kN, $W_3 = 75$ kN

6–125. $P = 283$ N, $B_x = D_x = 0$
$B_z = D_z = B_y = D_y = 283$ N

6–126. $M_{Ex} = 0.5$ kN $\cdot$ m, $M_{Ey} = 0$, $E_y = 0$, $E_x = 0$,
$F_{AB} = 1.56$ kN

6–127. $M_{Cx} = 0$, $C_x = 0$, $F_{BA} = 153.8$ kN,
$C_z = -18$ kN, $C_y = -117$ kN
$M_{C_z} = -414$ kN $\cdot$ m, $A_x = 0$,
$A_y = 144$ kN, $A_z = 54$ kN

6–129. $\theta = 16.1°$

6–130. $A_x = 1.40$ kN, $A_y = 250$ N,
$C_x = 500$ N, $C_y = 1.70$ kN

6–131. $\theta = 21.7°$

6–133. $B_x = B_y = 220$ N, $A_x = 300$ N, $A_y = 80.4$ N

6–134. Member AC: $A_x = 117$ N, $A_y = 397$ kN
Member CB: $B_x = 97.4$ N, $B_y = 97.4$ N

6–135. $P = \dfrac{kL}{2\tan\theta\sin\theta}(2 - \csc\theta)$

6–137. $F_{AD} = 2473.86$ kN (T), $F_{AC} = F_{AB} = 1220.9$ kN

Chapter 7

7–1. $V_A = 0$, $N_A = 12.0$ kN, $M_A = 0$, $V_B = 0$,
$N_B = 20.0$ kN, $M_B = 1.20$ kN $\cdot$ m

7–2. $N_A = 550$ N, $N_B = 250$ N, $N_C = 950$ N

7–3. $N_A = 5.00$ kN, $N_C = 4.00$ kN, $N_B = 3.00$ kN

7–5. $M_C = 5$ kN $\cdot$ m, $N_C = 0$, $V_C = 1.33$ kN,
$M_D = 0.98$ kN $\cdot$ m, $N_D = 0$, $V_D = 1.03$ kN

7–6. $N_C = 0$, $V_C = -4.0$ kN, $M_C = 8$ kN $\cdot$ m,
$N_D = 0$, $V_D = -4.0$ kN, $M_D = 48$ kN $\cdot$ m

7–7. $N_C = 0$, $V_C = -386$ kN, $M_C = -428.6$ kN $\cdot$ m,
$N_D = 0$, $V_D = 300$ kN, $M_D = -300$ kN $\cdot$ m

7–9. $N_D = -800$ N, $V_D = 0$, $M_D = 1.20$ kN $\cdot$ m

7–10. $w = 100$ N/m

7–11. $M_C = 5.33$ kN $\cdot$ m, $V_C = 2$ kN

7–13. $N_D = 0$, $V_D = 800$ N, $M_D = -1.60$ kN $\cdot$ m,
$N_C = 0$, $V_C = 0$, $M_D = 800$ N $\cdot$ m

7–14. $N_D = 1.92$ kN, $V_D = 100$ N, $M_D = 900$ N $\cdot$ m

7–15. $N_E = -1.92$ kN, $V_E = 800$ N, $M_E = 2.40$ kN $\cdot$ m

7–17. $N_C = -406$ N, $V_C = 903$ N, $M_C = 1.35$ kN $\cdot$ m

7–18. $N_D = -464$ N, $V_D = -203$ N, $M_D = 2.61$ kN $\cdot$ m

7–19. $N_C = -30$ kN, $V_C = -8$ kN, $M_C = 6$ kN $\cdot$ m

7–21. $N_B = 0$, $V_B = 9.6$ kN, $M_B = -12.8$ kN $\cdot$ m

7–22. $\dfrac{a}{b} = \dfrac{1}{4}$

7–23. $N_C = 20.0$ kN, $V_C = 70.6$ kN, $M_C = -302$ kN $\cdot$ m

7–25. $M_C = -17.8$ kN $\cdot$ m

7–26. $N_D = 0$, $V_D = 0.375$ kN, $M_D = 3.375$ kN $\cdot$ m,
$N_E = 0$, $V_E = -6$ kN, $M_E = -6$ kN $\cdot$ m

7–27. $N_D = 2.40$ kN, $V_D = 50$ N, $M_D = 1.35$ kN $\cdot$ m

7–29. $V_C = 2.49$ kN, $N_C = 2.49$ kN, $M_C = 4.97$ kN $\cdot$ m,
$N_D = 0$, $V_D = -2.49$ kN, $M_D = 16.5$ kN $\cdot$ m

7–30. $N_B = 59.8$ N, $V_B = -496$ N, $M_B = -480$ N $\cdot$ m,
$M_B = -480$ N $\cdot$ m, $N_C = -495$ N, $V_C = 70.7$ N,
$M_C = -1.59$ kN $\cdot$ m

7–31. $N_A = 86.6$ N, $V_A = 150$ N, $M_A = 45$ N $\cdot$ m

7–33. $N_D = 0$, $V_D = 0$, $M_D = 45$ kN $\cdot$ m,
$N_E = 0$, $V_E = -50$ kN, $M_E = -60$ kN $\cdot$ m

7–34. $V_E = 0$, $N_E = 894$ N, $M_E = 0$, $V_F = 447$ N,
$N_F = 224$ N, $M_F = 224$ N $\cdot$ m

7–35. $a = \dfrac{L}{3}$

7–37. $N = -0.866rw_0$, $V = -1.5rw_0$, $M = 1.23r^2w_0$

7–38. $C_x = -150$ N, $C_y = -350$ N, $C_z = 700$ N,
$M_{C_x} = 1.40$ kN $\cdot$ m, $M_{C_y} = -1.20$ kN $\cdot$ m,
$M_{C_z} = -750$ N $\cdot$ m

7–39. $C_x = -170$ N, $C_y = -50$ N, $C_z = 500$ N,
$M_{C_x} = 1$ kN $\cdot$ m, $M_{C_y} = -900$ N $\cdot$ m,
$M_{C_z} = -260$ N $\cdot$ m

7–41. $N_C = -350$ N, $(V_C)_y = 700$ N, $(V_C)_z = -150$ N,
$(M_C)_x = -1.20$ kN $\cdot$ m, $(M_C)_y = -750$ N $\cdot$ m,
$(M_C)_z = 1.40$ kN $\cdot$ m

7–42. $V = \dfrac{Pb}{a+b}$, $M = \dfrac{Pb}{a+b}x$, $V = -\dfrac{Pa}{a+b}$,
$M = Pa - \dfrac{Pa}{a+b}x$

7–43. For $0 \le x < 2$ m: $V = 100$, $M = 100x - 1200$
For $2 < x \le 4$ m: $V = 100$, $M = 100x - 400$

7–45. **a)** $V = 0$, $M = 0$, $V = 0$, $M = M_0$, $V = 0$, $M = 0$

b) $V = 0$, $M = 0$, $V = 0$, $M = 500$ N $\cdot$ m,
$V = 0$, $M = 0$

7–46. $M_{max} = M_0 = 2$ kN $\cdot$ m

7–47. $V = \dfrac{w}{2}(L - 2x)$, $M = \dfrac{w}{2}(Lx - x^2)$,

$w = 400$ N/m

7–49. $V = \{40 - 50x\}$ N
$M = \{40x - 25x^2\}$ N $\cdot$ m, $V = 0$, $M = -20$ N $\cdot$ m

7–50. $V = \dfrac{wL}{8}$, $M = \dfrac{wL}{8}x$, $V = \dfrac{w}{8}(5L - 8x)$,

$M = \dfrac{w}{8}(-L^2 + 5Lx - 4x^2)$

7–51. $V = 250(1 - x)$, $M = 25(10x - 5x^2 - 1)$

7–53. Member AB: $V = \{875 - 150x\}$ N,
$M = \{875x - 75.0x^2\}$ N $\cdot$ m,
$V = \{2100 - 150x\}$ N,
$M = \{-75.0x^2 + 2100x - 14700\}$ N $\cdot$ m
Member CBD: $V = 919$ N, $M = \{919x\}$ N $\cdot$ m,
$V = 306$ N, $M = \{2450 - 306x\}$ N $\cdot$ m

7–54. $V = \dfrac{w}{4}(3L - 4x)$, $M = \dfrac{w}{4}(3Lx - 2x^2 - L^2)$

7–55. $V = \dfrac{w}{18}(7L - 18x)$, $M = \dfrac{w}{18}(7Lx - 9x^2)$,

$V = \dfrac{w}{2}(3L - 2x)$,

$M = \dfrac{w}{18}(27Lx - 20L^2 - 9x^2)$,

$V = \dfrac{w}{18}(47L - 18x)$,

$M = \dfrac{w}{18}(47Lx - 9x^2 - 60L^2)$

7–57. $V = \dfrac{wx^2}{2L}$, $M = -\dfrac{wx^3}{6L}$, $w = 22.2$ N/m

7–58. $x = \dfrac{L}{2}$, $P = \dfrac{4M_{max}}{L}$

7–59. $V = \left\{4.8 - \dfrac{x^2}{6}\right\}$ kN,

$M = \left\{4.8x - \dfrac{x^3}{18} - 5.76\right\}$ kN $\cdot$ m,

$V = \left\{\dfrac{1}{6}(2.4 - x)^2\right\}$ kN,

$M = \left\{-\dfrac{1}{18}(2.4 - x)^3\right\}$ kN $\cdot$ m

7–61. $V = \dfrac{w}{12L}(4L^2 - 6Lx - 3x^2)$,

$M = \dfrac{w}{12L}(4L^2x - 3Lx^2 - x^3)$

7–62. **a)** $V = P$, $M = Px$, $V = 0$, $M = Pa$, $V = -P$,
$M = P(L - x)$

b) For $0 \le x \le 5$ m, $V = 800$ N, $M = 800x$ N $\cdot$ m
For 5 m $\le x \le 7$ m, $V = 0$, $M = 4000$ N $\cdot$ m
For 7 m $\le x \le 12$ m, $V = 800$ N,
$M = (9600 - 800x)$ N $\cdot$ m

7–63. $V_x = 1.5$ kN, $V_y = 0$, $V_z = 800(4 - y)$ N,
$M_x = 400(4 - y)^2$ N $\cdot$ m, $M_y = -3$ kN $\cdot$ m,
$M_z = -1500(4 - y)$ N $\cdot$ m

7–65. $V_x = 0$, $V_z = \{24.0 - 4y\}$ kN,
$M_x = \{2y^2 - 24y + 64.0\}$ kN $\cdot$ m,
$M_y = 8.00$ kN $\cdot$ m, $M_z = 0$

7–66. $B_y = 9.50$ kN, $A_y = 6.50$ kN

7–67. $E_y = 333.33$ N, $A_y = 166.67$ N

7–69. $F_C = 625$ N, $A_y = 625$ N

7–71. $B_y = 206.25$ kN, $A_y = 133.75$ kN

7–73. $B_y = 10.5$ kN, $A_y = 9.50$ kN

7–77. $D_y = 32.167$ kN, $A_y = 1.167$ kN

7–78. $w = 2$ kN/m

7–81. $C_y = 0.200$ kN, $B_y = 0.400$ kN,
$M_A = 10.4$ kN $\cdot$ m, $A_y = 1.10$ kN
$(M_{max})_{BC} = 0.462$ kN $\cdot$ m

7–82. $C_y = 3.00$ kN, $A_y = 3.00$ kN, $M = 3.00$ kN $\cdot$ m

7–83. $B_y = \dfrac{7w_0L}{6}$, $A_y = \dfrac{w_0L}{3}$

7–86. $B_y = 40.0$ kN, $A_y = 40.0$ kN, $M = -45.0$ kN $\cdot$ m

7–87. $V_{6-} = -9$ kN, $M_6 = -36$ kN $\cdot$ m

7–89. $F_{BC} = 46.7$ kN, $F_{BA} = 83.0$ kN, $F_{CD} = 88.1$ kN,
$y = 2.679$ m, $l = 20.2$ m

7–90. $F_{DC} = 43.7$ kN, $F_{DB} = 78.2$ kN, $F_{CA} = 74.7$ kN,
$y = 1.695$ m, $l = 15.7$ m

7–91. $y_B = 2.17$ m, $y_D = 1.76$ m

7–93. $x_B = 4.36$ m

7–94. $P = 71.4$ kN

7–95. $P = 2.50$ kN, $F_{max} = 12.5$ kN

7–97. $w = 51.9$ kN/m

7–98. $T_{max} = 14.4$ kN, $T_{min} = 13.0$ kN

7–99. $y = (38.5x^2 + 577x)(10^{-3})$ m, $T_{max} = 5.20$ kN

7–101. $y = 2.37(10^{-3})x^3$, $T_{max} = 4.42$ kN

7–102. $y = \dfrac{x^2}{7813}\left(75 - \dfrac{x^2}{200}\right)$ m, $T_{max} = 9.28$ MN

7–103. $h = 7.09$ m

7–105. $L = 302$ m

7–107. $\dfrac{h}{L} = 0.141$

7–109. $(T_{max})_B = 2.73$ kN, $(T_{max})_C = 2.99$ kN

7–110. $T_{min} = 832.2$ N, $h = 15.1$ m

7–111. $T_{max} = 170.5$ kN, $L = 150.44$ m

7–113. $T_{max} = 2.01$ kN, Total weight $= w_0 l = 4.00$ kN

7–114. $h = 1.863$ m, $N_m = 1377$ N

7–115. $l = 71.29$ m, $h = 28.12$ m

7–117. $a = 0.366L$

7–119. $V = 1.50$ kN, $M = \{1.50x\}$ kN $\cdot$ m,
$V = -4.50$ kN, $M = \{27.0 - 4.50x\}$ kN $\cdot$ m

7–121. Segment DC $N_C = 0$, $V_C = 9.00$ kN,
$M_C = -62.5$ kN $\cdot$ m,
Segment DB $N_B = 0$, $V_B = 27.5$ kN,
$M_B = -184.5$ kN $\cdot$ m

7–122. $T_{max} = 76.7$ kN

Chapter 8

8–1. $N_A = 16.5$ kN, $N_B = 42.3$ kN
When the wheels at A are locked, **the mine car moves**.
When both wheels at A and B are locked, **the mine car does not move**.

8–2. $F_C = 27.4$ N, $N_C = 309$ N

8–3. Yes, the pole will remain stationary.

8–5. $P = 75$ N

8–6. $P = 5$ N
The ladder will remain in contact with the wall.

8–7. $N_C = 4000$ N, $N_B = 4805$ N

8–9. The ladder will not slip.

8–10. $P = \dfrac{M_0}{\mu_s r a}(b - \mu_s c)$

8–11. $\mu_s \geq \dfrac{b}{c}$

8–13. **a)** $P = 30$ N < 39.8 N   No,
b) $P = 70$ N > 39.8 N   Yes

8–14. **a)** $P = 30$ N < 34.26 N   No,
b) $P = 70$ N > 34.26 N   Yes

8–15. Since $P_{Req'd} = 6000$ N < 7659.57 N
It is possible to pull the load without slipping or tipping.

8–17. $P = 416.67$ N

8–18. $P = 500$ N

8–19. $m = 54.9$ kg

8–21. **a)** $W = 1272$ N, **b)** $W = 1440$ N

8–22. Dresser: $F = 90$ N
Man: $\mu_m = 0.15$

8–23. Dresser: $F = 121.45$ N
Man: $\mu_m = 0.195$

8–25. $d = 72$ mm

8–26. $\theta = 16.7°$, $\phi = 42.6°$

8–27. $1 < 10.99$ Therefore car A will not move.

8–29. $\theta = 16.7°$, $P = 0.287W$

8–30. $(F_B)_{max} = 1157.06$ kN > 571.45 N
Slipping occurs at A.

8–31. $\theta = 11.0°$

8–33. $\mu_s = 0.268$

8–34. $L = 1.118$ m

8–35. $P = 450$ N, $d = 0.5$ m

8–37. $F_D = 184.6$ N, $A_y = 2341.34$ N, $B_x = 173.21$ N
$B_y = 1141.34$ N

8–38. $\theta = 10.6°$, $x = 0.184$ m

8–39. $\theta = 8.53°$, $F_A = 1.48$ N, $F_B = 0.890$ N

8–41. $P = 63.5$ N (**Control!**)

8–42. $\mu = 0.176$

8–43. $P = 225$ N $\mu_s' = 0.300$

8–45. $P = 66.67$ N

8–46. $M = 90.6$ N $\cdot$ m

8–47. $P = 355$ N

8–48. $\mu_C = 0.0734$, $\mu_B = 0.0964$

8–50. $\phi = \theta$, $P = W \sin(\alpha + \phi)$

8–51. $P = 107$ N

8–53. $P = 196$ N (**Control!**)

8–54. $P = 40.2$ N

8–57. $P_x = 95.4$ N The saw horse will start to slip.

8–58. $P_x = 74.57$ N The saw horse will start to slip.

8–59. $M = 77.3$ N $\cdot$ m

8–61. Cylinder tips, $P = 375$ N

8–62. $P = 2.39$ kN

8–63. $W = 66.6$ kN

8–65. $P = 34.5$ N

8–66. $P = 304$ N

8–67. $x = 32.9$ mm

8–69. $P_x = 69.4$ N

8–70. $P = 5.53$ kN

Since a force $P(> 0)$ is required to pull out the wedge, the wedge will be self-locking when $P = 0$.

8–71. $W = 7.19$ kN

8–73. $\mu_s = 0.0637$

8–74. $F = 620$ N, Since $\phi_s > \theta$, the screw is self-locking.

8–75. $M = 569.48$ N $\cdot$ mm

8–77. $P = 1.98$ kN

8–78. $M = 0.202$ N $\cdot$ m

8–79. $M = 48.3$ N $\cdot$ m

8–81. $F_E = 72.7$ N, $F_D = F_E = 72.7$ N

8–82. $A_x = 328.6$ N, $B_y = C_y = 164$ N

8–83. $F_{AB} = 1.38$ kN (T), $F_{BD} = 828$ N (C), $F_{BC} = 1.10$ kN (C), $F_{AC} = 828$ N (C), $F_{AD} = 1.10$ kN (C), $F_{CD} = 1.38$ kN (T)

8–85. $n = 2$ turns

8–86. Approx. 2 turns (695°)

8–87. On the verge of sliding up the plane, $W = 86.0$ N

On the verge of sliding down the plane, $W = 13.9$ N

8–89. **a)** $F = 4.60$ kN, **b)** $F = 16.2$ kN

8–90. $n = 3$ half turns, $N_m = 33.7$ N

8–91. $P = 42.3$ N

8–93. $W_D = 127$ N

8–94. $W_D = 127$ N

8–97. $M = 3.37$ N $\cdot$ m

8–98. $P = 133.6$ N

8–99. $P = 78.7$ N

8–101. $h = 2.48$ m

8–102. $M = 50.0$ N $\cdot$ m, $x = 286$ mm

8–103. $P = 223$ N

8–105. $m = 25.6$ kg

8–106. $m = 7.82$ kg

8–107. $M = 19$ N $\cdot$ m

8–109. $F_{sp} = 8.08$ kN

8–110. $M = 54.4$ N $\cdot$ m

8–111. $F = 106.67$ N

8–113. $M = 87.85$ N $\cdot$ m

8–114. $M = \dfrac{1}{2}\mu PR$

8–115. $M = 0.521\,P\mu R$

8–117. $M = 17.0$ N $\cdot$ m

8–118. $\theta = 68.2°$, $M = 0.0455$ N $\cdot$ m

8–119. $\mu = 0.215$, $\mu = 0.211$ (approx.) $F = 6$ N, $\mu = 0.215$

8–121. $T = 289$ N, $N = 479$ N, $F = 101$ N

8–122. $\mu_s = 0.0407$

8–123. $F = 18.9$ N

8–125. $T = 13.8$ N

8–126. $T = 29.0$ N

8–127. $(r_f)_A = 5$ mm, $(r_f)_B = 2$ mm

8–129. $P \approx 0.378$ kN

8–130. $P = 245$ N

8–131. Since ϕ_A and ϕ_B are very small, $\cos \phi_A \approx \cos \phi_B \approx 1$.

8–133. $F = 454.5$ N

8–134. $s = 0.750$ m

8–135. **a)** $W = 6.97$ kN, **b)** $W = 15.3$ kN

8–137. $F_{max} = 0.2$ N

The cam **cannot** support the broom.

8–138. $P = 300$ N for two cartons. $P = 450$ N for three cartons.

8–139. $M = 3.75$ kN $\cdot$ m

Chapter 9

9–1. $\bar{x} = 0.546$, $O_x = 0$, $O_y = 7.06$ N, $M_O = 3.85$ N $\cdot$ m

9–2. $\bar{x} = 0$, $\bar{y} = 1.82$ m

9–3. $\bar{x} = 124$ mm, $\bar{y} = 0$

9–5. $\bar{x} = 0.531$ m, $O_x = 0$, $O_y = 0.574$ N, $M_O = 0.305$ N $\cdot$ m

9–6. $\bar{y} = 0.183$ m

9–7. $\bar{x} = \dfrac{3}{8}b$, $\bar{y} = \dfrac{3}{5}h$

9–9. $\bar{x} = \dfrac{3}{4}b$, $\bar{y} = \dfrac{3}{10}h$

9–10. $\bar{x} = \dfrac{5a}{8}$

9–11. $\bar{x} = \dfrac{n+1}{2(n+2)}a$

9–13. $\bar{x} = 3.20$ cm, $\bar{y} = 3.20$ cm, $T_A = 1.707$ N
$T_C = 1.707$ N, $T_B = 5.12$ N

9–14. $\bar{x} = \dfrac{(n+1)}{2(n+2)}a$, $\bar{y} = \dfrac{n+1}{2(2n+1)}h$

9–15. $\bar{x} = \dfrac{n+1}{2(n+2)}a$, $\bar{y} = \dfrac{n}{2n+1}h$

9–17. $\bar{y} = \dfrac{4b}{3\pi}$, $\bar{x} = \dfrac{4a}{3\pi}$

9–18. $\bar{x} = \dfrac{\pi}{2}a$, $\bar{y} = \dfrac{\pi}{8}a$

9–19. $\bar{y} = 2.80$ m, $\bar{x} = 6.00$ m

9–21. $\bar{y} = 2.04$ m

9–22. $\bar{x} = 1.26$ m, $\bar{y} = 0.143$ m, $N_B = 47.9$ kN,
$A_x = 33.9$ kN, $A_y = 73.9$ kN

9–23. $\bar{x} = 0.4$ m

9–25. $\bar{x} = 0.45$ m

9–26. $\bar{y} = 0.45$ m

9–27. $\bar{x} = 1.08$ m

9–29. $\bar{x} = 1.61$ m

9–30. $\bar{y} = 1.33$ m

9–31. $\bar{r} = 0.833\,a$

9–33. $\bar{x} = \bar{y} = 0$, $\bar{z} = \dfrac{4}{3}$ m

9–34. $\bar{z} = \dfrac{3}{8}a$

9–35. $\bar{z} = \dfrac{5}{6}h$

9–37. $\bar{x} = 0.4a$

9–38. $\bar{z} = 2.50$ m

9–39. $\bar{y} = 2.67$ m

9–41. $\bar{z} = \dfrac{R^2 + 3r^2 + 2rR}{4(R^2 + r^2 + rR)}h$

9–42. $m = \dfrac{\pi k r^4}{4}$, $\bar{z} = \dfrac{8}{15}r$

9–43. $\bar{z} = \dfrac{c}{4}$

9–45. $\bar{x} = 1.30$ m, $\bar{y} = 2.30$ m

9–46. $\bar{x} = 34.4$ mm, $\bar{y} = 85.8$ mm

9–47. $\bar{x} = 179$ mm

9–49. $\bar{x} = 0$, $\bar{y} = 5.14$ cm

9–50. $\bar{x} = 0$, $\bar{y} = 58.3$ mm

9–51. $\bar{x} = 1.60$ m, $\bar{y} = 7.04$ m, $A_x = 0$,
$A_y = 149$ kN, $M_A = 502$ kN · m

9–53. $\bar{y} = 85.9$ mm

9–54. $\bar{x} = 2.22$ m, $\bar{y} = 1.41$ m

9–55. $\bar{y} = 53.0$ mm

9–57. $\bar{y} = 154$ mm

9–58. $\bar{x} = 4.62$ mm, $\bar{y} = 1.00$ mm

9–59. $\bar{x} = 3.00$ mm, $\bar{y} = 2.00$ mm

9–61. $\bar{y} = 11.9$ mm

9–62. $\bar{x} = \dfrac{\frac{2}{3}r \sin^3 \alpha}{\alpha - \frac{\sin 2\alpha}{2}}$

9–63. $\bar{y} = 2.00$ cm

9–65. $\bar{x} = 77.2$ mm, $\bar{y} = 31.7$ mm

9–66. $\bar{y} = 135$ mm

9–67. $\bar{y} = 102.4$ mm

9–69. $h = 323$ mm

9–70. $\bar{z} = 128$ mm

9–71. $\bar{x} = -1.14$ cm, $\bar{y} = 1.71$ cm, $\bar{z} = -0.857$ cm

9–73. $\bar{x} = 47.4$ mm, $\bar{y} = 29.9$ mm

9–74. $\bar{x} = 0.86$ m, $\bar{y} = 0.519$ m, $N_B = 354.75$ N,
$N_A = 445.5$ N

9–75. $\bar{x} = 19.0$ m, $\bar{y} = 11.0$ m

9–77. $\bar{z} = 0.70$ m

9–78. $h = 2.00$ m

9–79. $\bar{z} = 101$ mm

9–81. $\bar{z} = 1.625$ cm

9–82. $\bar{z} = 58.1$ mm

9–83. $\bar{x} = 82.3$ mm

9–84. $A = 1.33$ m^2, $\bar{x} = 0.6$ m, $V = 5.03$ m^3

9–85. $A = 118$ cm^2

9–86. $A = 3.33$ m^2, $\bar{y} = 1.2$ m, $V = 25.1$ m^3

9–87. $m = 138$ kg

9–89. $V = 3.49$ m^3

9–90. $V = 4.25(10^6)$ mm^3

9–91. $R = 1220.13$ kN

9–93. $V = 1.40(10^3)$ cm^3

9–94. $W = 13\,465.9$ kN

9–95. $A = 0.89(10^3)$ m^3

9–97. $W = 1696.5$ kN

9–98. 28.66 litres

9–99. $V = 207$ m^3, $A = 188$ m^2

9–101. $V = 28.7$ cm^3

9–102. $W = 1.57$ N

9–103. $h = 106$ mm

9–105. $A = 119(10^3)$ mm^2

9–106. $F = 1.41$ MN, $h = 4$ m

9–107. $F_{R_A} = 157$ kN, $F_{R_A} = 235$ kN, $d = 4.22$ m

9–109. $d = 2.68$ m

9–110. $F_{AB} = 77\,868$ kN

9–111. $F = 391$ kN/m

9–113. $F_{R_v} = 4334.2$ kN, $F_{R_h} = 81\,250$ kN

9–114. $A_y = 2.51$ MN, $B_x = 2.20$ MN, $B_y = 859$ kN

9–115. $F_R = 2768.26$ kN, $d = 5.22$ m, $F_R = 3024$ kN

9–117. $F_R = 39.0625$ kN, $\bar{z} = 2$ m

9–118. $F_{R_v} = 2.04$ kN, $F_{R_h} = 1.3$ kN

9–119. $F = 678$ N, $\bar{x} = 0.948$ m, $\bar{y} = 1.50$ m

9–121. $F_R = 4.00$ MN, $\bar{y} = -6.49$ m

9–122. $\bar{x} = 0$, $\bar{y} = 2.40$ m, $F_R = 42.7$ kN,
$B_y = C_y = 12.8$ kN, $A_y = 17.1$ kN

9–123. $F_R = 6.93$ kN, $\bar{y} = -0.125$ m

9–125. $A = 1.25$ m^2

9–126. $\bar{y} = 87.5$ mm

9–127. $\bar{x} = \bar{y} = 0$, $\bar{z} = \dfrac{2}{3}a$

9–129. $\theta = 37.8°$

9–130. $\bar{y} = 0.600$ cm

9–131. $\bar{x} = 1.22$ m, $\bar{y} = 0.778$ m, $\bar{z} = 0.778$ m,
$M_{A_x} = 240.0$ N · m, $M_{A_y} = 857.6$ N · m,
$M_{A_z} = 0$, $A_x = 0$, $A_y = 0$, $A_z = 308.5$ N

9–133. $F_R = 7.62$ kN, $\bar{x} = 2.74$ m, $\bar{y} = 3.00$ m

Chapter 10

10–1. $I_x = 39.0$ m^4

10–2. $I_y = 8.53$ m^4

10–3. **a)** $I_x = 23.8$ m^4 **b)** $I_x = 23.8$ m^4

10–5. **a)** $I_x = 1.07$ m^4 **b)** $I_x = 1.07$ m^4

10–6. $I_x = \dfrac{2}{15}bh^3$

10–7. $I_x = 1.54$ cm^4

10–9. $I_x = \dfrac{2}{7}bh^3$

10–10. $I_y = \dfrac{2}{15}hb^3$

10–11. $I_x = 10.7$ cm^4

10–13. $I_y = 2.44$ m^4

10–14. $I_x = 0.571$ cm^4

10–15. $I_y = 1.07$ cm^4

10–17. $I_x = 0.176$ m^4

10–18. $I_x = 9.05$ cm^4

10–19. $I_y = 30.9$ cm^4

10–21. $I_y = 10.7$ cm^4

10–22. $I_x = 3.20$ m^4

10–23. $I_y = 0.628$ m^4

10–25. $A = 2.44$ cm^2

10–26. $A = 11.1(10^3)$ mm^2

10–27. $I_x = 3.35(10^3)$ cm^4

10–28. $I_y = 832$ cm^4

10–29. $\bar{I}_x = 162(10^6)$ mm^4

10–30. $\bar{y} = 2.00$ cm, $I_{x'} = \Sigma(I_{x'})_i = 64.0$ cm^4

10–31. $\bar{x} = 3.00$ cm, $I_{y'} = \Sigma(I_{y'})_i = 136$ cm^4

10–33. $I_{x'} = 49.5(10^6)$ mm^4

10–34. $I_x = 1.217(10^3)$ cm^4, $I_y = 367.8$ cm^4

10–35. $I_{x'} = 95.9(10^6)$ mm^4

10–37. $\bar{y} = 80.7$ mm, $\bar{I}_{x'} = 67.6(10^6)$ mm^4

10–38. $\bar{x} = 61.6$ mm, $\bar{I}_{y'} = 41.2(10^6)$ mm^4

10–39. $\bar{y} = 0.181$ m, $I_{x'} = 4.23(10^{-3})$ m^4

10–41. $\bar{y} = 22.5$ mm, $I_{x'} = 34.4(10^6)$ mm^4

10–42. $I_{y'} = 122(10^6)$ mm^4

10–43. $I_x = 648$ cm^4

10–45. $\bar{y} = 2$ cm, $I_{x'} = 128$ cm^4

10–46. $I_x = 503$ cm^4

10–47. $\bar{I}_{x'} = \frac{1}{12}a^3b \sin^3 \theta$

10–49. $\bar{y} = 53.0$ mm, $I_{x'} = 3.67(10^6)$ mm^4

10–50. $I_{x'} = 30.2(10^6)$ mm^4

10–51. $\bar{y} = 91.7$ mm, $I_{x'} = 216(10^6)$ mm^4

10–53. $\bar{I}_{x'} = \frac{1}{36}bh^3$, $\bar{I}_{y'} = \frac{1}{36}hb(b^2 - ab + a^2)$

10–54. $I_{xy} = \displaystyle\int dI_{xy} = 0$

10–55. $I_{xy} = 0.667$ cm^4

10–57. $I_{xy} = \dfrac{1}{6}a^2b^2$

10–58. $I_{xy} = 48$ cm^4

10–59. $I_{xy} = 2.00 \text{ cm}^4$

10–61. $I_{xy} = \dfrac{3}{16}b^2h^2$

10–62. $I_{xy} = \dfrac{a^4}{280}$

10–63. $I_{xy} = \dfrac{a^2b^2}{4(n+1)}$

10–64. $I_{xy} = 1.33 \text{ m}^4$

10–66. $I_{xy} = \dfrac{1}{6}l^3t\sin 2\theta$

10–67. $I_{xy} = -28.1(10^3) \text{ mm}^4$

10–69. $I_{xy} = 36.0 \text{ cm}^4$

10–70. $I_{xy} = \dfrac{a^2c\sin^2\theta}{12}(4a\cos\theta + 3c)$

10–71. $I_{xy} = 98.4(10^6) \text{ mm}^4$

10–73. $I_{xy} = 0.740 \text{ cm}^4$

10–74. $I_{uv} = 135(10)^6 \text{ mm}^4$

10–75. $I_u = 114(10^6) \text{ mm}^4,\ I_v = 56.5(10^6) \text{ mm}^4$

10–77. $I_u = 15.75 \text{ cm}^4,\ I_v = 25.75 \text{ cm}^4$

10–78. $\theta = -22.5°,\ I_{max} = 250 \text{ cm}^4,\ I_{min} = 20.4 \text{ cm}^4$

10–79. $I_u = 3.47(10^3) \text{ cm}^4,\ I_v = 3.47(10^3) \text{ cm}^4,$
$I_{uv} = 2.05(10^3) \text{ cm}^4$

10–81. $I_{max} = 64.1 \text{ cm}^4,\ I_{min} = 5.33 \text{ cm}^4$

10–82. $I_{max} = 4.92(10^6) \text{ mm}^4,\ I_{min} = 1.36(10^6) \text{ mm}^4$

10–83. $I_{max} = 1.74(10^3) \text{ cm}^4,\ I_{min} = 435 \text{ cm}^4$

10–85. $I_{max} = 250 \text{ cm}^4,\ I_{min} = 20.4 \text{ cm}^4$

10–86. $I_{max} = 64.1 \text{ cm}^4,\ I_{min} = 5.33 \text{ cm}^4$

10–87. $I_{max} = 4.92(10^6) \text{ mm}^4,\ I_{min} = 1.36(10^6) \text{ mm}^4$

10–89. $I_{max} = 1.74(10^3) \text{ cm}^4,\ I_{min} = 435 \text{ cm}^4$

10–90. $I_y = \dfrac{1}{3}ml^2$

10–91. $I_z = mR^2$

10–93. $I_x = \dfrac{2}{5}mr^2$

10–94. $k_x = 57.7 \text{ mm}$

10–95. $I_x = \dfrac{2}{5}mb^2$

10–97. $I_x = \dfrac{2}{5}mb^2$

10–98. $I_y = \dfrac{m}{6}(a^2 + h^2)$

10–99. $I_y = 2.89(10^8) \text{ kg} \cdot \text{m}^2$

10–101. $I_z = 1.53 \text{ kg} \cdot \text{m}^2$

10–102. $I_G = 154.2 \text{ kg} \cdot \text{m}^2$

10–103. $I_O = 382.3 \text{ kg} \cdot \text{m}^2$

10–105. $I = 69.75 \text{ kg} \cdot \text{m}^2$

10–106. $I_z = 34.2 \text{ kg} \cdot \text{m}^2$

10–107. $I_A = 50.75 \text{ kg} \cdot \text{m}^2$

10–109. $I_G = 3.25(10^{-3}) \text{ kg} \cdot \text{m}^2$

10–110. $I_O = 7.20(10^{-3}) \text{ kg} \cdot \text{m}^2$

10–111. $I_G = 0.230 \text{ kg} \cdot \text{m}^2$

10–113. $I_y = 0.0954d^4$

10–114. $I_y = 0.187d^4$

10–115. $I_x = \frac{93}{70}mb^2$

10–117. $I_u = 5.09(10^6) \text{ mm}^4,\ I_v = 5.09(10^6) \text{ mm}^4,$
$I_{uv} = 0$

10–118. $I_y = 2.13 \text{ m}^4$

10–119. $I_x = 0.610 \text{ m}^4$

10–121. a) $I_x = \dfrac{bh^3}{12}$, b) $\bar{I}_{x'} = \dfrac{bh^3}{36}$

10–122. $I_{xy} = 0.1875 \text{ m}^4$

Chapter 11

11–1. $F_{AC} = 73.23 \text{ N}$

11–2. $\theta = 0°,\ \theta = 73.1°$

11–3. $F = 24.5 \text{ N}$

11–5. $k = 1.05 \text{ kN/m}$

11–6. $F = 512 \text{ N}$

11–7. $F = \dfrac{500\sqrt{0.04\cos^2\theta + 0.6}}{(0.2\cos\theta + \sqrt{0.04\cos^2\theta + 0.6})\sin\theta}$

11–9. $\theta = 13.9°,\ \theta = 90°$

11–10. $\theta = 0°,\ \theta = 36.9°$

11–11. $M = 1.37 \text{ kN} \cdot \text{m}$

11–13. $P = \dfrac{W}{2}\cot\theta$

11–14. $\theta = 16.6°,\ \theta = 35.8°$

11–15. $\theta = 15.5°,\ \theta = 85.4°$

11–17. $F = 4.62 \text{ kN}$

11–18. $k = 108 \text{ N/m}$

11–19. $F = 2P\cot\theta$

11–21. $M = 42.5 \text{ N} \cdot \text{m}$

11–22. $F = 21.6 \text{ N} \cdot \text{m}$

11–23. $\theta = \cos^{-1}\left(\dfrac{a}{2L}\right)^{\frac{1}{3}}$

11–25. $m = 100 \text{ kg}$

11–26. $x = 0,\ \dfrac{d^2V}{dx^2} = -4 < 0$ Unstable,

     $x = 0.167 \text{ m},\ \dfrac{d^2V}{dx^2} = 4 > 0$ Stable

11–27. $\theta = 34.6°,\ \dfrac{d^2V}{d\theta^2} = -57.2 < 0$ Unstable,

     $\theta = 145°,\ \dfrac{d^2V}{d\theta^2} = 57.2 > 0$ Stable

11–29. $(0, 0)$ is a position of equilibrium stable

11–30. $\theta = 0°,\ \theta = \cos^{-1}\left(\dfrac{W}{2KL}\right)$,

     neutral when $w = 2kL$

11–31. $k = 28.12 \text{ N/m}$

11–33. The truck is in unstable equilibrium at $\theta = 23.2°$

11–34. $k = 1000 \text{ N/m}$

11–35. The cylinder is in unstable equilibrium at $\theta = 0$.

11–37. $h = 0$

11–38. $\theta = 90°,\ \theta = \sin^{-1}\left(\dfrac{4W}{ka}\right)$

11–39. $m = 5.29 \text{ kg}$

11–41. $d = 87.9 \text{ mm}$

11–42. $b < 2r$

11–43. $d = \dfrac{h}{3}$

11–45. $\dfrac{\cos\theta}{\sin^3\theta} = \dfrac{a}{2r}$

11–46. $P = 5.28 \text{ N}$

11–47. The system is in stable equilibrium at $\theta = 37.8°$

11–49. The system is in stable equilibrium at $\theta = 90°$

     The system is in unstable equilibrium at $\theta = 9.47°$

11–50. $\theta = 90°,\ \theta = 30°$

11–51. $\theta = 90°,\ \dfrac{d^2V}{d\theta^2} = -10 < 0$ Unstable,

     $\theta = 30°,\ \dfrac{d^2V}{d\theta^2} = 15 > 0$ Stable

INDEX

Geometric Properties of Line and Area Elements

Centroid Location	Centroid Location	Area Moment of Inertia

Circular arc segment

Circular sector area

$$I_x = \tfrac{1}{4} r^4 \left(\theta - \tfrac{1}{2}\sin 2\theta\right)$$

$$I_x = \tfrac{1}{4} r^4 \left(\theta + \tfrac{1}{2}\sin 2\theta\right)$$

Quarter and semicircle arcs

Quarter circle area

$$I_x = \tfrac{1}{16}\pi r^4$$

$$I_y = \tfrac{1}{16}\pi r^4$$

Trapezoidal area

Semicircular area

$$I_x = \tfrac{1}{8}\pi r^4$$

$$I_y = \tfrac{1}{8}\pi r^4$$

Semiparabolic area

Circular area

$$I_x = \tfrac{1}{4}\pi r^4$$

$$I_y = \tfrac{1}{4}\pi r^4$$

Exparabolic area

Rectangular area

$$I_x = \tfrac{1}{12}bh^3$$

$$I_y = \tfrac{1}{12}hb^3$$

Parabolic area

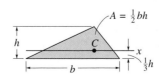

Triangular area

$$I_x = \tfrac{1}{36}bh^3$$

Center of Gravity and Mass Moment of Inertia of Homogeneous Solids

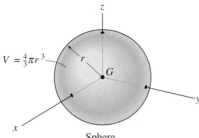

Sphere

$V = \frac{4}{3}\pi r^3$

$$I_{xx} = I_{yy} = I_{zz} = \frac{2}{5}mr^2$$

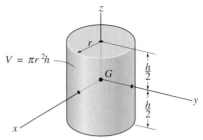

Cylinder

$V = \pi r^2 h$

$$I_{xx} = I_{yy} = \frac{1}{12}m(3r^2 + h^2) \quad I_{zz} = \frac{1}{2}mr^2$$

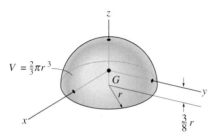

Hemisphere

$V = \frac{2}{3}\pi r^3$

$$I_{xx} = I_{yy} = 0.259mr^2 \quad I_{zz} = \frac{2}{5}mr^2$$

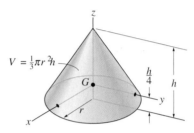

Cone

$V = \frac{1}{3}\pi r^2 h$

$$I_{xx} = I_{yy} = \frac{3}{80}m(4r^2 + h^2) \quad I_{zz} = \frac{3}{10}mr^2$$

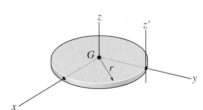

Thin Circular disk

$$I_{xx} = I_{yy} = \frac{1}{4}mr^2 \quad I_{zz} = \frac{1}{2}mr^2 \quad I_{z'z'} = \frac{3}{2}mr^2$$

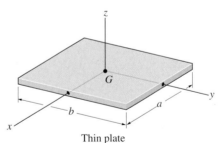

Thin plate

$$I_{xx} = \frac{1}{12}mb^2 \quad I_{yy} = \frac{1}{12}ma^2 \quad I_{zz} = \frac{1}{12}m(a^2 + b^2)$$

Thin ring

$$I_{xx} = I_{yy} = \frac{1}{2}mr^2 \quad I_{zz} = mr^2$$

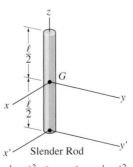

Slender Rod

$$I_{xx} = I_{yy} = \frac{1}{12}m\ell^2 \quad I_{x'x'} = I_{y'y'} = \frac{1}{3}m\ell^2 \quad I_{z'z'} = 0$$